PAPUA NEW GUINEA

MATHEMATICS

Grade

9

Ros O'Sullivan
Christine McRae

OXFORD

Oxford University Press is a department of the University of Oxford. It furthers the University's objective of excellence in research, scholarship, and education by publishing worldwide. Oxford is a registered trademark of Oxford University Press in the UK and in certain other countries.

Published in Australia by
Oxford University Press
253 Normanby Road, South Melbourne, Victoria 3205, Australia

First published 2009
Reprinted 2011, 2012, 2013, 2014, 2018, 2021, 2022

ISBN 978 0 19 556546 1

Edited by Scharlaine Cairns
Illustrated by Uramina and Nelson
Series design by Domenika Fairy
Typeset by Palmer Higgs
Printed in China by Golden Cup Printing Co. Ltd

Contents

Unit 1 Mathematics in our community 2

Unit 3 Working with data 164

Unit 1 Mathematics in our community

Unit summary

We use mathematics in our everyday lives and in most jobs. For example when:

- planning and cooking a family meal
- measuring to plot out garden beds, or to paint, or to put up pre-fabricated walls, or to make curtains or blinds
- working with money for a home or business budget
- working with interest rates, when we lend or borrow money
- looking at information presented as graphs and tables.

This unit focuses on everyday mathematics that students need for counting, measuring, comparing and presenting information numerically. This includes:

- basic operations with numbers, ratios and percentages
- estimation and calculations involving money
- simple costing/budgeting
- estimation and measurement of length and area
- simple organisation and graphical representation of data
- calculation of basic statistics.

Revision activities at the end of the unit allow students to revise concepts and skills and to apply these to individual or group investigations.

At the end of this unit students will consider how traditional mathematics is still used in people's everyday lives. This will encourage them to investigate how mathematics is used by the people of their area.

Syllabus references

9.1.1 Identify everyday situations where basic operations can be applied

9.1.2 Make calculations using a range of methods and be aware of whether or not the result is reasonable

9.1.3 Identify everyday situations where traditional operations can be applied

9.1.4 Communicate mathematical processes and results

9.1.5 Undertake investigations individually and cooperatively in which mathematics can be applied to solve problems

Mathematics in our community: Topics

NUMBERS AND OPERATIONS

Lesson 1

Comparing and ordering whole numbers and fractions

We often make comparisons between quantities in our everyday life. Which cereal costs less? Which coconut is heavier? Which runner ran the fastest?

As well as using such words as 'faster', 'higher' and 'smaller', we can show the relationship between two quantities using the signs < (is less than) and > (is greater than).

When comparing quantities, the bigger number is the one that would appear further to the right on a number line.

Fractions need to be written with the same denominator before we can compare them.

Example

Fill in the box with the correct sign (<, = or >).

$\frac{1}{2} \square \frac{3}{8}$

Answer

Using our knowledge of equivalent fractions, $\frac{1}{2} = \frac{4}{8}$.

$\frac{4}{8}$ is greater than $\frac{3}{8}$. So $\frac{1}{2} > \frac{3}{8}$.

Example

Which is bigger, $\frac{2}{3}$ or $\frac{4}{5}$?

Answer

We need to change both fractions into fifteenths.

$\frac{2}{3} = \frac{10}{15}$ and $\frac{4}{5} = \frac{12}{15}$

We can see that $\frac{12}{15} > \frac{10}{15}$. So $\frac{4}{5} > \frac{2}{3}$.

EXERCISE

1 Write 'True' or 'False' for each of the following statements:

a $7 > 4$ **b** $11 < 10$ **c** $2 > 0$ **d** $5 < 3$ **e** $24 > 27$

2 Copy the statements below and fill in the correct 'is less than' or 'is greater than' sign to make each statement true:

a 76 ☐ 74 **b** 1038 ☐ 1247 **c** 432 ☐ 427

d 789 ☐ 780 **e** 48 911 ☐ 48 916

Remember

- Numbers in **ascending** order go from smallest to largest.
- Numbers in **descending** order go from largest to smallest.

3 Arrange each of the following sets of numbers in ascending order:

a 15, 3, 8, 12, 4, 17 **b** 35, 28, 37, 29, 31, 42

c 134, 256, 93, 171, 249, 110 **d** 9824, 9801, 9832, 9846, 9819, 9857

e 7506, 689, 2314, 106, 3478, 732

4 Arrange each of the following sets of numbers in descending order:

a 18, 14, 9, 13, 11, 17 **b** 35, 28, 42, 27, 39, 29

c 315, 352, 381, 316, 395, 384 **d** 652, 499, 316, 530, 452, 601

e 12 025, 12 189, 12 547, 12 063, 12 822, 12 471

5 Write 'True' or 'False' for each of the following number statements:

a $\frac{1}{3} > \frac{1}{4}$ **b** $\frac{4}{5} < \frac{7}{10}$ **c** $\frac{2}{3} > \frac{1}{2}$ **d** $\frac{5}{6} < \frac{3}{4}$ **e** $\frac{2}{9} > \frac{2}{7}$

6 Arrange each of the following sets of fractions in ascending order:

a $\frac{1}{8}, \frac{1}{4}, \frac{1}{5}, \frac{1}{2}, \frac{1}{3}, \frac{1}{9}$ **b** $\frac{2}{17}, \frac{2}{7}, \frac{2}{9}, \frac{2}{3}, \frac{2}{11}, \frac{2}{5}$ **c** $\frac{1}{10}, \frac{3}{5}, \frac{9}{10}, \frac{7}{20}, \frac{1}{4}, \frac{11}{20}$

d $2\frac{1}{3}, 4\frac{1}{2}, 1\frac{1}{8}, 4\frac{2}{3}, 3\frac{2}{7}, 3\frac{1}{3}$ **e** $1\frac{7}{8}, 1\frac{3}{8}, 1\frac{4}{5}, 1\frac{1}{7}, 1\frac{5}{6}, 1\frac{1}{2}$

7 The 10 tallest mountains in the world have the following heights, in metres:

8167 8586 8125 8485 8848 8091 8163 8611 8516 8188

Arrange them in order, from highest to lowest.

8 At a weightlifting contest, the weights (in kilograms) lifted by the lifters in an Under-18 category were:

90, 105, 65, 108, 75

a Arrange these numbers in ascending order.
b What was the greatest weight lifted?
c The 2007 Pan American Games record for this category was 197 kg. How much bigger is this than the smallest weight lifted by the lifters in the Under-18 category?

9 Students ran laps of the running track for 30 minutes to raise money for an overseas trip. The best runners completed the most laps. Who were the top six runners, in order, from first place to sixth?

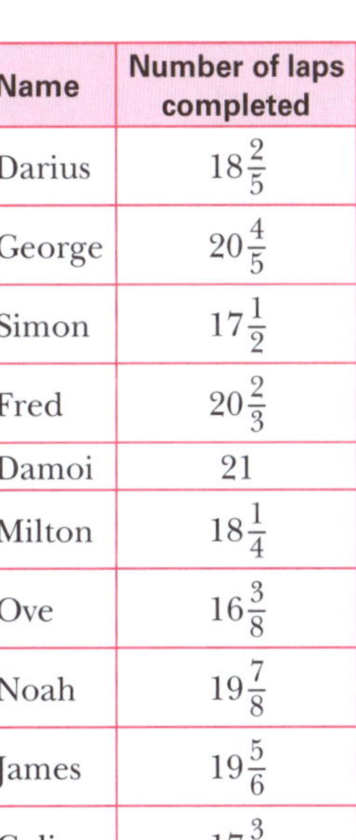

Name	Number of laps completed
Darius	$18\frac{2}{5}$
George	$20\frac{4}{5}$
Simon	$17\frac{1}{2}$
Fred	$20\frac{2}{3}$
Damoi	21
Milton	$18\frac{1}{4}$
Ove	$16\frac{3}{8}$
Noah	$19\frac{7}{8}$
James	$19\frac{5}{6}$
Colin	$17\frac{3}{5}$

10 Two groups of students set out on a long trek.

a At the end of the day, Paul's group had walked $\frac{2}{5}$ of the total distance and Gai's group had walked $\frac{4}{7}$ of the total distance. Which group had walked the farthest?
b After two days, Percy had eaten $\frac{7}{15}$ of his food rations and Riat had eaten half of his. How much more of his rations does Percy have left than Riat has left?

Discussion

Describe the counting system used by the people from your area. How similar to or different from the decimal system is it?

Did you know?

In PNG and Oceania, there are few counting systems with a regular base in which numerals are used for powers of the base as in the metric system. The counting systems are described in terms of cycles and patterns within a counting system, for instance the pattern in the ways the number words are constructed.

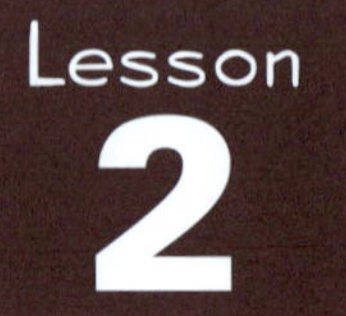
Lesson 2

Operations with fractions

In our previous work with fractions, we learned that fractions can only be added or subtracted if they have the same denominator.

Example

Find $\frac{4}{9} - \frac{2}{5}$.

Answer

$\frac{4}{9} - \frac{2}{5}$ — We need to find the lowest common denominator (LCD).

$= \frac{20}{45} - \frac{18}{45}$ — LCD is 45, so we make equivalent fractions with denominator of 45.

$= \frac{20 - 18}{45}$

$= \frac{2}{45}$

Example

Find $5\frac{1}{6} + 1\frac{4}{9}$.

Answer

$5\frac{1}{6} + 1\frac{4}{9}$ — Rearrange the sum so that the whole numbers are together and the fractions are together.

$= 5 + 1 + \frac{1}{6} + \frac{4}{9}$ — Find the LCD, which is 18.

$= 6 + \frac{3}{18} + \frac{8}{18}$ — Add the whole numbers; and write equivalent fractions.

$= 6 + \frac{11}{18}$

$= 6\frac{11}{18}$

Example

Find $2\frac{1}{2} - 1\frac{3}{4}$.

Answer

$2\frac{1}{2} - 1\frac{3}{4}$

$= \frac{5}{2} - \frac{7}{4}$ — First change the mixed numbers to improper fractions and find the LCD.

$= \frac{10}{4} - \frac{7}{4}$ — LCD is 4.

$= \frac{3}{4}$

Help box

To change a **mixed number** to an **improper fraction**, we multiply the denominator and whole number together and add that to the numerator.

$2\frac{1}{6} = \frac{6 \times 2 + 1}{6}$

$= \frac{13}{6}$

EXERCISE

1 Find:

a $\frac{3}{4} + \frac{2}{3}$ b $\frac{5}{6} - \frac{2}{3}$ c $\frac{7}{9} - \frac{1}{6}$

d $\frac{1}{5} + \frac{3}{10} + \frac{3}{20}$ e $\frac{11}{24} + \frac{5}{6} - \frac{2}{3} - \frac{5}{8}$

2 Find:

a $1\frac{5}{6} + 1\frac{1}{4}$ b $3\frac{7}{12} + 4\frac{5}{9}$ c $8\frac{1}{2} - 2\frac{1}{3}$

d $2\frac{3}{5} + 6\frac{1}{2}$ e $3\frac{4}{9} - 1\frac{1}{3}$

3 Mae's shopping bag contained $2\frac{1}{2}$ kg of rice, $\frac{3}{4}$ kg of salt and $1\frac{1}{3}$ kg of taro. What was the total weight of the contents of the shopping bag?

4 Daniel had $7\frac{9}{10}$ litres of petrol in his van. He used $5\frac{1}{2}$ litres making a delivery, then added $6\frac{1}{5}$ litres at the petrol station. How much petrol was then in the tank?

5 Mr Gen needs 10 metres of wire for his fence. He has pieces measuring $3\frac{1}{5}$ m, $3\frac{11}{15}$ m and $3\frac{1}{30}$ m. Does he have enough wire? (Show your calculations.)

6 Mirou mixed $2\frac{2}{3}$ litres of mango juice with $3\frac{3}{4}$ litres of pineapple juice, then drank $\frac{1}{8}$ litre of the mixture. How much was left?

7 Find the simplest answer each time:

a $\frac{3}{4} \times \frac{5}{6}$ b $\frac{5}{8} \times \frac{2}{5}$ c $2\frac{1}{4} \times 1\frac{1}{2}$

d $1\frac{1}{9} \times 2\frac{1}{3}$ e $\frac{4}{7} \times \frac{2}{5} \times \frac{1}{3}$ f $1\frac{1}{4} \times \frac{5}{8} \times \frac{4}{5}$

8 In the store, Lena stacks up eight cans each weighing $\frac{1}{3}$ kg. What is the total weight of the cans?

9 A tank holds $25\frac{1}{4}$ litres of water. How much water is in the tank if it is only $\frac{2}{3}$ full?

10 Find $\frac{3}{5}$ of $7\frac{1}{9}$.

11 Find the simplest answer to each of the following:

a $6 \div \frac{1}{4}$ b $\frac{3}{4} \div \frac{2}{5}$ c $1\frac{3}{4} \div \frac{7}{8}$ d $3\frac{3}{7} \div 1\frac{1}{14}$ e $12\frac{1}{2} \div 1\frac{1}{4}$

12 Piliki needs to feed his chickens $\frac{1}{3}$ of a cup of corn each day. The bag of corn holds $5\frac{2}{3}$ cups. For how many days will the bag of corn last?

13 Four children shared a $\frac{2}{3}$ litre bottle of drink equally. How much did each child drink?

Remember

To multiply fractions

Step 1: Write mixed numbers and whole numbers as improper fractions.

Step 2: Multiply numerators, and multiply denominators.

Step 3: Simplify the answer, if possible.

Example

$$1\frac{1}{5} \times \frac{7}{12} = \frac{6}{5} \times \frac{7}{12} = \frac{6 \times 7}{5 \times 12} = \frac{42}{60} = \frac{7}{10}$$

or simplify by cancelling:

$$\frac{{}^{1}\cancel{6} \times 7}{5 \times \cancel{12}_{2}} = \frac{7}{10}$$

Remember

To divide fractions

Step 1: Change any whole or mixed numbers to improper fractions.

Step 2: Invert the divisor (the number after the division sign) and change ÷ to ×.

Step 3: Complete the multiplication.

Example

$$3\frac{1}{3} \div \frac{3}{4} = \frac{10}{3} \times \frac{4}{3} = \frac{40}{9} = 4\frac{4}{9}$$

Decimals, fractions and estimations

To compare decimal numbers, we need to think of the place value of each digit.

$$100 \quad 10 \quad 1 \quad \bullet \quad \frac{1}{10} \quad \frac{1}{100} \quad \frac{1}{1000} \quad \frac{1}{10\,000}$$

Example

Which is greater, 12.349 or 12.36?

Answer

Comparing digits, the whole numbers (12) are equal. Both have $\frac{3}{10}$ (a 3 in the tenths place), so we compare the hundredths place.
$\frac{4}{100}$ is less than $\frac{6}{100}$, so 12.36 is greater than 12.349.

Note that the decimal number with more digits is not necessarily greater in value.

Example

Find the decimal equivalent for $\frac{5}{6}$.

Answer

$\frac{5}{8} = 5 \div 8$

```
   0.83
6)5.000
  48
   20
   18
    20
    18
     2
```

So $\frac{5}{6} = 0.8$ (repeating).

We sometimes need to convert fractions to their decimal form.

For example:

$\frac{3}{10} = 0.3$

$\frac{3}{100} = 0.03$

$\frac{3}{1000} = 0.003$

The rule for changing any fraction to a decimal is:
Divide the denominator into the numerator.

EXERCISE

1 Write the fraction represented by the digit 9 in each of the following decimal numbers:

a 34.98   b 25.809   c 2.35469

d 0.093   e 0.0009

2 Write 'True' or 'False' for each of the following number statements:

a $18.09 > 18.13$ b $0.671 < 0.68$

c $4.56 > 4.513$ d $67.14 < 67.2$

e $80.05 > 80.008$

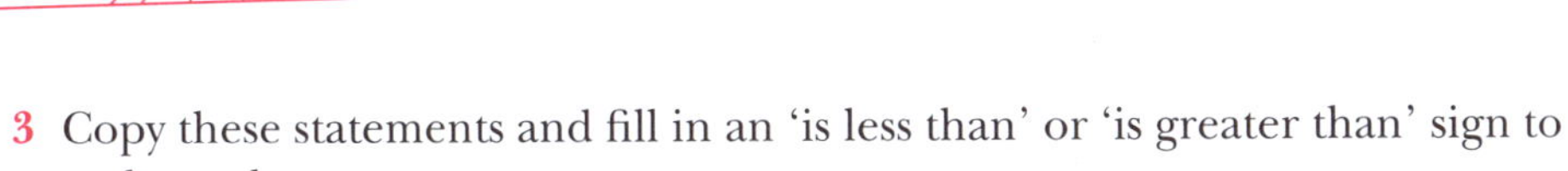

3 Copy these statements and fill in an 'is less than' or 'is greater than' sign to make each statement true:

a 0.05 ☐ 0.047 b 7.23 ☐ 7.2 c 5.38 ☐ 5.386

d 0.059 ☐ 0.07 e 10.3 ☐ 10.286

4 Arrange these decimal numbers in order, from smallest to largest:

a 11.01, 12.03, 11.34 b 6.87, 6.556, 6.9

c 1.02, 1.202, 12.02 d 5.72, 5.701, 5.7

e 91.002, 91, 91.202

5 Write each of these decimal numbers as a simple fraction:

a 0.6 b 0.25 c 0.71

d 0.004 e 0.125

6 Write each of these fractions as a decimal:

a $\frac{3}{8}$ b $\frac{7}{10}$ c $\frac{3}{25}$

d $\frac{5}{6}$ e $\frac{7}{9}$

7 Copy these statements and fill in an 'is less than' or 'is greater than' sign to make each statement true:

a 0.05 ☐ $\frac{1}{2}$ b $\frac{3}{4}$ ☐ 0.74 c 0.8 ☐ $\frac{3}{5}$

d $\frac{9}{10}$ ☐ 0.93 e 0.7 ☐ $\frac{7}{9}$

8 Three students were asked to time themselves while they each completed a problem. The times they took are shown below.

David: $\frac{1}{2}$ minute

Paulus: 0.55 minute

Gai: 0.505 minute

a Who took the longest time?

b How much quicker than the slowest person was the fastest person?

9 In which of these situations would an accurate answer be needed rather than an **estimate**? Explain your answer.

a Calculating your pay for working 6 hours in the local store

b Deciding the length of string to tie up six bean plants

c Deciding the amount of water to add to baby formula

d Deciding the number of bottles of cola to buy for a party given for 10 friends

e Deciding the number of litres of paint needed to paint the walls of a room

10 Round each number to the nearest unit, 10 or 100 (as appropriate) to find an estimate of the answer. Show your working.

a 0.8×37 b $893 \div 19$

c 56.4×2.1 d $23.75 \div 4.8$

e $15.8 \times 6.3 \div 11.7$

Help box

$\frac{1}{2} = \frac{5}{10} = 0.5$

$\frac{3}{4} = \frac{75}{100} = 0.75$

Remember

An estimate is an approximate value worked out by completing a simpler calculation. Estimates can be made in a number of ways. For example, to add 78 and 123:

- we could use **rounded numbers** (Rounded to the nearest 10, $78 + 123 \approx 80 + 120$, so one estimate is 200.)
- we could use **front-end estimation**, using only the digit with the highest value. So, $78 + 123 \approx 70 + 100 = 170$

Lesson 4

Operations with decimals

Decimal numbers are used much more commonly than fractions. They are used in such areas as banking, construction, measurement, and research, as well as in everyday life.

Discussion

- How do people where you live use fractions? What do they measure and calculate using fractions?
- Suggest why decimals instead of fractions are used in such occupations as mining, building, nursing or banking.

1 Adding and subtracting decimals

To add or subtract decimal numbers, we need to 'line up' the decimal points.

Example

Find 9.284 + 18.3.

Answer

$$\begin{array}{r} 9.284 \\ +\ 18.\mathbf{300} \\ \hline 27.584 \end{array}$$

← Insert zeros as place holders.

Example

Calculate 356.7 – 39.97.

Answer

$$\begin{array}{r} 356.7\mathbf{0} \\ -\ 39.97 \\ \hline 316.73 \end{array}$$

← Insert zero as a place holder.

2 Multiplying and dividing by powers of 10

Because decimal numbers are based on powers of 10, it is very easy to multiply or divide by 10, 100, 1000, and so on. The answer will contain the same digits as when we multiply or divide by 1, but the decimal point will be in a different position.

Example

Calculate 6.75 × 10.

Answer

Step 1: Ignoring the decimal point and multiplying by 10 gives 6750.

Step 2: There are two decimal places in the numbers in the question, so there must be two decimal places in the answer.

$$\begin{aligned} 6.75 \times 10 &= 67.50 \\ &= 67.5 \end{aligned}$$

We see that, when we **multiply** by 10, the decimal point moves one place to the **right**.

When we **divide** by 10 (or 100, 1000, etc.) the decimal point moves to the **left**.

Example

342.7 ÷ 10 = 34.27

3 Multiplying decimal numbers together

To multiply decimals, we:

- multiply the numbers ignoring the decimal point(s)
- count the total number of decimal places in both numbers being multiplied in the question
- insert the decimal point in the answer, making sure the answer has the same number of decimal places as the total number of decimal places in the question.

(We should also always work out an estimate of the answer as a check.)

Example

Find 85.3 × 4.2.

Answer

85.3 × 4.2

There is a total of two decimal places.
An estimate is 80 × 4 = 320.

$$\begin{array}{r} 853 \\ \times \quad 42 \\ \hline 1706 \\ 34120 \\ \hline 35826 \\ \hline \end{array}$$

The answer is 358.26 (two decimal places).

4 Dividing decimals

When dividing by a whole number, the decimal point stays fixed.

Example

Find 75.5 ÷ 5.

Answer

When dividing by a whole number, the decimal point stays fixed.

$$5\overline{)75.5} = 15.1$$

So 75.5 ÷ 5 = 15.1

Example

Find 81.04 ÷ 0.5.

Answer

When dividing by a decimal number, we must change the numbers so that we are dividing by a whole number.

First, multiply both numbers by 10, so the new question is 810.4 ÷ 5.

Note that $\frac{81.04}{0.5}$ is equivalent to $\frac{810.4}{5}$.

$$5\overline{)810.40} = 162.08$$

So 81.04 ÷ 0.5 = 162.08

EXERCISE

1 Find:

- a 2.34 + 6.9
- b 23.4 + 8.15
- c 6.23 + 11.29 + 4.1
- d 135.6 + 214.75 + 98.34
- e 26.321 + 8.5 + 63.92

2 Find:

- a 34.4 – 5.3
- b 156.09 – 67.89
- c 10.8 – 7
- d 13.445 – 9.68
- e 10 – 8.97

3 Find:

a $45 + 9.03 - 2.46$ b $12.2 - 0.1 + 22.5$ c $11 - 0.098 + 26.78$
d $67 - 18.97 - 6.45$ e $0.453 + 34 - 0.8$

4 Calculate:

a 78.3×1000 b $3.56 \div 10$ c 1.24×1000
d $45.7 \div 1000$ e 8.22×10 f $1.45 \div 100$

5 Given that $85 \times 41 = 3485$, write the answer to each of these products:

a 8.5×41 b 8.5×4.1 c 85×0.41
d 0.085×4.1 e 8.5×0.0041

6 Calculate:

a 3.4×4 b 54.78×5 c 23×2.5
d 6.2×0.4 e 1.7×0.2 f 0.023×12

Remember

A **repeating** or **recurring** decimal is one in which one or more of the digits repeat.

For example $\frac{1}{3}$ means $1 \div 3$.

$$3\overline{)1.000000} = 0.333333\ldots$$

$0.333\,333\ldots = 0.\dot{3}$

$\frac{1}{7} = 1 \div 7 = 0.142857142857\ldots = 0.\overline{142857}$

7 Find:

a $61.5 \div 3$ b $1.428 \div 7$ c $0.582 \div 3$
d $7.8 \div 0.2$ e $0.63 \div 0.9$ f $1.44 \div 1.2$

8 **Families of recurring decimals.**

You will have noticed that the decimal equivalent of some fractions does not terminate (does not come to an end).

Copy and complete the table of fraction families on the right. The members of these families of fractions form recurring decimals. See if you can find the pattern in each family and use it to predict the next answer.

Fraction	Recurring decimals
Thirds family ($\frac{1}{3}$ and $\frac{2}{3}$)	$0.\dot{3}$
Sixths family ($\frac{1}{6}, \frac{2}{6}$, etc.)	
Sevenths family ($\frac{1}{7}, \frac{2}{7}$, etc.)	
Ninths family ($\frac{1}{9}, \frac{2}{9}$, etc.)	

Lesson 5

Practical calculations

Now that we have practised all of the number operations, let us see how we use these in everyday life.

EXERCISE

Show your working out for each of these problems.

1 A store owner has 10 baskets of mangoes for sale at K5.25 a basket. How much will he receive in total when he sells them?

2 How much does Simeon pay for supplies to build a shed if he buys 3 packets of nails costing K4.26 each, 8 brackets costing 65 toea each and 4 lengths of timber costing K3.12 each?

3 Imelda's garden has 300 plants. Half of the plants are peanuts, one-third of the plants are beans and the rest are yams. How many yam plants does she have?

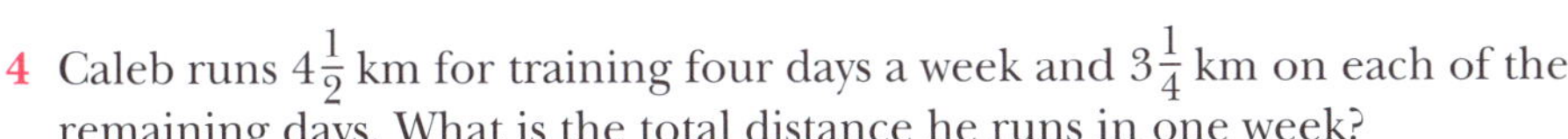

4 Caleb runs $4\frac{1}{2}$ km for training four days a week and $3\frac{1}{4}$ km on each of the remaining days. What is the total distance he runs in one week?

5 **a** How many bottles containing 1.25 L can be filled from a barrel containing 25 L?

b When eight bottles have been filled from a full barrel, how many litres remain in the barrel?

6 Mr Pao needs to move 135 kg of soil using a bucket that can carry 7.5 kg. How many times does he need to fill the bucket to move all the soil?

7 The fishermen on one fishing boat caught $6\frac{1}{3}$ kg of fish and the fishermen on another boat caught $9\frac{1}{2}$ kg fish. They shared the total catch between five families. How much was each family's share?

8 A tank holding 1000 L of water is leaking at a rate of 4.5 L per hour.

a How many litres are in the tank after one day?

b How long before the tank is empty?

9 The rainfall recorded for four months in a city was:

April: 56.7 mm
May: 19 mm
June: 76.5 mm
July: 45 mm

a What was the total rainfall over the four months?

b How much more rain fell in June than in May?

c What was the average rain over the four months?

d If the average monthly rainfall for the year for this city was 52.3 mm, what was the total rainfall of the other eight months?

10 Times for the US Nationals 100 m junior backstroke championships are shown in the table.

a Who was the fastest swimmer? (*Hint*: Will that person's time be the smallest or largest number?)

b Who was the second slowest swimmer?

c Which swimmer had a time nearest to 54 seconds?

d Which two swimmers were closest in finishing time?

e What is the average swimming time, correct to two decimal places?

Name	Time (seconds)
Lucy	54.63
Petra	54.35
Stephanie	53.17
Christine	54.49
Barbara	53.82
Tiffany	52.52
Ruth	53.45
Sonya	54.45

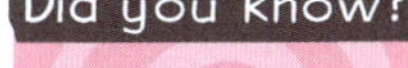

Did you know?

Enga province has a universal language that is spoken throughout the whole province.

Words are used in place of figures and counting is only used when counting real-life materials or items. This traditional counting system incorporates the four major mathematical operations and is known and used widely throughout the province.

RATIO AND PERCENTAGE

Recognising and using percentages

A percentage is a special type of fraction with a denominator of 100.

$x\% = \frac{x}{100}$ or x parts out of 100.

So 7% is 7 parts out of 100, or $\frac{7}{100}$, or 0.07.

Discussion

- From your previous work with percentages, explain the meaning of 0%, 50% and 100%, and give an example of each.
- Explain the meaning of percentages greater than 100%, such as 200% and 420%.
- Describe some situations in which you see figures expressed as percentages (in school, in the newspaper, in the store, etc.).

EXERCISE

Choose the correct answer (**A**, **B**, **C** or **D**) by estimating the percentages required.

1 How much drink is in the glass?

A 10% **B** 25%
C 50% **D** 80%

2 How much charge is left in the battery?

A 20% **B** 40%
C 60% **D** 80%

Charge left

3 How much of the floor still needs to be tiled?

A 30%
B 45%
C 65%
D 90%

4 How much of the pizza is left?

A 10% **B** 25%
C 50% **D** 80%

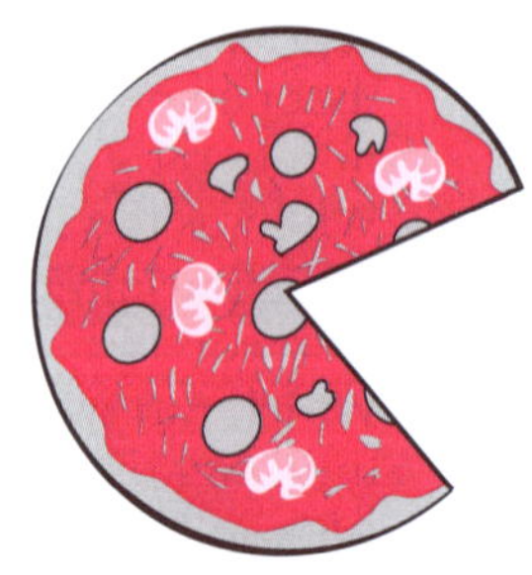

5 Four people are running in a race as shown below.

a Which runner has over 90% of the distance still to run?
b Which runner is nearly 50% of the way to the finish?
c Which runner has covered 95% of the distance?

6 If the illustration on the right represents a full lemonade bottle, which of the illustrations below shows the same bottle after 85% of the lemonade has been drunk?

A
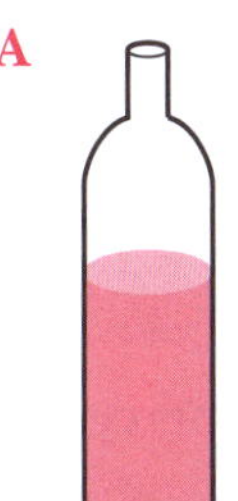

B
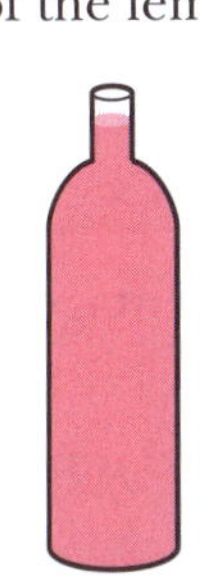

C

D

7 Copy and complete the table of commonly used percentages on the right.

Percentage	Fraction	Decimal
5%		
10%		
20%		
25%		
50%		
60%		
75%		
80%		
100%		

8 Convert each of the following percentages to a simple fraction:

a 3% **b** 30%
c 300% **d** 40%
e 55% **f** 68%
g 33% **h** 110%
i 4%

9 Express each of the following percentages as a decimal:

a 43% **b** 8% **c** 104% **d** 48% **e** 21%
f 35% **g** 66% **h** 125% **i** 1%

Remember

Every percentage can be written as a fraction or decimal.

- To change a percentage into a fraction, write it with 100 as the denominator and cancel the fraction to its simplest form.
  For example:
  $80\% = \frac{80}{100} = \frac{4}{5}$
- To change a percentage into a decimal, divide it by 100.
  For example:
  $35\% = 0.35$
  $2\% = 0.02$

Changing fractions and decimals to percentages

Lesson 2

To convert a fraction or a decimal to a percentage, multiply by 100.

Example

Convert to a percentage:

a $\frac{3}{5}$ **b** 0.68 **c** $\frac{2}{3}$

Answer

a $\frac{3}{5} \times 100 = \frac{300}{5}\% = 60\%$

b $0.68 \times 100 = 68\%$

c $\frac{2}{3} \times 100 = \frac{200}{3}\% = 66\frac{2}{3}\%$

To convert a decimal percentage to a fraction:

- write it with a denominator of 100
- multiply both the numerator and denominator by a power of 10 to remove the decimal point
- cancel to the simplest fraction.

Example

Express 4.55% as a fraction.

Answer

$$4.55\% = \frac{4.55}{100} = \frac{4.55}{100} \times \frac{100}{100} = \frac{455}{10\,000} = \frac{91}{2000}$$

EXERCISE

1 Express each of the following amounts as a percentage:
 a 0.34 b 0.465 c 0.050 d 0.002 e 2 f 5.7

2 Express each of the following fractions and mixed numbers as a percentage:
 a $\frac{7}{100}$ b $\frac{7}{50}$ c $\frac{3}{40}$ d $\frac{1}{25}$ e $1\frac{1}{4}$ f $3\frac{1}{50}$

3 Express each of the following values as a percentage:
 a 0.14 b 2.3 c $\frac{2}{50}$ d 0.66 e $\frac{7}{20}$
 f 24.5 g $\frac{9}{40}$ h 1.00 i $3\frac{8}{25}$

4 Convert each of the following percentages into a:
 i decimal ii fraction.
 a 22.5% b 6.5% c 0.25% d 120.5% e 87.5%

5 Of the students at Banare's school, 66% are girls.
 a What percentage of the students are boys?
 b What fraction of the students are girls?
 c What fraction of the students are boys?

6 Three students were asked how much of their project they had completed. Siri said 62%, Aaron said $\frac{3}{5}$ and Jemima said 0.64. Who had completed:
 a the most of their project? b the least of their project?

7 Samuel had 24 answers correct out of 30 questions on his science test. What percentage mark did he get?

8 Milla has 95% of her ice-lolly left and Lini has eaten $\frac{1}{27}$ of hers. Who has the most of their ice-lolly left?

Finding a percentage of an amount

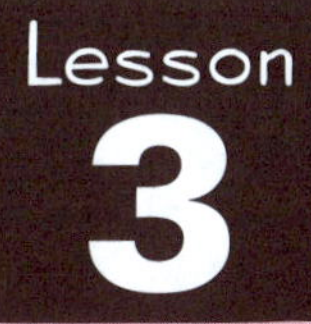

Finding a percentage of an amount is the same as finding a fraction of the amount.

Example

Find 15% of 300.

Answer

15% is equivalent to $\frac{15}{100}$, so we need to calculate $\frac{15}{100} \times 300$.

$$\frac{15}{100} \times 300 = \frac{15}{{}_1\cancel{100}} \times \frac{\cancel{300}^3}{1} = 45$$

Remember

Show all working out, not just the answer.

EXERCISE

1 Find:

a 20% of 75	b 25% of 8000	c 16% of 200
d 30% of 9	e 80% of 12.5	f 40% of 3.5
g 65% of 20	h 15% of 20	i 75% of 48

2 Nikki earns K15 at the store and decides to put 30% of it in the bank.
   a How much does she bank? b How much does she have left to spend?

3 15% of the passengers on a PMV are school children. If the PMV has 60 passengers, how many are school children?

4 Ismene's vegetable garden has 240 plants. Of the plants, 30% are kaukau, 45% are taro and the rest are yams. How many of each type of plant does she have?

5 A village has a population of 380 people. Of the villagers, 70% have visited Rabaul. How many villagers have not visited Rabaul?

6 There are 80 boys and 64 girls in a school. If 5% of the boys are left-handed and 12.5% of the girls are left-handed, how many students altogether are left-handed?

7 The coffee bean crop in one area was 1750 kilograms. Unfortunately, 4% of the crop was damaged before it was sold and the rest was sold at K8 per kilogram. What was the total of the money raised from the sale?

8 A contest has prize money of K10 000, 75% of which goes to the first-prize winner. The runner-up receives 15% and the rest is divided equally between the other five losing contestants.
   a What percentage does each of the five losing contestants receive?
   b How much money does the first-prize winner receive?
   c How much does the runner-up receive?
   d How much more money does the winner receive than each of the five losing contestants?

9 In 2003, a certain village had a population of 1850. By 2008, the population had increased by 8%. How large was the population in 2008?

10 At the last big athletics carnival in a region there were 230 entrants. It is expected that there will be 10% more entrants at the next carnival. How many entrants will there be altogether at the next carnival?

Lesson 4 Expressing one quantity as a percentage of another

When expressing a quantity as a fraction or percentage of another quantity we must make sure that we are using the same unit of measurement for both the quantities. Then we write the two quantities as a fraction and multiply by 100.

Example

Express 1 gram as a percentage of 8 grams.

Answer

Percentage required $= \frac{1}{8} \times 100\%$

$= 12.5\%$ (or $12\frac{1}{2}\%$)

Example

Express 16 mm as a percentage of 20 cm.

Answer

First convert: 20 cm = 200 mm

Percentage required $= \frac{16}{200} \times 100\%$

$= 8\%$

Help box

To express one quantity as a fraction of another:

- make sure both quantities are expressed in the same unit of measurement
- express as a fraction
- multiply by 100 and add the percentage sign.

EXERCISE

1 a Express each of the following as a percentage:
   i 4 out of 5 ii 4 out of 50
   iii 4 out of 500 iv 4 out of 5000
  b Can you see a pattern? Use the pattern to help you write 4 out of 500 000 as a percentage.

2 Write each of the following as a percentage:
  a 13 out of 40 b 20 out of 80 c 145 out of 200
  d 17.5 out of 20 e 25 out of 125 f 18 out of 72
  g 18 out of 50 h 1 out of 8 i 13 out of 26

3 Express the given number as a percentage of the number in brackets each time, giving your answers correct to two decimal places:
  a 12 (34) b 6 (9) c 564 (24)

4 Write the given amount as a percentage of the amount in brackets each time, giving your answers correct to two decimal places:
  a 3 cm (4 cm)
  b 4 mm (5 cm)
  c 25 g (2 kg)
  d 4 hours (30 hours)
  e 5 seconds (4 minutes)

5 Dorcas obtained a mark of 67 out of 79 for her mathematics test. What was her percentage score?

6 Mino purchased a second-hand car for K800. Six months later, after repairing it, he sold it for K1000. What percentage of his purchase price did he sell it for?

7 For their science tests, Kila scored 54 out of 86 and Mirou scored 45 out of 67. Change these to percentages, correct to one decimal place, to determine who had the better percentage, and by how much.

Using ratios

Lesson 5

From our previous maths work, we know that a ratio is a comparison between two quantities. For example, if there are 8 boys and 12 girls in your class, we can say that the ratio of **boys to girls** is 8 to 12, or the ratio of **girls to boys** is 12 to 8.

boys : girls = 8 : 12

girls : boys = 12 : 8

Notice that the ratio is written in a particular order matching the order in which it is stated.

A ratio can also be expressed as a fraction, so the ratio of boys to girls in the example just discussed is $\frac{8}{12}$. This can be simplified to $\frac{2}{3}$.

8 : 12 and 2 : 3 are called **equivalent ratios**.

One application of ratio is when goods are exchanged in bartering. For example, in the Sepik provinces, food materials and the traditional money for pigs and for bride price are exchanged proportionally, according to a ratio.

Did you know?

The traditional method of ratio and proportion can be observed in the barter system. For example throughout the Sepik provinces, shell money is used for bride price and pig payment. The value of these shells corresponds to the size of the shells. Therefore the values range from the smallest to the biggest.

Discussion

- How are items exchanged using ratio in your area?
- Imagine that two families wanted to exchange pigs and chickens. Which of the following do you think would be a reasonable ratio to use?
  1 pig for 1 chicken?
  1 pig for 5 chickens?
  1 pig for 10 chickens?
  Or some other ratio?
  Explain how you would determine the fairest proportion in this case.

Remember

Both quantities in a ratio must be expressed in the same unit before simplifying.

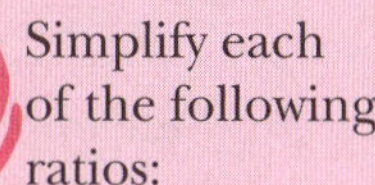

Example

Simplify each of the following ratios:

a 50t to K2

b 8 : 10 : 4

c 0.4 : 1.8

d $\frac{1}{2} : \frac{3}{4} : \frac{2}{5}$

Answer

a First, change K2 to toea, so we are comparing the same units.
50 : 200 = 1 : 4

b 8 : 10 : 4 = 4 : 5 : 2 Dividing each part of the ratio by 2.

c 0.4 : 1.8 = 4 : 18 Multiplying both parts by 10 to give whole numbers.
= 2 : 9 Dividing both parts by 2.

d $\frac{1}{2} : \frac{3}{4} : \frac{2}{5} = \frac{10}{20} : \frac{15}{20} : \frac{8}{20}$ Writing the fractions with a common denominator.
= 10 : 15 : 8 Multiplying by 20, to give whole numbers.

Remember

A ratio is:

- a comparison of two quantities measured in the same units
- expressed in a specific order
- written in the form $a : b$ or $\frac{a}{b}$
- expressed in its simplest form.

Help box

1 litre = 1000 millilitres

1 gram = 1000 milligrams

1 kilometre = 1000 metres

EXERCISE

1 Write each of the following ratios in its simplest form:

a 3 : 6 **b** 3 : 9 **c** 4 : 2 **d** 12 : 3
e 13 : 39 **f** 2 000 000 : 5000 **g** 256 : 48 **h** $\frac{2}{3} : \frac{7}{9}$
i 4 : 2.5 **j** 33 : 60 **k** 0.4 : 46 **l** 4.8 : 1.2

2 Write each of the following fractions as a ratio in its simplest form:

a $\frac{2}{4}$ **b** $\frac{3}{9}$ **c** $\frac{12}{18}$ **d** $\frac{25}{75}$ **e** $\frac{13}{195}$ **f** $\frac{45}{27}$

3 Write two equivalent ratios for each of the ratios below.

a 1 : 3 **b** 4 : 5 **c** 1 : 3 : 4

4 Express each of the following ratios in its simplest form:

a 1 : 0.2 : 0.75 **b** $2.3 : \frac{6}{5} : 0.01$ **c** $\frac{5}{8} : \frac{7}{16} : \frac{11}{32}$ **d** $1\frac{1}{4} : 3\frac{1}{2} : 2\frac{2}{3}$

5 Find each of the following ratios, in simplest form:

a 1 hour to 20 minutes **b** 240 millilitres to 3 litres
c 80 milligrams to 2 grams **d** 75 metres to 50 centimetres
e 30 toea to K2.70 **f** 40 seconds to 20 minutes
g 1.6 kilometres to 800 centimetres **h** 2.4 kilograms to 300 grams
i 300 days to 4 years

6 A gardener spends 6 hours working. He spends $\frac{1}{4}$ of the time digging, $\frac{3}{8}$ of the time weeding, $\frac{1}{4}$ of the time planting and $\frac{1}{8}$ of the time cleaning up.

a What is the ratio of time spent digging to time spent weeding?
b What is the ratio of time spent weeding to time spent cleaning up?
c What is the ratio of time spent planting to time spent weeding?
d What time, in hours and minutes, was spent on each activity?

Lesson 6

Finding a quantity when the ratio is known

If we know one part of the ratio, we can easily work out the other part.

Example

The weights of two watermelons are in the ratio of 3 : 4. If the smaller watermelon has a weight of 6 kg, what is the weight of the larger watermelon?

Answer

small to large = 3 : 4
= 6 : ?

We can see that 3 × 2 = 6, so we need to multiply 4 by 2 to find the missing value. The equivalent ratio for 3 : 4 is 6 : 8. So the weight of the larger watermelon is 8 kg.

EXERCISE

1 Find x in each of the following pairs of equivalent ratios:

a $1 : 2 = x : 6$ b $2 : 5 = x : 15$ c $4 : 9 = 16 : x$

d $x : 8 = 6 : 16$ e $12 : 5 = 4 : x$ f $x : 2 = 1.5 : 4$

g $80 : 16 = 10 : x$ h $x : 3.4 = 22 : 17$

2 The weights of two boxes are in the ratio 7 : 12. The bigger box has a weight of 36 kg.

a What is the weight of the smaller box?

b What is the combined weight of the boxes?

3 The lengths of two sticks are in the ratio 11 : 3. The smaller stick is 15 cm long. How long is the longer stick?

4 The ratio of the weight of a baby pig to its mother is 3 : 17. The baby pig has a weight of 6 kg. What is the weight of the mother pig?

5 A soft drink is made from concentrate and water, mixed in the ratio 1 : 6. How many litres of water should be mixed with 500 mL of concentrate?

6 The ratio of boys to girls in a class is 6 : 5. If there are 15 girls in the class, what is the total number of students?

7 The ratio of cement to sand to gravel when making concrete is 1 : 3 : 5. If Tanua has 12 buckets of sand, how much cement and gravel will be needed to make concrete?

8 Brass is made of a mixture of copper, tin and zinc in the ratio 8 : 2 : 3.
How much copper and tin will be required if 1.5 kilograms of zinc is available to make brass?

9 A glider has a glide ratio of 1 in 10, which means it descends 1 unit for every 10 units travelled forward.

a How far has the glider travelled forward when it has descended 200 metres?

b If the glider is 900 metres above the ground, how much farther will it travel forward at this rate before reaching the ground? Give your answer in kilometres.

10 The proportions of an iceberg above and below the water is described by the ratio 1 : 9. If 80 cubic metres of the iceberg are visible above the water:

a how much of the iceberg is below the water?

b what is the total volume of the iceberg?

Challenge

a If 200 mL of medicine contains 150 mg of the active ingredient, how much medicine should be taken to provide 7.5 mg of the active ingredient?

b Two gliders start at the same height. One has a glide ratio of 0.3 and the other has a glide ratio of $\frac{2}{7}$. When they have dropped 1200 metres, which one will have travelled the farther distance, and by how far?

c There are three different mango fruit juice concentrates in the store: Fresh M, Go-mango and Mighty Mango. The recipe for Fresh M uses 2 cups of concentrate to 5 cups of water, while Go-mango uses 3 cups of concentrate to 8 cups of water and Mighty Mango uses 4 cups of concentrate to 9 cups of water.

i Write the ratio of 'concentrate : water' for each brand.

ii Work out which uses the most concentrate for 1 cup of water.

Lesson 7

Dividing a quantity in a given ratio

Example

Suppose that five men from two families went on a fishing trip and caught a total of 140 fish. How should they divide the fish fairly if there are three men from one family and two from the other?

Answer

They should share the fish in the same proportion as their contribution to the fishing.

There are five people altogether, so the family with three men fishing should receive $\frac{3}{5}$ of the fish.

$\frac{3}{5} \times 140 = 84$

The family with two men fishing should receive $\frac{2}{5}$ of the fish.

$\frac{2}{5} \times 140 = 56$

Check: Adding these two amounts together gives 140, which is the total number of fish caught.

Example

Share 30 coconuts in the ratio 4 : 11.

Answer

Total number of shares, $t = 4 + 11$
$= 15$

Share 1 $= 30 \times \frac{4}{15}$
$= 8$

Share 2 $= 30 \times \frac{11}{15}$
$= 22$

So the shares are 8 coconuts and 22 coconuts.

Example

Divide K2000 in the ratio 2 : 3 : 5.

Answer

Total number of shares, $t = 2 + 3 + 5$
$= 10$

Share 1 $= \frac{2}{10} \times 2000$
$= 400$

Share 2 $= \frac{3}{10} \times 2000$
$= 600$

Share 3 $= \frac{5}{10} \times 2000$
$= 1000$

So the shares are K400, K600 and K1000.

EXERCISE

Remember

To divide a quantity (n) into shares in the ratio $x : y$

- the total number of shares, t, equals $x + y$
- share 1 $= n \times \frac{x}{t}$
- share 2 $= n \times \frac{y}{t}$

This rule can be extended to calculate more than two shares.

1 Share K50 in each of the following ratios:

a 2 : 3 b 4 : 1 c 1 : 9 d 12 : 13
e 7 : 3 f 3 : 17 g 65 : 35 h 3 : 5

2 Divide a 50 cm piece of string in each of the following ratios:

a 1 : 1 b 4 : 1 c 7 : 13 d 3 : 2
e 2 : 3 f 3 : 17 g 17 : 8 h 9 : 11

3 A machine uses a mixture of petrol and oil in the ratio 5 : 1. How much petrol and oil will there be in 3 L of the mixture?

4 The ratio of pigs to chickens in a village is 2 : 7. If there are 126 animals altogether, how many pigs and how many chickens are there?

5 On a fishing trip, two boys catch 48 fish. Tao catches three times as many fish as Kaupa.

a What is the ratio of Tao's fish to Kaupa's fish?
b How many of the 48 fish should each boy take home for his share?

6 The ratio of flour : sugar : cocoa in a chocolate cake recipe is 4 : 3 : 1. If I want 12 kg of the mixture, how much of each ingredient do I need?

7 The ratio of children : teenagers : adults in a survey sample is 2 : 11 : 17.

a Which is the largest group represented?
b If there are 180 people surveyed, how many are:

i adults? ii not teenagers?

RATES

What are rates used for?

A **rate** is a comparison between two related quantities. Speed is the most commonly used rate. Speed compares the distance travelled with the time taken. We can also talk about a rate of pay, the rate at which water is rising in the river, the rate of population growth, and so on. What other rates can you name?

The two quantities related in a rate can be multiplied or divided by the same number.

Examples

- A rate of K4.60 per 2 kilograms is equal to a rate of K2.30 per kilogram (dividing both measures by 2).
- Travelling at 60 km per hour you would cover 240 km in 4 hours (multiplying both measures by 4).

EXERCISE

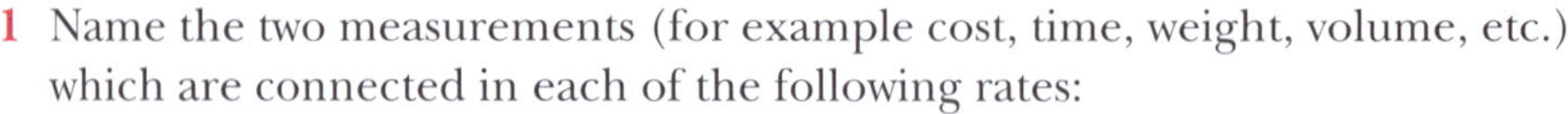

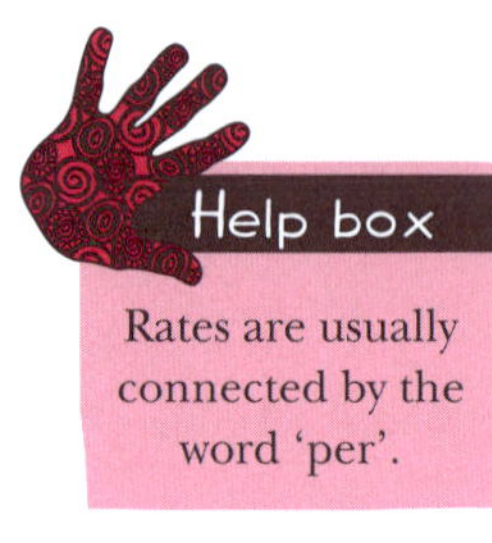

Rates are usually connected by the word 'per'.

1 Name the two measurements (for example cost, time, weight, volume, etc.) which are connected in each of the following rates:
  - a candlenuts are K2 per kilogram
  - b a medicine dose is 5 mL per day
  - c the rainfall is 238 mm per year
  - d a hire car charges K3.50 per kilometre
  - e a mobile phone charge is 31 toea per 30 seconds of call

2 Copy and complete each of the following rate statements:
  - a 20 metres of irrigation pipe costing a total of K4.60 means it is sold at a rate of ___ per metre.
  - b A 2 litre bottle of soft drink costing K3.70 means it is sold at a rate of ___ per litre.
  - c A township which has a population of 480 people and an area of 10 square kilometres has a population density of ___ people per square kilometre.
  - d A tanker pumping 350 litres of water in five minutes is pumping at a rate of ___ litres per minute.
  - e A painter working for six hours and earning K42 is earning at the rate of ___ kina per hour.

3 Describe a situation in which each of the following rates might be used:
  - a kina per kilogram
  - b milligrams per day
  - c litres per day
  - d kina per kilometre
  - e millilitres per day
  - f milligrams per kilogram
  - g kina per tonne
  - h kina per square metre

4
  - a A breadmaking machine pours out 50 kilograms of mixture in 8 minutes. At what rate is the mixture being poured?
  - b A paper mill wasted 30 centimetres of the paper roll in producing every 150 metres of paper. What is the rate of wastage per metre?
  - c A commercial laundry used 6300 litres of water in six days. What is the rate of water use per day?
  - d Grace is paid K45 for 9 hours work. At what rate is she being paid?
  - e A PMV travels 4 km in 5 minutes. At what rate is it travelling (in km/h)?

5 A jogger covers 5 kilometres in one hour. At this rate, how far will she jog in:

a 3 h? b $1\frac{1}{2}$ h? c $2\frac{1}{4}$ h? d 45 min? e 1 h 10 min?

6 Paint solvent costs 80 toea per litre. What is the cost of:

a 10 litres? b 5 litres? c 4.5 litres? d 23 litres? e $7\frac{1}{4}$ litres?

7 A shower uses 20 litres of water per minute. How many litres will be used in a shower lasting:

a 2 min? b 3 min? c 3.5 min? d 45 seconds? e $2\frac{1}{4}$ min?

8 A fast food restaurant makes 350 hamburgers a day. At this rate, how many hamburgers would be made in:

a 2 days? b 5 days? c 1 week? d 3 weeks? e 1 year?

9 Orim spreads 30 grams of fertiliser per square metre on his garden. How much fertiliser does he use if the garden measures:

a 4 m^2? b 10 m^2? c 25 m^2? d 12.3 m^2? e 18.6 m^2?

10 Laskam's truck uses 28 litres of petrol per 100 kilometres. How many litres would each of the following trips use?

a 10 km b 25 km c 40 km d 250 km e 35.6 km

Rate tables and graphs

Lesson 2

Information is often recorded using tables or graphs. These allow us to work out the rates involved or to calculate other information based on the table or graph.

EXERCISE

1 The table on the right shows the charge rates for different mobile phone calls with UCall phone company. Rates are per 30 seconds, but are charged by the second.

Type of call	Peak	Off peak
National	K0.80	K0.30
International	K1.35	K0.90
SMS message	K0.30	K0.30

Calculate the cost of each of the following:

a an off peak national call lasting 40 seconds
b a peak international call of 3 minutes
c five SMS messages, off peak
d a peak national call for 2 minutes 15 seconds
e an off peak international call for 25 minutes

2 A builder is paid K60 for 6 hours work. Copy and complete the table below and plot the points on grid paper.

Number of hours worked	0	1	2	3	4
Kina earned					

a How much is the builder paid per hour?
b Using your graph, or otherwise, find how much the builder earns in:
 i an hour-and-a-half ii 7 hours
c Copy and complete this statement:
 'For every additional ________ he works, the builder earns ________.'

Remember

- Graphs should be drawn using pencil.
- Axes should be ruled and labelled.
- Each graph should have a title.

3 At an agriculture research farm, watermelons were grown in different plots which were all the same size. The watermelons were given different amounts of fertiliser to test its effectiveness. The results are shown in this table:

Amount of fertiliser used (cups)	0	1	2	3	4
Average length of watermelon (cm)	30	32.5	35	37.5	40

a Copy the table and graph the points.

b Using your graph, or otherwise, calculate the expected length of a watermelon if each of the following amounts of fertiliser was used:

i $1\frac{1}{2}$ cups **ii** 5 cups

c Copy and complete the statement:
'As the amount of fertiliser increased by 1 cup, the length of the watermelon ______________.'

d Do you think it would make sense to extend the graph indefinitely? Explain.

4 A large corrugated iron tank, used for drinking water, has sprung a leak right at the bottom of the tank. Mr Gip checked the water level above the leak daily and found these results:

Number of days	0	2	4	6	8
Height of water (cm)	400	296	192		

a Copy and complete this statement:
'The water level is dropping at a rate of ____ cm per ____.'

b Plot the points on grid paper.

c Using your graph, or otherwise, find the height of the water after 6 days if it continues leaking at the same rate.

d The tank will be mended on day 8. Will there still be water in the tank?

5 The graph on the right shows the temperature (in degrees Celsius) of a mixture over a number of minutes.

a How can you see from the graph that the mixture is cooling?

b What temperature is the mixture to start with?

c By how many degrees has the mixture cooled in 40 minutes?

d What is the rate of cooling?

e What is the temperature of the mixture after 50 minutes?

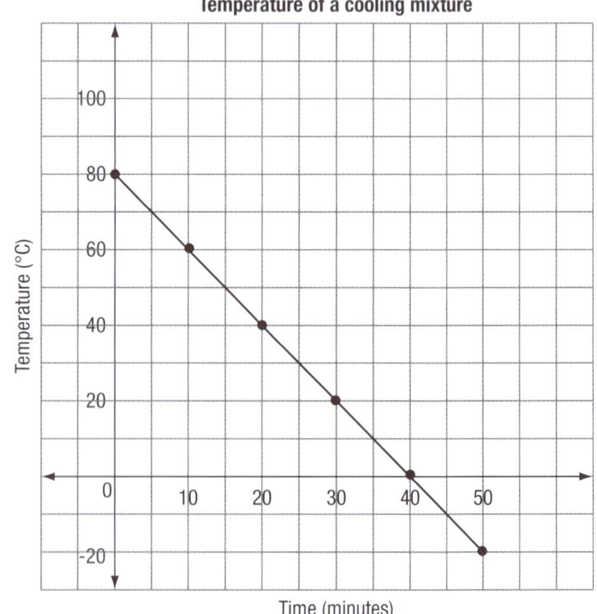

6 At a fish breeding farm, a large pond is being filled with water. The graph on the right shows the relationship between the number of minutes and the number of litres of water pumped into the pond.

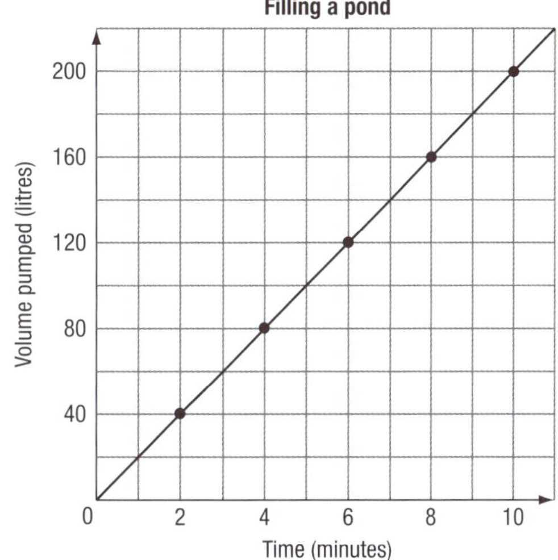

a How many litres of water have been pumped into the pond in 5 minutes?

b At what rate (in litres per minute) is the pond being filled?

c How many litres will be in the pond after 10 minutes?

d At this rate, how many litres will be in the pond after 1 hour?

e If the pond has a capacity of 1 000 000 L, how long will it take to fill at this rate?

7 A town has two freight services, Zippy Vans and Top Trucks. Their charges for delivering parcels are shown in the table and graph below.

Freight service	Fixed cost	Rate per km
Zippy Vans (*Z*)	K5.00	K0.80
Top Trucks (*T*)	K9.00	K0.60

a Copy the graphs and use the information given to label the graphs (*Z*) and (*T*).

b Copy and complete the table below:

Kilometres travelled	Total Cost	
	Zippy Vans	Top Trucks
5		
10		
15		
20		
25		
30		
35		
40		

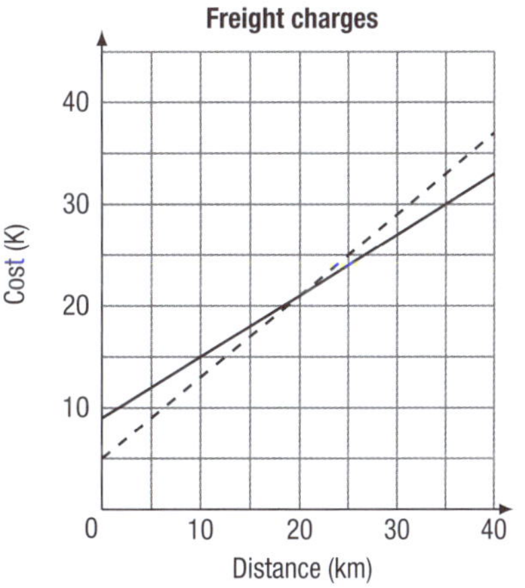

c For what length of trip are the total costs the same?

d A company wants to use the freight services for different trips. Write a simple set of instructions that will help the company choose the most economical service.

Lesson 3

Speed and distance tables and graphs

Travel graphs show the distance an object or person has moved over a particular time period. If they are travelling at a constant rate, the graph will be a straight line. There can be more than one rate of travel as the object speeds up, slows down or stops.

Remember
Average speed equals distance divided by time.

EXERCISE

1 **a** Copy and complete these tables showing the speeds at which different animals can travel:

Animal	Rate (km/h)	Distance travelled in 2 hours
Rabbit	56.3	
Pig	17.7	
Chicken	14.5	
Tortoise	0.27	
Snail	0.05	

Animal	Rate (km/h)	Distance travelled in 15 minutes
Cheetah	112	
Lion	80	
Zebra	64	
Giraffe	51.5	
Elephant	40.4	

b Considering both tables above, would it make sense to use these figures to calculate how far each animal could travel in 4 hours? Explain your answer.

2 Calculate the average speed, given each of the following distances and times:

a 200 km in 4 hours
b 150 km in $1\frac{2}{3}$ hours
c 450 km in 30 minutes
d 10 km in 2 hours
e 56 km in 1 h 45 min

3 Match each speed in Question **2** with the correct means of transport below.

a a jet aeroplane
b a person walking
c a PMV
d a car on a freeway
e a heavily loaded truck

4 Think about how distance changes over time when an object is moving. Match each distance–time graph below to one of the descriptions **A** to **E**.

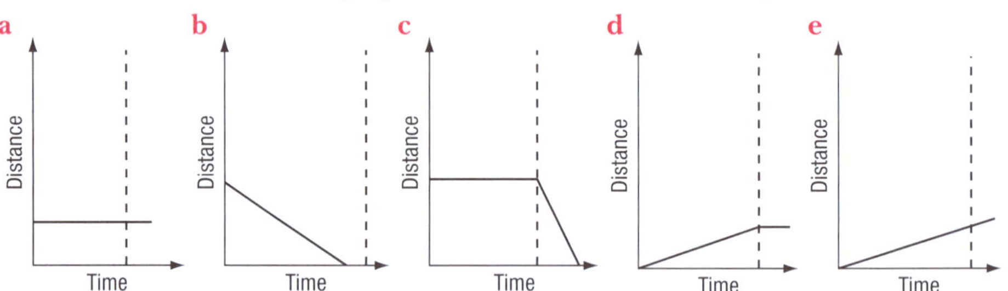

A The object is stationary.
B The object is moving away at a constant rate.
C The object first moves away at a constant rate then stops.
D The object is stationary then starts moving at a constant rate towards you.
E The object moves towards you and stops when it reaches you.

Challenge
Can you give another explanation for graph **a** in Question **4**?

5 Leo's mother walks quickly to the store, spends 15 minutes there, then walks home carrying the shopping. Which graph below best matches this information? Copy the graph and explain why you chose it.

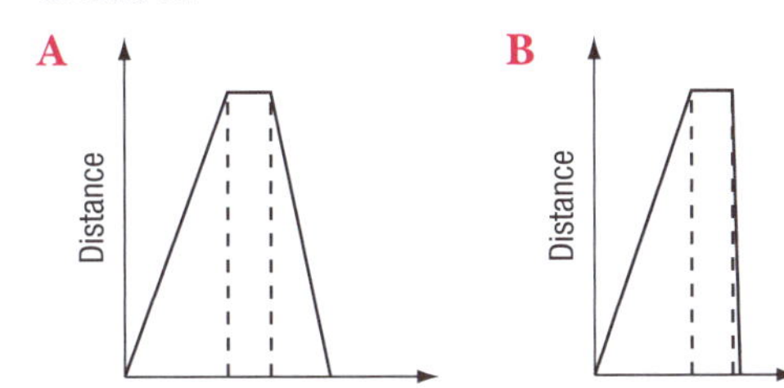

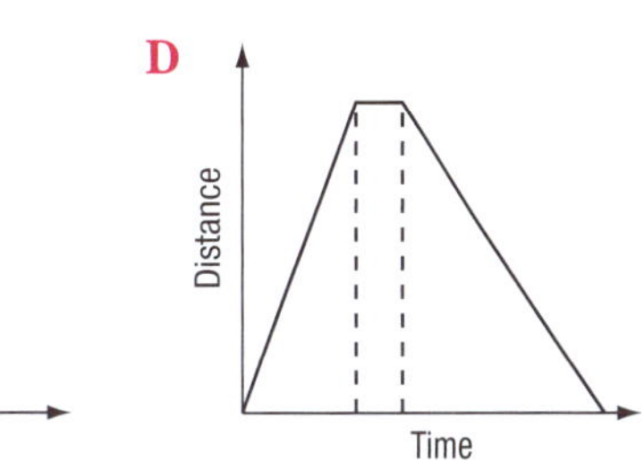

6 Amos works at a stationery supply store in Rabaul and delivers orders on his motorbike. The graph below shows one of his trips.

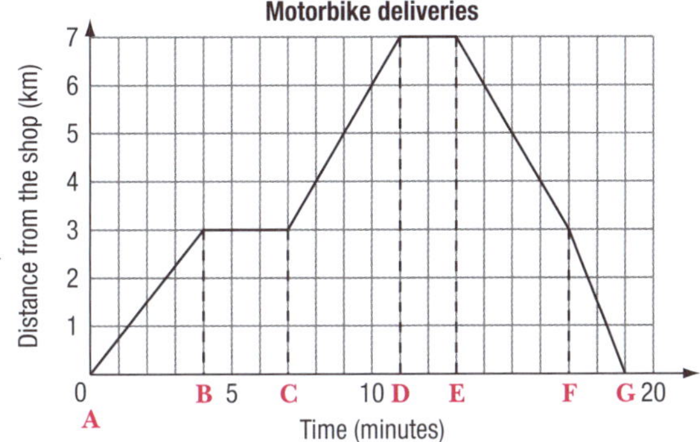

a How long did it take Amos to reach his first stop?
b How far did Amos travel to get to his first stop?
c What was his average speed (in km/h) for this first part of his trip?
d When did he travel faster: from time **A** to time **B** or from time **C** to time **D**? Give a reason for your answer.
e How far was the farthest point Amos travelled from the shop?
f How long did it take Amos to return to the shop?

7 The two graphs below illustrate Kenai's trip home from school. Graph A is what Kenai told his parents. Graph B is what actually happened. Write a story to match the different version of events.

a

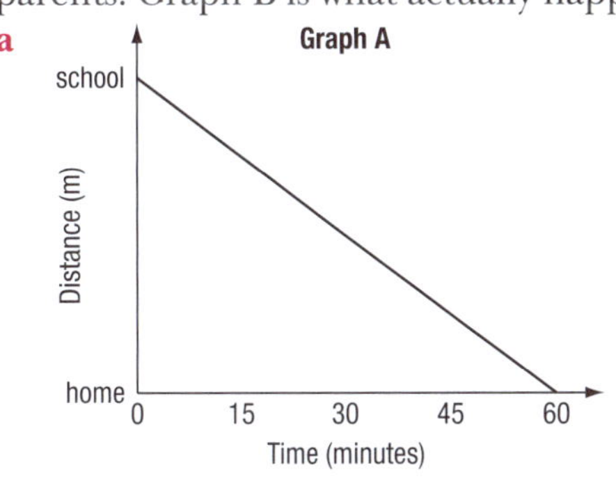

b

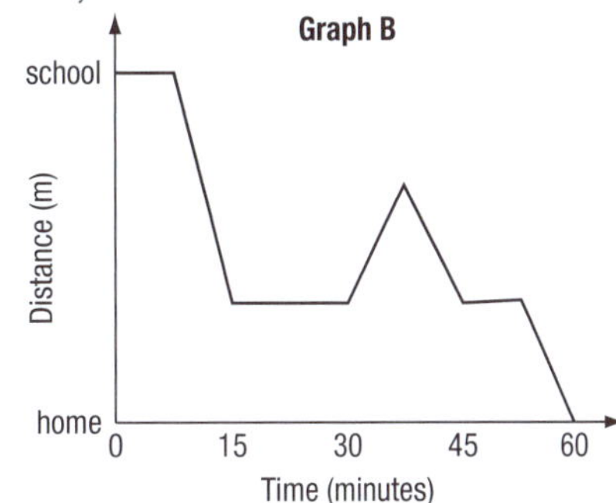

8 Create a graph that represents your trip home from school on a typical day and write a paragraph explaining it. Remember to label the axes for distance and time and show approximate values matching your trip.

Interpreting graphs

We have all seen temperature or rainfall graphs before, but temperature and rainfall are only two of the many practical situations graphs can be used to display. Can you suggest some advantages graphs have that sometimes make them better than written descriptions?

EXERCISE

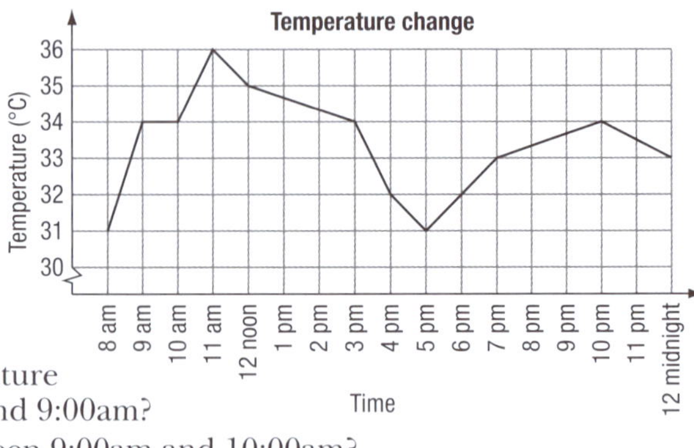

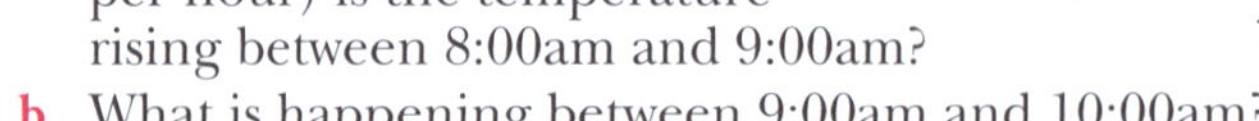

1. This graph shows the change in temperature in a tropical town between 8:00am and midnight.

   Use the graph to answer the following questions.

   a. At what rate (in degrees per hour) is the temperature rising between 8:00am and 9:00am?
   b. What is happening between 9:00am and 10:00am?
   c. What is the maximum temperature for the day?
   d. When does the maximum temperature occur?
   e. For how many hours is the temperature rising? How does the graph show this?
   f. When is the temperature rising the fastest? How does the graph show this?
   g. When does a cool change hit the town.
   h. Between what times was the temperature dropping the quickest?
   i. At what two times was the temperature falling at the same rate?

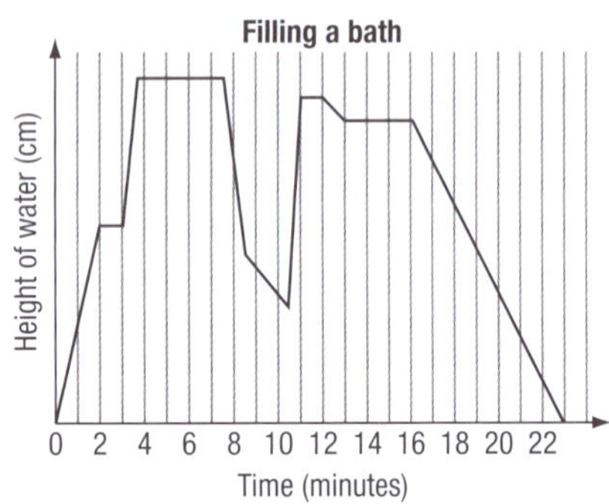

2. A zookeeper needed to bath a gorilla. The graph on the right shows the height of the bath water over time.

   a. For how long was there water in the bath?
   b. What happened about 8 minutes after water started being run in the bath?
   c. Write a story describing the gorilla's bath time.

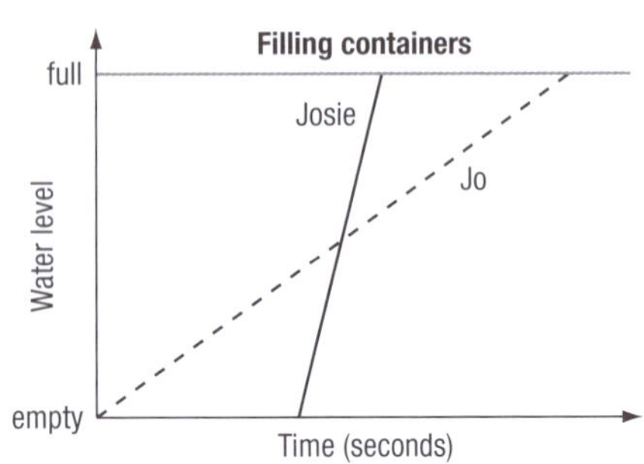

3. Jo and Josie are filling identical water containers from a tank. Use the graph on the right to answer these questions.

   a. Who starts filling their container first?
   b. Whose container was filled first?
   c. Who fills their container more quickly? How do you know?
   d. If a third person, Peter, started filling another container at the same time as Josie and at the same rate as Jo, sketch graphs of the filling of all three containers.

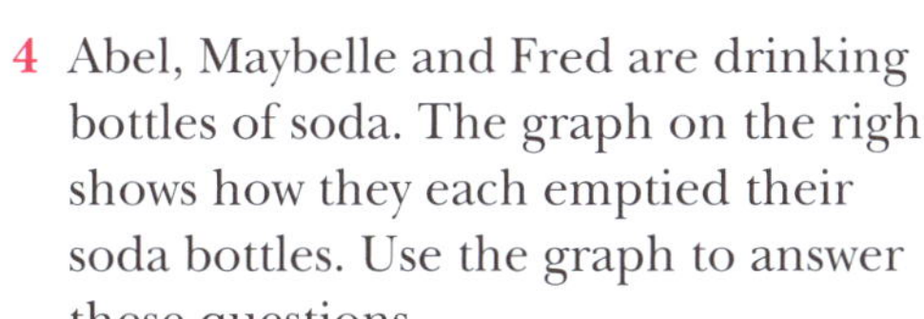

4 Abel, Maybelle and Fred are drinking bottles of soda. The graph on the right shows how they each emptied their soda bottles. Use the graph to answer these questions.
   a Who finished first?
   b Who drank their bottle without stopping? How can you tell?
   c Who had the longest rest in the middle of drinking their bottle? How can you tell?

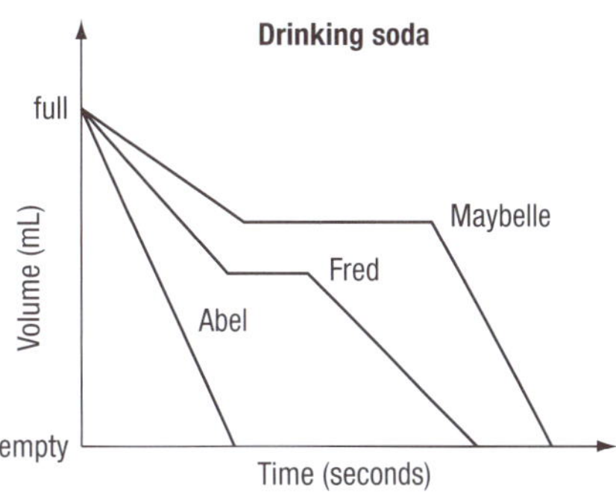

5 Sung, Kate and Jake have agreed to dig a vegetable garden. The graph on the right shows their rate of work.
   a Copy and complete this statement: 'The graph shows the rate of digging in ____ per ________.'
   b Who stopped in the middle of digging to talk to a friend?
   c Who finished their third of the digging first?
   d Who dug at the slowest rate?

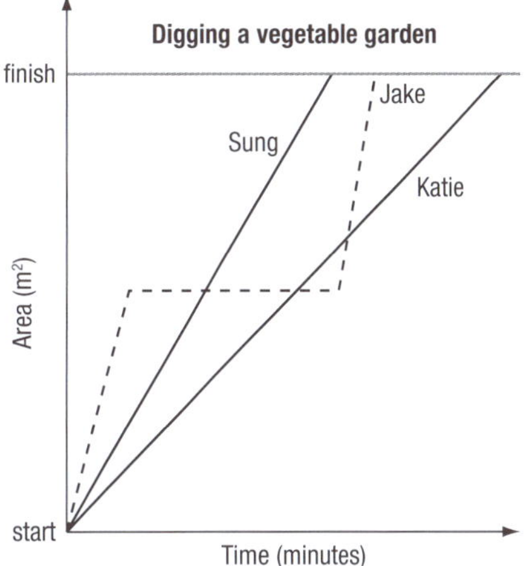

Graphs showing variable rates

Some of the graphs we have seen earlier involved constant rates, where the rate remains the same throughout. In such cases, the graph is a straight line. In other cases, several constant rates can be shown by different straight line segments.

We are now going to look at these variable rates.

Consider filling two containers with water (which is flowing from the tap at a steady rate). In the first container (a cylinder), the water rises steadily and the graph showing the height of water versus time is a straight line.

Container 1

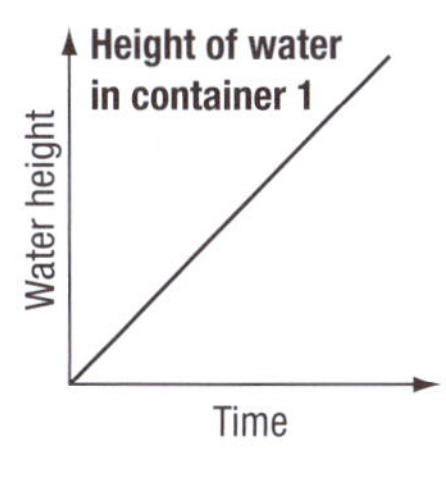

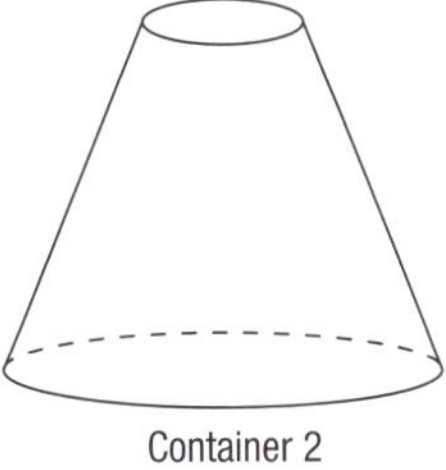
Container 2

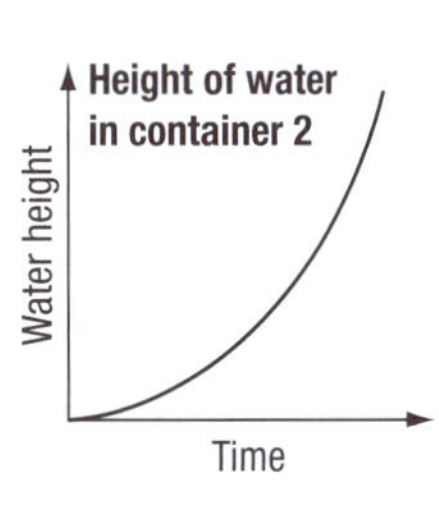

In the second container, the rate at which the water level rises changes because of changes in the container's shape. The water level increases slowly at first, then more and more quickly as the container narrows and fills.

EXERCISE

1 Three different containers of the same height are being filled with water flowing at a constant rate. The graph shows the change in water level over time for container A.

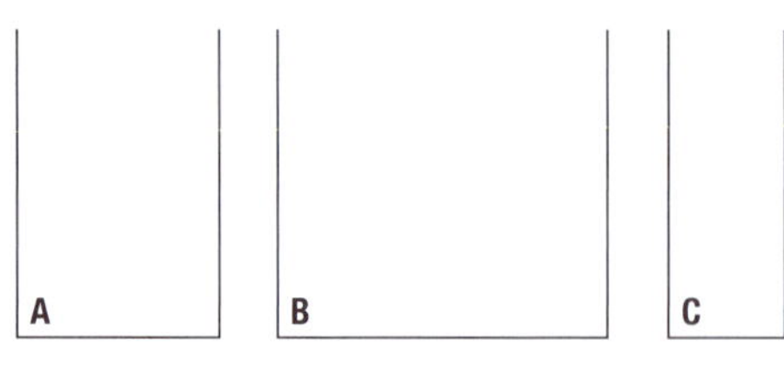

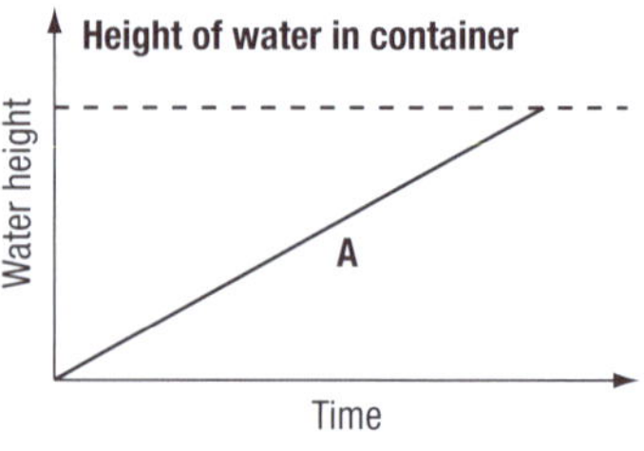

a Explain which container will fill the most quickly.

b Copy the graph for container A and sketch the graphs for containers B and C on the same set of axes.

2 Another three different containers of the same height are being filled with water flowing at a constant rate. The graph shows the change in water level over time for container A.

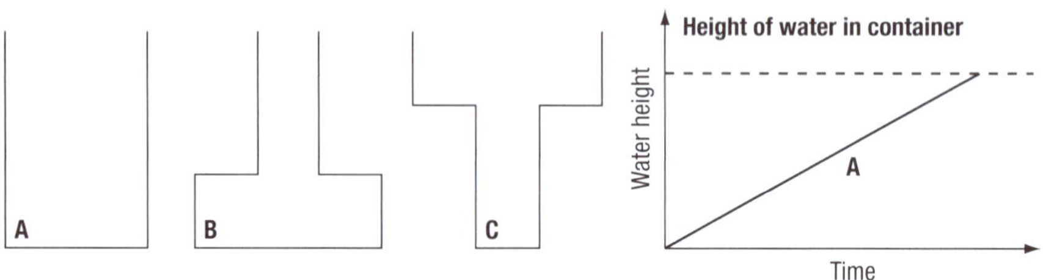

a Describe how the water level will change in:
- **i** container B
- **ii** container C.

b Copy the graph for container A and sketch the graphs for containers B and C on the same set of axes.

3 Match the graphs showing the change in water level over time to the different shaped vases, all of which are being filled at a constant rate.

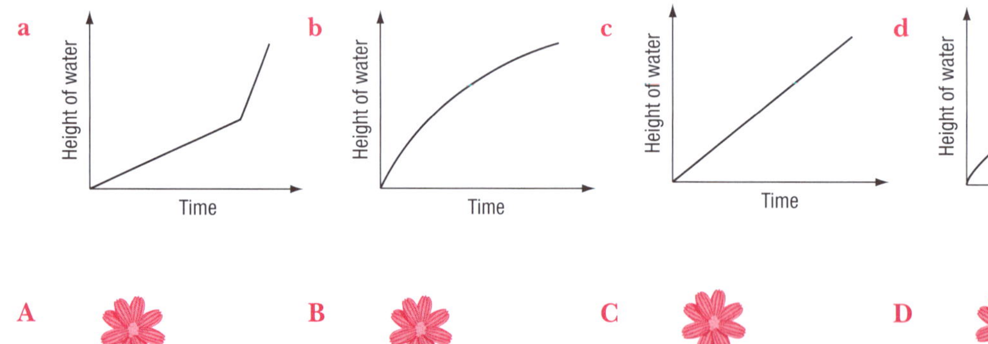

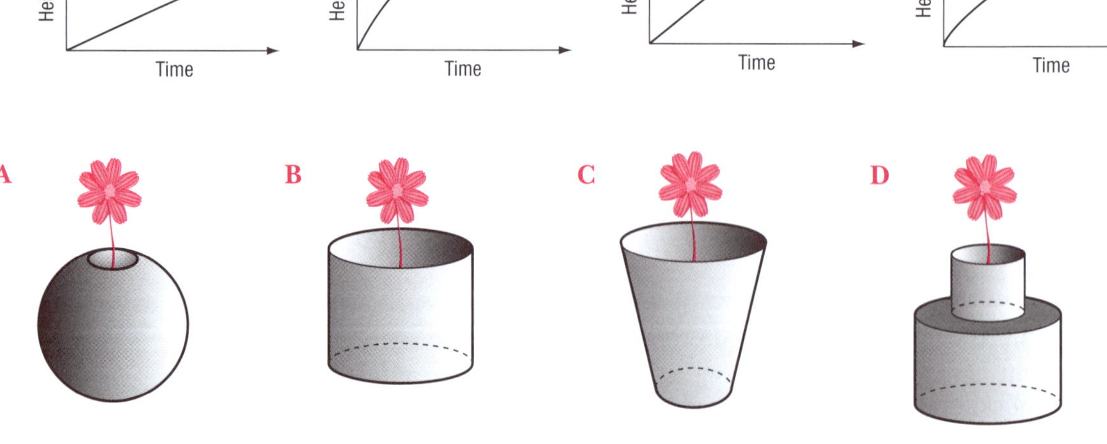

4 Draw the height–time graph for the water level if these two containers are filled with water flowing at a constant rate. Show time on the horizontal axis and height on the vertical axis.

a

b

5 The graph on the right shows the speed of a formula 1 racing car at certain distances from the starting line.

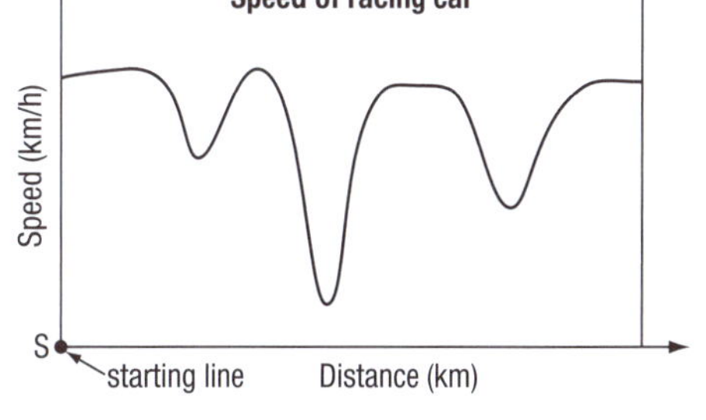

- a How can you tell that the graph does not represent the first lap?
- b How many corners are there on the circuit? (*Hint*: When turning a corner, the driver must slow down. The sharper the corner, the slower the driver will go.)
- c Copy the graph and put a cross on the point where you believe the driver turns the sharpest corner.
- d Which of the diagrams below shows the circuit matching the graph above? Explain your choice.

A

B

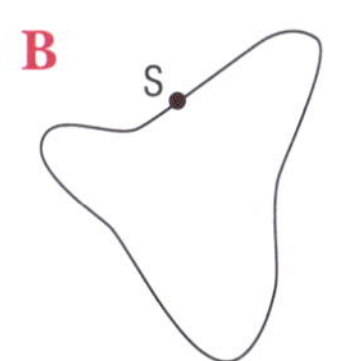

C

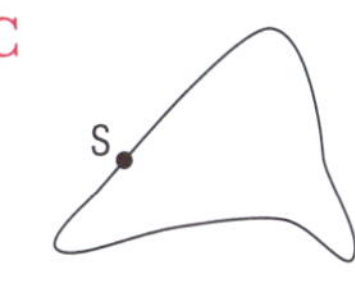

D

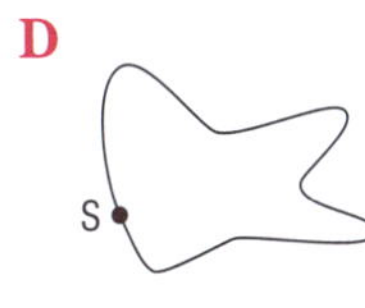

Challenge

Sketch the formula 1 circuits that would produce these speed–distance graphs.

a

b

c

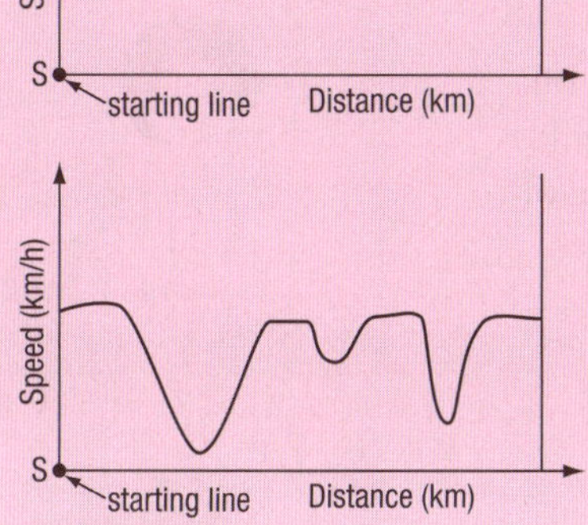

d

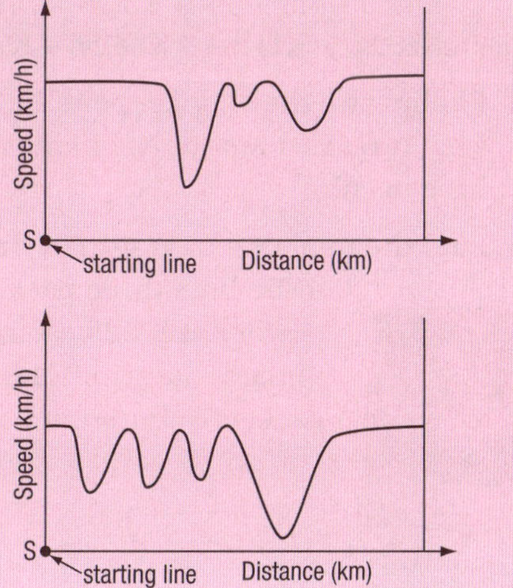

Weeks 5 and 6

MONEY

Lesson 1

Everyday money calculations

We all need to be able to manage money transactions, particularly when we are shopping.

Most of our buying and selling nowadays is done with currency (coins and notes). Coins have been used for trading around the world for more than 2000 years, but other forms of payment have also been used, such as precious metals (gold and silver), shells, gems, and even horses.

Trading can also be done by bartering, when goods or services are exchanged for other goods and/or services.

Some traditional forms of payment used in Papua New Guinea are pigs, salt and shell money.

Did you know?

The Tolai people on the Gazelle Peninsula use shell money called Tabu as bride price, and as payment for land, pigs and certain items in the market. The shell money is convertible into kina and toea at the rate of 10 fathoms of shell money to between K30 and K40 (depending on the current value of the kina). Papua New Guinea's first legal shell money exchange was opened in the Rabaul district.

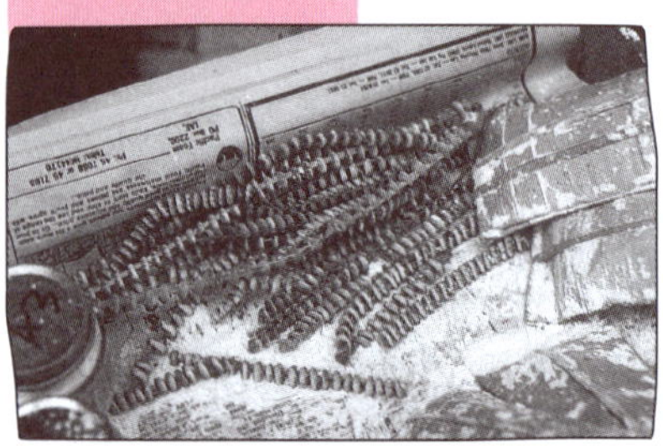

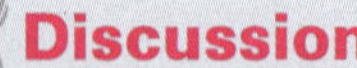

Discussion

- What other forms of payment are used, or have been used in the past, where you live?
- What are the advantages and disadvantages of trading using:
  - **a** currency?
  - **b** bartering?

EXERCISE

1 Round each of the following amounts to the nearest kina:

a K1.86 **b** K56.34 **c** K4.50
d K67.49 **e** K99.89 **f** K120.07

2 Round each of the following amounts to the nearest 5 toea:

a 87t **b** K3.72 **c** K72.38
d K9.11 **e** K13.49 **f** K66.23

3 Calculate each of these amounts:

a 95t + 84t + K1.29 + 73t + K2.16
b K7.23 + 95t + K4.99 + 23t + K1.33
c K4.15 + 85t + 39t + K5.58 + 74t
d 55t + 93t + K1.83 + K7.29 + K3.57
e K23.90 + K11.84 + K34.56 + K9.12
f K103.45 + K94.22 + K210.39

4 Phillip gave the shopkeeper a K20 note and received K13.50 change. Between what two exact amounts, in kina and toea, could the total of his shopping bill have been?

5 Find the total cost of Waisina's shopping bill, if she buys three packets of milk powder at K4.25 each, a bottle of tomato sauce for K4.85 and two packets of butter at K2.15.

6 Find the cost of five bottles of soft drink if:
a 1 bottle costs K1.25
b 10 bottles cost K8.50
c 3 bottles cost K2.88
d 8 bottles cost K8.64
e 12 bottles cost K14.76
f 7 bottles cost K6.09.

7 The table below is a storekeeper's bank record, showing deposits and withdrawals for one week.
a Copy the table and complete the missing values **A**, **B**, **C**, **D** and **E**.

Date	Deposit	Withdrawal	Balance
Friday, 12 March			**A**
Monday, 15 March	K320.60		K9076.60
Tuesday, 16 March	K215.45		**B**
Wednesday, 17 March	K189.50	K100	**C**
Thursday, 18 March	K348.65		**D**
Friday, 19 March	**E**		K10 000.00

b What was the minimum balance?
c Which day of the week appears to be the best for business in this store? Give a possible reason for this.

8 The 2007 PNG budget was around K7.6 million and the population at that time was approximately 6 million. How much per capita does this represent, to the nearest 5 toea?

Percentages of money

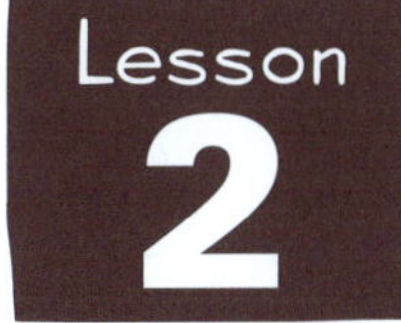

Value added tax (VAT), taxes, commissions, price rises and other charges are often expressed as percentages. We need to be able to find 10% of any sum of money and use that to calculate other percentages.

To **find a percentage of an amount**, we express the percentage as a fraction of 100 and multiply.

Example

Find 10% of K46.

Answer

$$\frac{10}{100} \times 46 = \frac{46}{10}$$
$$= 4.6$$

Help box

A shortcut for finding 10% of a sum of money is to move the decimal point one place to the left.

10% of K235.60
= K23.560
= K23.56

This rounds to K23.55.

EXERCISE

1 Write the simple fraction equivalent to each of the following percentages:
a 10% b 5% c 25% d 50% e 80% f 20%

2 Use your answers for Question 1 to find:
a 20% of K95
b 5% of K6000
c 80% of K35
d 10% of K66.50
e 50% of K4.66
f 25% of K34 600

3 Find 10% of each of the following amounts, giving your answer to the nearest 5t:
a K34.50
b K1.74
c K3.95
d K601.18
e K299.95
f K526.90

Help box

Salespeople are sometimes paid a **commission** based on the value of the sales they make.

Example

Albert was paid 5% commission when he sold a car for K12 000. How much commission did he earn?

Answer

Commission

$= \frac{5}{100} \times 12\,000$

$= 600$

So Albert earns K600.

4 Find each of the following percentages by first finding 10%:

a 5% of K78 b 30% of K60 c 90% of K550

d 60% of K5.30 e 40% of K12 500 f 70% of K7100

5 What does a salesperson earn for:

a a sale of K1500 at 3% commission?

b a sale of K600 at 2% commission?

c a sale of K8000 at 2.5% commission?

d a sale of K12 500 at 3% commission?

e a sale of K4260 at 2% commission?

f a sale of K9210 at 2.5% commission?

6 An agent is promised 2.5% commission. He sells a house for K240 000.

a How much commission does the agent earn?

b How much does the person selling the house receive?

Help box

To **increase** or **decrease** an amount by a percentage, we find the percentage and **add** or **subtract** as appropriate.

Example

Increase K450 by 20%.

Answer

$\frac{20}{100} \times 450 = 90$

So: New amount = 450 + 90 = 540

Another, shorter, method would be to calculate (100 + increase)% in one step:

New amount = (100 + 20)% of K450

= 120% × 450

= 1.**20** × 450 (0.20 represents the 20%)

= 540

To **decrease** K450 by 20% using this method, we would calculate:

New amount = (100 − 20)% of K450

= 80% × 450

= 0.80 × 450

= K360

7 Increase K460 by:

a 10% b 20% c 5% d 25% e 8% f 200%

8 Decrease K95 by:

a 10% b 5% c 75% d 60% e 30% f 15%

9 Find the value of the following goods after 1 year:
  a a car that cost K14 000, but depreciated 10%
  b land that cost K25 000, and appreciated 5%
  c a rare coin that cost K600, and appreciated 3%
  d a tractor that cost K8600, but depreciated 15%

10 The Tsonga family owns a collection of masks which is valued at K1200. After 1 year the collection is worth K1260.
  a How much has the collection appreciated?
  b By what percentage has the collection appreciated?
  c If the collection appreciates at the same rate each year, what is its value at the end of:
    i year 2?
    ii year 3?

Help box

Appreciation and depreciation

Most goods that we buy, such as fishing gear, a van or a guitar, lose value or **depreciate** as they get older. Some things, such as a block of land, a rare coin or a ceremonial mask, increase in value, or **appreciate**, as they age.

The rate of appreciation or depreciation is a percentage of the item's value.

Budgets

Lesson 3

Budgets are an important tool for managing our money. They help us plan for our expenses and our savings.

Discussion

In your class, who gets an allowance or earns wages? Who plans their spending and saving? Why do you think it is important to save?

EXERCISE

1 We can separate our spending into two categories: essential items and non-essential items. Name four essential items you believe families would need to allow for in their weekly budget and four non-essential items they might commonly spend money on.

2 Copy and complete this table showing the Rau family's weekly budget:

Income	Kina	Expenses	Kina
Mr Rau's pay	445	Food	160
Mrs Rau's pay	255	Rent	350
		Power, phone and other services	55
		Transport	15
		Clothes	32
		Household goods	18
		School fees	26
		Lollies	7
		Magazines and papers	8
		Savings	20
		Entertainment	9
Total income:		**Total expenses:**	

To express one quantity as a percentage of another, we write the quantities as a fraction and multiply by 100.

Example

What percentage is K4.5 of K25?

Answer

$\frac{4.5}{25} \times 100$

$= \frac{45}{_{10}\cancel{250}} \times \cancel{100}^{4}$

$= \frac{18\cancel{0}}{1\cancel{0}}$

$= 18$

a What are the: **i** total income? **ii** total expenses?

b Which of the Rau family's expenses do you consider to be non-essential?

c What fraction of the family's income is spent on rent?

d Which item uses nearly 5% of the family's income?

e If the family stopped buying magazines and papers, how much could it save in one year?

f Which two items, together, use $\frac{1}{10}$ of the family's income?

3 This pie graph shows how the Tahli family spends its fortnightly budget of K750.

a How much does the family spend on entertainment?

b Which item is close to half the family's expenses?

c What fraction of the family's income is spent on school fees?

d What fraction of the family's income is spent on food?

e What percentage of the family's income goes into savings?

f What percentage of the family's income is spent on clothes?

4 Tao has an allowance of K10 a week for doing work around the house and yard. He is saving his money for a bicycle which costs K250.

a How long will Tao need to save before he can buy the bicycle?

b If Tao's allowance increases to K12.50, how much sooner can he buy the bicycle?

5 Mrs Pamina is being sent to Brisbane for an important conference. She will be away for 3 days and expects her expenses while she is away to be as follows:

- Airfare (return): K1200
- Hotel accommodation (2 nights): K75 per night
- Food: K56 per day for 3 days
- Transport: K12 per day for 3 days

a What is the total of her expenses?

b Which of her expenses is close to 10% of the total?

c If she plans to take an extra 5% for emergencies, how much will she need to budget for altogether?

d Mrs Pamina's flight home is delayed and she needs to spend an extra night at the hotel. Will the extra 5% included in part **c** cover her costs?

Sales and shopping

To get the best value when we are shopping, we need to be able to compare the prices for different goods. We can also purchase bargains if we shop at sales, when prices are reduced.

To compare prices, we need to work out the cost of the same quantities.

Example

Which is the better buy, a 100 g packet of peanuts for K1.25 or a 150 g packet for K1.75?

Answer

The 150 g packet contains half as much again as the 100 g packet.

Half of K1.25 is 62.5t, so 150 g at the rate for the 100 g packet would cost K1.25 + K0.625 which is more than K1.75. So the 150 g packet is the better buy.

EXERCISE

1 Wal's Outdoor Shop is having a sale and all the prices have been reduced by 10%. What would a shopper pay for:
- a a fishing rod, normally priced at K24?
- b a sports bag, normally priced at K12.50?
- c a folding chair, normally priced at K29.95?
- d a cooler, normally priced at K175?
- e a lantern, normally priced at K7.95?

2 Which is the better buy from each of the following pairs of options, and by how much?
- a 6 bananas for K3.06 or 9 bananas for K4.68
- b 3 kg of taro for K7.50 or 5 kg for K12
- c 2 L of sauce for K3.75 or 500 mL for 95t
- d 5 kg of rice for K4.55 or 12 kg for K11.95
- e 1.5 L of drink for K2.25 or 500 mL for 75t

3 A CD player, normally priced at K199, is on sale at one store for K180 and at another store for 10% off. Which is the better deal and by how much?

4 Steak is marked down from K16.50 per kg to K15 per kg. What percentage price reduction is this?

5 The store is advertising milk powder for K4.25 per 200 g pack which is a saving of 75t.
- a What is the usual price?
- b By what percentage has the usual price been reduced?

6 A 14-inch television on sale usually sold for K495 but was on sale for K450.
- a How much would I save if I bought the television in the sale?
- b By what percentage (approximately) has the price been reduced: 8%, 9% or 10%?

Remember

To work out the percentage savings, we use:

$$\frac{\text{amount saved}}{\text{original price}} \times 100$$

Example

A can of fish, originally cost K2.60, but now it costs K2.21. Find the:

a amount saved on the original price

b percentage savings (to the nearest whole %) on the original price.

Answer

a Amount saved
= 2.60 – 2.21
= 0.39

b % savings
$= \frac{0.39}{2.60} \times 100$
= 15%

7 For each of the following sale items, find:
- **i** the amount saved
- **ii** the percentage saving on the original price, to the nearest whole percentage.

a shampoo: was K10.95, is now K9.50
b corn flakes: was K7.60, is now K6.95
c packet of biscuits: was K8.45, is now K7.95
d 250 g pack of Devon roll: was K3.85, is now K2.95
e butter: was K2.25, is now K1.95
f packet of cake mix: was K5.95, is now K4.70

8 In a clearance sale, all prices are reduced by 25%. What would you pay for each of these items at the sale?

a shorts, normally priced at K16
b a top, normally priced at K12.50
c shoes, normally priced at K18.60
d a sports bag, normally priced at K14.90
e jeans, normally priced at K19.99
f a jumper, normally priced at K23.45

9 Mya works in a food store and is given a staff discount of 5% on goods she buys. What does Mya pay for each of the following items?

a deodorant, normally priced at K9.40
b bread, normally priced at K4.20
c 2 kg of candlenuts, normally priced at K1.50 per kilogram
d 3 kg of sausages, normally priced at K6.95 per kilogram
e a laundry basket, normally priced at 24.95
f a mop, normally priced at K9.95

10 Aaron travels to Sydney to visit relatives. He takes K50 with him to buy souvenirs.

a He exchanges his money and gets 0.45 Australian dollars for each kina. How many Australian dollars does he receive?

b He buys souvenirs costing 20 Australian dollars. How much money does he have left?

Sydney, Australia.

Running a business

Many businesses operate by buying goods from the manufacturer then adding an amount before selling the goods to customers.

The amount the business owner pays for the goods is called the **cost price**; the amount added to this price is called the **mark-up**, and the amount the customer pays is the **selling price**.

The **profit** made by the business operator is equal to the mark-up. It is the difference between the **selling price** and the **cost price**.

Lesson 5

Discussion

- One of a business operator's expenses or costs is wages. What are some of the other things he or she needs to pay for out of the profit?
- Can you explain how a business works if the operator makes or grows goods to sell, rather than buying them from a manufacturer?

Remember

- A business owner calculates the percentage profit for the business by subtracting the running costs, or expenses, from the income.

Profit = income − costs

Percentage profit $= \frac{\text{profit}}{\text{costs}} \times 100$

- For a particular item on sale, the profit is the difference between the cost price and the selling price.

EXERCISE

1 Mr Abal owns a tradestore. His income for July was K2100. His expenses for July were:

- Wages K450
- Rent K65
- Electricity K20
- Goods purchased K1200
- Insurance and other K15

a What were Mr Abal's total expenses for the month?

b How much profit did he make?

c Express Mr Abal's profit as a percentage.

d If July is a typical month, what is the lowest income Mr Abal could make in order to cover his expenses (that is, to **break even**)?

e If Mr Abal wants to make at least 15% profit, what would his income need to be?

2 Mrs Kambla runs a food stall. Her income and expenses for three months are shown in this table:

Month	Income	Expenses	Profit	% profit
May	520	480		
June	490	460		
July	510	470		

a Copy and complete the table, giving the percentage profit to one decimal place.

b In which of these months did Mrs Kambla earn the most income?

c In which month did she earn the greatest percentage profit?

d Give an explanation why the answers for parts **b** and **c** are not the same.

e If Mrs Kambla's expenses for August are K500, what income will she need to make a profit of 10%?

3 Piliki and Rose start a small business selling chickens. They buy materials for the chicken house and pay other start-up costs, which total K65. They plan to sell the chickens at K5 each.

a How many chickens do they need to sell to cover their start-up costs?
b If they sell 20 chickens, what profit will they make?
c What is the profit from part **b** as a percentage?

4 **a** Philomena's shop has a CD player which cost her K40 wholesale. She calculated her selling price by adding 25%. What is Philomena's selling price for the CD player?
b At sale time, the selling price of the CD player in part **a** is marked down by 25%. What is Philomena's sale price?
c If Philomena sells the CD player at sale time, does she make a profit or a loss? Can you explain why the two 25% amounts had different values?

5 A small mining company keeps a record of its income and costs each month. The table below shows the records for six months.

Month	Income (thousand kina)	Costs (thousand kina)	Profit (thousand kina)
January	4.8	3.9	
February	4.5		1.3
March	3.6	3.4	
April	2.9		0.1
May	0.75	2.75	−2.0
June	2.8	2.75	

a Copy and complete the table, using the formula:
Profit = income – costs
b In which month did the company make the greatest profit? Write the profit in figures.
c Because the costs were higher than the income in May, the company made a loss. What could be one explanation for the result?
d What is the company's total income for this 6 months? Write the total income in figures.

6 A gift shop receives a shipment of goods from a wholesaler. What is the selling price of each of these items given the cost price and the percentage mark-up?

a mirror: cost price K25, mark-up 100%
b statue: cost price K15, mark-up 75%
c glass vase: cost price K40, mark-up 200%
d bowl: cost price K18, mark-up 150%
e plate: cost price K24, mark-up 120%

7 What percentage profit does a shopkeeper make on each of the following goods, to the nearest whole number?

a radio: cost price K70, selling price K99
b girl's watch: cost price K25, selling price K49
c rice cooker: cost price K65, selling price K99.95
d video camera: cost price K600, selling price K1199
e fridge: cost price K490, selling price K660

Borrowing and investing money

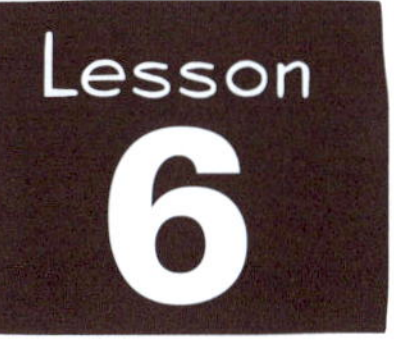

Simple interest

When we borrow money, we must repay what we have borrowed (the **principal**) as well as an additional percentage (**interest**) over the time of the loan.

If we **invest** money (the **principal**), for example in a bank savings account, we earn **interest**, which we can also take out of the account.

Simple interest = principal × rate p.a. (%) × number of years

$I = P \times \frac{r}{100} \times n$

Remember

p.a. means 'per annum,' or 'each year'.

Example

Find the interest on a **loan** of K500, borrowed for 2 years at 8% p.a.

Answer

$I = 500 \times \frac{8}{100} \times 2$
$= 80$

We would pay K80 interest.
After 2 years, we would owe K580.

If we **invested** K500 for 2 years at 8% p.a., we would earn K80 interest and our investment would be worth K580. In reverse, if we know how much in total has been paid in interest and we know the principal, we can calculate the **rate of interest**.

Example

To start her business, Kendra borrowed K1000 for 4 years. At the end of the 4 years she had paid K1800. What rate of interest was she charged?

Answer

Kendra paid an extra K800 over 4 years:

Total repaid – principal = 1800 – 1000
= 800

This is K200 per year.

The rate of interest is $\frac{200}{1000} \times 100 = 20\%$

Discussion

- What sort of loans can you get where you live, or in the nearest large town?
- What choices do you have for investing your money?

EXERCISE

1 This table shows amounts, interest rates and numbers of years. Copy and complete this table to show the amount of interest each time.

Principal	Interest rate	Number of years	Interest
K200	10%	3	
K150	20%	4	
K40	15%	2	
K3000	8%	5	
K600	12%	2	

2 Calculate the interest paid on K3000 at:

a 10% p.a. for 5 years
b 8% p.a. for $2\frac{1}{2}$ years
c 12% p.a. for 6 months
d 4% p.a. for 9 months
e 15% p.a. for 40 months
f 8.5% p.a. for $3\frac{1}{2}$ years

3 Mr Amalio invested K100 for 8 years in an account paying 6.5% p.a.

a How much interest did he earn?
b What is the total value of Mr Amalio's investment at the end of 8 years?

4 If you borrowed K500 for 2 years and repaid K570 at the end of that time, what percentage interest were you being charged?

5 Samara was charged K30 for a loan of K500 for 6 months. What rate of interest was she being charged?

6 Joshua invested K200 in a savings account. At the end of 5 years he had a balance of K450. What percentage interest was his investment earning?

7 Banks pay higher interest on a fixed-term deposit than for a savings account. Why do you think this is so?

8 For a person investing in a fixed-term account, describe:

a one advantage
b one disadvantage.

9 Mrs Lau won K10 000 in a competition and put it in a 3-month short-term deposit account, paying 4.5% interest per year. What was her prize worth at the end of 3 months?

10 Copy and complete this table showing the interest earned on the following fixed-term deposits.

	Term (months)	Interest rate (% p.a.)	Amount deposited	Interest
a	6	4.5	K1000	
b	24	6	K400	
c	3	4	K8000	
d	36	6.5	K550	
e	9	5.5	K15 000	

Lesson 7

Paying in instalments

Hire-purchase is an arrangement for paying off goods over a period of time. After we have paid a deposit, we are able to take the goods home and use them.

Discussion

- What are the advantages and disadvantages of buying goods on hire-purchase?
- What other options are there for buying goods, and what are the advantages and disadvantages of these?

EXERCISE

1 Lucas pays a deposit of K31 on a mobile phone worth K281, and pays monthly instalments of K10 for 3 years.
   - a What is the balance owing on the phone after the deposit has been paid?
   - b How much does Lucas pay in instalments over the 3 years?
   - c How much in total does Lucas pay?
   - d How much interest does Lucas pay?

2 The Romero family buys a refrigerator costing K663, for a deposit of K63 and instalments of K20 a month for 5 years.
   - a What does the Romero family pay in total?
   - b How much interest does the Romero family pay?
   - c What percentage interest per annum are the Romero family being charged?

3 I can buy a laptop priced at K2899 for a K299 deposit and 24 monthly instalments of K130. How much interest am I paying over the 2 years?

4 A hi-fi system is advertised at K1629, or K129 deposit and K55 fortnightly for 18 months.
   - a How many fortnightly payments will be made?
   - b What does the hi-fi cost in total if we pay in instalments?
   - c How much interest are we being charged, to the nearest per cent?

5 Philomena bought a guitar for K350 with a cash deposit of K50. She paid the balance plus 10% interest per annum in equal monthly instalments over 2 years.
   - a How much was the interest?
   - b How much in total did Philomena need to repay?
   - c How much was each monthly instalment?

6 Mr Tan is paying off K70 a week on a truck for use in his business. He needs to repay K10 000 and is being charged 15% interest per annum on the whole of the balance.
   - a How much will he have repaid in 1 year?
   - b How much will he have repaid in 3 years?
   - c What is the interest on K10 000 for 3 years?
   - d Will Mr Tan have finished paying off the truck at the end of 3 years? Give reasons for your answer.

7 Rufus is setting up a small business selling timber. He buys an industrial saw worth K4500 for K100 deposit and pays it off over 3 years. He is charged 12% p.a. interest on the whole of the balance.
   - a What is the balance owing?
   - b How much interest does he pay in total?
   - c If the payments are K167 for the first 35 months, what final payment must Rufus make?

MEASUREMENT

Lesson 1

Traditional measurement of length

In almost all cultures, traditional measurement of length was based on body parts. Many people in local communities in Papua New Guinea still do not use standard units, such as the units of the metric system, to measure length. They use concrete measures such as arm spans, hand spans, string, sticks, bamboo, rope and paces, as well as verbal descriptions/comparisons which everyone in the community understands and uses to measure length. Specific methods are used for specific jobs. Items that are long, such as beams or wall structures, are measured by pacing, while shorter items, such as windows and doorways, are measured using sticks, ropes, arm spans and hand spans.

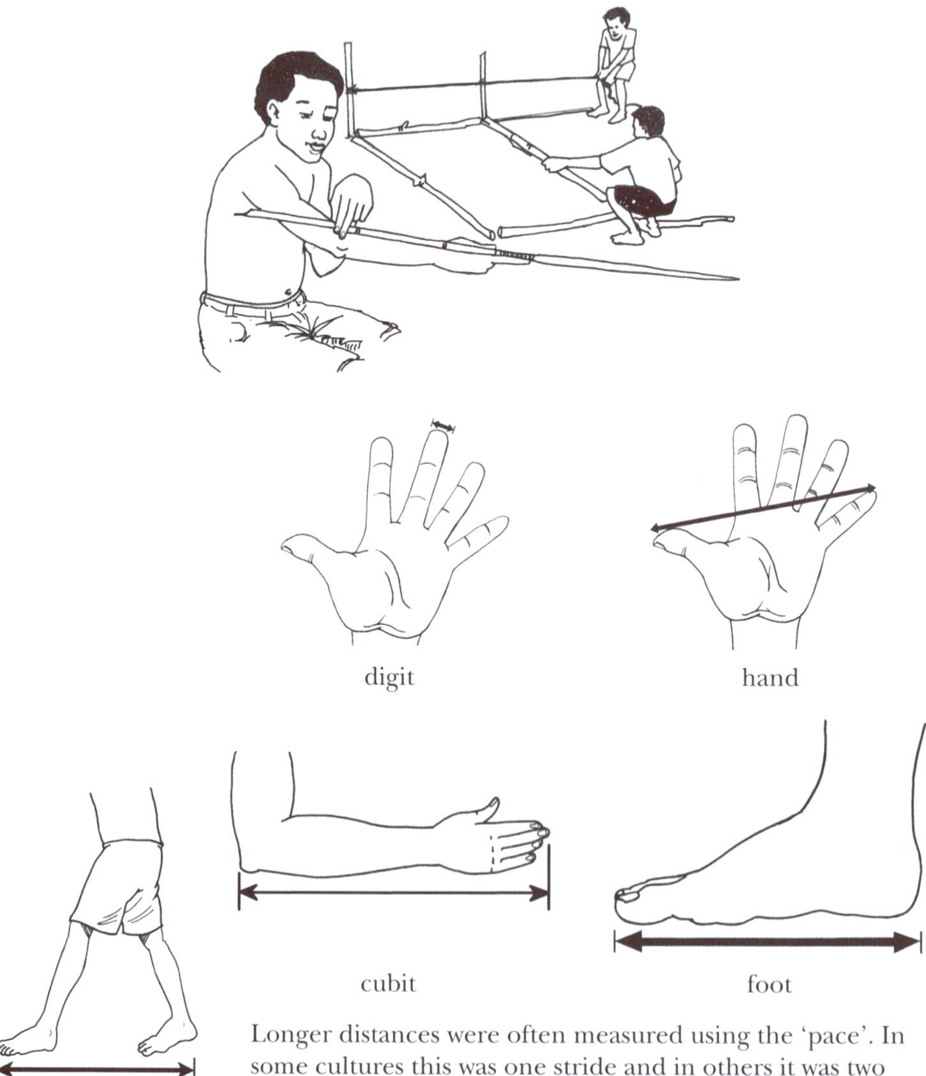

Longer distances were often measured using the 'pace'. In some cultures this was one stride and in others it was two strides.

EXERCISE

1 Which of the body part units, **cubit**, **digit**, **hand**, **foot** or **pace** would you use to measure each of the following?*

   a the length of your classroom
   b the length of your school desk/table.
   c the width of this textbook
   d the perimeter of a sportsfield
   e the height of a doorway

2 a Use the body parts you chose in Question 1 to measure each of the given objects. You may need to use mixed body parts. For example the width of the textbook may be '1 hand and 4 digits'.

   b On a new, full page of your workbook construct the following table to record your measurements. (A row has been completed as an example for you.) At this stage, complete only the first three columns for all the objects you measured.

   c Compare your measurements with those made by some other class members. Why do you think they differ?

Object	Body part used	Measurement in body units	Length of my body parts (in cm or m)	Converted body part measurements (in cm or m)	Actual measurements (in cm or m)	Percentage error
Textbook length	Hand, digit	1 hand, 4 digits	Hand 19 cm Digit 1.5 cm	$1 \times 19 + 4 \times 1.5$ = 25.0 cm	25.3 cm	$\frac{25.3 - 25}{25.3} \times \frac{100}{1}$ = 1.19%

   d Working in pairs, use a metric ruler or tape measure to measure your body parts, recording your answers in centimetres (cm) or metres (m) in the fourth column of your table.

   e Convert your body part measurement of each of the objects to lengths in centimetres or metres, by multiplying the entries in the third column of your table by those in the fourth column. Record the answers in the fifth column.

   f Obtain the accurate actual length of each of the objects (in centimetres or metres) from your teacher. Record these values in the sixth column of your table.

   g Calculate the accuracy of your body part measurements by calculating 'percentage error' where:

$$\text{Percentage error} = \frac{\text{actual measurement} - \text{converted body part measurement}}{\text{actual measurement}} \times \frac{100}{1}$$

   h Comment on the accuracy of the measurements you found using body parts.

3 Why do you think most cultures eventually changed to standard units for measuring length?

4 Papua New Guinea and many other countries use the metric system of measurement. The USA and the UK still use the Imperial system, in which long distances are measured in miles.

   1 mile ≈ 1.6093 kilometres.

   The mile was originally based on the Roman soldier's 'pace', which was two strides. One thousand of these 'paces' was the basis for 1 mile.

   a Working in pairs and using a tape measure, calculate the length of your 'Roman pace', in metres. It is usually more accurate to measure several 'paces' and then average the length.

   b Calculate the length of 1000 of your 'paces', in metres, and convert this to kilometres.

   c How does the length you calculated compare with 1.6093 km? Calculate the percentage error for your 'Roman mile':

$$\text{Percentage error} = \frac{1.6093 - \text{your measurement in km}}{1.6093} \times \frac{100}{1}$$

* Teachers can also choose their own list of available objects to measure.

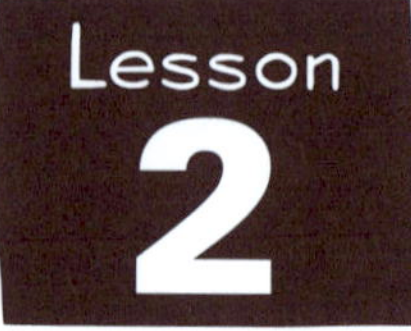

Estimating lengths, conversions and perimeter

The metric system of measurement is based on the **metre** (**m**). Other commonly used units in the metric system are the **centimetre** (**cm**) and the **millimetre** (**mm**); both used to measure small lengths or where accuracy is important. Long distances in the metric system are measured in **kilometres** (**km**).

1 metre = 100 centimetres
= 1000 millimetres
1000 metres = 1 kilometre

Converting units in the metric system is relatively simple because it involves multiplying or dividing by 10, 100 or 1000 which, in our decimal system of numbers, means moving the decimal point the appropriate number of places.

EXERCISE

1 a How long is one centimetre (1 cm)? Draw a line 1 cm long in your workbook.
  b Name something you would measure in centimetres.
  c How many centimetres are there in one metre?

2 a How long is one millimetre (1 mm)? Draw a line 3 mm long in your workbook.
  b Name something you would measure in millimetres.
  c How many millimetres are there in:
    i 1 centimetre? ii 1 metre?

3 a How long is one metre (1 m)? Name something that is about 1 metre long.
  b How many metres are there in one kilometre?

Remember
To estimate you need to find an approximate value.

4 Estimate the length of each of the following, recording each answer in the appropriate unit.
  a the width of your classroom
  b the length of the blackboard in your classroom
  c your height
  d your arm span; fingertip to fingertip
  e the diameter of your pen
  f the thickness of this textbook
  g the distance from your home to your school
  (Compare your estimates with the actual lengths where possible.)

5 Copy and complete the following conversion diagram in your workbook.

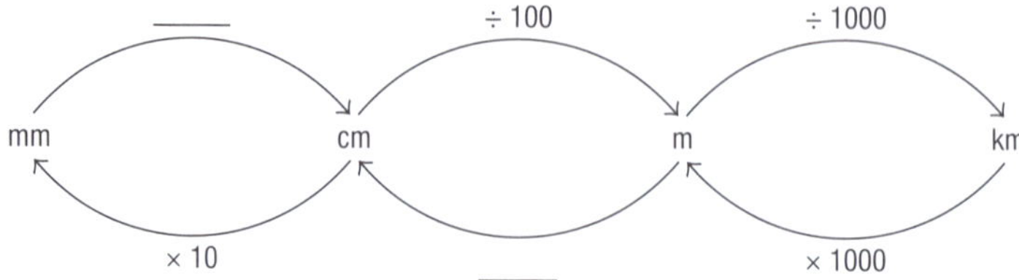

6 Convert each of the following to the required units:

a 15 cm to millimetres
b 1860 m to kilometres
c 23 km to centimetres
d 4050 mm to metres
e 23.5 m to centimetres
f 1.025 m to millimetres
g 658 mm to centimetres
h 4.75 km to metres
i 88 cm to metres

7 Find the perimeter of each of the following closed figures:

a
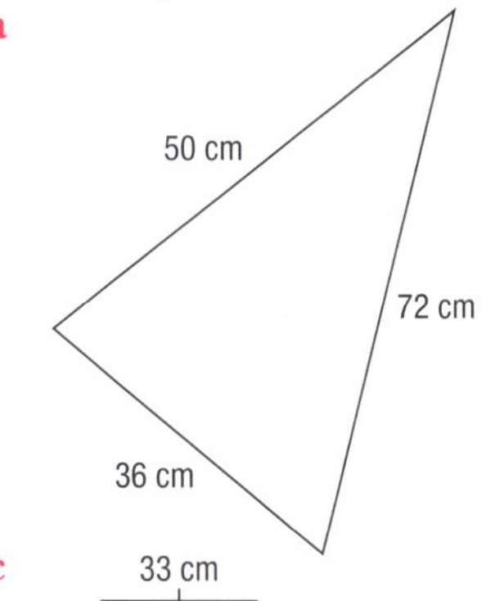

b
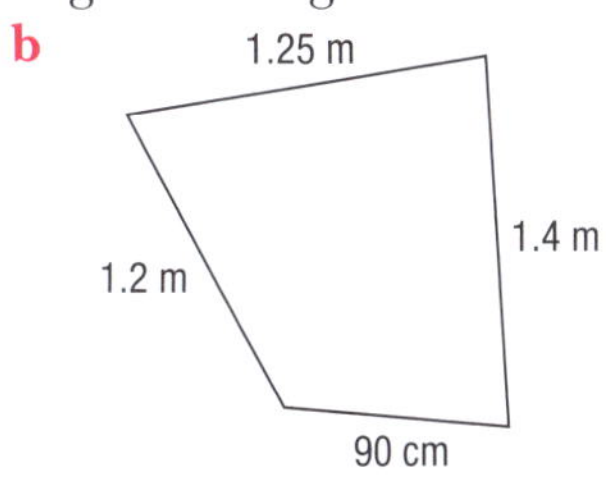

c
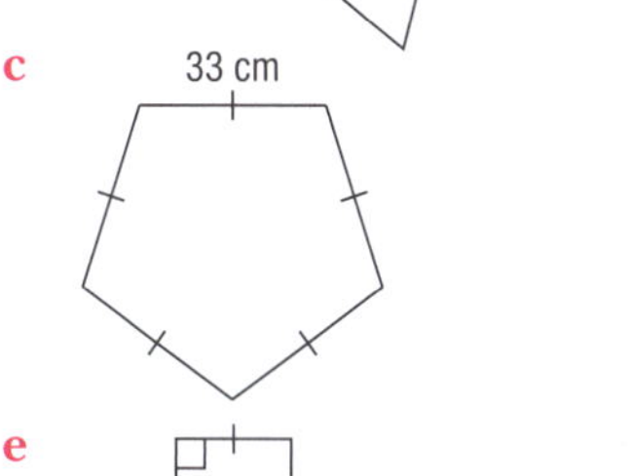

d
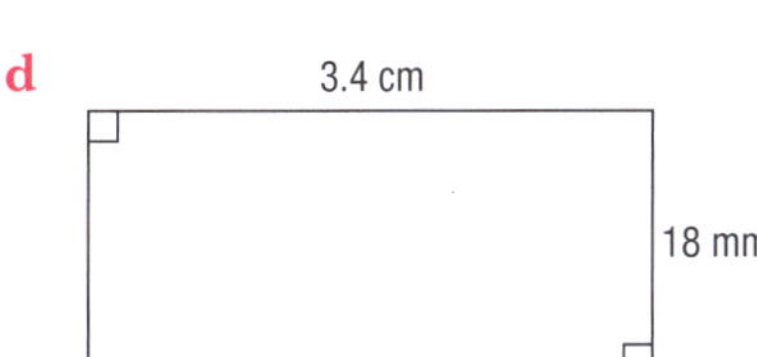

e
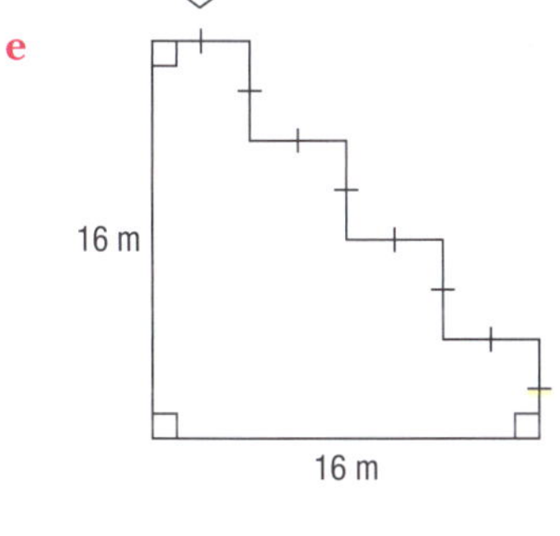

f
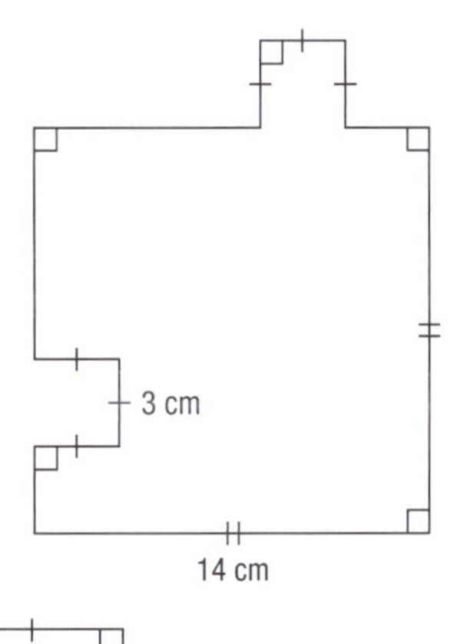

g
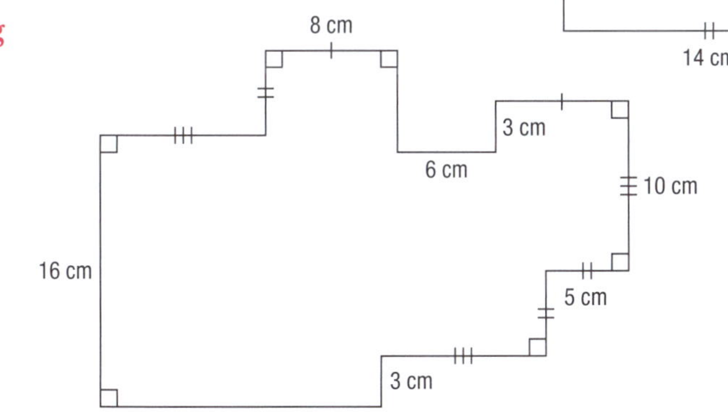

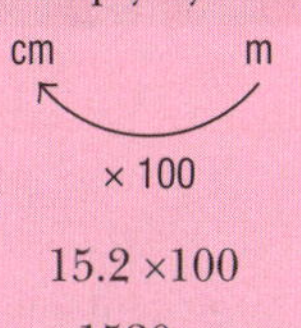

Help box

To convert 15.2 m to centimetres, we multiply by 100:

cm ← m × 100

15.2×100
$= 1520$ cm

Remember

- The perimeter of a closed figure is the distance around the outside of the figure.
- Lengths must be in the same unit before they can be added.

Remember

Sides marked in the same way are the same length.

Lesson 3

Remember

- π has the value 3.141 592 …
- Approximations for π are 3 or $\frac{22}{7}$.

Perimeter by rule

Rectangle

width, w

length, l

Perimeter = $2(l + w)$

Circle

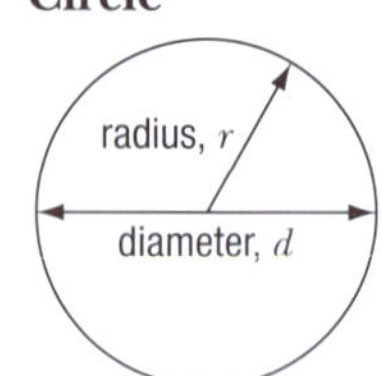

The perimeter of a circle is called the **circumference**. The ratio between the circumference and the diameter of a circle is always equal to pi (π).

Circumference $= \pi d$

$= 2\pi r$

EXERCISE

1 Find the perimeter of each the following rectangles:

a

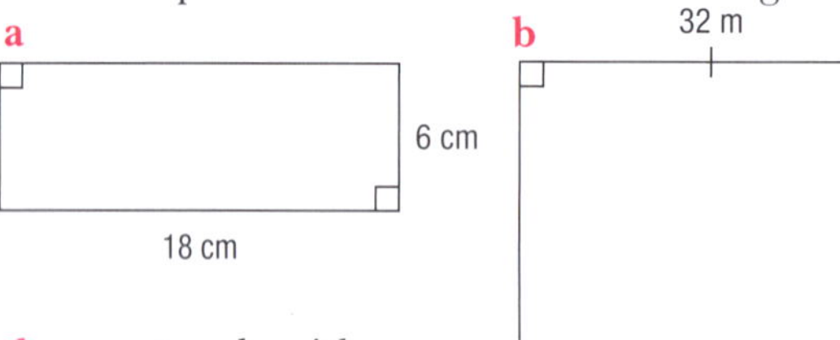

b 32 m

c 30 mm, 1.25 cm

d a rectangle with dimensions 12 metres × 8.4 metres

2 a Estimate the circumference of each of the following circles, using 3 as an approximation for π.

i

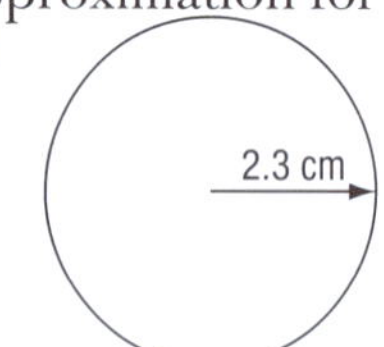

ii

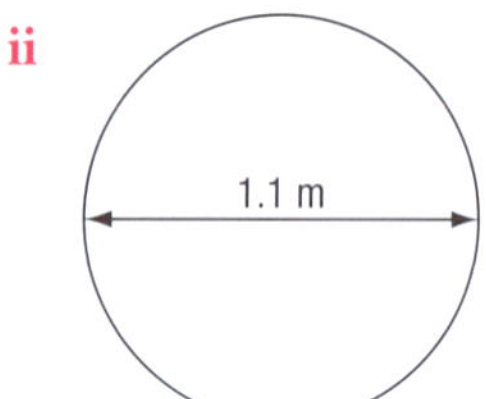

b The diameter of the Earth at the equator is 12 756.1 km. Estimate the circumference of the Earth at the equator to the nearest kilometre, using 3 as an approximation for π.

3 Calculate the circumference of the following circles using $\pi = \frac{22}{7}$.

a

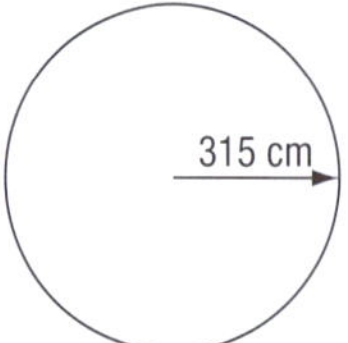

b

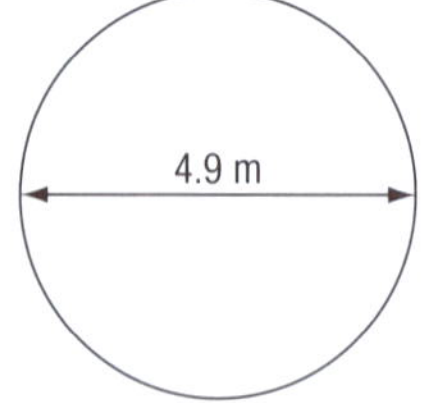

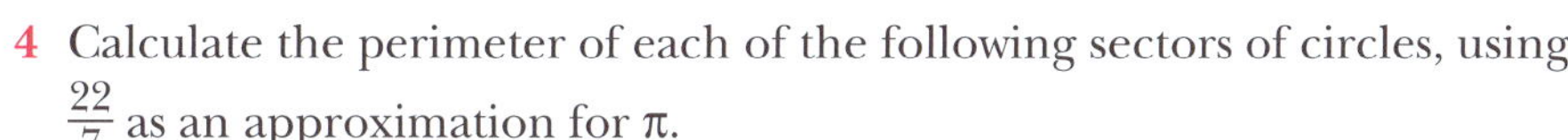

4 Calculate the perimeter of each of the following sectors of circles, using $\frac{22}{7}$ as an approximation for π.

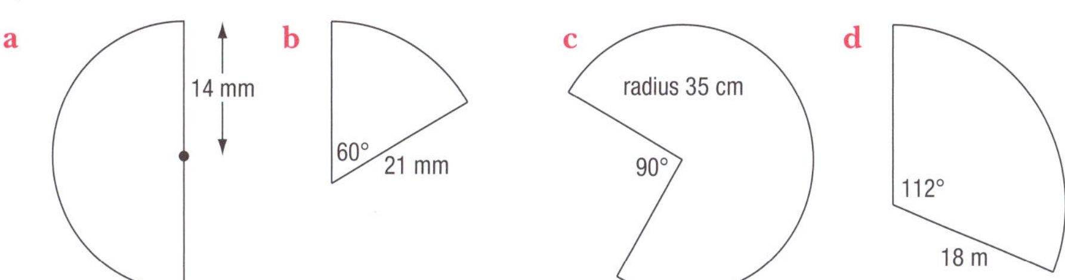

5 Measure the appropriate dimensions for each of these closed figures and calculate the perimeter.

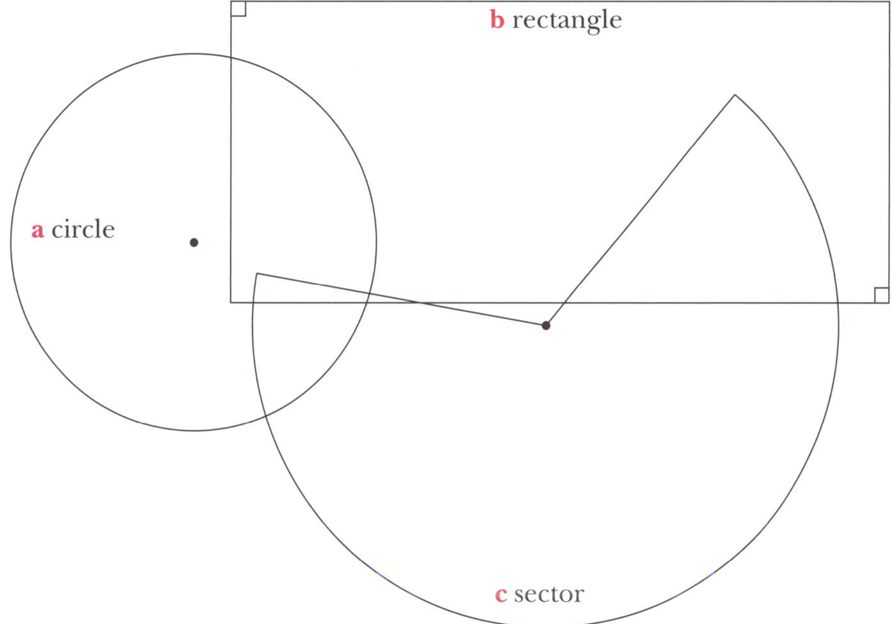

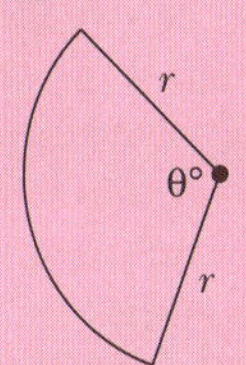

An arc is a fraction of a circle's circumference, so the length of an arc with angle θ at the centre of the circle is:

$$\frac{\theta}{360^\circ} \times 2\pi r$$

The perimeter of the sector equals the length of the arc plus two radii:

$$\frac{\theta}{360^\circ} \times 2\pi r + 2r$$

Area

The area of a closed figure is the amount of surface covered by the closed figure. In the metric system, the units of area are **square metres** ($\mathbf{m^2}$), **square centimetres** ($\mathbf{cm^2}$), **square millimetres** ($\mathbf{mm^2}$) and **square kilometres** ($\mathbf{km^2}$).

The area of some common figures

Rectangle

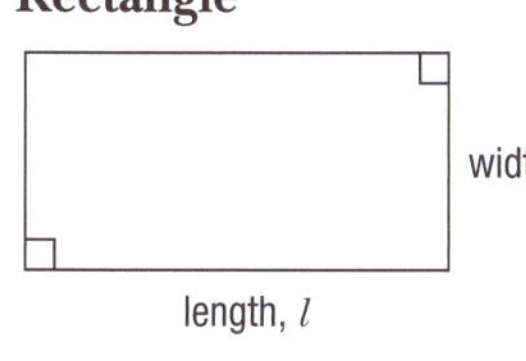

Area = $l \times w$

Parallelogram

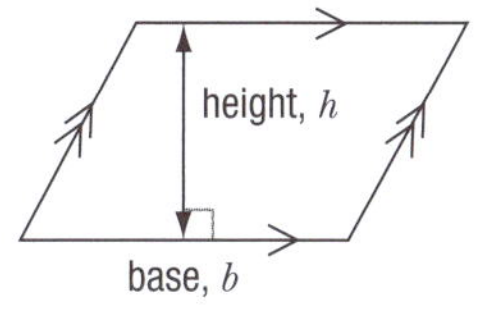

Area = $b \times h$

Trapezium

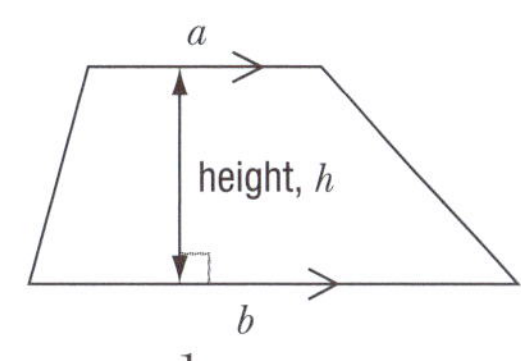

Area = $\frac{1}{2}(a + b) \times h$

Did you know?

Many people in the highlands use a length measure to determine area, such as measuring garden plots. It is as if the width of area unfolds in the mind as the length is paced out. Many garden plots are a certain width, so the total size (area) of the garden could be determined by pacing out lengths.

Triangle

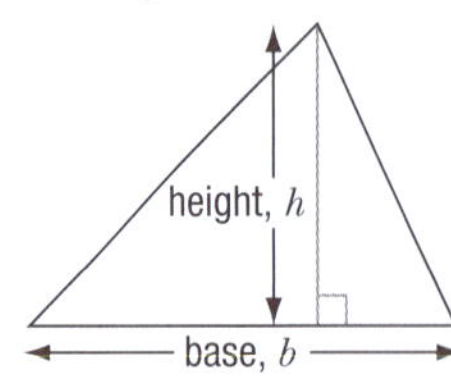

Area = $\frac{1}{2}(b \times h)$

Circle

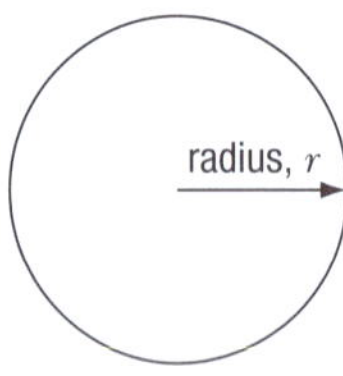

Area = πr^2

Sector

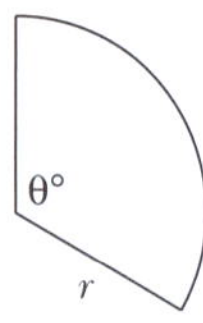

Area = $\frac{\theta}{360} \times \pi r^2$

EXERCISE

1 a In your workbook, draw a square that has side length 1 cm. This square has an area of one square centimetre (1 cm^2).

b Draw a rectangle that has an area of:

i 2 cm^2

ii 12 cm^2

iii 11 cm^2

c How many square millimetres (mm^2) in 1 cm^2?

2 a How large is a square metre? Imagine (or draw on the board or the ground) a square that has sides of length 1 m. This square has an area of one square metre (1 m^2).

b Estimate the area of your classroom, in square metres. How would you obtain an accurate measure of the area of your classroom?

c How many square centimetres are there in one square metre?

3 Copy and complete the following conversion diagram in your workbook.

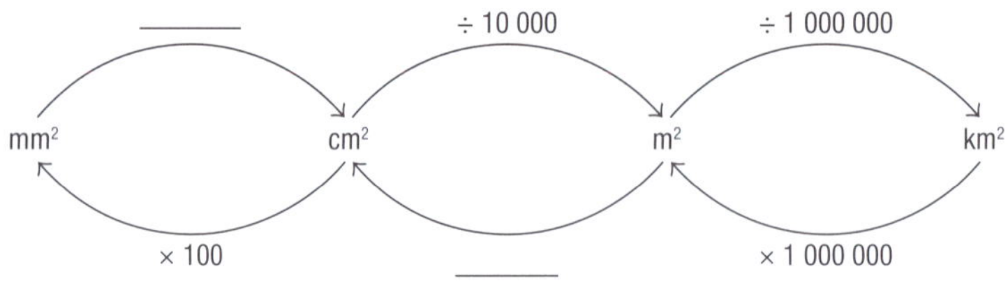

Help box

To convert 3.2 m^2 to cm^2, multiply by 10 000.

$3.2 \times 10000 = 32000$ cm^2

4 Convert each of the following to the required units:

a 12 500 cm^2 to m^2

b 0.563 km^2 to m^2

c 5430 mm^2 to cm^2

d 13 456 500 m^2 to km^2

5 Calculate the area of each of the following figures. Use $\pi \approx \frac{22}{7}$.

a

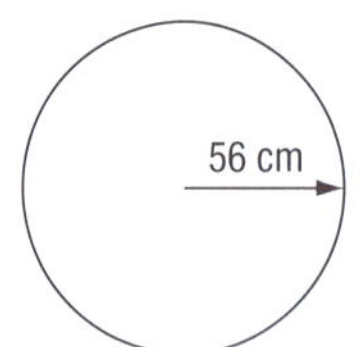

b

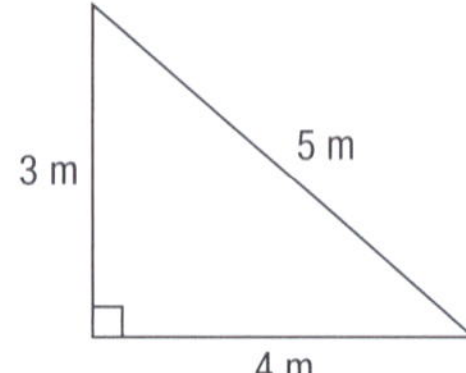

c

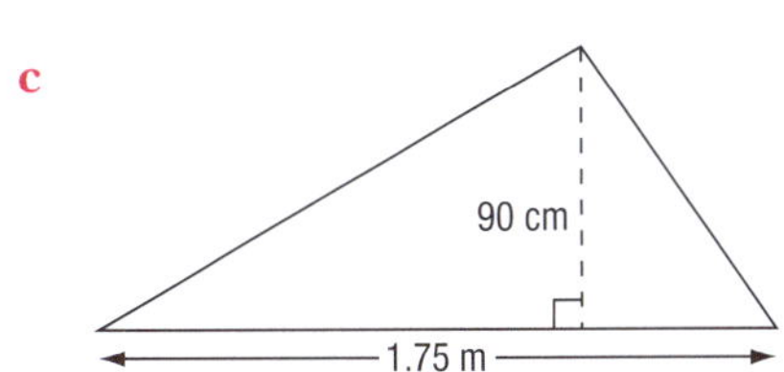

d

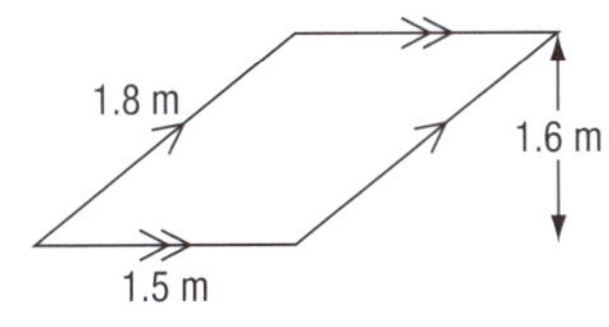

e

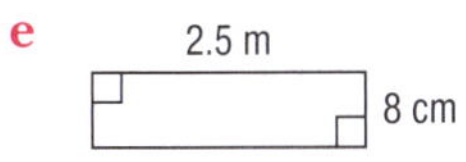

f

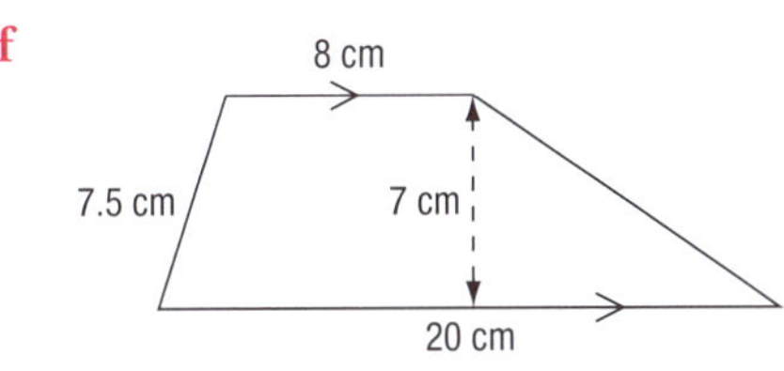

g

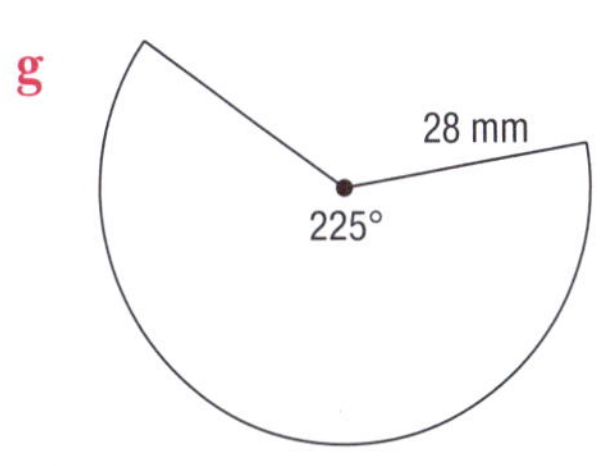

h

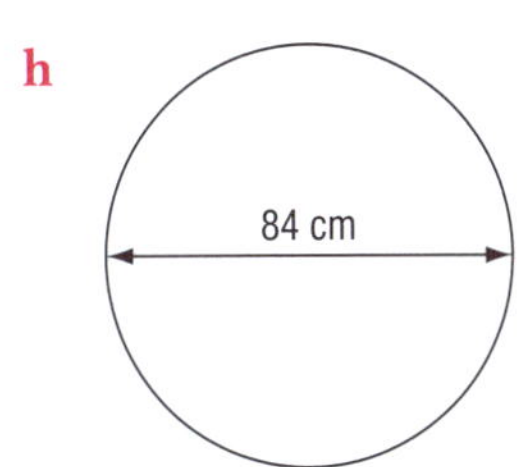

6 Measure the appropriate dimensions for each of these closed figures and calculate the area.

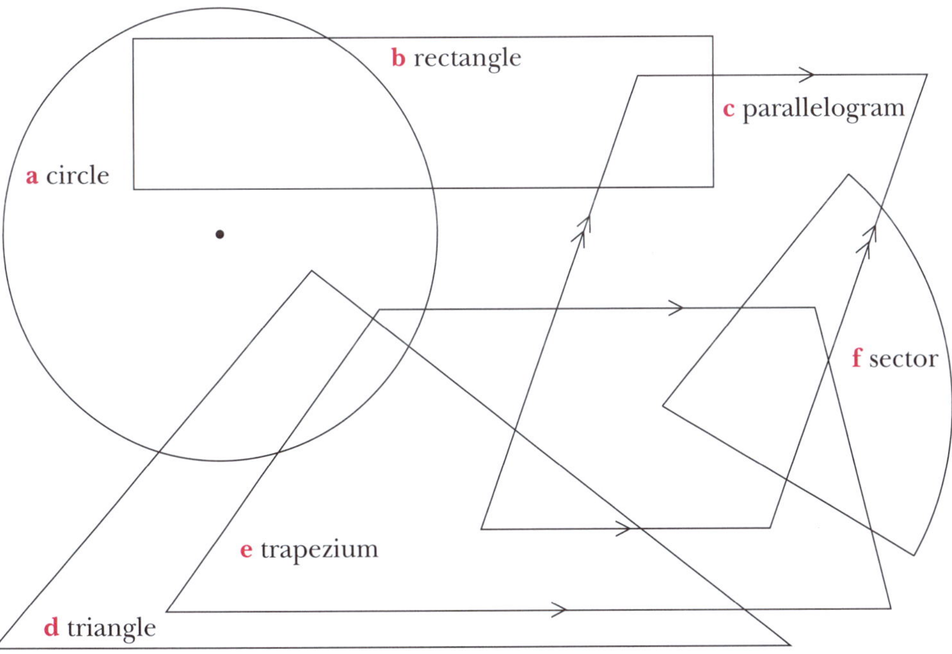

Area of composite figures, and total surface area

Many figures are constructed from combinations of the basic shapes: rectangles, triangles and circles. They are called **composite shapes**.

Example

Find the shaded area of this composite shape.

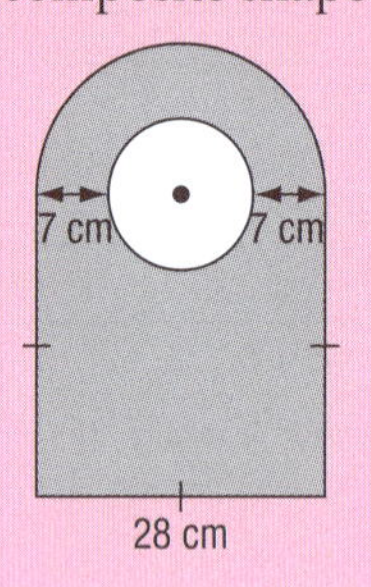

Answer

The shaded area is a composite of a square, a semi-circle and a smaller circle.

The area of the composite figure equals:

the area of a 28 cm square + the area of a semi-circle of radius 14 cm – the area of a circle of radius 7 cm.

$= 28 \times 28 + \frac{1}{2} \times \frac{22}{7} \times 14^2 - \frac{22}{7} \times 7^2$

$= 784 + 308 - 154$

$= 938 \text{ cm}^2$

The **total surface area** (TSA) of a solid is the sum of all the surface areas of the solid. For example the TSA of a cube is the sum of the areas of all six faces of the cube. It is often helpful to draw a net of all the faces that make-up the solid.

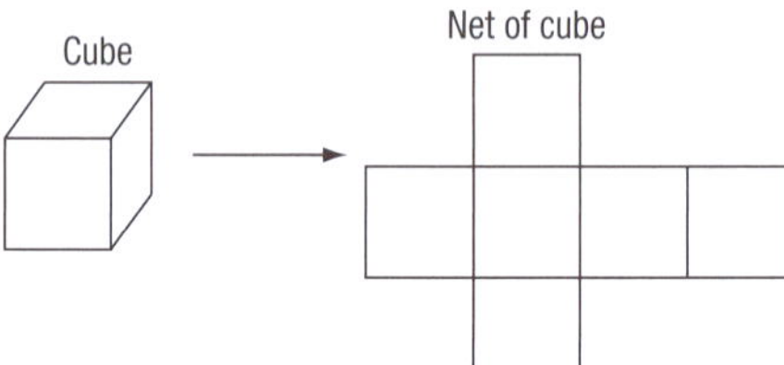

EXERCISE

Use $\pi \approx \frac{22}{7}$, where necessary.

1 Find the area of each of the following composite figures:

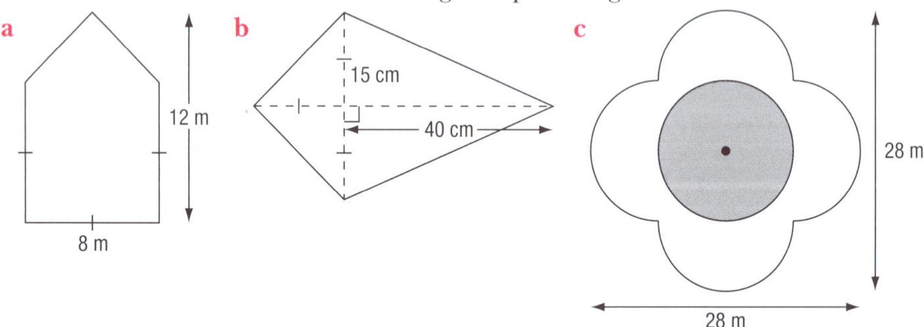

2 Find the perimeter and the area of each of the following composite figures:

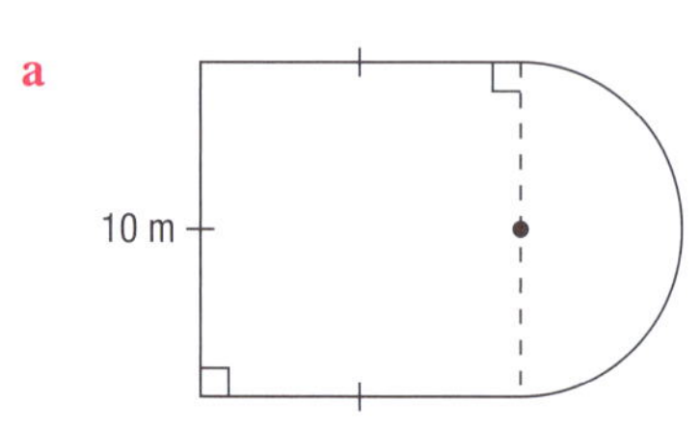

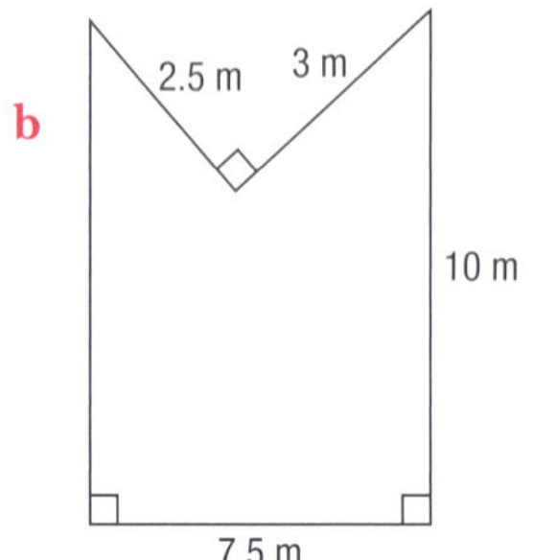

c

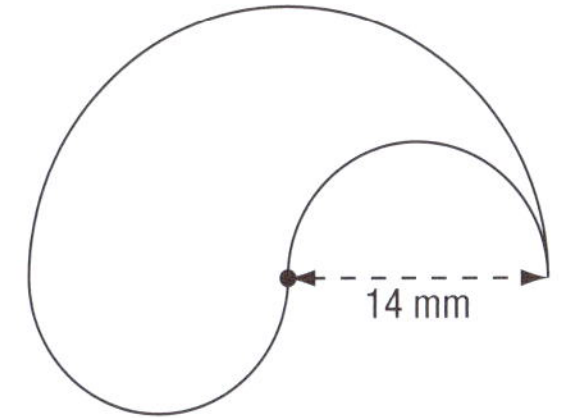

d

14 m
30 m

3 Find the total surface area of each of the following solids:

a

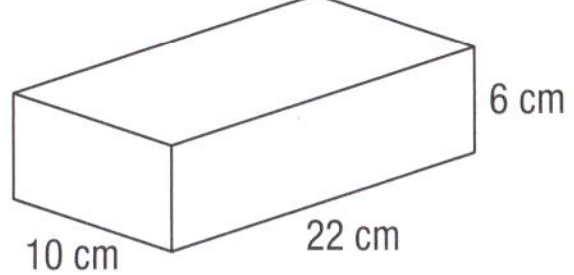

b

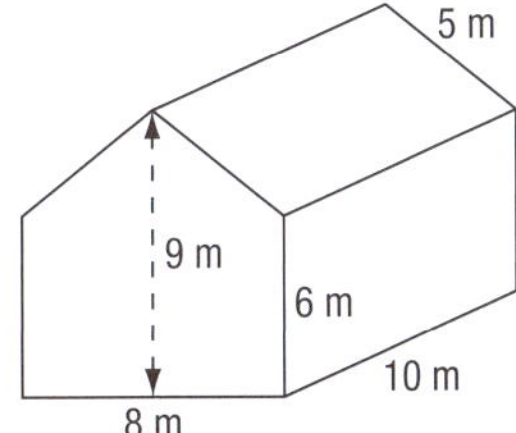

c

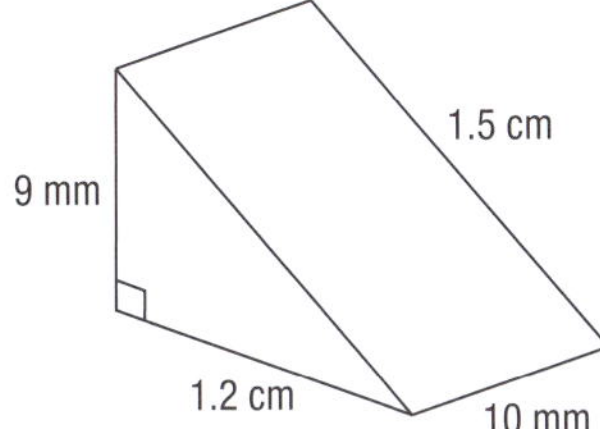

d

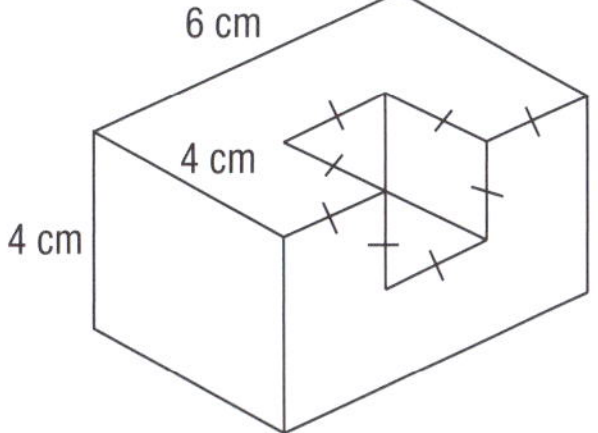

4 Nesting boxes are made from large cylindrical metal containers that have their circular ends removed, as shown in the diagram below.

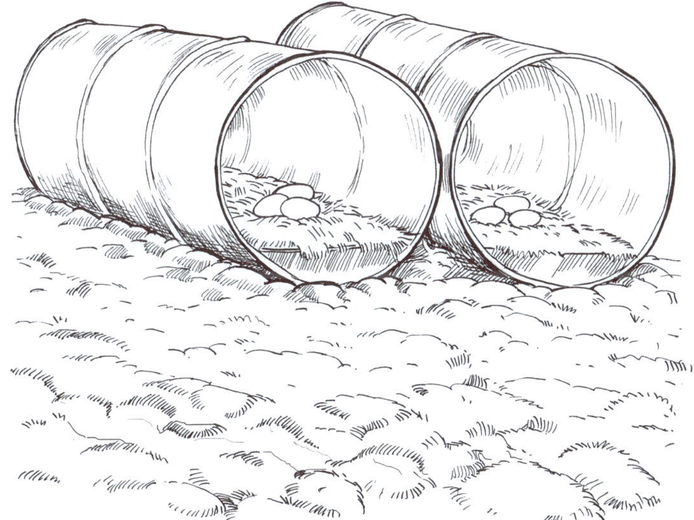

If each original metal container had circular ends of diameter 570 mm and the height of the container was 850 mm, find the amount of metal, in square metres, used for one nesting box.

DATA

When information for a statistical investigation is collected and recorded this is referred to as data.

Types of data

i Data can have a numerical value (number). This is called **numerical data**.

For example: If the investigation is about the number of people in a household, the data that is recorded will have the values 1, 2, 3, 4 …

ii Data can be assigned a quality or category. This type of data is called **categorical data**.

For example: If the investigation is about the sport played by a group of students, the data categories that could be recorded are rugby, basketball, volleyball, soccer, hockey …

Organisation and display of categorical data

An investigation into the favourite sport of 20 students was made and each student was asked to nominate one of the following categories as their favourite: rugby, soccer, basketball or volleyball.

Their answers were recorded and a tally and frequency table was constructed to organise the results:

Favourite sport	Tally	Frequency
rugby	𝍸 |||	8
soccer	𝍸	5
basketball	||||	4
volleyball	|||	3
	Total:	**20**

Remember

𝍸 is a tally of five.

A column graph, a bar chart or a pie graph can be used to display the results.

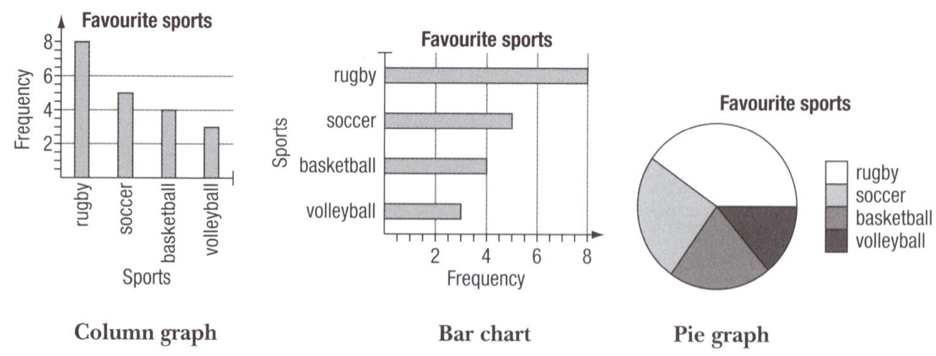

Column graph **Bar chart** **Pie graph**

EXERCISE

1 What type of data, categorical or numerical, would be collected if each of the following was investigated?

- **a** the time taken to travel to school
- **b** the number of cousins each person has
- **c** voting intention at the next election
- **d** the speed of cars on a particular stretch of road
- **e** favourite type of fruit
- **f** the weights of three-year-old children

2 a i If you were asked, 'Which region of PNG do you live in?' what would you write?

ii If a number of people were asked the region of PNG in which they lived, would the data collected be categorical data or numerical data?

b In the 2000 census, the number of people living in each of the regions of PNG was counted and the percentage of the whole population living in each of the regions was calculated. The results are shown in the table below.

A pie chart is a suitable graph for displaying this data. Copy the table and complete the last column by calculating the angles for each sector of the pie chart. Then draw a circle and use a protractor to construct the pie chart. Label each sector, and use a different colour to shade each sector.

Region	Percentage of population	Sector of circle
Southern	20	$\frac{20}{100} \times \frac{360}{1} = 72°$
Highlands	38	
Momase	28	
Islands	14	
Total:	**100**	**360°**

3 Eighty randomly selected Year 8 students were asked to nominate their favourite school subject. The results of the survey are displayed in the bar chart on the right.

Favourite subject

Subject: English, Mathematics, Science, Social science, Personal development

Frequency: 2 4 6 8 10 12 14 16 18 20

- **a** Which subject was the most favoured subject?
- **b** How many students chose English as their favourite subject?
- **c** What percentage of the students nominated mathematics as their favourite subject?
- **d** What percentage of the students chose either personal development or social science as their favourite subject?

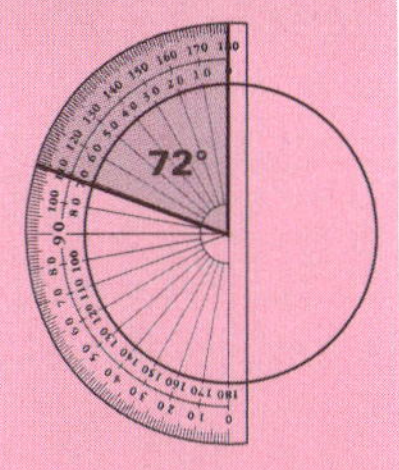

To construct a pie chart:

- use a compass to draw a circle, then draw a radius
- using a protractor with the drawn radius as the base line, mark the angle for the first sector by drawing another radius.
- move the protractor around the circle, using each new radius as the base line to draw the angles for the other sectors.

4 Children from a particular primary school were asked to nominate their favourite fruit. A pie chart of the results is shown on the right.

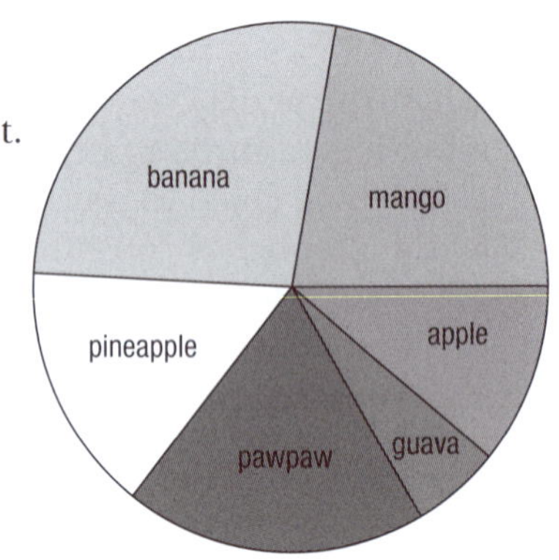

a For this group of students:
  i which was the most favoured fruit?
  ii which was the least favoured fruit?

b If there were 45 students in this survey, work out how many chose mango as their favourite fruit?

c i What is your favourite fruit?
  ii Is data recorded about favourite fruit numerical data or categorical data?

d Another group of primary school students was asked to nominate their favourite fruit. Their responses are listed below:
banana, banana, mango, pawpaw, mango, apple, apple, guava, mango, pineapple, banana, mango, apple, pineapple, pawpaw, banana, guava, mango, pineapple, guava, mango, banana, banana, apple, pineapple, pineapple, guava, apple, pineapple, banana
  i Construct a tally and frequency table for this data.
  ii Construct a bar chart to illustrate this data.
  iii What is the most favoured fruit in this group of students?
  iv What is the least favoured fruit in this group of students?
  v What percentage of these students nominated mango as their favourite fruit?

Class exercise

Collect data about the favourite fruit of the members of your class and construct a suitable graph to display the results.

5 A town is interested to know the type of vehicle using a particular road. They collected the following data on one day.

Type of vehicle	Frequency
Bike	32
Motor bike	12
Private car	56
Light truck	37
Heavy truck	13

a How many vehicles travelled along this road on the day the data was collected?

b What percentage of the vehicles were trucks?

c Construct a column graph to display this data.

Numerical data

Lesson 2

When the data collected involves numbers it is called **numerical data**.

There are two types of numerical data:

- **discrete numerical data**: The data recorded will be distinct values. Often they are whole number values. Discrete data is often **counted** so that the data can be recorded. For example the number of children in a family may be 0, 1, 2, 3, 4 …
- **continuous numerical data**: The data recorded can be any values on a number line, including the values between whole numbers, and is obtained by **measuring**. For example the heights of Year 9 students may be any values between about 140 cm and 200 cm.

Organisation and display of discrete numerical data

A farmer has a crop of peas and wishes to investigate the number of peas in the pods. He takes a sample of 50 pods and counts the number of peas in each pod.

To organise the data the farmer could use a **tally and frequency table**, similar to the one shown below.

Number of peas in pod	Tally	Frequency
0		0
1	|	1
2	||	2
3	||	2
4	||||	4
5	~~||||~~ |	6
6	~~||||~~ ||||	9
7	~~||||~~ ~~||||~~ |||	13
8	~~||||~~ ~~||||~~	10
9	|||	3
	Total:	**50**

A **column graph** could be used to display the results:

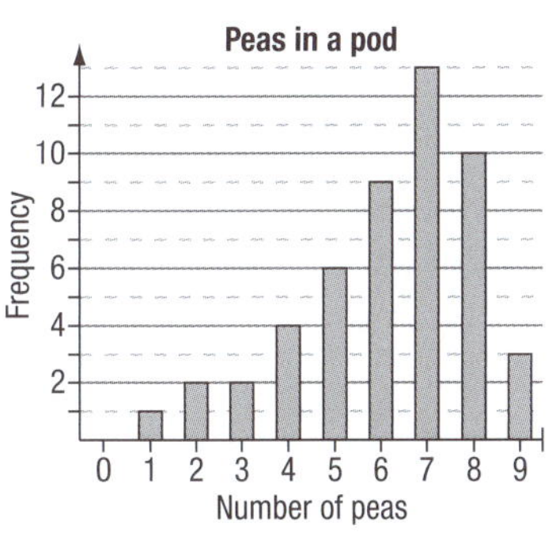

Remember

The columns of the graph are the same width and are separated, **not** joined together.

EXERCISE

1 When each of the following is investigated would the numerical data collected be discrete or continuous?

- **a** the times taken by athletes to run 100 metres
- **b** the maximum temperatures reached each day in January
- **c** the number of matches in matchboxes
- **d** the weight of luggage taken on an aircraft
- **e** the time taken for batteries to run down
- **f** the number of goals scored by each player in a soccer match
- **g** the number of passengers on buses
- **h** the heights of five-year-old children

2 A randomly selected sample of households was asked, 'How many people live in your household?' The column graph on the right has been constructed using the results.

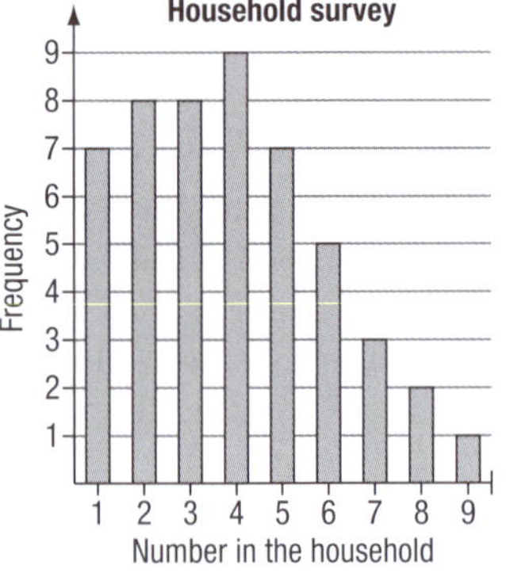

- **a** How many households were in the survey?
- **b** How many of the households had two occupants?
- **c** What percentage of the households had five occupants?
- **d** What percentage of the households had six or more occupants?

3 A class of 20 students was asked, 'How many pets do you have in your household?' The following set of data was collected:

0 1 2 2 1 3 4 3 1 2 0 0 1 0 2 1 0 1 0 1

- **a** Is the numerical data collected in this investigation discrete or continuous? Why?
- **b** Construct a tally and frequency table to organise the data.
- **c** Construct a column graph for the data. Be sure to give your graph a title and scale and to carefully label the axes.
- **d** What is the most common number of pets in a household for these 20 students.
- **e** What percentage of the households had no pets?
- **f** What percentage of the households had three or more pets?

4 The number of matches in a matchbox is stated on the label as 50 but the actual number of matches has been found to vary. To investigate this, the number of matches in a sample of 60 boxes has been counted. The number counted in each of the 60 boxes is recorded below.

50 50 51 52 49 50 48 51 50 47 50 52 48 50 49 51 50 50 52
51 50 50 52 50 53 48 50 51 50 50 49 48 51 49 52 50 49 50
52 50 51 49 52 52 50 49 50 49 51 50 50 51 50 53 48 49 49

- **a** Is the numerical data collected in this investigation, continuous or discrete? Why?
- **b** Organise the data by constructing a tally and frequency table.
- **c** Display the data using a column graph, taking care to label both axes and to give the graph a title.
- **d** What percentage of the boxes contained exactly 50 matches?
- **e** What percentage of the boxes contained fewer than 50 matches?

5 The number of letters per word in a magazine article was investigated. A paragraph was randomly chosen and the number of letters in each word of that paragraph was counted:

3 8 6 2 10 3 3 5 2 4 9 4 3 5 3 12 4 3 9 3
8 4 4 4 2 9 2 4 3 2 2 2 6 2 3 6 11 4 9 4
2 8 2 4 3 2 4 2 2 6 2 3 6 10 10 3 4 2 4 6
4 8 9 2 9 2 11 11 2 6 4 5 7 2 3 5 4 7 7 7

- **a** How many words were in the paragraph?
- **b** Explain why the set of data is discrete numerical data.
- **c** Construct a column graph for the data, labelling the axis and using a suitable title for the graph.
- **d** Copy and complete the following:
  - **i** Words with ____ letters were the most frequently occurring words in the paragraph.
  - **ii** ____% of the words in the paragraph had eight or more letters.
  - **iii** ____% of the words in the paragraph were two-letter or three-letter words.

Organisation and display of continuous numerical data

Lesson 3

The data in a set of continuous numerical data could all be different values. Therefore, we usually group continuous numerical data into **class intervals** to help us organise and graph it.

A **histogram**, is used to display continuous numerical data. A histogram is similar to a column graph, but the 'columns' are joined together.

When the class intervals are the same size, the frequency is represented by the height of the 'columns'.

Suppose the heights of 14-year-old children are being investigated. The data collected is **continuous numerical data** because the values recorded could be any value on the number line.

The heights of thirty children are measured, and recorded to the nearest centimetre:

163	154	152	160	148	149	154	172	171	162
165	160	166	175	143	174	180	162	167	158
159	164	163	183	150	163	181	158	165	156

The data can be organised using **class intervals** and a tally and frequency table. First, we need to decide the class intervals we will use.

The shortest height is 143 cm and the tallest is 183 cm. We can use five class intervals from 140 cm to 190 cm. The first class interval includes 140 cm and extends to 150 cm but does not include 150 cm. This is written as '140–'.

The other class intervals are:

- 150–
- 160–
- 170–
- 180–

The tally and frequency table for this data looks like the one below.

Height (cm)	Tally	Frequency
140–	|||	3
150–	~~||||~~ |||	8
160–	~~||||~~ ~~||||~~ ||	12
170–	||||	4
180–	|||	3
	Total:	**30**

A **histogram** can be constructed for this data:

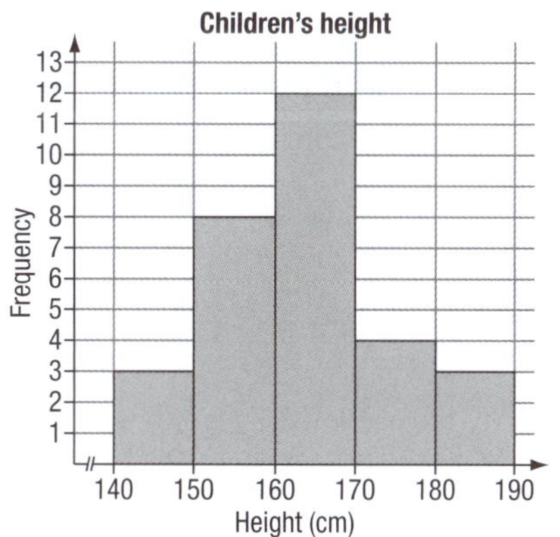

Remember

The marks // on the horizontal axis tell us that there is a break in the scale on the axis of the graph and that, in this case, it does not start at zero.

EXERCISE

Weight (kg)	Frequency
30–	2
40–	14
50–	24
60–	22
70–	8
80–	2

1 The weights of 72 students were recorded and the results were recorded in the frequency table on the right. Draw a histogram from the frequency table, taking care to give the graph a title and to use an appropriate scale and label the axes.

2 The monthly rainfall, in millimetres, was recorded for a town over 30 months. This is the set of data collected:

0.6	1.5	5.7	10.6	34.2	25.3	7.2	15.4	23.6	34.8
14.6	18.5	22.3	28.6	3.4	11.6	5.8	38.9	19.4	39.9
30.0	10.8	15.0	26.4	7.9	20.6	11.2	35.8	19.9	33.3

a State reasons why this data is continuous numerical data.

b Construct a tally and frequency table using the class intervals of 0–, 5–, 10–, 15–, 20–, 25–, 30– and 35–.

c Draw a histogram for the data, taking care to give the graph a title and to use an appropriate scale and label the axes.

3 Residents have complained about cars speeding along their street where the speed limit is 60 km/h. The speed of cars on this street was recorded for a 24-hour period and the histogram on the right was drawn of the results.

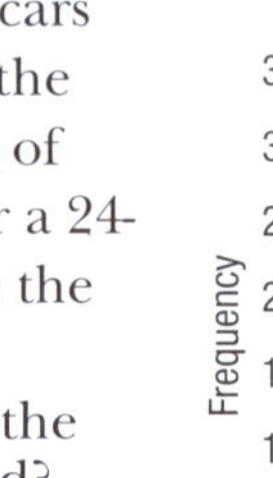

a How many cars travelled along the street during the 24-hour period?

b i If the speed limit is 60 km/h, how many of the drivers were travelling below the speed limit?

ii If only one car was travelling at exactly 60 km/h, what percentage of the drivers were travelling above the speed limit?

4 The ages of people in an extended family group have been rounded to one decimal place and recorded. The data is listed below:

75.3	54.1	56.4	48.6	43.6	36.6	45.4	16.7	15.1
13.8	9.8	7.3	34.1	39.3	15.1	13.7	11.3	11.3
8.5	28.5	29.3	9.6	6.3	4.4	2.1	0.4	24.6
28.5	4.2	2.8	0.7	22.7	20.6			

a Construct a tally and frequency table for this data, using class intervals of 0–, 10–, 20–, 30–, 40–, 50–, 60– and 70–.

b Construct a histogram for the data.

c The class interval with the highest frequency is called the **modal class**. Which is the modal class?

d What percentage of the family is aged:

i less than 20? ii 50 or more?

5 A school conducted a survey of 60 of their students to investigate the time it takes for students to travel to school. The following data shows the students' travel times to the nearest minute:

12	15	16	8	10	17	25	34	42	18	24	18	45	33	38
45	40	3	20	12	10	10	27	16	37	45	15	16	26	32
35	8	14	18	15	27	19	32	6	12	14	20	10	16	14
28	31	21	25	8	32	46	14	15	20	18	8	10	25	22

a Is this data collected about travelling time to school discrete or continuous numerical data? Give a reason for your answer.

b Construct a frequency table for the data using class intervals of five minutes.

c Construct a histogram from your frequency table.

d Calculate the percentage of the students whose travelling time to school is:

i less than 15 minutes

ii more than 30 minutes.

Class exercise

Investigate the time taken for the students in your class (or in the whole school) to travel to school.

Measures of centre

Lesson 4

A quick understanding of a numerical data set can be obtained if we have an indication of the **centre** of the data.

There are three **statistics** that provide a **measure of the centre** of a set of numerical data.

1 The mean

- The **mean** is the statistical name for the 'average'.
- Mean = $\frac{\text{sum of the data values}}{\text{the number of data values}}$
- The mean does not need to be one of the data values.

2 The median

- The **median** is the **middle value** of an **ordered set of data**. An ordered set of data is the data listed in order, from the smallest value to the largest value.
- If there is an even number of data, then the median is the average of the two middle values.
- Half of the values in a set of data are less than the median and half of the values are more than the median.
- The median does not need to be one of the data values.

3 The mode

- The **mode** is the most frequently occurring value in the data set. If there are two modes in a data set then the data can be described as **bimodal**.
- The mode can be found easily from a frequency table or column graph. On a histogram, we call the class interval that occurs most frequently **the modal class**.

Example

For the set of data 7, 6, 6, 8, 5, 8, 6, 9, calculate:

a the mean

b the median

c the mode.

Answer

a The **mean** is the average of the eight numbers:

$$\text{Mean} = \frac{7+6+6+8+5+8+6+9}{8} = \frac{55}{8} = 6.875$$

b The data needs to be ordered so that the **median** can be found:

5 6 6 6 7 8 8 9

There are eight data values, so the median is the average of the two middle numbers:

$$\text{Median} = \frac{6+7}{2} = 6.5$$

c The **mode** is the most frequently occurring data value so the mode is 6 (because 6 appears three times in the data set while the other values each appear only one or two times).

EXERCISE

1 For each of the following data sets, find:

i the mean **ii** the median **iii** the mode.

a 2 3 3 3 4 4 4 5 5 5 5 6
6 6 6 6 7 7 8 8 8 9 9

b 10 12 15 15 16 16 17
18 18 18 18 19 20 21

c 22.4 24.6 21.8 26.4 24.9 25.0 23.5 26.1 25.3 29.5 23.5

d 127 123 115 105 145 133 142 115 135 148 129
127 103 130 146 140 125 124 119 128 141

2 The test scores, out of 30 marks, for a class of 22 students are:

15 16 18 23 22 28 29 25 25 24 27
18 11 20 23 26 26 30 25 18 15 17

a Look at the data and write an estimate of the 'central value'.

b Calculate:

i the mean **ii** the median **iii** the mode.

c One of the data values (11) is particularly low. Which of the measures of centre (mean, median or mode) is most affected by this low value?

d Give a reason why the mode is not the most suitable measure of centre for this set of data.

e Noah scored 23 on this test. What do the mean and median of the data tell us about Noah's score?

3 In 2007, the median age for residents of Papua New Guinea was 21.4 years. Explain what this means. Do you think that the median age will be different now?

4 Consider the following two ordered data sets.

Data set A:	3	4	4	5	6	6	7	7	7	8	8	9	10
Data set B:	3	4	4	5	6	6	7	7	7	8	8	9	15

a Find the mean for both Data set A and Data set B.

b Find the median of both Data set A and Data set B.

c Find the mode of both Data set A and Data set B.
d Explain why the mean of Data set A is less than the mean of Data set B, but the median and mode are the same for both data sets.

5 The members of a rugby team scored a mean of 15.4 points in their first 10 games. They scored 24 points and 16 points in their next two games.
a What was the total of the points scored by the team in the first 10 games.
b What is the total of the points scored in all 12 games.
c What is the team's mean score for the 12 games?

6 A basketball team scored 43, 55, 41 and 37 goals in the first four matches of the season.
a What is the mean number of goals scored for the first four matches?
b What score will the team need to shoot in the next match to maintain the same mean score?
c The team shoots only 25 goals in the fifth match. What is the mean number of goals scored for the five matches?
d The team shoots 41 goals in the sixth and final match.
i Will this increase or decrease the previous mean score?
ii What is the mean score for all six matches?

Interpreting graphs and estimating the mean, median and mode

Consider the data: 5 5 7 3 8 2 3 4 6 5 7 6 4

A column graph for the data is shown on the right.

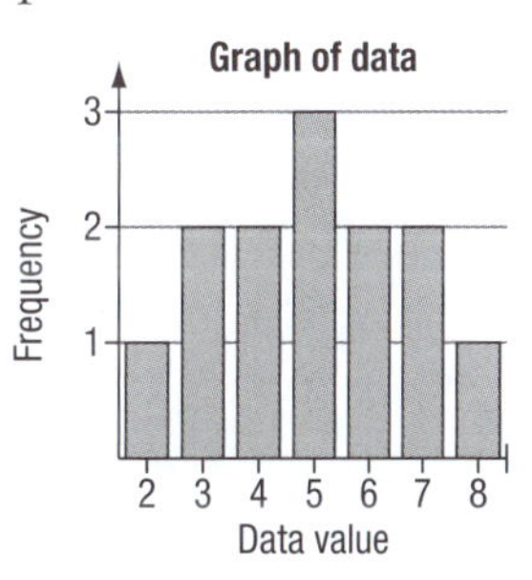

For this data, the mean, median and mode all have the same value: 5.

We say that the **distribution of data** for this graph is **symmetric** (equally spread on either side of the 'central' value 5).

When the distribution of data is not symmetric the mean, median and mode can have different values.

The distribution of data graphed on the right is described as **negatively skewed** (it is stretched to the left).

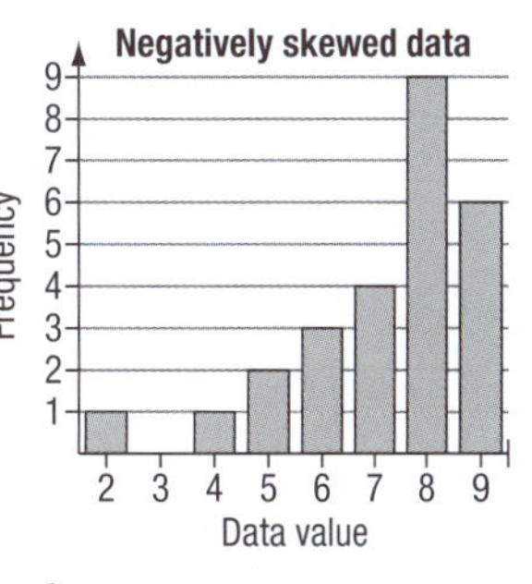

In this example, the **mean** (7.23) is less than the median (8) and the mode (8) because there are some extremely small values (2 and 4) in the data set. Because the mean depends on the actual values of the data, it has been 'dragged' towards these extreme small values.

The **median** is not influenced by extreme values because it depends on the position of data rather than their value.

The distribution of a **positively skewed** set of data would appear to be stretched to the right and the mean would be greater than the median and mode.

The mean for the data graphed on the right is 19, but the median is 17.5 and the **modal class** is the 10–15 minute class interval.

The mean has been dragged towards the extreme value in the 60–65 minute class interval.

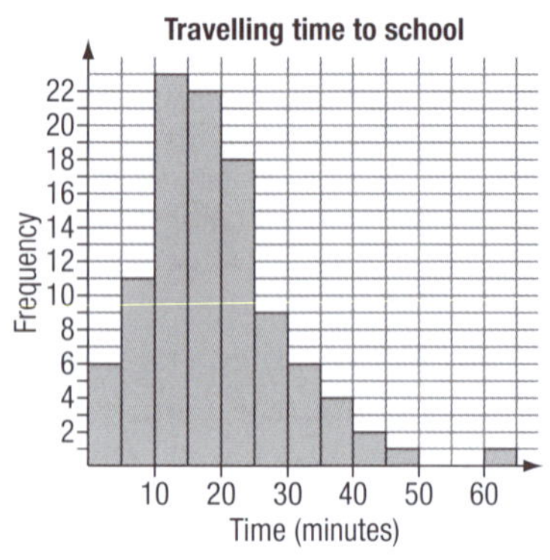

Remember

To estimate, we need to find an approximate value.

EXERCISE

1 Look at the numbers in each of the following ordered sets of data and estimate the value of:

i the mean ii the median iii the mode.

a 2 2 3 3 4 4 4 5 5 5 5 6
6 7 7 7 8 8 8 9 9 10 10

b 4 5 6 7 8 9 10 10 11 12 13 14

c 0 0 0 0 0 0 1 1 1 1 1 1 1 2

d 0 0 1 1 2 2 2 3 3 4 4 5 5 6 6 12

2 For each of the data sets graphed below:

i describe the distribution of the data as either symmetric, positively skewed or negatively skewed.

ii estimate the mean

iii estimate the median

iv estimate the mode (or modal class).

a

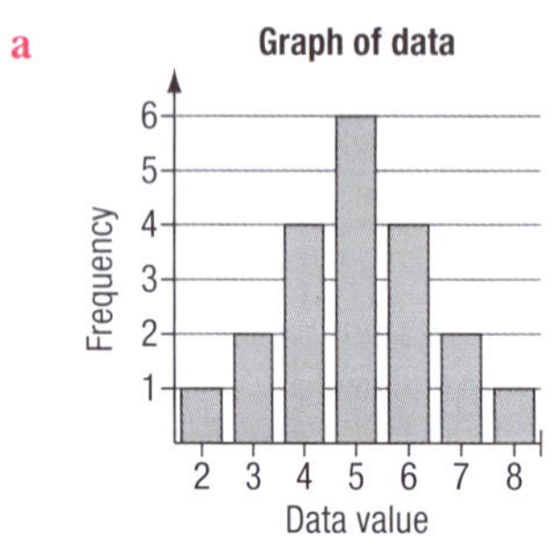

b

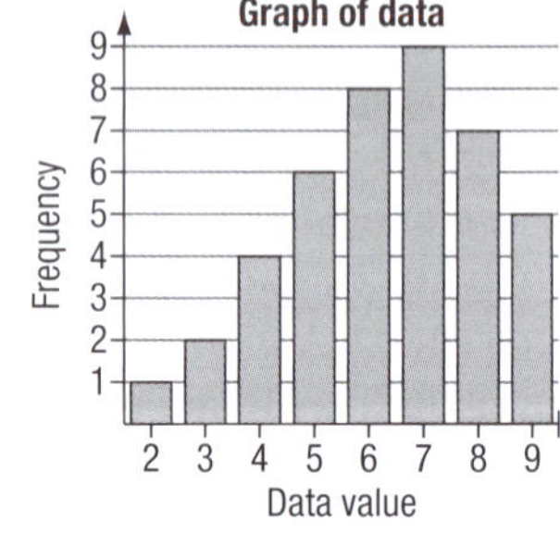

c

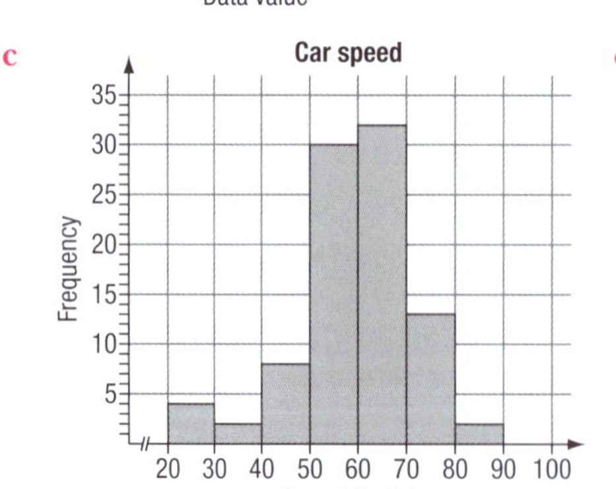

d

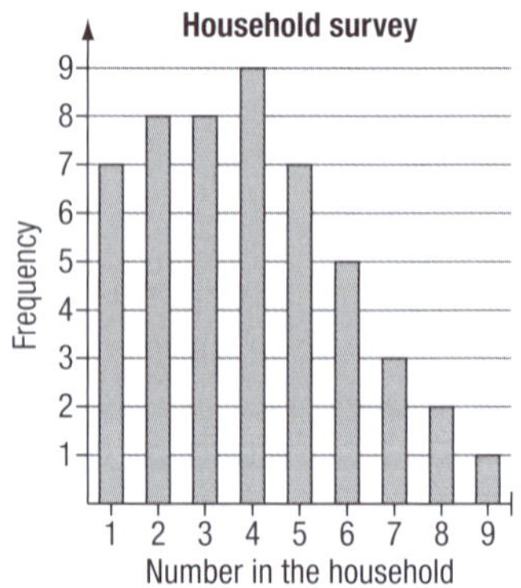

e

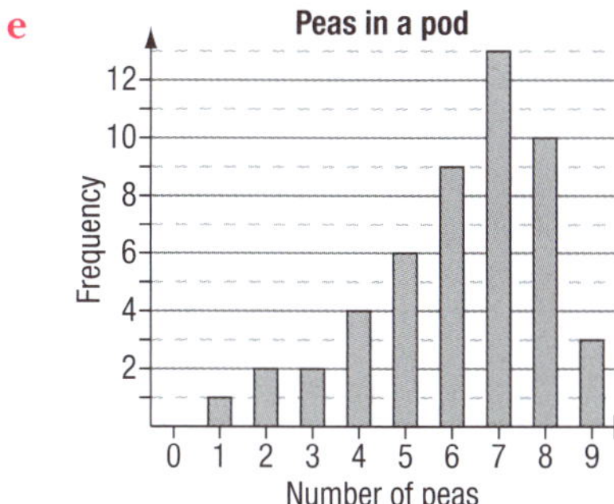

f

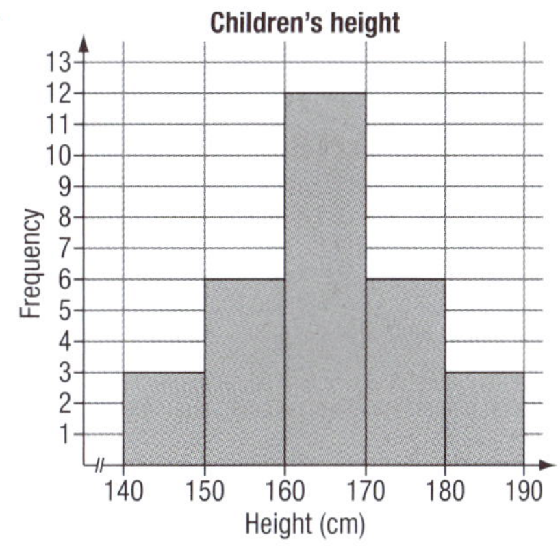

g

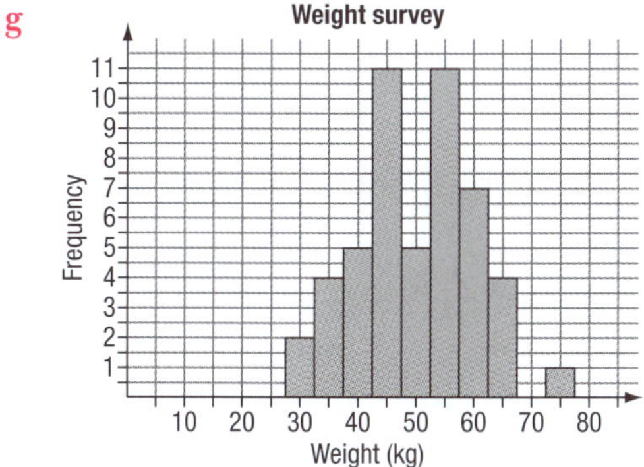

h

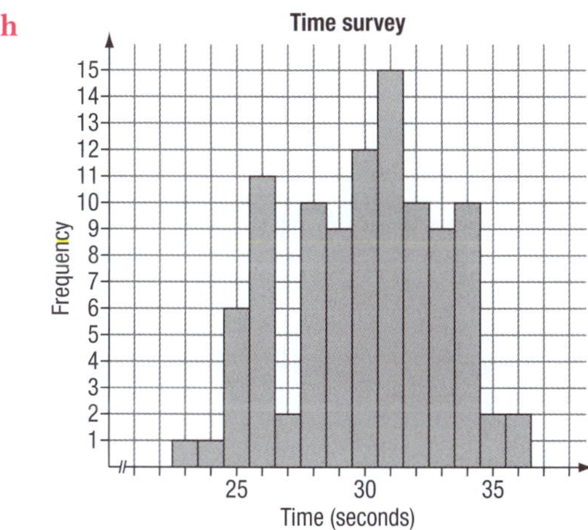

REVISION AND ASSESSMENT

Numbers and operations

1 Write 'True' or 'False' for each of the following statements:

a $\frac{3}{4} + \frac{1}{6} = \frac{11}{12}$ b $16 \div \frac{1}{4} = 4$ c $5\frac{1}{4} + 11\frac{2}{5} = 16\frac{1}{3}$ d $5.75 > 5\frac{2}{3}$

e $86.3 \times 1000 = 8630$ f $17.45 \div 10 = 1.745$ g $\frac{2}{5} = 0.4$ h $5.6 = 5\frac{1}{6}$

2 Calculate each of the following, showing all working:

a $4\frac{2}{3} + 1\frac{1}{5} + 6\frac{3}{4}$ b $3\frac{2}{5} \div 2\frac{3}{10}$ c $6\frac{3}{4} \times \frac{5}{9}$ d $\frac{2}{3}$ of $1\frac{1}{5}$

3 Given $762 \times 19 = 14\,478$, write the answers for:

a 762×1.9 b 0.762×1.9 c 76.2×0.19

4 Calculate each of the following, showing all working:

a $41.67 + 9.2 + 86.19$ b 86.5×1.1 c $845.24 \div 4$ d $380 \div 0.4$

5 Selma has 4 kg of sago. She gives $\frac{2}{3}$ kg to her aunt and $1\frac{1}{4}$ kg to her friend. How much sago does she have left?

6 Raoul has 153 metres of string to make bird traps which require $2\frac{1}{4}$ metres each. How many traps can he make?

7 A storage bin of grain holds 5.8 cubic metres. It is to be emptied into containers holding 0.4 cubic metres each. How many containers will it fill?

8 What is the cost of 35 items at K4.20 each?

9 Amos is training for the Pacific Games. He runs $5\frac{1}{2}$ kilometres each weekday and $8\frac{1}{4}$ kilometres on both Saturday and Sunday. How far does he run in the 7 days?

10 a What is the total weight of four piglets if their weights are 3.3 kg, 3.25 kg, 2.95 kg and 3.62 kg?

b What is the piglets' average weight?

c If another two piglets weighing 4.5 kg and 4.82 kg are put in the yard with the original four, by how much does the average weight of all the piglets increase?

Ratio and percentage

1 Choose the correct answer (**A**, **B**, **C** or **D**) for each of the following:

a Which of the following is $\frac{1}{4}$ expressed as a percentage?

A 50% **B** 25% **C** 75% **D** 100%

b Which of the following is 12.5% expressed as a fraction in its simplest form:

A $\frac{125}{1000}$ **B** $\frac{1}{8}$ **C** $\frac{12.5}{100}$ **D** $\frac{1}{125}$

c Which of the following is 75% of 34.40?

A 8.60 **B** 24.08 **C** 25.80 **D** 28.50

d Which of the following is K5 as a percentage of K20?

A 5% **B** 10% **C** 20% **D** 25%

e A class is made up of 12 girls and 14 boys. Which of the following is the ratio of boys to girls?

A 12 : 26 **B** 14 : 26 **C** 7 : 6 **D** 6 : 7

f Which of the following is the ratio 2 : 3 expressed as a fraction?

A $\frac{3}{5}$ **B** $\frac{2}{5}$ **C** $\frac{3}{2}$ **D** $\frac{2}{3}$

g Which of the following is **not** an equivalent ratio for 4 : 5?
A 0.2 : 2.5 B 12 : 15 C 0.8 : 1.0 D 16: 20

h A box contains 24 tins of fish. The ratio of mackerel to tuna is 3 : 5. How many tins of tuna are there in the box?
A 3 B 12 C 9 D 15

2 a What percentage is 15 of 75?
b What percentage is K2.40 of K7.20?
c What percentage is 5 g of 2 kg?
d What percentage is 72 000 of 1 000 000
e What percentage is $\frac{2}{3}$ of $\frac{7}{12}$?

3 Of the paw paw trees planted around the perimeter of the school grounds, 72% became diseased and bore no fruit. If 150 trees were planted:
a how many trees were diseased?
b how many trees bore fruit?

4 Write each of the following as a simplified ratio:
a 18 : 27 b $\frac{2}{3} : \frac{3}{5}$ c 1.25 : 3.75 d 420 : 30 : 1500

5 Express each of the following fractions as a ratio in its simplest form:
a $\frac{3}{9}$ b $\frac{12}{18}$ c $\frac{72}{56}$

6 A length of string is divided in the ratio 7 : 5. If the smaller piece measures 15.25 cm, how long is the other piece?

7 Divide 48 cm in the ratio:
a 1 : 2 b 1 : 2 : 3 c 2 : 11 : 5 : 6

8 Two families share a bag of fertiliser for their garden in the ratio 3 : 2. If the smaller share is 8 kg, how much was in the bag originally?

9 Paint colours are identified by code numbers. To make O014 orange paint, R012 red paint is mixed with Y022 yellow paint in the ratio 7 : 5. If 350 litres of R012 red paint is used by the paint mixer:
a how much Y022 yellow paint is used?
b how much O014 orange paint will there be?
c and 5% of the paint is accidentally spilled during the mixing, how much paint is actually produced?
d how many 4 litre tins can be filled with orange paint?

10 9-carat gold contains 37.5% pure gold, with the rest being a mixture of other metals.
a What is the ratio of pure gold to other metals in 9-carat gold?
b How much pure gold would be used to make 24 grams of 9-carat gold?

Rates

1 What is the hourly rate for each of the following?
a A factory producing 6356 cans of food in 7 hours.
b A painter using 12 litres of paint in 8 hours.
c A miner earning K33 for 6 hours work.
d A cook making 3 pizzas every 12 minutes.
e A boat travelling 336 kilometres in 1 day.

2 a Solomon spreads fertiliser on his garden at the rate of 5 grams per square metre. How much will he use if the garden measures 24 square metres?
  b A store sells carrots for K7.30 per bag. Find the cost of 12 bags.
  c Mata is planting seedlings at the rate of 40 per square metre. How many will she plant in an area of 8.5 square metres?
  d A coffee growing region produces 750 kilograms of coffee per hectare. How much coffee will be produced from a plot measuring 0.2 hectares?
  e Song rides 23.5 kilometres in half an hour. What is his speed in km/h?

3 Two friends, Kala and Kashmir, set off on a 500 metre cross country race. Kala can run faster than Kashmir, but did not win the race.
  a Match the graphs on the right to each of the two runners.
  b Describe the progress of the race, as if you were a commentator watching from a helicopter.

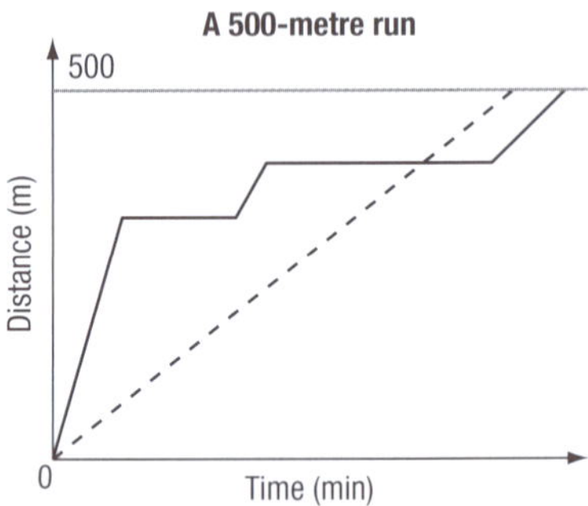

4 Yawie rides a motorbike from home to the store and draws this distance–time graph of his journey.
  a When did he stop for people crossing the road?
  b How long did he stop for the people crossing the road?
  c How long did he spend at the store?
  d How far is the store from Yawie's home?
  e When was his rate of travel the greatest? What was his speed then?

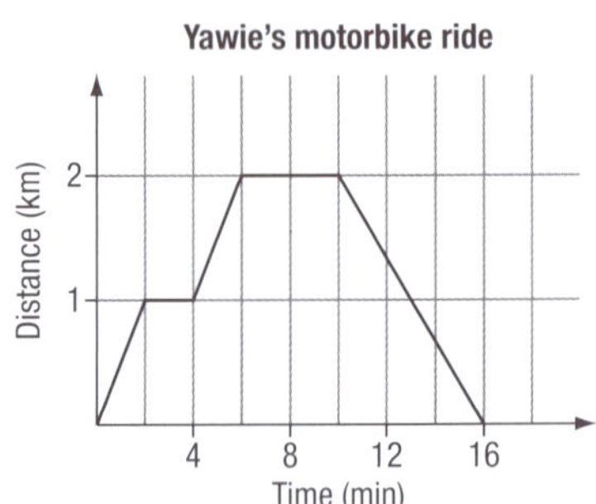

5 Gerard plotted the amount of petrol in his car's tank over a 24-hour period.
  a When was the car not being driven?
  b When did Gerard fill the tank?
  c When was Gerard driving in heavy traffic? (*Hint*: You use more petrol in stop/start traffic.)
  d When did Gerard stop for lunch?

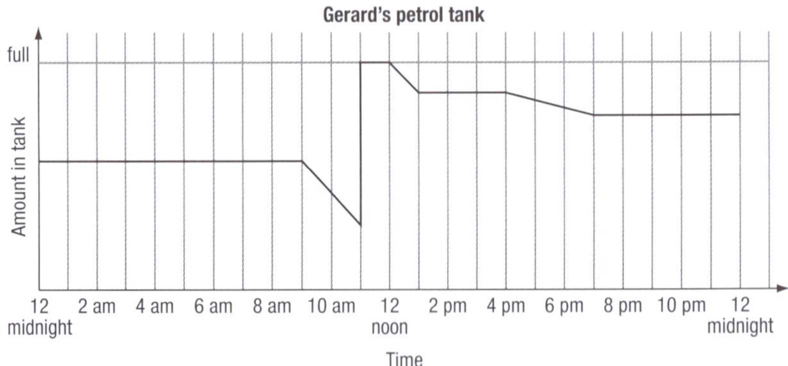

6 A pond is being filled with water, and the height of the water in the pond each minute is recorded in the table below.

Number of minutes	0	1	2	3	4	5	6	7	8	9	10
Height of water (cm)	0	6.5	13	19.5	26	32.5	39	45.5	52	58.5	65

  a Draw a graph, on grid paper, showing the information in the table.
  b At what rate is the water level rising?
  c Find the height of the water in the pond after 4.5 minutes.

7 Water is leaking out of a tank. The table below shows how much is left in the tank every day for 5 days.

Number of days	0	1	2	3	4	5
Number of kilolitres remaining	500	440	380	320	260	200

a How much water is in the tank to start with?
b How much water has been lost after 2 days?
c Is the water leaking at a constant rate? Explain.
d At what rate is the tank leaking?
e When will the tank be empty (assuming it keeps leaking at this rate and no water is added)?

8 The following graphs each represent a container being filled with water at a constant rate. Explain the shape of each container and what is happening in each case.

a
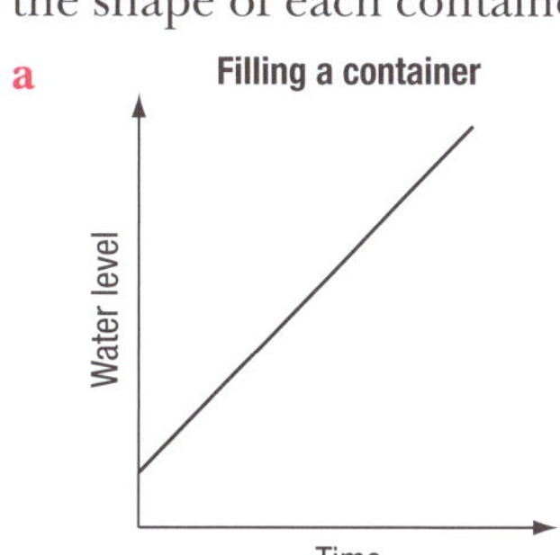

b
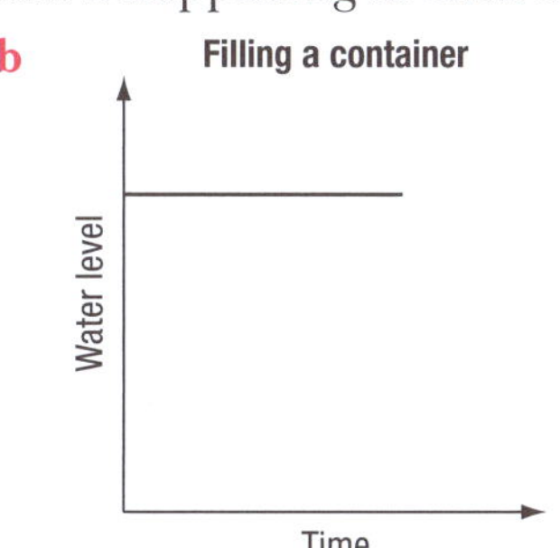

c
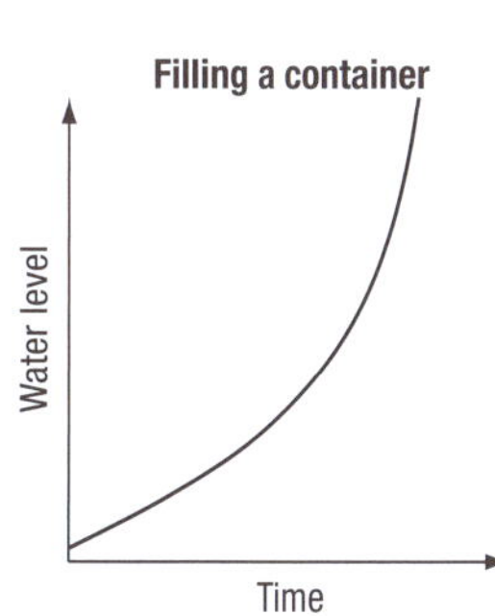

d
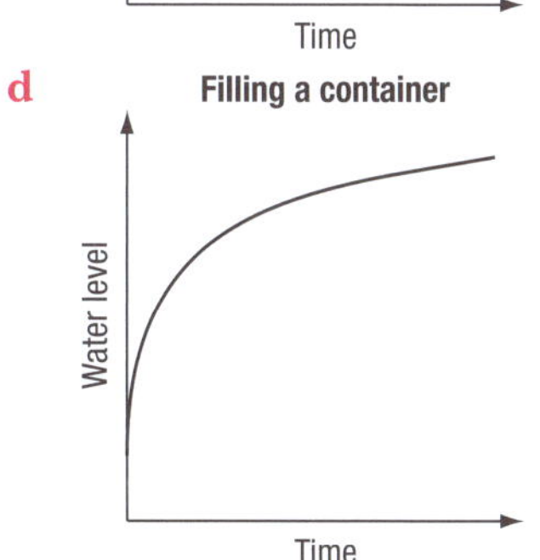

e
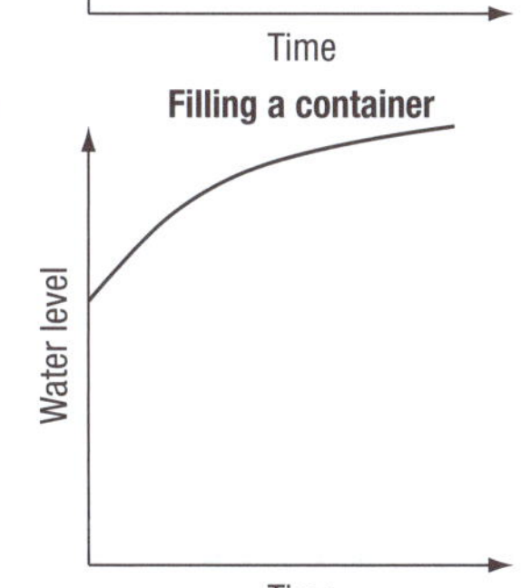

Money

1 Kisam buys goods costing K4.58, K11.07, K2.13 and 89t. How much change will she get from K20?

2 Find each of the following amounts, to the nearest toea:
a 10% of K58
b 5% of K45 000
c 15% of K23.90
d 85% of K5000
e 130% of K650

3 Solomon is paid K1950 per week. If he is given a pay rise of 5%, what is his new monthly pay? (*Hint*: Use 4 weeks = 1 month.)

4 A second-hand tractor, bought for K12 500, depreciates at the rate of 12% p.a.
a Find the amount of the first year's depreciation.
b What is the tractor's value after one year?

5 In one month, an agent sold goods totalling K25 300. If the agent is paid 5% commission, how much will he earn for the month?

6 Caleb earns K1200 per month. His budget for one month is shown on the right.
   a Which category accounts for nearly 5% of Caleb's expenses?
   b How much can Caleb save each month?
   c Which categories account for more than 10% of his expenses?
   d What percentage of Caleb's earnings is spent on rent (to the nearest whole number)?

Expenses	Amount budgeted
Rent	K550
Food	K480
Services	K58
Transport	K32
Clothes and incidentals	K40
Savings	

7 A store is advertising peanuts at K2.45 per packet, which is a saving of 45t.
   a What is the usual price of a packet of peanuts?
   b By what percentage has the price been reduced?

8 Which item in each of the following is the best value for money?
   a canned fish: a 95 g tin for K1.20; a 185 g tin for K2.10; or a 425 g tin for K4.92
   b paint: a 4 L tin on sale for K19.90 less 10%, or a 10 L tin for K46.60

9 All goods are reduced at a clearance sale. What is the sale price of these items after the percentage reduction stated? Give your answer correct to 5 toea.
   a towels costing K12.90 5% off
   b jeans costing K25.40 10% off
   c CD box costing K28.70 25% off
   d cupboard costing K179 15% off
   e guitar costing K450 30% off

10 At a sale, a CD player is marked down from K135 to K109. What is the percentage mark-down on the full price? Give your answer to the nearest whole number.

11 Philomena buys goods for her store from the importer at wholesale prices and marks them up by 45%. What price tag will she put on a chair with a wholesale cost of K78?

12 A washing machine is advertised for sale at K499. If the profit is 20%, calculated on the selling price, how much money does the storekeeper make on the sale?

13 Find the simple interest on the following amounts:
   a K4000 invested at 6% p.a. for 4 years
   b K23 000 invested at 6.5% p.a. for 8 years
   c K580 borrowed at 5% p.a. for 9 months

14 Domenica puts K450 in a fixed term deposit for 6 months. If the account pays 3.50% p. a. interest, how much in total will she have in the account at the end of 6 months?

15 Kema is charged K18 interest on a loan of K300 for six months. What percentage interest is being charged?

16 Tau is paying off a motor scooter which costs K1575. He pays a deposit of K75 and monthly payments of K90 for two years.
   a What does Tau pay in total?
   b How much interest does he pay over two years?
   c What percentage interest is this?

Measurement

1 Convert each of the following lengths to the required units:
   a 345 cm to metres
   b 1.654 km to metres
   c 1200 mm to metres
   d 2.564 m to centimetres

2 Convert each of the following areas to the required units:

a 345 mm^2 to cm^2 b 1.2 km^2 to m^2 c 2 865 000 cm^2 to m^2

3 Find the perimeter of each of the following figures:

a

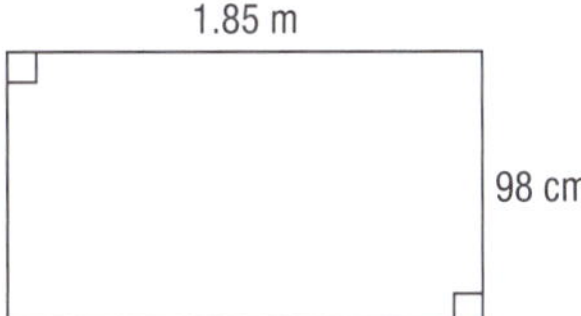

b

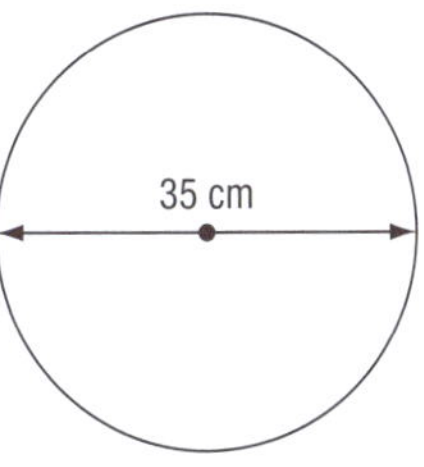

c

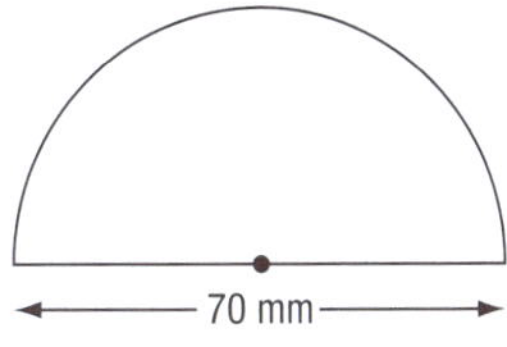

4 Find the area of each of the following figures:

a

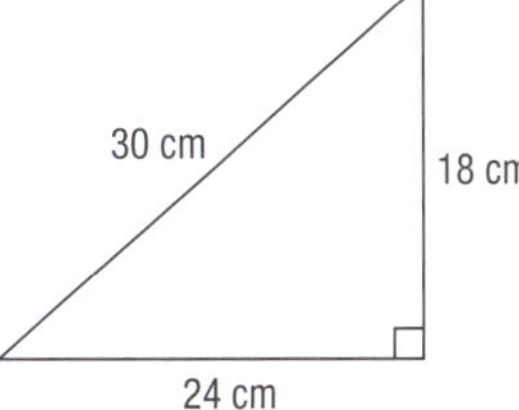

b

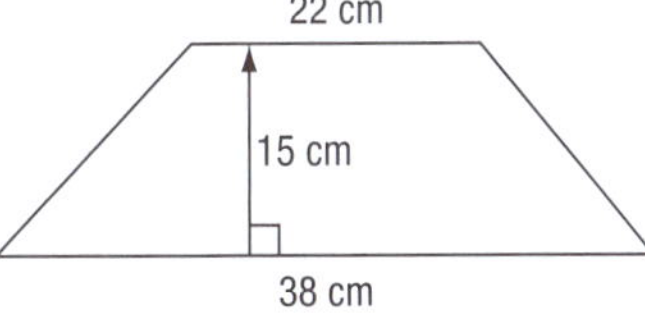

c

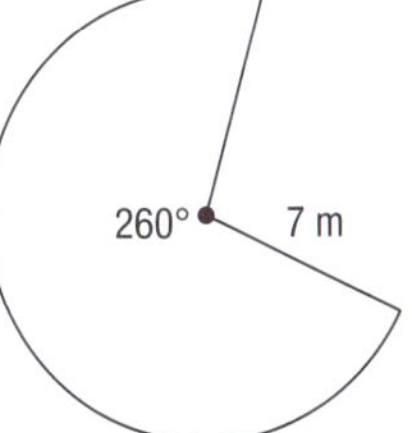

5 Calculate the area of each of these composite figures:

a

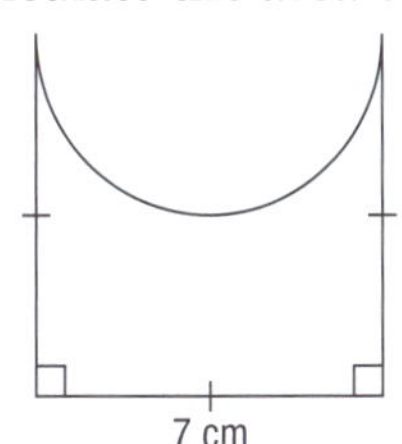

b

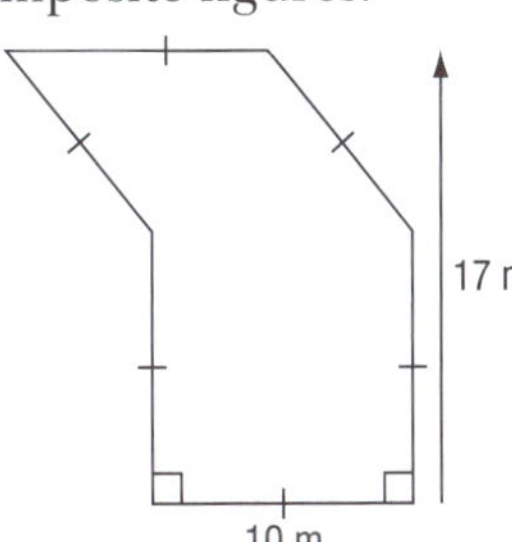

c

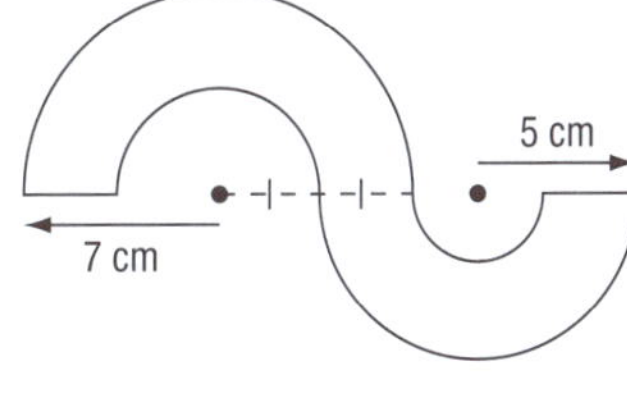

6 Calculate the total surface area of each of the following solids:

a

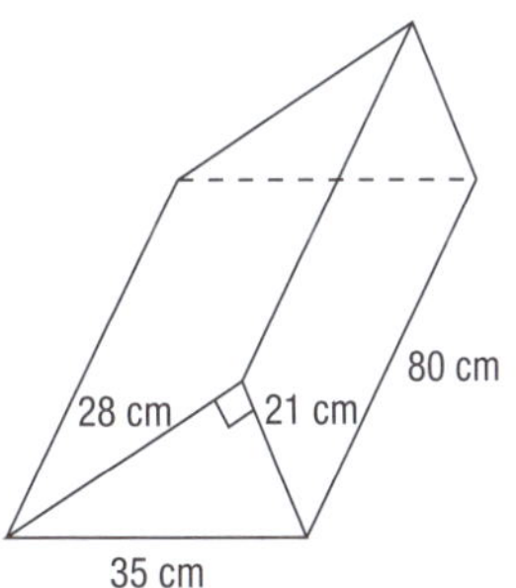

b

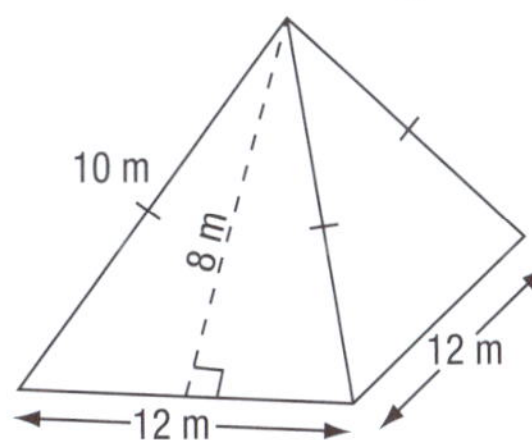

c

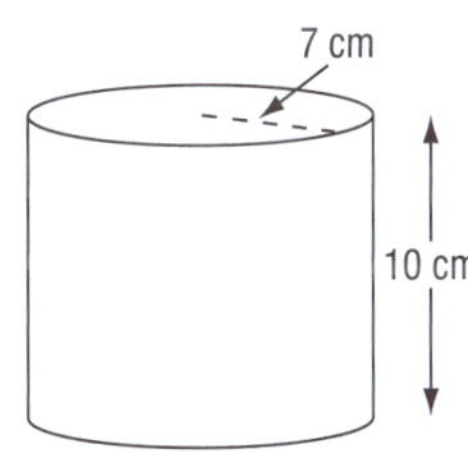

Data

1 If each of the following was investigated which investigation would produce categorical data?

A the time spent by customers in a bank queue

B the number of pigs owned by each family in a village

C the area of land that each family in a village farms

D the soccer teams which members of your class support

2 Which one of the following is suitable for displaying continuous numerical data?
A a histogram B a bar chart C a pie graph D a column graph

3 Data about the favourite football teams of children in a village has been collected and organised into the frequency table shown on the right.

Football team	Frequency
Tigers	11
Bears	8
Eagles	24
Demons	30
Broncos	15
Bulldogs	20

a Is the data collected categorical data or numerical data?
b Which team is the team most favoured by the people surveyed?
c Construct a bar chart for the data. Remember to give your graph a title, use an appropriate scale and label the axes.
d What percentage, of those that were surveyed, supported the Eagles?

4 For the following set of data, find:
a the mean b the median c the mode.

2 3 3 3 4 4 5 6 6 6 6 7 7 8

5 In five matches, a cricketer scored the following numbers of runs:

23 38 21 7 78

a Calculate the cricketer's mean number of runs per match.
b If the cricketer scores 0 runs and 56 runs in the next two matches, calculate his average for the seven matches.

6 The graph on the right displays the results of a survey of 90 women into the number of children that they have had.

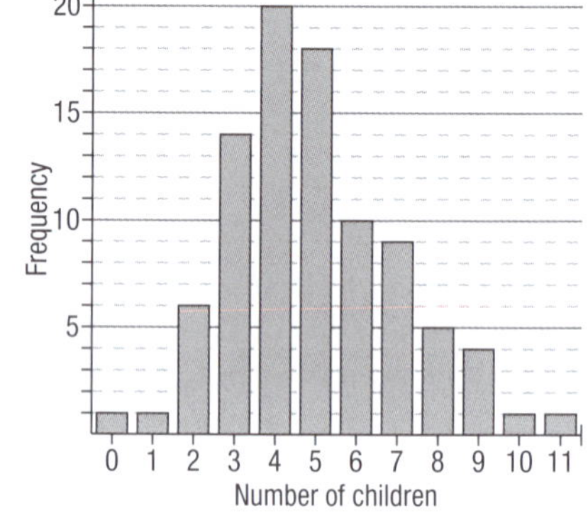

a Explain why this data is discrete numerical data.
b How many of the women had fewer than four children?
c What percentage of the women had six or more children?
d For this data, find:
i the mode ii the median.
e Describe the distribution of the data.
f Would you expect the mean to be larger or smaller than the mode? Why?

7 The following graph displays the heights of sixty 24-month-old boys.
a What is the modal class for this data?
b What percentage of the boys had a height less than 80 cm?
c What percentage of the boys had a height of 100 cm or more?
d Is the numerical data displayed discrete or continuous?
e From the graph, estimate the mean height of 24-month-old boys.

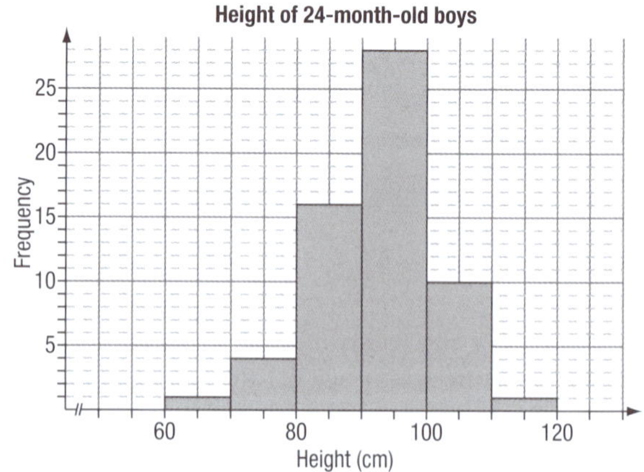

PROJECTS

Numbers and operations: Working with numbers

Your teacher will instruct you about which of these activities to complete, and whether to do them individually or with others in a small group.

Project 1: Demonstrate your understanding

Using your knowledge of number operations and design, produce a wall poster for younger students, showing how to carry out one of the operations (addition, subtraction, multiplication or division) with either fractions or decimals. Your poster should explain the operation, set out any steps and show examples relating to students' everyday life.

Project 2: Explain your local counting system

Many groups in Papua New Guinea use counting systems based on non-decimal cycles. Cycles of 4, 6 and 8 are some examples, as well as cycles of 5 and 20. Cycles of 2 are common in many language groups as well.

What counting system is used where you live? List the numbers from 1 to 50 in your counting system and give their names. If numbers can be shown on the body, for example by touching the shoulder or nose, draw a diagram to show this.

In some places, different objects such as coconuts and fish are counted in different sets. Is this true where you live? If so, describe two different examples of counting objects and how they are used.

a evutu two(2)

a varivi four (4)

a kurene six (6)

a kurene ma varivi ten (10)

Project 3: Working with another counting system

The ancient Mayans, who lived more than 3000 years ago in the area that is now Mexico, had their own system of counting based on the numbers 5 and 20. Their counting system used three symbols: a dot for one, a bar for 5, and a shell for 0. As in our decimal system, they used place value and the zero symbol was used if a place was empty.

1 Here are some Mayan numbers. Use this information to show the symbols for the numbers from 1 to 100.

Symbol		Value
shell	(shell symbol)	zero (0)
dot	●	one (1)
bar	▬	five (5)

Instead of writing their symbols from left to right as we do in our base-10 number system, the Mayans wrote their numbers from bottom to top, in rows, and used a base-20 system.

In the examples on the right, the dot in the top row in each of the numbers between 20 and 29 stands for 20 (signifying 1×20). The symbols in this row can go as high as 19×20 (or 380). With 19 in the bottom row added to this, it would be possible to show numbers up to 399.

If numbers as large as 400, or larger, are required, a third row (the 400s) would be added, standing for 20×20, or 20^2.

Here is they way that some larger numbers were written by the Mayans:

Example

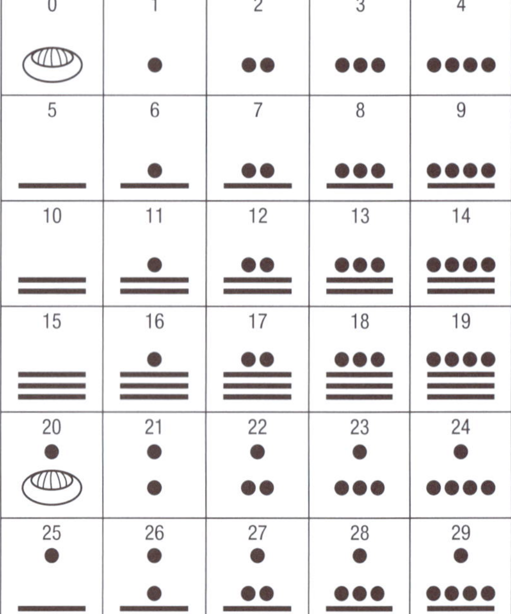

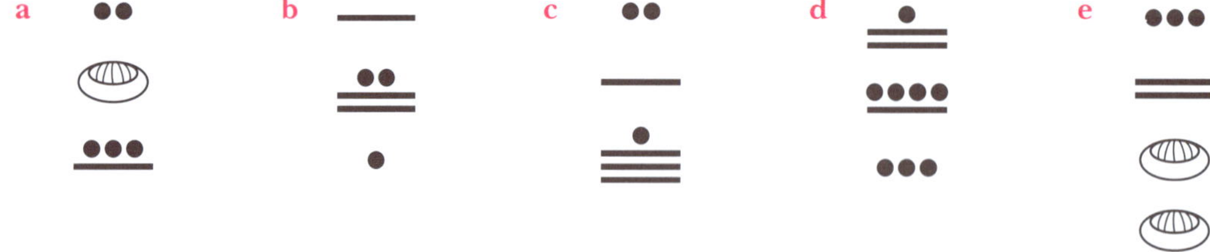

2 Test your understanding by writing the decimal value of each of these number representations.

a b c d e

Ratio and percentage: Applications of ratio

Project 1: Scale drawings

Maps and building plans are examples of scale drawings. They are a scaled down version of the original, with all the measurements in proportion to the real object.

We can also use scale drawings to create an enlargement of a drawing.

Your turn

Use these steps to create an enlargement of the drawing on the right.

Step 1: Trace the lizard onto your page and rule a 1 cm grid on it in pencil.

Step 2: Count how many squares there are across and down the grid.

Step 3: On another piece of paper, rule a grid with the same number of squares across and down, using 2 cm squares. Draw the grid **lightly in pencil** because you will rub these lines out later.

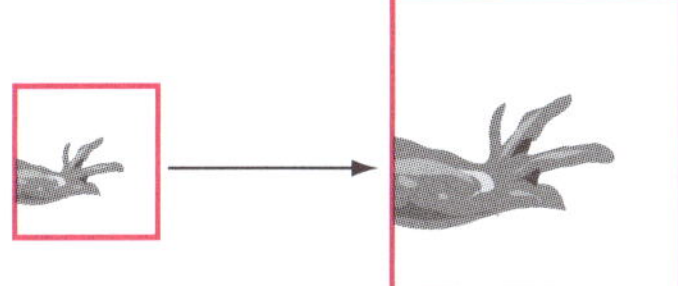

Step 4: Look at each box in the smaller grid and copy the shape into the box in the same position in the larger grid. Draw it lightly in pencil.

Step 5: When you are satisfied with the enlarged shape, draw over the shape more heavily with pencil and rub out the grid lines.

1 What is the ratio of the original drawing to your scale drawing?

Project 2: Map scales

On a map, the map scale or ratio scale is the ratio of map distance to actual distance.

For example, if the scale on the map is 1 : 1 000 000, this means that 1 unit measured on the map represents 1 000 000 units in actual distance. So 1 cm on the map is equivalent to 1 000 000 cm actual distance.

We can get a better idea of what this means by changing the larger figure into kilometres.

1 000 000 cm = 10 000 metres
= 10 kilometres

So 1 cm on the map is equivalent to 10 km in actual distance.

A scale can be written using units, such as: 1 cm = 10 km

or in numbers only, such as: 1 : 1 000 000

Your turn

1 Using the scale above, how far apart are two towns in actual distance when they are 3.5 cm apart on the map?

2 Two mountain peaks are 180 km apart. What will be the distance between them on a map using the scale above?

3 Two towns that are known to be 20 kilometres apart, measure 5 cm apart on a map. What is the ratio scale of the map?

Project 3: Gear ratios

A gear is a wheel with a number of cogs or 'teeth' that fit into the teeth on a second wheel. We see gears in a lot of machinery with moving parts, such as bicycles, car engines and wind-up clocks.

The speeds of the gear wheels depend on the number of teeth they have. In the diagram on the right, when the larger wheel has turned once, the smaller wheel will have turned twice. The speed of the smaller wheel will be twice that of the larger wheel.

The gear ratio equals:

number of teeth on large wheel : number of teeth of small wheel = 12 : 6
= 2 : 1

Notice that the ratio of the speeds of the wheels is the opposite of this:

speed of large wheel : speed of small wheel = 1 : 2

Your turn

1 State the gear ratio A : B for each of the following gear systems:

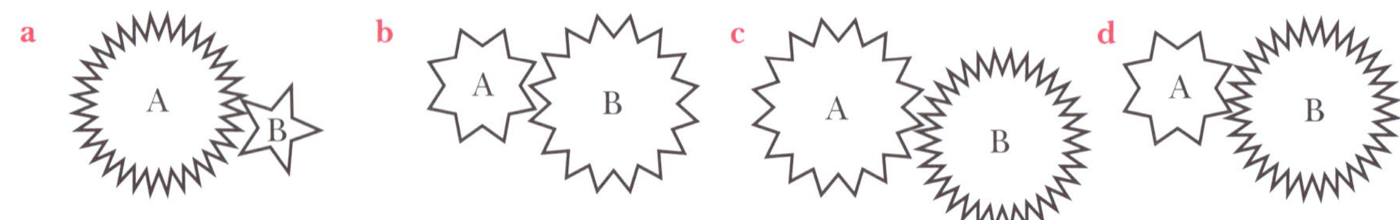

2 For each example in Question 1, describe how the wheel with fewer teeth moves when the other wheel rotates once clockwise.

3 For each example in Question 1, if the wheel with more teeth has rotated twice, how many times will the other wheel have rotated?

4 Trace the diagram below.

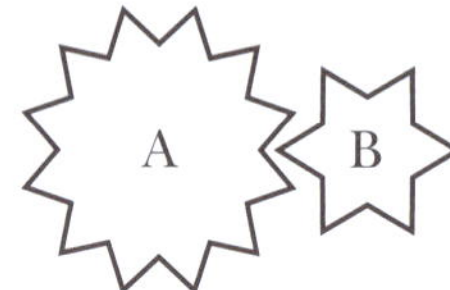

Sketch the two wheels after gear A has turned $\frac{1}{4}$ of a revolution, carefully numbering the teeth.

5 Find two examples of gear wheels around your school or town (for example on a bicycle, or a machine) and draw a diagram of the wheels. Write the gear ratio.

Money: Catering for a sports carnival

Imagine there is going to be a sports carnival in your town and your class has decided to carry out some fundraising. At least 500 people are expected to attend the carnival. The profits from the fundraising will be used to buy some sports equipment for your school.

Work with two or three other students on the following activities.

Project 1: A raffle

You are going to run a fundraising raffle, for which the prize is a mobile phone with a value of K150. (As an alternative, you can go to the store or look through newspaper advertisements to choose a phone for the prize yourself, so each group will have different calculations. Round the price of the phone to the nearest kina.)

1 How many 50t tickets would you need to sell to cover the cost of the phone?

2 How many 40t tickets would you need to sell to cover the cost of the phone?

3 If you sold 400 tickets at 50t, would you make a profit or a loss? How much would the profit or loss be?

4 Which do you think would be the better price for selling the tickets, 40t or 50t? Give reasons for your answer.

5 If you raised K75 from selling tickets (after paying for the phone), how many 50t tickets would you have sold?

Project 2: A soft drink stall

1 From your local store, or from advertisements in the newspaper, write the cost of a 1 L bottle of your favourite soft drink.

2 Estimate how many cups can be filled from a 1 L bottle of soft drink. Check this using 1 L of water.

3 If all 500 people attending the carnival want to buy a cup of drink, how many bottles of soft drink should the organisers of the stall buy and what would the cost be?

4 How much should the drinks be sold for to make a profit of 20%, assuming you sell all 500 drinks?

5 What is the smallest number of drinks you would need to sell to break even?

Project 3: A pizza snack stall

1 Find out the size and cost of the largest size of pizza available in your town.

2 Sketch the size of the pizza on a sheet of paper and estimate how many snack-size slices you would get from one pizza.

3 Assuming 400 of the people at the carnival wanted to buy your snack, how many pizzas should you order, and what would that cost you in total?

4 How much should you charge for each slice if you were to make no profit at all?

5 If you wanted to make 100% profit, what should you charge for each pizza slice?

Measurement: Using measurement

Project 1: Constructing a feed-box

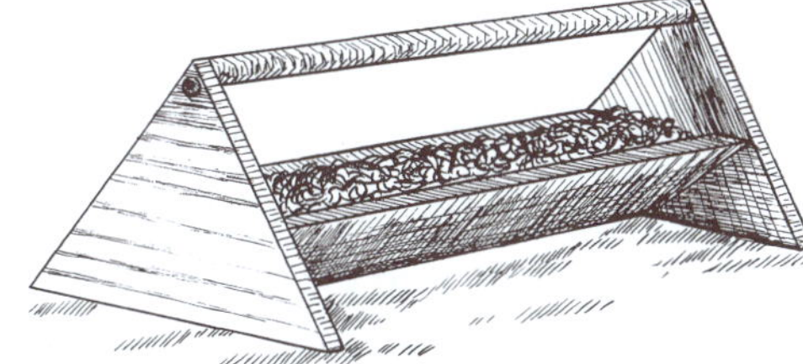

Suppose you are constructing a wooden feed-box in the shape shown on the right.
The feed-box is to be made from 2.5 cm thick wood.

1 Decide on the size of the feed-box you want to make and then draw the feed-box, labelling it with its overall dimensions (length, width and height).

2 Draw the shape of each of the components that make up the feed-box, labelling each shape with its dimensions and angles.

3 Calculate the total area of wood that would be required to make the feed-box.

Project 2: Designing a feed-box

You are to design a wooden feed-box with a shape different from the one in Project 1 above. It could be the shape of any feed-box that you use at your place or that you have seen. It could be constructed from 2.5 cm thick wooden planks, cylindrical logs or pieces of wood with any other dimension.

1 Draw the feed-box you would like to build, labelling it with its overall height, length and width.

2 Draw the shape of all the components that the feed-box is constructed from, labelling each of them with their dimensions and angles.

3 Calculate the total surface area of the inside of the feed-box, where the feed is placed.

Data: Collecting and analysing data

Project 1: Carrying out your own investigation

You are to conduct a small investigation into an aspect of your everyday life.

- The data that you collect could be from your class, your school, your family or your immediate neighbourhood.
- The data that you collect could result from a direct question, from measuring something, or from counting something.

Some suggestions of things that you may like to investigate are:

i the heights of students/boys/girls in your class
ii the lengths of the hand spans of students/boys/girls in your class
iii the number of siblings/cousins/people living in each of the households of students in your class
iv the favourite sports/drinks/vegetables/school subjects of students in your class
v the number of each type of vehicle travelling along a road in a particular time interval
vi the number of chickens/pigs/rabbits/dogs kept by the families of the students in your class
vii the types of vegetables grown in the garden plots of students in your class
viii the number/types of fruit trees in the garden plots of students in your class
ix the times that students in your class spend travelling to school/working in the family garden plot/doing schoolwork at home
x the times that students arrive at school

When you have chosen the subject of your investigation, work through the questions below.

1 Decide how, when and where you are going to collect the data for your investigation.

2 a Carry out your investigation and record the data.
b Establish what type of data you have recorded.

3 Construct an appropriate table and graph of your data.

4 If appropriate, find the mean, median and mode of the data.

5 Comment on the graph you constructed and any other findings from your investigation.

6 Write a report of your investigation. Be sure to include references to all of Questions 1 to 5 above.

ENRICHMENT

Traditional methods of measurement

Many people in local communities had no idea of the metric system. They did not use rulers and tape measures to measure distances, or clocks to measure time, or scales or balances to measure weights.

Scales being used to weigh coffee using the metric system of measurement.

In traditional mathematics, people used simple materials, such as natural objects and their body parts, to measure distances of length, or time or weight. For example, when measuring distances, they used ropes, arm lengths and pacing. They measured the weight of something by lifting it and comparing it with another object which was heavy or light. Estimation was also commonly used for comparing lengths and areas. Because this traditional system is based on estimation, the accuracy of measurement with these informal units could not be guaranteed.

Many people in traditional communities continue to use traditional mathematical methods for a variety of essential everyday activities. Most of these traditional methods can be related to modern mathematics as taught in schools of today.

Examples

- In the district of Bugia in Madang province, one full sago bag is used as the standard unit of measurement to measure volumes, capacities and quantities of different items, and is accepted throughout the whole province.
  For example, a full bag of fish is determined by comparison to a full bag of sago. Half a bag of fish is measured by comparing it to half a bag of sago, and so on. The fisherman compares amounts of fish to either a full bag, $\frac{3}{4}$ of a bag, or $\frac{1}{4}$ of a bag of sago.

- In the Southern Highlands, mounds of kaukau were made equal by using a rope for the circumference of the base and a stick of fixed size for the height.
- In the New Britain province of New Guinea Islands, fathoms (arm spans) were used to measure lengths of shell money.

- Lengths were estimated when selecting trees and saplings for building bridges and houses. House builders had a good idea of how much bush rope, bamboo and palm fronds would be needed for a particular house. Lengths were paced out for houses and gardens.

- In the Highland provinces, paces, foot-sized steps and ropes were commonly used (and continue to be used) for measuring lengths. There are a wide range of activities that require the measurement of lengths, and some indicate links to area.
These activities include building houses, drains, and gardens. Many village communities use a length measure to determine area. For example, garden plots are generally a certain width, so the total size could be determined by pacing out the other dimension, the length.
Measuring activities also include the construction of smaller three-dimensional objects, such as wigs, in which small lengths of string and finger parts are used to ensure symmetry and a good fit on a person's head. The wig makers (Huli people) have words to indicate the exact finger-width units that could be used to mark off lengths on a wig.

Huli wig maker

Huli wig man

- Sometimes a long rope is used to measure lengths. The length of rope may or may not be equal to a fixed number, but it is common for it to equal 20 paces or arm spans, since most languages have 20 cycle counting systems.
- In traditional Enga society, mathematics was very important, especially in the field of trade and barter. The measurement system was used to measure things that were of importance to the people in this society. There were no standard units but concrete objects and verbal descriptions/comparisons were used as standard units which were understood and used by everyone in the province. There are no mathematical figures in traditional Enga language. Instead of 1, 2, 3, etc., people just used Enga language words to count concrete materials.

The table below summarises some of the traditional methods of measurement:

Modern (metric) system	Traditional system
Time: seconds	**Time**: Smoking period (one smoking period is the amount of time taken to smoke one traditional cigarette). This unit measures lengths of time within a day.
Length: metres	**Length**: Pace, ropes, arm spans or hand spans, and verbal descriptions (specific methods were used for specific jobs). Things that were very long were measured using paces, while shorter items were measured using sticks, ropes, arm spans and hand spans. Pacing was very useful when designing houses. Hand spans were used in trade and barter to measure long bamboo containers of oil, strings of shells, etc. Measuring high trees physically was difficult, so verbal description was used.
Weight: kilograms	**Weight**: Verbal comparison in relation to pandanus nuts was used. A method of verbal description was used to make comparison between the weight of a specific item and the weight of a pandanus nut. The weight of a kaukau bag was also used for making comparisons, but this was not used as commonly as the pandanus method. Objects that weighed only slightly more than a pandanus nut were described as more than 1 pandanus nut but less than 2 pandanus nuts.  Pandanus fruit  Pandanus nuts

In general the use and understanding of traditional measurement methods is simple as long as one knows the language.

For example time measurement in Simbu, expressed in words of the Konda language, involves using words derived from names of different stages of growth (for humans, animals and plants), special events and the weather.

There is no measurement equivalent to hours, minutes and seconds. A day is defined as the period of daylight only. Time within the day is worked out and explained by the position of the Sun.

A new month is shown by the Moon.

Time is also measured by looking at the growth of animals and plants. For example, when a person is born a tree is planted. The person's age is later judged not in years but in the growth of the tree. Dates are related to the stages of human growth, or to events. A new year or Christmas is related to the time of an important pig-killing ceremony.

Centuries, as in hundreds of years, are not counted, but long periods of time are related to family ancestry. So, measuring time is easy and not a problem for Simbu people because they know the language. An outsider would find it easy too, if they knew the language.

FIND OUT

1. Describe how people in your community carry out measurement activities. What are the situations in which they need to compare or measure? Do these situations involve length, area, volume (size) or weight?
2. Describe how people in your community compare or measure. What units do they use, and what do they do with these units or tools?
3. Do people use a unit that combines smaller units? If so, how many smaller units are combined and how do they join them? Why might this be done?
4. Do people in your community use body parts to measure? Give an example of how each body part is used.
5. Do people in your community have a standard unit for comparison that is widely used by all? Give an example.
6. How important is 'accuracy' to people in your community? How is 'accuracy' considered in specific activities that involve measurement, such as planning and making garden plots, buildings drains and houses, making handicraft items, and so on?

Unit 2

Patterns of change

Unit summary

This topic extends our knowledge of number operations and shows how patterns can be used to find general rules and formulas. It focuses on the mathematical techniques needed to work both with numbers and with algebraic symbols in practical situations. It also explores how algebra can be applied to solve everyday problems by posing questions, translating these questions into symbols, and using logical steps to arrive at the answer.

There are many examples around the world of people in ancient civilisations applying these methods to discover rules, which could be used for counting, measuring, constructing, and predicting, in an attempt to find the answers to problems in the world around them.

For example, surviving clay tablets show that, more than 3500 years ago, the Babylonians discovered and used formulas to solve mathematical problems. The Babylonians also developed algebraic methods of solving equations. The ancient Egyptians also used algebra and we can see examples of this in documents like the Rhind Papyrus.

A Babylonian clay tablet.

Syllabus references

9.2.1 Identify and create representations of patterns to solve equations

9.2.2 Create, investigate and interpret equations, explain the effect of order of operations, and justify solutions to equations

9.2.3 Communicate mathematical processes and results in writing

9.2.4 Undertake investigations individually in which mathematics can be applied to solve problems

Patterns of change: Topics

Week	Content	Page
1 and 2	**Directed numbers** • Using positive and negative numbers in addition, subtraction, multiplication and division • Solving everyday problems with directed numbers • Carrying out mixed operations	88
3, 4 and 5	**Indices** • Estimating and finding squares and square roots • Using square root tables • Expressing numbers in index form • Evaluating expressions involving index notation • Observing patterns leading to the index rules for multiplication, division and powers of powers and the meaning of zero and negative indices • Simplifying algebraic expressions • Using scientific notation	100
6, 7, 8 and 9	**Equations** • Completing and describing number patterns • Special sequences • Finding terms and writing rules for arithmetic sequences and series • Using the t_n formula • Writing algebraic expressions and equations • Removing grouping symbols • Forming and solving equations • Checking solutions to equations • Using equations to solve simple problems • Using mathematical formulas to find values • Changing the subject of a formula	121
9 and 10	Revision and assessment, and project work	150

Weeks 1 and 2

DIRECTED NUMBERS

Why are there negative numbers?

Negative numbers occur in many practical situations. For example, $^{-}10$ could represent a diver's position 10 metres below sea level. In this situation, what would $^{+}10$ represent?

Can you describe three other situations that could give a result of $^{-}10$?

Positive numbers and **negative numbers** together with **zero** make up the set of **directed numbers**. Directed numbers have both size and direction and can be illustrated on a number line.

We have used number lines before, to place numbers in order. On a positive number line, by convention (or agreement), zero is on the left and numbers are marked off in equal intervals to the right:

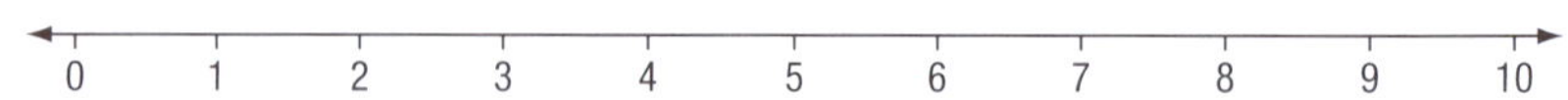

A number line showing positive and negative numbers looks like this:

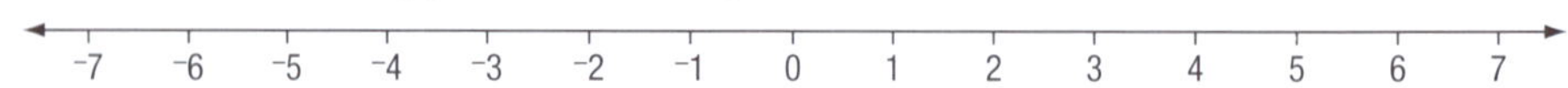

EXERCISE

1 If right is positive and left is negative, write the numbers that would appear at positions **A**, **B**, **C**, **D** and **E** of the number line below, using zero as the reference position.

D E B A C

−7 −6 −5 −4 −3 −2 −1 0 1 2 3 4 5 6 7

2 Write each of these temperatures as a positive or negative number. Zero degrees is the reference point.

a 7° above zero **b** 5° below zero **c** 9° below zero
d 25° above zero **e** 13° below zero

3 Write these gains or losses as positive or negative numbers, if zero kina is the reference point.

a a K40 loss **b** a K300 gain **c** a K24 loss
d a K13 loss **e** a K89 gain

4 If north is the positive direction, write these directions as positive or negative numbers.

a 6 metres north **b** 10 metres south **c** 385 metres south
d 250 metres north **e** 1.8 metres south

5 If the ground floor (street level) is regarded as zero, write a directed number for the following positions in a skyscraper.

a 4 floors above ground level **b** 2 floors below ground level
c 21 floors above ground level **d** 6 floors below ground level
e 1 floor below ground level

6 If right is positive, write a number for the position from zero which is:

a 7 units left **b** 5 units right

c 12 units left **d** 9 units right

e 23 units left

7 A baby boy weighed 365 grams at birth. The record of his weight for the first five days showed the following:

Day 1: 25 g loss
Day 2: 10 g loss
Day 3: 15 g gain
Day 4: 20 g gain
Day 5: 25 g gain

a Write each day's gain or loss as a positive number or a negative number.

b Use the number line below to help you calculate the baby's weight at the end of the five days?

340 g 350 g 360 g 370 g 380 g 390 g 400 g 410 g

8 Carlos had K55 in the bank. How much will be in his account after the following transactions?

Week 1: deposit K5
Week 2: withdraw K13
Week 3: withdraw K10
Week 4: withdraw K7
Week 5: deposit K20

9 Copy the two columns of numbers below and draw lines to connect the pairs of opposites.

$^+9$	-18
-45	$^+\frac{1}{2}$
$^+3$	$^+47$
$-\frac{1}{2}$	-9
$^+18$	$^+45$
-47	-3

10 Write the opposite of these numbers:

a $^+7$ **b** -3 **c** 0 **d** 5 **e** -1

f $^+5.4$ **g** $-2\frac{1}{2}$ **h** 21 **i** -16 **j** -15

Remember

Every number on the *right* of zero on the number line has a 'partner' on the *left* of zero. (Zero has no partner because it lies at the centre of the number line.) The pairs of numbers are called **opposites**.

Remember

Positive numbers can be written with a positive sign (+) before the number, or with no sign at all (when we assume the number is positive).

Weeks 1 and 2

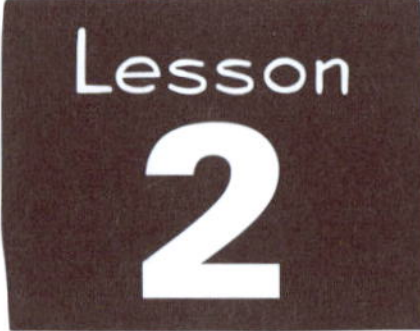

Using directed numbers

Thinking of the position of numbers on a number line makes it easy to compare their size and arrange them in order. Remember, the number farther to the right is the larger number.

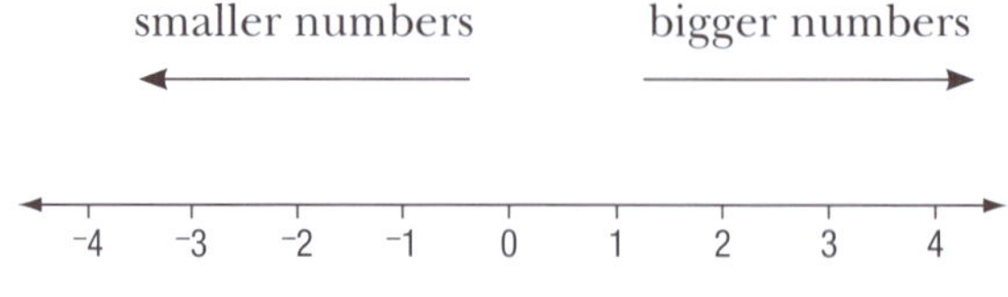

$^{+}3$ is bigger than $^{-}1$ because it is farther to the right. In symbols, we can write $^{+}3 > {}^{-}1$.

EXERCISE

Draw a number line showing the numbers from $^{-}12$ to $^{+}12$ at the top of the page of your workbook to help you answer these questions.

1 Show $^{+}5$ and $^{-}4$ on a number line and write a sentence comparing their size.

2 Show $^{+}3$ and $^{-}2$ on a number line and mark in two numbers smaller than both.

3 Show $^{-}1$ and $^{+}5$ on a number line and mark in two numbers larger than both.

4 Show $^{-}6$ and $^{-}3$ on a number line and mark in two numbers between them.

5 Write the statement $^{-}7 > {}^{-}4$ in words, then state whether it is true or false.

6 Which is the larger number in each of the following pairs of numbers?
a $^{+}6$ or $^{+}9$ **b** $^{+}5$ or $^{-}2$ **c** $^{-}3$ or $^{+}6$
d $^{+}6$ or $^{-}8$ **e** $^{-}9$ or $^{-}7$ **f** $^{-}4$ or $^{-}10$

7 Which is the smaller number in each of the following pairs of numbers?
a 11 or 5 **b** 7 or $^{-}4$ **c** $^{-}5$ or 5
d $^{-}9$ or $^{-}8$ **e** $^{-}12$ or $^{-}7$ **f** $^{-}3$ or $^{-}8$

8 Write 'True' or 'False' for each of the following.
a $5 < {}^{-}2$ **b** $11 > {}^{-}3$ **c** $0 > {}^{-}6$ **d** $5 < {}^{-}8$
e $9 > {}^{-}7$ **f** $^{-}6 > {}^{-}4$ **g** $^{-}5 > {}^{-}4$ **h** $^{-}10 < 3$
i $^{-}6 > {}^{-}9$ **j** $^{-}11 < {}^{-}10$

9 Insert < or > in the box to make each statement true.
a $3 \square {}^{-}2$ **b** $8 \square {}^{-}5$ **c** $7 \square {}^{-}7$ **d** $^{-}3 \square {}^{-}6$
e $^{-}5 \square {}^{-}9$ **f** $^{-}12 \square {}^{-}11$ **g** $0 \square {}^{-}4$ **h** $^{-}8 \square 0$
i $^{-}6 \square {}^{-}2$ **j** $^{-}1 \square 4$

10 Arrange the numbers in each of these sets in ascending order:
a $^{-}5, 8, {}^{-}2$ **b** $4, {}^{-}3, {}^{-}4, 0$ **c** $^{-}6, {}^{-}5, {}^{-}11, {}^{-}10$
d $7, {}^{-}9, {}^{-}6, 6$ **e** $^{-}2, 3, 12, {}^{-}5$ **f** $1, {}^{-}1, 7, {}^{-}12, 9, 0$
g $2.5, {}^{-}1.2, 4, {}^{-}3.1$ **h** $^{-}9.5, {}^{-}8.9, {}^{-}10, {}^{-}9.7$ **i** $3\frac{1}{2}, {}^{-}2\frac{1}{4}, 1, {}^{-}1\frac{1}{5}$
j $^{-}\frac{1}{8}, {}^{-}\frac{7}{8}, \frac{5}{8}, {}^{-}\frac{3}{8}, {}^{-}\frac{5}{8}$

11 Arrange the numbers in each of these sets in descending order:
a $3, {}^{-}8, 10$ **b** $^{-}6, 0, {}^{-}1, 4$ **c** $9, {}^{-}8, {}^{-}3, 3$
d $^{-}6, {}^{-}5, {}^{-}2, {}^{-}9$ **e** $^{-}11, 10, {}^{-}2, 8, {}^{-}12$ **f** $3, 7, {}^{-}1, {}^{-}4, 11$
g $^{-}6.3, 5.2, 7.1, {}^{-}4.9$ **h** $^{-}1.1, {}^{-}5.7, {}^{-}1.8, {}^{-}3.9$ **i** $^{-}5\frac{2}{3}, 7\frac{1}{3}, {}^{-}1\frac{1}{3}, 1\frac{2}{3}$
j $^{-}\frac{11}{12}, \frac{5}{12}, {}^{-}\frac{7}{12}, \frac{1}{12}$

Remember

- Numbers in **ascending** order go from smallest to largest.
- Numbers in **descending** order go from largest to smallest.

12 The number line on the right is vertical. As we go up the number line, the numbers increase, and as we go down the numbers decrease. Write the directed number for each of the points marked on the number line, and write 'True' or 'False' for each of the statements below.

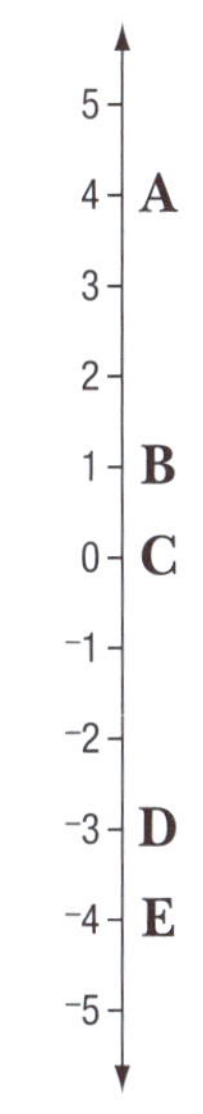

- **a** **B** is greater than **D**.
- **b** **A** < **E**
- **c** **D** is less than **A**.
- **d** **B** < **C**
- **e** **C** > **E**
- **f** **C** < **B**
- **g** **B** and **D** are opposites.
- **h** **A** and **E** are opposites.

13 Write the next three numbers for each of these number patterns.

a 8, 6, 4, …	**b** 15, 10, 5, …	**c** 20, 10, 0, …	**d** 4, 0, ⁻4, …
e 7, 3, ⁻1, …	**f** ⁻6, ⁻8, ⁻10, …	**g** 1, ⁻1, ⁻3, …	**h** ⁻8, ⁻13, ⁻18, …

14 **a** Which number in each of the following pairs of numbers is farthest from 7?

i 3 or 15 **ii** 10 or ⁻1 **iii** ⁻20 or 28

b Which number in each of the following pairs of numbers is farthest from ⁻3?

i 5 or ⁻8 **ii** ⁻10 or 6 **iii** 32 or ⁻28

Adding and subtracting negative numbers

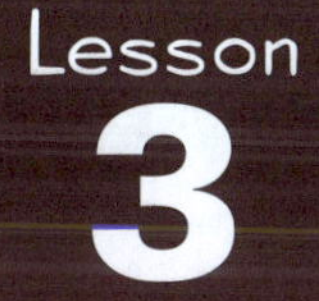

We can recall from Year 8 that adding a positive number is the same as '+' and adding a negative number is the same as '–'.

Example	**Answer**
Find each of the following:	
a ⁻6 + ⁺5	**a** ⁻6 + ⁺5 = ⁻6 + 5 = ⁻1
b ⁻3 + ⁻5	**b** ⁻3 + ⁻5 = ⁻3 – 5 = ⁻8
c ⁺6 + ⁻4	**c** ⁺6 + ⁻4 = 6 – 4 = 2

Also, subtracting a negative number is the same as adding its opposite.

Example	**Answer**
Find each of the following:	
a ⁺9 – ⁻5	**a** ⁺9 – ⁻5 = ⁺9 + 5 = 9 + 5 = 14
b ⁻3 – ⁻4	**b** –3 – ⁻4 = ⁻3 + 4 = 1
c ⁺8 – ⁺6	**c** ⁺8 – ⁺6 = 8 – 6 = 2

When adding and subtracting directed numbers:

$+\,^{+}$ is the same as $+$	$-\,^{-}$ is the same as $+$
$+\,^{-}$ is the same as $-$	$-\,^{+}$ is the same as $-$

EXERCISE

1 Find:

a ${}^{+}9 + {}^{+}3 = {}^{+}9 + 3 =$ **b** ${}^{-}4 + {}^{-}5$ **c** ${}^{+}2 + {}^{+}3$

d ${}^{-}8 + {}^{-}2$ **e** ${}^{-}1 + {}^{+}9$ **f** ${}^{-}3 + {}^{-}6$

g ${}^{-}11 + {}^{-}9$ **h** ${}^{-}15 + {}^{+}6$ **i** ${}^{-}2 + {}^{-}8$

2 Find the sum of:

a ${}^{+}12$ and ${}^{+}4$ **b** ${}^{-}10$ and ${}^{+}6$ **c** ${}^{-}1$ and ${}^{+}3$

d ${}^{-}5$ and ${}^{+}9$ **e** ${}^{-}8$ and ${}^{+}11$ **f** ${}^{+}3$ and ${}^{-}7$

g ${}^{+}5$ and ${}^{-}5$ **h** ${}^{-}7$ and ${}^{-}4$ **i** ${}^{-}1$ and ${}^{-}13$

3 Find the answer each time, first replacing pairs of signs with a single sign, as in the first example.

a ${}^{+}9 + {}^{-}2 + {}^{+}3 = {}^{+}9 - 2 + 3 =$ **b** ${}^{-}7 + {}^{-}4 + {}^{+}5$

c ${}^{+}1 + {}^{-}8 + {}^{+}4$ **d** ${}^{-}2 + {}^{-}7 + {}^{+}6$

e $9 + {}^{+}7 + {}^{-}2$ **f** ${}^{-}1 + {}^{+}10 + {}^{-}7$

g ${}^{+}12 + {}^{-}9 + {}^{+}5$ **h** ${}^{-}11 + {}^{-}3 + {}^{+}14$

i $15 + {}^{-}7 + {}^{-}9$

4 Change each subtraction into an addition, then find the answer.

a ${}^{+}8 - {}^{+}4 = {}^{+}8 + {}^{-}4 =$ **b** ${}^{+}3 - {}^{+}6 = {}^{+}3 +$ __ =

c ${}^{-}4 - {}^{+}1 = {}^{-}4 +$ __ = **d** ${}^{-}7 - {}^{+}3 = {}^{-}7 +$ __ =

e ${}^{+}7 - {}^{+}8 = {}^{+}7 +$ __ = **f** ${}^{+}4 - {}^{-}9 = {}^{+}4 + {}^{+}$__ =

g ${}^{-}8 - {}^{-}3 = {}^{-}8 +$ __ = __ **h** ${}^{+}7 - {}^{-}11 = {}^{+}7 +$ __ = __

i ${}^{-}10 - {}^{-}5 = {}^{-}10 +$ __ = **j** ${}^{-}13 - {}^{-}2 = {}^{-}13 +$ __ =

Help box

'Evaluate' means 'find the value of'.

5 Change each of the following into an addition, then evaluate:

a ${}^{+}9 - {}^{+}3$ **b** ${}^{-}5 - {}^{+}6$ **c** ${}^{-}11 - {}^{+}5$

d ${}^{+}4 - {}^{-}3$ **e** ${}^{-}7 - {}^{-}4$ **f** ${}^{-}5 - {}^{+}9$

g ${}^{+}8 - {}^{-}6$ **h** ${}^{-}9 - {}^{-}1$ **i** ${}^{-}11 - {}^{-}4$

j ${}^{+}13 - {}^{-}9$ **k** ${}^{+}5 - {}^{+}16$ **l** ${}^{+}20 - {}^{-}8$

m ${}^{-}15 - {}^{-}4$ **n** ${}^{+}14 - {}^{-}16$

6 Copy the following into your book then write 'True' or 'False' for each.

a $7 + {}^{-}5 = 7 - 5$ **b** ${}^{-}4 + {}^{-}3 = 4 - 3$ **c** ${}^{-}11 - {}^{+}3 = {}^{-}11 - 3$

d $15 + {}^{-}7 = 15 - 7$ **e** ${}^{-}8 - {}^{+}11 = {}^{-}8 + 11$ **f** $21 - {}^{-}12 = 21 + 12$

7 Simplify then evaluate:

a $7 - {}^{+}9 + {}^{+}3$ **b** ${}^{-}6 + {}^{-}7 + {}^{-}4$ **c** $9 + {}^{-}4 - {}^{+}3$

d ${}^{-}8 + {}^{+}4 - {}^{+}7$ **e** $11 - {}^{-}5 - {}^{+}3$ **f** ${}^{+}15 + {}^{-}3 + {}^{-}8$

g ${}^{-}2 - {}^{-}1 - {}^{-}9$ **h** ${}^{+}7 + {}^{-}6 + {}^{-}5$ **i** ${}^{-}16 + {}^{-}9 + {}^{-}4$

j ${}^{-}18 - {}^{-}6 + {}^{-}12$ **k** ${}^{+}20 - {}^{+}15 - {}^{+}11$ **l** ${}^{+}31 - {}^{+}27 + {}^{-}11$

m ${}^{+}16 - {}^{-}24 - {}^{+}10$ **n** ${}^{-}17 + {}^{-}13 + {}^{-}6$ **o** ${}^{-}9 + {}^{-}17 + {}^{-}22 + {}^{-}6$

Multiplying and dividing directed numbers

Rosa has forgotten her purse, so she borrows K2 from each of her four friends in order to do some shopping. It is easy to see that she has borrowed K8 altogether, so she owes K8.

Owing 8 kina can be written as $^{-}8$. Below, we will look at some ways of arriving at the answer of $^{-}8$.

Method 1

$4 \times {}^{-}2$ means '4 lots of $^{-}2$', or $^{-}2 + {}^{-}2 + {}^{-}2 + {}^{-}2 = {}^{-}2 - 2 - 2 - 2$
$= {}^{-}8$

Method 2

If we look at a list of multiples of 4, we can work backwards down the list to arrive at the answer for 4×-2, because it will appear in the same pattern (decreasing by 4 each time).

$4 \times 3 = 12$
$4 \times 2 = 8$
$4 \times 1 = 4$
$4 \times 0 = 0$
$4 \times {}^{-}1 = {}^{-}4$
$4 \times {}^{-}2 = {}^{-}8$
and so on …

Whether we use Method 1 or Method 2 above, we can see that multiplying a positive number (4) by a negative number ($^{-}2$) gives a negative result ($^{-}8$).

We can use the same methods described above to find what happens when the signs on the two numbers are reversed, that is to find that $2 \times {}^{-}4 = {}^{-}8$.

So, **multiplying a positive number by a negative number gives a negative answer**:

$^{+} \times {}^{-} = {}^{-}$ and $^{-} \times {}^{+} = {}^{-}$

Turning the question around, we can see that this rule also applies to division. Rosa owes a total of K8 to her four friends. She owes each friend the same amount. How much does she owe each friend?

To find the answer, we would be dividing $^{-}8$ by 4. From the above explanation, we know that the answer is $^{-}2$, so $^{-}8 \div 4 = {}^{-}2$.

So, **dividing a positive number by a negative number gives a negative answer**.

To see what happens when we multiply two negative numbers together, look at the following list of multiples of $^-3$.

Multiples of $^-3$

$5 \times {}^-3 = {}^-15$ | $0 \times {}^-3 = 0$
$4 \times {}^-3 = {}^-12$ | ${}^-1 \times {}^-3 = 3$
$3 \times {}^-3 = {}^-9$ | ${}^-2 \times {}^-3 = 6$
$2 \times {}^-3 = {}^-6$ | ${}^-3 \times {}^-3 = 9$
$1 \times {}^-3 = {}^-3$

We can see that the answers follow a pattern, increasing by 3 each time.

From this pattern, we can also see that the product of two negative numbers is a positive number.

To investigate the situation for division, we can use the first line in the list of multiples of $^-3$:

$5 \times {}^-3 = {}^-15$

If we ask 'How many $^-3$s are there in $^-15$?' we can work backwards to see that the answer must be 5:

${}^-15 \div {}^-3 = 5$

So, dividing a negative number by a negative number gives a positive number

We can see that the same rules for positive and negative signs apply to both multiplication and division.

EXERCISE

1 Copy into your workbook these rules for multiplying and dividing with directed numbers.

When multiplying or dividing two numbers, numbers with the **same** signs give a positive answer:

$^+ \times {}^+ = {}^+$ $\quad$ $^+ \div {}^+ = {}^+$
$^- \times {}^- = {}^+$ $\quad$ $^- \div {}^- = {}^+$

When multiplying or dividing two numbers, number with **different** signs give a negative answer:

$^+ \times {}^- = {}^-$ $\quad$ $^+ \div {}^- = {}^-$
$^- \times {}^+ = {}^-$ $\quad$ $^- \div {}^+ = {}^-$

2 Find the product of:

a $^+5$ and $^-7$ b $^-9$ and 2 c $^-6$ and $^-1$ d $^-4$ and $^-7$ e $^-8$ and 5

3 Calculate:

a ${}^+4 \times {}^-2$ b ${}^-7 \times {}^+1$ c ${}^-6 \times {}^+8$ d ${}^+4 \times {}^+9$
e ${}^+2 \times {}^+8$ f ${}^+10 \times {}^-4$ g ${}^-4 \times {}^-3$ h ${}^-7 \times {}^-9$
i ${}^+12 \times {}^-7$ j ${}^-11 \times {}^+6$

4 Copy and complete these multiplication grids.

a

×	+8	−7	−4	+5
−3				
+6				
+9				
−5				

b

×	−9	+5	−6	+7
−1				
−8				
+3				
−11				

5 Evaluate:

a	${}^{-}20 \times {}^{+}11$	**b**	${}^{+}15 \times {}^{-}16$	**c**	${}^{-}9 \times {}^{-}34$	**d**	${}^{+}12 \times {}^{-}19$
e	${}^{-}23 \times {}^{-}18$	**f**	${}^{+}16 \times {}^{-}13$	**g**	${}^{-}100 \times {}^{-}4$	**h**	${}^{+}11 \times {}^{-}35$
i	${}^{-}15 \times {}^{-}21$	**j**	${}^{+}30 \times {}^{-}42$				

6 Find the value of:

a	${}^{-}1 \times {}^{+}2 \times {}^{-}3$	**b**	${}^{+}7 \times {}^{-}3 \times {}^{+}2$	**c**	${}^{-}4 \times {}^{+}3 \times {}^{-}1$
d	${}^{+}5 \times {}^{+}2 \times {}^{-}3$	**e**	${}^{-}6 \times {}^{-}1 \times {}^{+}5$	**f**	${}^{+}2 \times {}^{-}7 \times {}^{-}3$
g	${}^{-}8 \times {}^{-}2 \times {}^{+}5$	**h**	${}^{-}4 \times {}^{-}2 \times {}^{-}5$	**i**	${}^{+}6 \times {}^{-}1 \times {}^{-}3$
j	${}^{+}3 \times {}^{-}5 \times {}^{-}4$				

7 Find the result of dividing:

a	${}^{-}18$ by ${}^{+}3$	**b**	12 by ${}^{-}6$	**c**	10 by ${}^{-}2$	**d**	${}^{-}3$ by 3
e	5 by ${}^{-}5$	**f**	${}^{-}20$ by ${}^{-}2$	**g**	${}^{-}15$ by ${}^{+}5$	**h**	${}^{-}14$ by ${}^{-}7$
i	${}^{-}8$ by ${}^{+}4$	**j**	${}^{-}6$ by 3				

8 Calculate:

a	${}^{+}25 \div 5$	**b**	${}^{-}16 \div {}^{+}8$	**c**	${}^{+}36 \div {}^{-}4$	**d**	${}^{-}12 \div {}^{-}3$
e	${}^{-}48 \div {}^{-}8$	**f**	${}^{+}100 \div {}^{-}5$	**g**	${}^{-}50 \div {}^{+}10$	**h**	$144 \div {}^{-}6$
i	${}^{-}28 \div {}^{-}7$	**j**	${}^{-}56 \div {}^{-}2$				

9 Find the value of:

a	$\frac{{}^{+}12}{{}^{-}2}$	**b**	$\frac{{}^{-}16}{{}^{-}4}$	**c**	$\frac{{}^{-}9}{{}^{+}3}$	**d**	$\frac{{}^{-}8}{{}^{+}2}$	**e**	$\frac{{}^{+}20}{{}^{-}5}$
f	$\frac{{}^{-}30}{{}^{-}6}$	**g**	$\frac{{}^{-}14}{{}^{+}7}$	**h**	$\frac{6}{{}^{-}3}$	**i**	$\frac{{}^{-}24}{8}$	**j**	$\frac{36}{{}^{-}9}$

10 Divide:

a	${}^{-}100$ by ${}^{-}20$	**b**	144 by ${}^{-}12$	**c**	${}^{-}250$ by ${}^{-}5$	**d**	140 by ${}^{-}20$
e	${}^{-}360$ by 4	**f**	${}^{-}900$ by ${}^{-}90$	**g**	${}^{+}450$ by ${}^{-}50$	**h**	${}^{-}198$ by ${}^{-}6$
i	72 by ${}^{-}24$	**j**	${}^{-}1000$ by ${}^{-}40$				

Mixed operations with directed numbers

Remember that operations are carried out in this order:

- **brackets** first
- **multiplication** and **division**, from left to right
- **addition** and **subtraction**, from left to right.

Example

Find:

a ${}^{-}7 + 5 \times {}^{-}3$

b ${}^{-}2 \times {}^{+}1 + 7 \times {}^{-}3$

c ${}^{+}8 - 2 \times {}^{-}3$

d $({}^{-}1 + 3) \times {}^{-}4 \div {}^{-}2$

Answer

a ${}^{-}7 + 5 \times {}^{-}3 = {}^{-}7 - 15$
$= {}^{-}22$
First, multiply $5 \times {}^{-}3$.

b ${}^{-}2 \times {}^{+}1 + 7 \times {}^{-}3 = {}^{-}2 - 21$
$= {}^{-}23$
First, multiply ${}^{-}2 \times {}^{+}1$, and $7 \times {}^{-}3$.

c ${}^{+}8 - 2 \times {}^{-}3 = {}^{+}8 + 6$
$= {}^{+}14$
$= 14$
First, multiply $2 \times {}^{-}3$.

d $({}^{-}1 + 3) \times {}^{-}4 \div {}^{-}2 = {}^{+}2 \times {}^{-}4 \div {}^{-}2$
$= {}^{-}8 \div {}^{-}2$
$= {}^{+}4$
$= 4$
First, work out brackets. Then work from left to right, so multiply ${}^{+}2 \times {}^{-}4$.

EXERCISE

1 Use the correct order of operations to evaluate each of the following, showing all working.

a $^-3 + 2 \times {}^-1$	b $^+12 \div {}^-6 + 5$	c $^+8 + 12 \div {}^-4$
d $^-11 + 3 \times {}^-2$	e $^+5 \times {}^-3 - 2 \times 4$	f $(7 - 12) \times {}^-2$
g $^-1 \times 6 - 8$	h $13 + {}^-3 \times 6$	i $20 \div {}^-5 + 7$
j $^+5 - 12 \div {}^-4$	k $9 + 15 \div {}^-3$	l $({}^-4 + 7) \times (3 - 8)$
m $18 \div {}^-6 - 12 \div {}^-4$	n $^-2 \times {}^-6 + 5 \times {}^-1$	o $18 \div {}^-2 - (3 \times {}^-4)$

2 Use the correct order of operations to evaluate each of the following, showing all working.

a $5 - 6 \times {}^-8$	b $10 \div {}^-5 + 4$	c $7 + 14 \div {}^-2$
d $^-12 \times {}^-3 + 16$	e $(9 - 11) \times {}^-5$	f $^-1 \times {}^-8 \div {}^-4$
g $^-20 \div 4 \div {}^-2 \times 3$	h $11 - (5 \times {}^-3 + 7)$	i $2 - 6 \times {}^-9 + 8$
j $7 + {}^-4 \times {}^-9 + 11$	k $12 - (8 - 9) \times {}^-5$	l $9 \times {}^-3 \div {}^-1 + 7$
m $35 \div {}^-7 + 18 \div {}^-6$	n $4 \times {}^-8 - 3 \times {}^-2$	o $(8 - 15) \times (9 + {}^-4)$

Challenge

Use the digits 1, 2, 3 and 4, together with $^+$ and $^-$ signs and the operations of addition, subtraction, multiplication and division to make as many number statements as you can that have answers from 1 to 10.

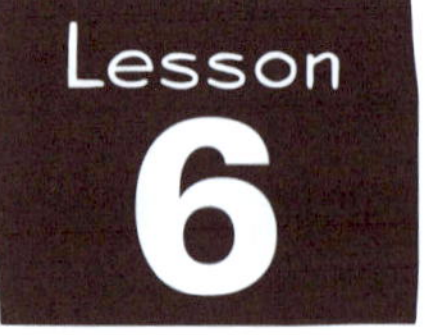

Everyday problems with directed numbers

There are many everyday situations in which positive and negative numbers need to be combined in order to solve a problem, using some or all of the steps below.

Step 1: Carefully identify which of the numbers in the problem should be written as positive numbers and which should be written as negative numbers.

Step 2: Decide which operation to use: addition, subtraction, multiplication, division, or more than one of these.

Step 3: Write the calculation(s) necessary, labelling the steps so that someone else can follow the working.

Step 4: Carry out the calculation(s) and write the answer to the problem.

Step 5: Make an estimate of the answer, and think about whether the answer is reasonable. This step will help to identify errors in the working.

Example

Goodfoods manufactures frozen dinners. The food is cooked and partly cooled to a temperature of 25°C before it is placed in a freezer. The freezer cools the food by 6°C per hour until it is reduced to the temperature in the freezer cabinet, which is $^-8$°C.

What will the temperature of the food be after:

a 2 hours?

b 5 hours?

c 6 hours?

Answer

a After 2 hours, the temperature will be:
$25 - 2 \times 6 = 25 - 12$
$= 13$°C

b After 5 hours, the temperature will be:
$25 - 5 \times 6 = 25 - 30$
$= {}^-5$°C

c After 6 hours, the temperature will be:
$25 - 6 \times 6 = 25 - 36$
$= {}^-11$°C

But the food will have reached the temperature of the freezer, which is $^-8$°C, and cannot be cooled any more. So the temperature of the food will be $^-8$°C.

EXERCISE

1 Declan's bank passbook had a balance of K83 at the end of May and showed the following transactions for June:

3 June	K40 deposit
7 June	K23 withdrawal
14 June	K28 deposit
22 June	K15 withdrawal
26 June	K69 withdrawal

a Write each deposit or withdrawal as a directed number.

b Find the balance of Declan's account at the end of June.

2 A submarine measures its position as a negative number, using sea level as zero. It is 360 metres below sea level when it detects a shipwreck 435 metres below it. Use directed numbers to state the position of:

a the submarine **b** the shipwreck.

3 Four friends equally invest a total of K200 in a syndicate to buy lottery tickets. What is each person's profit or loss if the syndicate wins each of the amounts below, and the winnings are shared equally:

a K1 000 000 **b** K1000 **c** K50 **d** K10

4 A company keeps a record of each week's profit, in thousands of dollars, as shown in the table on the right.

Week	Profit (000s of K)
1	12.5
2	7.2
3	$^-1.4$
4	$^-2$
5	1.2
6	3.8

a Which weeks ran at a loss?

b Which week had:

i the highest profit?

ii the greatest loss?

c Write in figures the loss made in week 4.

d What was the total profit after 6 weeks?

e If the company had a balance of K40 000 before week 1, what was its balance after the six weeks shown in the table?

5 The temperatures in major cities around the world on a particular day were as shown in the table on the right.

City	Minimum °C	Maximum °C
Amsterdam	5	9
Bahrain	28	39
Cairo	24	36
Helsinki	⁻4	4
London	1	9
Melbourne	19	33
New York	⁻2	7
Oslo	⁻1	5
Stockholm	⁻9	3

a List the cities with the four highest maximum temperatures in order, starting with the hottest.

b List the cities with the four lowest minimum temperatures in order, starting with the coldest.

c How many degrees did the temperature rise in:

i Cairo? ii London? iii New York?
iv Oslo? v Stockholm?

d Find the minimum temperature in Helsinki the next day if it was 2°C lower than the minimum shown in the table.

e On the same day, the coldest temperature recorded worldwide was ⁻62°C, at Oimyakon in Siberia. How much colder was this than the maximum temperature in Melbourne?

f On the same day, the hottest temperature recorded worldwide was 44°C, at Eneabba in Western Australia. How much hotter was this than the temperature in Oimyakon?

6 Vagi and Segana have invented their own scoring system for darts. The bullseye scores 10, the rest of the board scores 4, and a shot that misses the board scores ⁻5.

Vagi and Segana each throw three darts at the board when it is their turn. Show how it is possible for a player to score each of the following scores with their three darts:

a 30 b 18
c 12 d 9
e 3 f 0
g ⁻6 h ⁻15

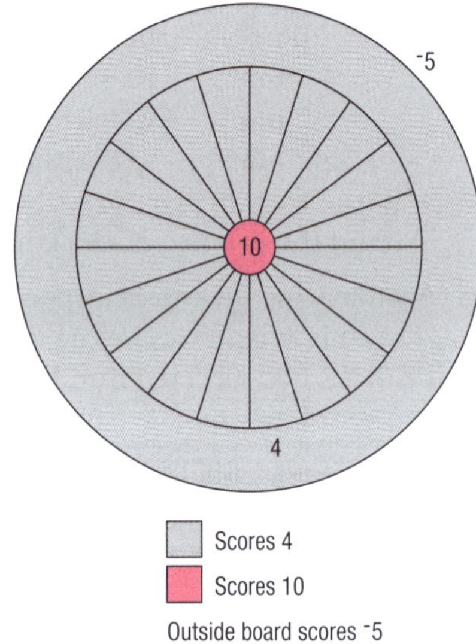

7 A marker is placed in a river with marks to show the water level, in metres, above and below flood level.

a After a dry season, the river level is ⁻3 metres. How many metres would the river need to rise to reach flood level?

b If the level rose from ⁻3 m to ⁻2 m, how much did it rise?

c In the rainy season, the level rose from ⁻2 m to 1.5 m. By how much did it rise?

d When some of the water had drained away, the depth fell from 1.5 m to ⁻0.6 m. By how many metres had it fallen?

Directed numbers on a calculator

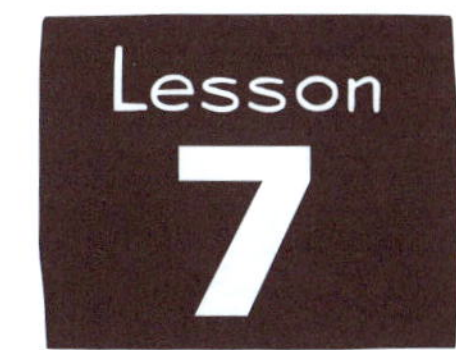

Calculators have a key for addition (+), a key for subtraction (−), and a key for inserting a positive or negative sign before numbers (+/−). On some calculators, the + or − sign is entered **before** we enter the number while, on other calculators, the number is entered first and the + or − sign is entered **after** (but appears on the screen in front of the number).

Test the order in which you should enter directed numbers on your calculator.

Enter $^{-}5$ using:

- [5] then [+/−]
- [+/−] then [5]

Which one of these alternatives shows −5 on your calculator display?

Example

a Use a calculator to work out $^{-}255 \div {}^{-}15$.

b Evaluate $^{-}198 \times (^{-}19 - 13)$.

c Calculate $^{-}789 \div {}^{-}14$, giving the answer correct to one decimal place.

Answer

a 255 [+/−] [÷] 15 [+/−] [=] 17

b 198 [+/−] [×] [(] 19 [+/−] [−] 13 [)] [=] 6336

c 789 [+/−] [÷] 14 [+/−] [=] 56.357 — Because the number in the second decimal place is 5, we round up.

= 56.4 — To one decimal place.

EXERCISE

1 Use your calculator to find answers for the following:

a $^{+}103 - 2089$	b $^{-}75 \times {}^{-}43$
c $(240 - 658) \times 14$	d $^{-}89 - 271 - 347$
e $^{-}900 \div {}^{-}25 + 74$	f $(541 - 932) \times {}^{-}67$
g $2450 \div {}^{-}70 - 31$	h $64 \times {}^{-}83 \times {}^{-}11$
i $^{+}876 - (24 \times 78)$	j $319 - 876 - 24 \times 11$

2 Find the answers, correct to one decimal place:

a $^{+}25.2 \times {}^{-}13.4$	b $^{-}785 \div {}^{-}8$
c $^{+}450 \div {}^{-}16$	d $^{-}239 \times {}^{-}11.4$
e $415 \times {}^{-}17 \div {}^{-}3 + 815$	f $1295 - (^{-}23 \times {}^{-}7) + 2843$
g $6701 \div {}^{-}56 \div {}^{-}21$	h $(4039 - {}^{-}652) \times 433$
i $236 \div {}^{-}19 \times (^{-}37 - 421)$	j $^{-}345 \div {}^{-}29 + {}^{-}217$

INDICES

Squares and square roots

In Year 8, we saw that **square numbers** result from multiplying a number by itself. Square numbers can be represented by a pattern of squares made from dots:

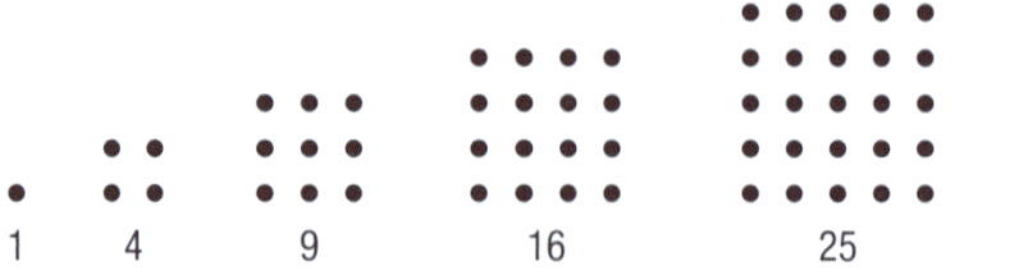

Square numbers are also called **perfect squares**.
We can see that $5 \times 5 = 25$.

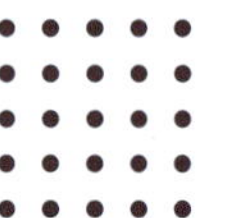

That means that 25 is a square number.
It is the result of multiplying 5 by itself.
We can write this as $5^2 = 25$ and we can say '5 squared equals 25' or '5 to the power 2 is 25'.
Similarly, '8 squared' = $8^2 = 8 \times 8 = 64$. Therefore 64 is a square number.
Finding the square of any number (whole number, fraction or decimal) is very easy. We just multiply the number by itself.

Example

Find:

a 0.5^2
b $(1\frac{2}{3})^2$
c 300^2

Answer

a $0.5^2 = 0.5 \times 0.5$
$= 0.25$

b $(1\frac{2}{3})^2 = 1\frac{2}{3} \times 1\frac{2}{3}$
$= \frac{5}{3} \times \frac{5}{3}$
$= \frac{5 \times 5}{3 \times 3}$
$= \frac{25}{9}$
$= 2\frac{7}{9}$

c $300^2 = 300 \times 300$
$= 90\,000$

Did you know?

There is a simple shortcut for finding the squares of numbers ending in 5.
$75^2 = 5625$
Multiply 7 by the next number, that is 7×8, and place that answer in front of 25.
This works with decimals too:
$9.5^2 = 90.25$
Try these:
$45^2 = \square 25$
The $\square$ equals 4×5.
$115^2 = \square 25$
The $\square$ equals 11 by 12.

The opposite operation to squaring is finding the **square root**. The square root of a number is another number which, when multiplied by itself, gives the original number.

The square root of 25 is 5 because, when 5 is multiplied by itself, the result is 25:

$$5 \times 5 = 25 \text{ so } \sqrt{25} = 5$$

The square of 9 is 81, so the square root of 81 is 9.
That is:

$$9^2 = 81 \quad \text{and} \quad \sqrt{81} = 9$$

Remember

The symbol $\sqrt{}$ means 'find the square root of'.

Finding the square root of a number is not straightforward because very few numbers are perfect squares. For instance, the only square numbers between 1 and 10 are 1, 4 and 9. The only square numbers between 20 and 40 are 25 and 36.

So, what if we want to find the square root of 8, or we want to find $\sqrt{37}$? First, we will look at how we can use an integer (whole number) estimate. Later, we will see how to find more accurate values.

Using estimates

To find the square root of a number which is not a perfect square, identify the perfect squares it lies between.

Example

What is the closest whole number estimate of:

a $\sqrt{8}$?

b $\sqrt{37}$?

Answer

a The number 8 lies between the perfect squares 4 and 9. It is closer to 9 than it is to 4. We know that $\sqrt{9} = 3$ (because $3 \times 3 = 9$), so an estimate for $\sqrt{8}$ would be 3.

b The number 37 is between the perfect squares 36 and 49, and is closer to 36 than to 49. We know that $\sqrt{36} = 6$, so an estimate for $\sqrt{37}$ would be 6.

EXERCISE

n	n^2
1	
2	
3	
4	
5	
6	
7	
8	
9	
10	
11	
12	
13	
14	
15	
16	
17	
18	
19	
20	

1 Copy the table on the right down the side of a page in your workbook and complete the column of square numbers. Use your completed table to help you answer the questions in this exercise.

2 From the table you completed in Question **1**, or otherwise, find the square of each of these numbers:

a 11 **b** 60 **c** 15

d 40 **e** 1.9 **f** 80

g 0.9 **h** $\frac{6}{7}$ **i** 0.4

j 300 **k** $2\frac{1}{4}$ **l** $\frac{1}{25}$

m 9000 **n** 4.9 **o** $5\frac{1}{2}$

3 What number do we square to get:

a 64? **b** 100? **c** 16?

d $\frac{4}{49}$? **e** 1.21? **f** $\frac{81}{121}$?

4 Using your table from Question **1**, or your answers to Question **2**, find:

a $\sqrt{64}$ **b** $\sqrt{0.81}$ **c** $\sqrt{144}$

d $\sqrt{90\,000}$ **e** $\sqrt{1600}$ **f** $\sqrt{1}$

g $\sqrt{225}$ **h** $\sqrt{49}$ **i** $\sqrt{1.44}$

j $\sqrt{6400}$

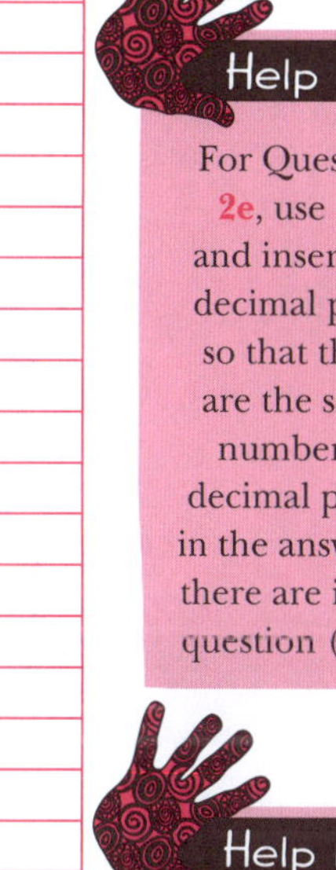

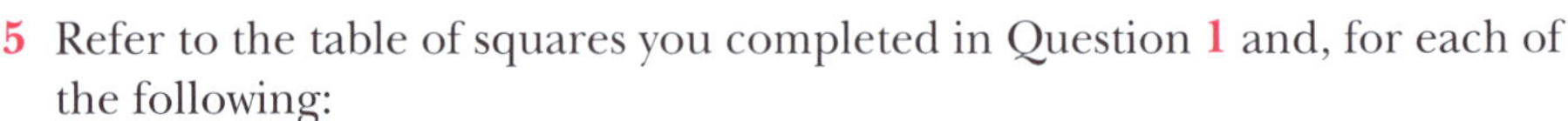

5 Refer to the table of squares you completed in Question **1** and, for each of the following:

- **i** write the two whole numbers between which the answer lies
- **ii** give the closest whole number estimate of the answer.

a $\sqrt{12}$ **b** $\sqrt{90}$ **c** $\sqrt{20}$ **d** $\sqrt{38}$ **e** $\sqrt{45.6}$

f $\sqrt{130}$ **g** $\sqrt{7\frac{1}{2}}$ **h** $\sqrt{99.1}$ **i** $\sqrt{60.45}$ **j** $\sqrt{3.8}$

6 Evaluate:

a $8^2 + \sqrt{49}$ **b** $13^2 - 8^2$ **c** $\sqrt{225} - \sqrt{100}$

d $\sqrt{\frac{64}{81}}$ **e** $0.7^2 + \sqrt{\frac{1}{25}}$

Lesson 2

Finding square roots using tables or calculators

Imagine the following problem:
Ilikis wants to make a square garden with an area of 40 square metres. How long should the sides be?

$z^2 = 40$

$z = \sqrt{40}$

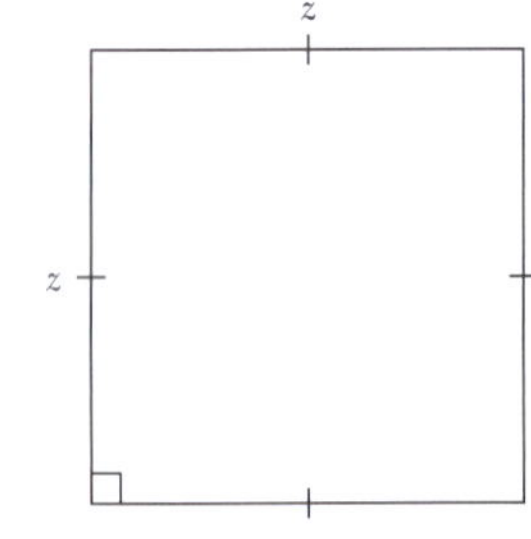

Because 40 is not a square number, Ilikis is uncertain how to calculate the answer.

Following are four methods that could be used.

1 Calculator

If we have a calculator with a √ key, this is the easiest method. We simply enter 40 √ and the calculator will show 6.324 555 32.

2 Square root tables (see pages 391–2)

To use the table for a two-digit number, we locate the first digit of the number in the left-hand column and move across that row to the column headed by the second digit. Doing this for $\sqrt{40}$ gives the value 6.3246.

×	0	1	2	3	4	5	6	7	8	9
0	0.0000	1.0000	1.4142	1.7321	2.0000	2.2361	2.4495	2.6458	2.8284	3.0000
1	3.1623	3.3166	3.4641	3.6056	3.7417	3.8730	4.0000	4.1231	4.2426	4.3589
2	4.4721	4.5826	4.6904	4.7958	4.8990	5.0000	5.0990	5.1962	5.2915	5.3852
3	5.4772	5.5678	5.6569	5.7446	5.8310	5.9161	6.0000	6.0828	6.1644	6.2450
4	6.3246	6.4031	6.4807	6.5574	6.6332	6.7082	6.7823	6.8557	6.9282	7.0000
5	7.0711	7.1414	7.2111	7.2801	7.3485	7.4162	7.4833	7.5498	7.6158	7.6811

For a one-digit number, we go to 0 in the left-hand column and across that row to find the digit required in the column headings. For example $\sqrt{8} = 2.8284$.

×	0	1	2	3	4	5	6	7	8	9
0	0.0000	1.0000	1.4142	1.7321	2.0000	2.2361	2.4495	2.6458	2.8284	3.0000
1	3.1623	3.3166	3.4641	3.6056	3.7417	3.8730	4.0000	4.1231	4.2426	4.3589
2	4.4721	4.5826	4.6904	4.7958	4.8990	5.0000	5.0990	5.1962	5.2915	5.3852
3	5.4772	5.5678	5.6569	5.7446	5.8310	5.9161	6.0000	6.0828	6.1644	6.2450
4	6.3246	6.4031	6.4807	6.5574	6.6332	6.7082	6.7823	6.8557	6.9282	7.0000
5	7.0711	7.1414	7.2111	7.2801	7.3485	7.4162	7.4833	7.5498	7.6158	7.6811

For a three-digit number, we find the first two digits down the left-hand column and move across the row to the third digit. For example $\sqrt{285} = 16.8819$.

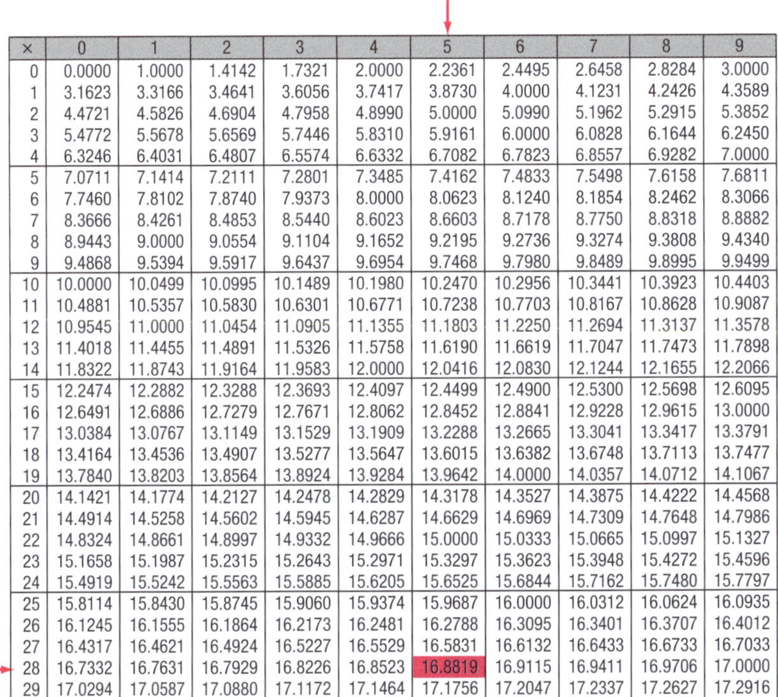

×	0	1	2	3	4	5	6	7	8	9
0	0.0000	1.0000	1.4142	1.7321	2.0000	2.2361	2.4495	2.6458	2.8284	3.0000
1	3.1623	3.3166	3.4641	3.6056	3.7417	3.8730	4.0000	4.1231	4.2426	4.3589
2	4.4721	4.5826	4.6904	4.7958	4.8990	5.0000	5.0990	5.1962	5.2915	5.3852
3	5.4772	5.5678	5.6569	5.7446	5.8310	5.9161	6.0000	6.0828	6.1644	6.2450
4	6.3246	6.4031	6.4807	6.5574	6.6332	6.7082	6.7823	6.8557	6.9282	7.0000
5	7.0711	7.1414	7.2111	7.2801	7.3485	7.4162	7.4833	7.5498	7.6158	7.6811
6	7.7460	7.8102	7.8740	7.9373	8.0000	8.0623	8.1240	8.1854	8.2462	8.3066
7	8.3666	8.4261	8.4853	8.5440	8.6023	8.6603	8.7178	8.7750	8.8318	8.8882
8	8.9443	9.0000	9.0554	9.1104	9.1652	9.2195	9.2736	9.3274	9.3808	9.4340
9	9.4868	9.5394	9.5917	9.6437	9.6954	9.7468	9.7980	9.8489	9.8995	9.9499
10	10.0000	10.0499	10.0995	10.1489	10.1980	10.2470	10.2956	10.3441	10.3923	10.4403
11	10.4881	10.5357	10.5830	10.6301	10.6771	10.7238	10.7703	10.8167	10.8628	10.9087
12	10.9545	11.0000	11.0454	11.0905	11.1355	11.1803	11.2250	11.2694	11.3137	11.3578
13	11.4018	11.4455	11.4891	11.5326	11.5758	11.6190	11.6619	11.7047	11.7473	11.7898
14	11.8322	11.8743	11.9164	11.9583	12.0000	12.0416	12.0830	12.1244	12.1655	12.2066
15	12.2474	12.2882	12.3288	12.3693	12.4097	12.4499	12.4900	12.5300	12.5698	12.6095
16	12.6491	12.6886	12.7279	12.7671	12.8062	12.8452	12.8841	12.9228	12.9615	13.0000
17	13.0384	13.0767	13.1149	13.1529	13.1909	13.2288	13.2665	13.3041	13.3417	13.3791
18	13.4164	13.4536	13.4907	13.5277	13.5647	13.6015	13.6382	13.6748	13.7113	13.7477
19	13.7840	13.8203	13.8564	13.8924	13.9284	13.9642	14.0000	14.0357	14.0712	14.1067
20	14.1421	14.1774	14.2127	14.2478	14.2829	14.3178	14.3527	14.3875	14.4222	14.4568
21	14.4914	14.5258	14.5602	14.5945	14.6287	14.6629	14.6969	14.7309	14.7648	14.7986
22	14.8324	14.8661	14.8997	14.9332	14.9666	15.0000	15.0333	15.0665	15.0997	15.1327
23	15.1658	15.1987	15.2315	15.2643	15.2971	15.3297	15.3623	15.3948	15.4272	15.4596
24	15.4919	15.5242	15.5563	15.5885	15.6205	15.6525	15.6844	15.7162	15.7480	15.7797
25	15.8114	15.8430	15.8745	15.9060	15.9374	15.9687	16.0000	16.0312	16.0624	16.0935
26	16.1245	16.1555	16.1864	16.2173	16.2481	16.2788	16.3095	16.3401	16.3707	16.4012
27	16.4317	16.4621	16.4924	16.5227	16.5529	16.5831	16.6132	16.6433	16.6733	16.7033
28	16.7332	16.7631	16.7929	16.8226	16.8523	16.8819	16.9115	16.9411	16.9706	17.0000
29	17.0294	17.0587	17.0880	17.1172	17.1464	17.1756	17.2047	17.2337	17.2627	17.2916

3 Estimate and check

Looking at the table of square numbers we constructed in Question 1 on page 101, we can see that 40 lies between 36 and 49. So $\sqrt{40}$ will lie between $\sqrt{36}$ and $\sqrt{49}$ (that is between 6 and 7). We could make an estimate and then check to see how close our estimate is. Suppose we chose an estimate of 6.3:

$6.3^2 = 39.69$ This is close.

We could then try an estimate of 6.4.

$6.4^2 = 40.96$ Which estimate is closer?

Depending on how accurate a measurement is needed, we could use the closer of these two estimates, or we could add a second decimal place to this guess and then check again:

$6.34^2 = 40.1956$ This is a little high. We could try 6.33, and so on.

4 The Babylonian method

This method involves three steps:

Step 1: Estimate the square root of the number to at least one digit.

Step 2: Divide this into the number to give at least one decimal place.

Step 3: Average the answers found in Step 1 and Step 2.

Repeat this process, finding more decimal places on each step.

Below is how this works for $\sqrt{40}$:

Step 1: Try 6.3 again.

Step 2: $40 \div 6.3 = 6.34$

Step 3: $(6.3 + 6.34) \div 2 = 6.32$

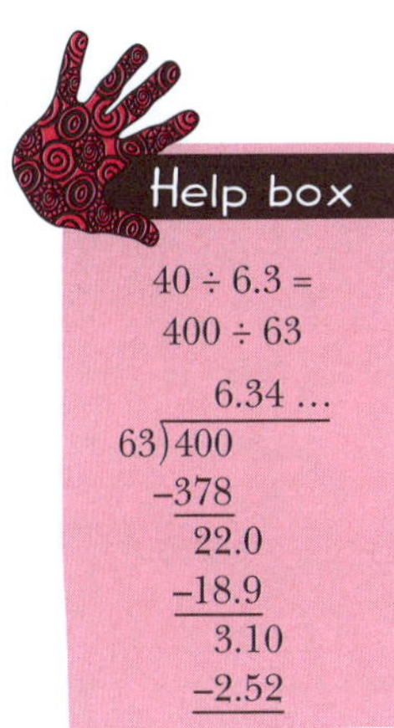

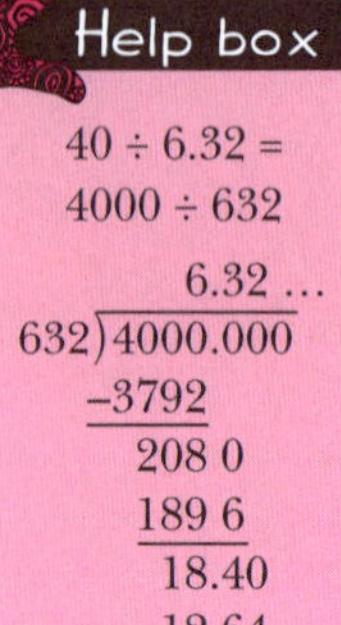

40 ÷ 6.32 =
4000 ÷ 632

```
       6.32 ...
632)4000.000
   -3792
     208 0
     189 6
      18.40
      12.64
```

Start again:

Step 1: 6.32

Step 2: 40 ÷ 6.32 = 6.32

Step 3: The average of the numbers in Steps 1 and 2 is 6.32, so that means 6.32 is the square root of 40, correct to two decimal places.

5 An old algorithm (a step-by-step procedure)

This process works a little like long division and is also called the 'iteration method'. Each step gives us the next digit/decimal place of the answer to the square root. The following shows how this works for $\sqrt{40}$.

Step 1: We mark off the digits in pairs to the left and right of the decimal point, and add some pairs of zeros on the right: $\sqrt{40.00\,00\,00}$

Step 2: We find a number whose square is less than or equal to the first pair of digits. Multiply that number by itself, as shown on the right, write the product, and subtract to find the remainder. Bring down the next pair of digits from above.

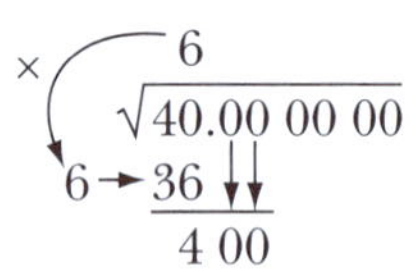

Step 3: Double the last digit of the number you found to square in Step 2 and write that with a space beside it as shown.

Step 4: Work out a number that can be used in both marked spaces as shown on the right.

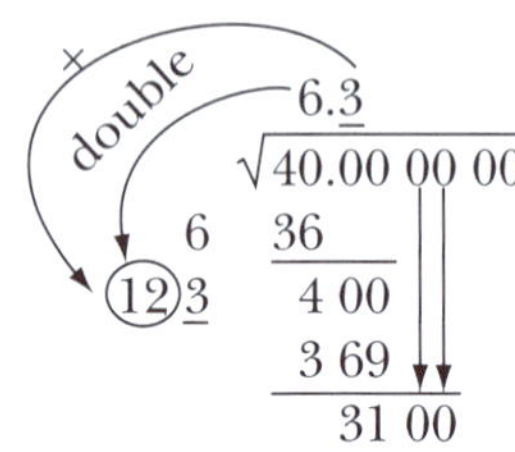

Step 5: Double the last two digits of the number above the square root symbol and write that with a space beside it. Continue as shown on the right.

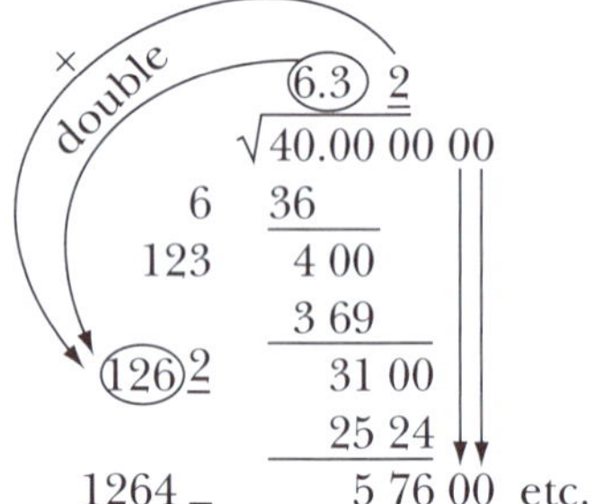

EXERCISE

1 **i** Between which two integers does each of the following square roots lie?
ii Use the square root table (see pages 391–2) to find the square root of each of the following, correct to four decimal places.

a $\sqrt{78}$ **b** $\sqrt{59}$ **c** $\sqrt{7}$ **d** $\sqrt{22}$ **e** $\sqrt{160}$ **f** $\sqrt{904}$

2 Use an 'estimate and check' method to find $\sqrt{31}$ to two decimal places.

3 Use the Babylonian method to find each of the following, to two decimal places.

a $\sqrt{2}$ **b** $\sqrt{95}$

4 Use a calculator to find each of the following, giving your answers correct to three decimal places.

a $\sqrt{5}$ **b** $\sqrt{78}$ **c** $\sqrt{241}$ **d** $\sqrt{3099}$ **e** $\sqrt{523}$

5 Use the iteration method, described above, to find $\sqrt{2521}$, to two decimal places.

Using indices

Lesson 3

We can find the prime factors of a number using a factor tree like this:

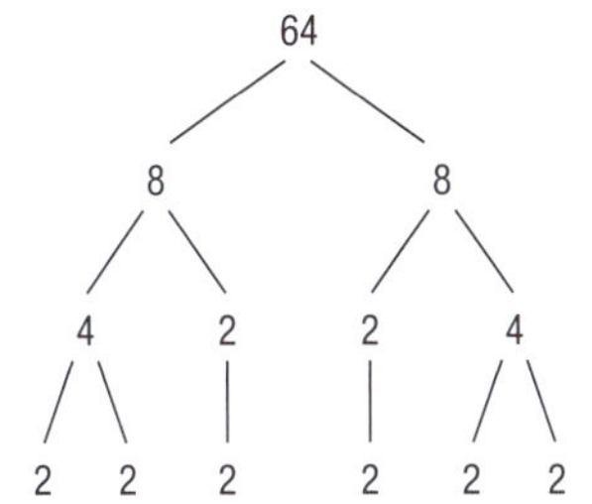

We can then express 64 as a product of its prime factors: $64 = 2 \times 2 \times 2 \times 2 \times 2 \times 2$

A simpler way of writing this is:

$$64 = 2^6$$

(6 — the index; 2 — the base)

This says 'sixty-four equals two to the power 6'. The number 2 is called the **base** and 6 is called the **index**. (The plural of 'index' is '**indices**').

Another word for 'index' is 'power' or 'exponent'.

When 64 is written as 2^6, it has been written in **index form**.

When 64 is written as $2 \times 2 \times 2 \times 2 \times 2 \times 2$, it has been written in **expanded form**.

Help box

Another way of obtaining the prime factors of a number is to use a factor ladder as shown below.

To illustrate this, we will express 360 in index form, as a product of prime factors.

Step 1: Divide by prime numbers until the number one is reached. Start with the smallest prime number (2) until that will not divide exactly, then go on to 3, 5, 7, 11 … (whichever divides with no remainder).

$360 \div 2 = 180$
$180 \div 2 = 90$
$90 \div 2 = 45$
$45 \div 3 = 15$
$15 \div 3 = 5$
$5 \div 5 = 1$

Step 2: Write the number as a product of its prime factors. $360 = 2 \times 2 \times 2 \times 3 \times 3 \times 5$

Step 3: Write in index form. $= 2^3 \times 3^2 \times 5^1$

A special case: The index '1'

Since the index shows the number of factors that are to be multiplied together, 6^1 must mean there is one factor of six, which is 6.

In the same way: $8^1 = 8$ $5^1 = 5$ $23^1 = 23$ and so on.

EXERCISE

1 Match each of these expressions with one of the given descriptions.

a 5^2 b 3^4 c 5^3 d 2^5 e 4^3

A the power is 5
B the base is 3
C the index is 3
D it reads as '5 squared'
E it reads as '5 cubed'

2 Write in index form:

a $6 \times 6 \times 6 \times 6 \times 6$ b 3×3 c 8
d $7 \times 7 \times 7 \times 7 \times 7 \times 7 \times 7 \times 7$ e $12 \times 12 \times 12$

Did you know?

Only the powers 2 and 3 have special names (squared and cubed). Can you explain where these might come from?

3 Write in words:

a 9^3 b 6^5 c 2^3 d 11^7 e 8^6

4 Write the expanded form of each of the following terms and evaluate:

a 1^6 b 2^3 c 3^3 d 4^2 e 5^4
f 10^1 g 1^9 h 3^5 i 2^6 j 8^3

5 Find the prime factors of each of these numbers, and write each number in simplest index form.

a 32 b 40 c 27 d 48 e 75
f 45 g 160 h 500 i 192 j 10 000
k 6468 l 17 875 m 10 400

Challenge: Tower of Hanoi

In this activity, you will be working out the pattern involved in solving a traditional puzzle. The Tower of Hanoi (also known as The Towers of Brahma) was invented in 1883 by a French mathematician. It consists of three vertical pegs fixed to a board and eight circular discs of different sizes with a hole in the centre. The discs are stacked, from the largest to the smallest, so that they look like a tower.

You need to rebuild the tower on one of the other pegs. However, you can only move one disc at a time and you cannot stack a larger disc on a smaller disc.

You can solve this puzzle using pencil and paper, or you can make a model of the tower in the classroom or outdoors.

Start by solving a smaller version of the puzzle, say using 3 or 4 discs. Once you are familiar with how the puzzle works, aim to restack the tower using the least number of moves. Even if you have trouble solving the puzzle with 8 discs, you will be able to carry on with this investigation.

Count the moves and complete the table below.

Number of discs in the tower	1	2	3	4	5	6	7	8
Least number of moves required								

Can you identify a pattern in your answers? Determine the least number of moves for any given number of discs, n. Write this as a statement in words or as a rule.

Least number of moves = _____

Indices in algebra

In general, if the base is any real number, a, and the index (power or exponent) is a natural (counting) number, n, then:

$a^n = a \times a \times a \times \ldots \times a$ (n times)

Understanding the meaning of indices is important for many algebraic operations. One example is working out the value of algebraic expressions.

Example

If $m = 3$ and $n = ^-2$, evaluate:

a m^4 **b** n^3
c $m^2 + n$ **d** $(mn)^3$
e mn^3 **f** $4n^2$

Answer

a $m^4 = m \times m \times m \times m$
$= 3 \times 3 \times 3 \times 3$
$= 81$

b $n^3 = n \times n \times n$
$= ^-2 \times ^-2 \times ^-2$
$= ^-8$

c $m^2 + n = m \times m + n$
$= 3 \times 3 + ^-2$
$= 9 - 2$
$= 7$

d $(mn)^3 = (3 \times ^-2)^3$
$= (^-6^3)$
$= ^-216$

e $mn^3 = m \times n \times n \times n$
$= 3 \times ^-2 \times ^-2 \times ^-2$
$= ^-24$

f $4n^2 = 4 \times n \times n$
$= 4 \times ^-2 \times ^-2$
$= ^-8 \times ^-2$
$= 16$

Remember

Use the order of operations:
1 brackets
2 multiplication and division (from left to right)
3 then addition and subtraction (from left to right).

EXERCISE

1 Copy and complete this table:

	Index form	Expanded form
a	k^6	
b		$y \times y \times y$
c	$d^4 \times g^2$	
d		$t \times r \times t \times t \times r \times t$
e	p^3q^4	
f		$w \times w \times w \times w \times w \times q$
g	yz^6	
h		$b \times b \times z \times w \times b$
i	$a^2b^3c^2$	
j		$g \times g \times f \times f \times f \times f \times f \times f$

2 If $t = 2$ evaluate:
a t^5 **b** t^2 **c** t^6 **d** t^3 **e** t **f** t^4

3 If $b = 4$, find the value of:
a $3b^2$ **b** $2b^2$ **c** $(b^2)^2$ **d** $b^3 \div 8$ **e** $4b^2 \times 6b$

4 If $t = \frac{1}{2}$, find the value of:
a t^2 **b** $2t^3$ **c** $3t^4$ **d** $-5t$ **e** $6t^2$

5 If $p = 2$ and $q = 5$, find the value of:
a $p^2 \times q$ **b** $q^2 \times p$ **c** $(pq)^3$ **d** q^3p **e** pq^4 **f** p^5q

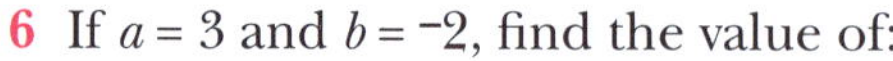

6 If $a = 3$ and $b = {}^{-}2$, find the value of:

a a^2	**b** b^4	**c** $3a^3$	**d** $(ab)^2$
e $4b^3$	**f** $(a + b)^2$	**g** $5b^2$	**h** $a^4 + 5$
i $(ab)^3 \div 4$	**j** ab^2	**k** $a^2b \div 3$	**l** $b^4 \div a$
m $a^2 \div b$	**n** $b^5 \div a$	**o** $\frac{1}{a^2}$	**p** $(\frac{b}{a})^3$

7 **a** Find the value of each of these expressions when $g = 1$:

i g^3 **ii** g^6 **iii** g^1 **iv** g^7 **v** g^{11}

b Write a sentence giving your conclusion about the number 1 raised to any power.

Lesson 5

The index rule for multiplying

Index challenge 1

a Copy and complete this table, following the completed example shown in the first row. This will help you to work out the rule for multiplication using indices.

Index notation	Factor form	Simplified index form
$x^3 \times x^2$	$(x \times x \times x) \times (x \times x)$	x^5
$2^4 \times 2^3$		$2^{\square}$
$3^1 \times 3^5$		
$y^3 \times y^3$		
$5^2 \times 5^3 \times 5^1$		
$a^2b \times ab^3 \times a^3b^2$	$(a \times a \times b) \times (a \times b \times b \times b) \times (a \times a \times a \times b \times b)$ $= a \times a \times a \times a \times a \times a \times b \times b \times b \times b \times b \times b$	
$pq^3 \times p^2q \times pq^3$		
$2^2 \times 2^n$	$(2 \times 2) \times (2 \times \ldots \times 2)$ n times	$2^{\square}$
$3^x \times 3^y$	$(3 \times \ldots \times 3) \times (3 \times \ldots \times 3)$ x times y times	$3^{\square}$

b Compare your answers for the last column of the table with those of other members of the class and complete this statement:

When using index notation, if like bases are multiplied then their indices are ______________.

c Copy into your book this rule for multiplication of indices:

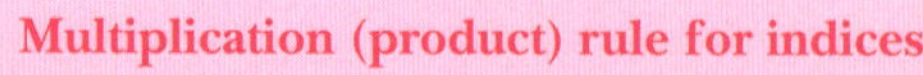

Multiplication (product) rule for indices

To multiply powers which have the same base, we add the indices:

$$a^m \times a^n = a^{m+n}$$

Example

Simplify, giving your answers in index notation:

a $t^4 \times t^2$ **b** $a^x \times a$ **c** $5^3 \times 5^4$

d $5ab^2 \times 2a^6b$ **e** $2w^3 \times 4w^2 \times 8$

Answer

a $t^4 \times t^2 = t^{4+2}$
$= t^6$

b $a^x \times a = a^x \times a^1$
$= a^{x+1}$

c $5^3 \times 5^4 = 5^{3+4}$
$= 5^7$

d $5ab^2 \times 2a^6b = 5 \times a^1 \times b^2 \times 2 \times a^6 \times b^1$
$= 5 \times 2 \times a^{1+6} \times b^{2+1}$
$= 10a^7b^3$

e $2w^3 \times 4w^2 \times 8 = 2w^3 \times 2^2w^2 \times 2^3$
$= 2^{1+2+3}w^{3+2}$
$= 2^6w^5$

2, 4 and 8 can all be written with the **same base** as powers of 2.

EXERCISE

1 Copy the following and write the correct numbers or pronumerals in the boxes:

a $d^{10} \times d^2 \times d^3 = d^{\square}$ **b** $e^{\square} \times e^3 = e^6$

c $6^2 \times 6 = 6^{\square}$ **d** $k^{\square} \times k^2 = k^{b+2}$

e $3j \times 3j^2k \times j^3k^4 = 3^2j^{\square}k^5$ **f** $3d^3 \times 9d^2 = 3^{\square}d^{\square}$

g $2^4 \times 7^3 \times 2^3 \times 7 = 2^{\square} \times 7^{\square}$ **h** $5^3 \times 3^{\square} \times 5 \times 5^2 \times 3^n = 5^{\square} \times 3^{m+n}$

i $125rt \times 5r^3t^5 \times 25r^7t^4 \times r^2t = 5^{\square}r^{\square}t^{\square}$

2 Use the 'Multiplication rule for indices' to help you simplify each of the following:

a $t^5 \times t^3$ **b** $10^5 \times 10^3$ **c** $y^4 \times y^2 \times y^5$

d $x^3 \times x \times x^9$ **e** $a^2 \times a^3 \times a^5$ **f** $5w^2 \times 4w$

g $p^7 \times q^5 \times p$ **h** $2a^4 \times 3a$ **i** $z^5 \times 2y^3 \times y^4 \times z^2$

j $2^n \times 2^{3n}$ **k** $f^2 \times f^x \times f$ **l** $7^2 \times 7^4 \times 7^5$

m $a^3b^2 \times 4ab^7 \times 8a^2$

3 Simplify:

a $-3u^4 \times -11u^9$ **b** $6z^2 \times -3z^7$ **c** $10c^4 \times 2c^3 \times -c^2$

d $-4y \times y^3 \times -5y^2 \times 3y^4$ **e** $6a^2 \times \frac{1}{12}a^3$ **f** $\frac{1}{2}a^4 \times \frac{1}{3}a^5$

4 Evaluate:

a $3^2 + 3$ **b** $3^2 \times 3$ **c** $(-6)^3$ **d** -6^3

e $4^3 \times 4^2$ **f** $4^3 + 4^2$ **g** $-2^3 + 2^2$ **h** $-(2^3 + 2^2)$

Remember

'Evaluate' means 'work out the value of'.

The index rule for dividing

Index challenge 2

a Copy and complete this table, following the completed example shown in the first row. This will help you to work out the rule for division using indices.

Index notation	Factor form	Simplified index form
$x^5 \div x^3$	$\dfrac{x \times x \times x \times x \times x}{x \times x \times x}$	$x^{\square}$
$4^4 \div 4^3$		$4^{\square}$
$\dfrac{3^7}{3^4}$		
$\dfrac{a^6b^4}{a^5b^2}$		$a^{\square}b^{\square}$
$x^4 \div x^2 \div x$	$\dfrac{x \times x \times x \times x \times x}{x \times x \times x} \div x$ $= \dfrac{x \times x}{x}$ $=$	$x^{\square}$
$\dfrac{y^4}{y^2}$		$y^{\square}$

b Copy and complete this sentence:

When using index notation, if *like* bases are *divided* then their indices are ____________.

c Copy into your book this rule for division of indices:

Division (quotient) rule for indices

To divide powers which have the same base, we subtract the indices:

$a^m \div a^n = a^{m-n}$ where $a \neq 0$

Example

Simplify, giving your answers in index form:

a $7^4 \div 7^2$

b $5^x \div 5^3$

c $2x^7 \div 4x^3$

d $m^6n^5 \div m^5n^2$

e $\dfrac{5a^3b^4 \times 4a^6b^2}{10a^7b^3}$

Answer

a $7^4 \div 7^2 = 7^{4-2}$
$= 7^2$

b $5^x \div 5^3 = 5^{x-3}$

c $2x^7 \div 4x^3 = \dfrac{{}^1\cancel{2}x^7}{{}_2\cancel{4}x^3}$
$= \dfrac{x^{7-3}}{2}$
$= \dfrac{x^4}{2}$

d $m^6n^5 \div m^5n^2 = m^{6-5}n^{5-2}$
$= mn^3$

e $\dfrac{5a^3b^4 \times 4a^6b^2}{10a^7b^3} = \dfrac{5 \times 4 \times a^{3+6} \times b^{4+2}}{10a^7b^3}$
$= \dfrac{20a^9b^6}{10a^7b^3}$
$= \dfrac{{}^2\cancel{20}a^{9-7}b^{6-3}}{\cancel{10}_1}$
$= 2a^2b^3$

EXERCISE

1 Copy the following and write the correct numbers or pronumerals in the boxes:

a $t^{10} \div t^2 = t^{\square}$

b $w^{\square} \div w^3 = w^7$

c $9^4 \div 9 = 9^{\square}$

d $x^{\square} \div x^2 = x^{b-3}$

e $j^5k^8 \div j^2k = j^{\square}k^7$

f $3d^3 \div 9d^2 = \dfrac{d^{\square}}{3}$

2 Simplify each of the following, giving your answers in index form:

a $d^{11} \div d^4$

b $h^7 \div h^5$

c $20e^5 \div 5e^2$

d $\dfrac{9d^{15}}{18d^{13}}$

e $16f^3g^2 \div 18fg$

f $g^9h^5 \div g^5h^3$

g $-12j^3 \div -20j^2k$

h $\dfrac{a^7b^6c^3}{a^6b^4c}$

i $\dfrac{2r^2 \times 3r^5}{r^4}$

j $\dfrac{6p^2 \times 4p^4}{12p^3}$

k $\dfrac{5t^4 \times 6t^3}{15t^6}$

l $\dfrac{5r^8 \times 21s^3}{15r^6s^2}$

m $\dfrac{12x^2y \times 6x^4y^5}{18xy^4 \times 8x^3}$

n $\dfrac{-2q \times 18q^5 \times -q^8}{-6q^7 \times q^6}$

o $\dfrac{64r^7t^4 \times 2r^3t^5}{4rt \times 10r^2t^3}$

p $\dfrac{15r^2s \times 20r^2s^7}{r^3s^2 \times 6rs^4}$

3 Simplify and evaluate:

a $\dfrac{4^5 \times 10^3}{4^3 \times 10^2}$

b $\dfrac{3^4 \times 5^2}{5 \times 3^2}$

c $(7^9 \times 3^3) \div (7^4 \times 3^2)$

d $(10^5 \times 2^8) \div (10^4 \times 2^6)$

4 Evaluate:

a $6^2 \div 6$

b $6^2 \times 6$

c $6^2 + 6$

d $6^2 - 6$

e $4^3 \times 4^2$

f $4^3 + 4^2$

g $4^3 \div 4^2$

h $4^3 - 4^2$

i $-2^3 \div 2^2$

j $-2^3 \times 2^2$

Lesson 7

The index rule for raising a power to a power

Index challenge 3

a Copy and complete this table.

Index notation	Factor form	Simplified index form
$(x^3)^2$	$x^3 \times x^3 = (x \times x \times x) \times (x \times x \times x)$	$x^{\square}$
$(g^4)^3$	$g^4 \times g^4 \times g^4$	$g^{\square}$
$(3a^2)^5$	$(3^1a^2) \times (3^1a^2) \times (3^1a^2) \times (3^1a^2) \times (3^1a^2)$ $= 3^1 \times a^2 \times 3^1 \times a^2 \times 3^1 \times a^2 \times 3^1 \times a^2 \times 3^1 \times a^2$	$3^{\square}a^{\square}$
$(y^{\square})^3$	$y^{\square} \times y^{\square} \times y^{\square}$	$y^{\square}$
$(5^7d^2)^3$	$5^7d^2 \times 5^7d^2 \times 5^7d^2 = 5^7 \times d^2 \times 5^7 \times d^2 \times 5^7 \times d^2$	$5^{\square}d^{\square}$
$(2^2)^3$	$2^2 \times 2^2 \times 2^2$	$2^{\square}$
$(6ef^3)^4$	$(6^1e^1f^3)^4 = 6^1e^1f^3 \times 6^1e^1f^3 \times 6^1e^1f^3 \times 6^1e^1f^3$ $= 6^1 \times e^1 \times f^3 \times 6^1 \times e^1 \times f^3 \times 6^1 \times e^1 \times f^3 \times 6^1 \times e^1 \times f^3$	$6^{\square}e^{\square}f^{\square}$

b Copy into your book this index rule for raising a power to a power:

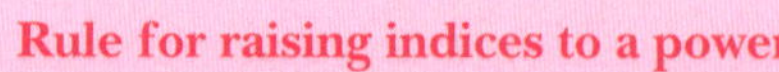

Rule for raising indices to a power

To raise a power to a power, we multiply the indices:

$$(a^m)^n = a^{m \times n} = a^{mn}$$

Example

Simplify:

a $(k^5)^3$

b $(7a^4)^2$

c $\dfrac{(e^3)^2f^8}{e^3(f^2)^3}$

Answer

a $(k^5)^3 = k^{5 \times 3} = k^{15}$

b $(7a^4)^2 = (7^1a^4)^2 = 7^{1 \times 2}a^{4 \times 2} = 7^2a^8$

c $\dfrac{(e^3)^2f^8}{e^3(f^2)^3} = \dfrac{e^{3 \times 2}f^8}{e^3f^{2 \times 3}} = \dfrac{e^6f^8}{e^3f^6} = e^{6-3}f^{8-6} = e^3f^2$

EXERCISE

1 Copy the following and write the correct numbers or pronumerals in the boxes:

a $(z^4)^{\square} = z^{24}$ **b** $(y^{\square})^3 = y^{27}$ **c** $(5^{\square})^2 = 5^6$

d $(3^2)^{\square} = 3^{2t}$ **e** $(x^{\square})^3 = x^{12n}$ **f** $(w^{2-x})^{\square} = w^{4-2x}$

2 Simplify each of the following, giving your answers in index form:

a $(a^4)^2$ **b** $(c^4)^6$ **c** $(b^3)^3$ **d** $(11^2)^n$ **e** $(7^3)^t$

3 **a** Arrange in ascending order: $3(5^2)^3$ 3^35^2 $(3 \times 5^2)^3$

b Arrange in descending order: 3×5^2 $3^2 \times 5$ $(3 \times 5)^2$

4 Simplify:

a $(g^2)^5 \times g^2$ **b** $\dfrac{(h^3)^5}{h^{12}}$ **c** $(5^4)^3 \times (5^6)^2$ **d** $(3^3)^5 \times (3^2)^7$

e $\dfrac{j^{20}}{(j^3)^5}$ **f** $\dfrac{(m^6)^3}{(m^2)^7}$ **g** $\dfrac{n^8 \times (n^3)^4}{(n^7)^2}$ **h** $\dfrac{2p^3 \times (p^5)^3}{4(p^4)^3 \times p^5}$

i $\dfrac{x^6yz^5 \times x^3(y^5z^2)^4 \times xz}{(xz)^2 \times (x^3y^2z)^2 \times y^4}$ **j** $\dfrac{3^4 \times (3s^2t)^5 \times st^2}{s^3t^2 \times 3^7(s^3t^2)^2}$

Negative or zero indices

We can find the meaning of a **negative index** by simplifying it in two ways, as shown below for $\frac{7^2}{7^4}$:

Using the 'Division rule for indices':

$$\frac{7^2}{7^4} = 7^{2-4}$$
$$= 7^{-2}$$

Using factors:

$$\frac{7^2}{7^4} = \frac{{}^1\cancel{7 \times 7}}{7 \times 7 \times \cancel{7 \times 7}_1}$$
$$= \frac{1}{7^2}$$

So we can see that: $7^{-2} = \frac{1}{7^2}$

Note that 7^{-2} is the **reciprocal** of 7^2.

Remember

The **reciprocal** of a fraction is its inverse (the fraction upside down).
Examples of pairs of reciprocals are:
- $\frac{1}{9}$ and $\frac{9}{1}$ (which is 9)
- $\frac{4}{5}$ and $\frac{5}{4}$
- -3 (which is $-\frac{3}{1}$) and $-\frac{1}{3}$
- t^4 and $\frac{1}{t^4}$

The product of a pair of reciprocals is always 1.

Example

Simplify $\frac{x}{x^4}$.

Answer

Using the 'Division rule for indices'

$$\frac{x}{x^4} = \frac{x^1}{x^4}$$
$$= x^{1-4}$$
$$= x^{-3}$$

Using factors

$$\frac{x}{x^4} = \frac{x^1}{x^4}$$
$$= \frac{x}{x \times x \times x \times x}$$
$$= \frac{1}{x^3}$$

Therefore: $x^{-3} = \frac{1}{x^3}$

Again, note that x^{-3} is the **reciprocal** of x^3.

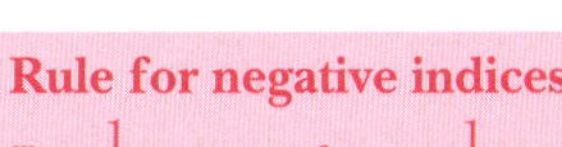

Rule for negative indices

$$a^{-m} = \frac{1}{a^m} \quad \text{and} \quad \frac{1}{a^{-m}} = a^m$$

Note: A factor can be moved from the denominator to the numerator (or vice versa) by changing the sign of its index.

Example

Simplify each of the following and write your answer with a positive index each time:

a $\frac{m^3}{m^5}$ **b** $(p^2)^{-1}$

c $\frac{1}{m^{-6}}$ **d** $\frac{3a^5}{a^{-3}}$

e $5p^{-3}$ **f** $(2m)^{-1}$

Answer

a $m^{3-5} = m^{-2} = \frac{1}{m^2}$

b $p^{2 \times -1} = p^{-2} = \frac{1}{p^2}$

c $\frac{1}{m^{-6}} = m^6$

d $\frac{3a^5}{a^{-3}} = 3a^5 \times a^3 = 3a^8$

Moving a^{-3} in the denominator to a^3 in the numerator.

e $5p^{-3} = 5 \times p^{-3}$
$= \frac{5}{p^3}$

f $(2m)^{-1} = \frac{1}{(2m)^1}$
$= \frac{1}{2m}$

Zero index

To investigate the meaning of a **zero index**, we will simplify $3^2 \div 3^2$ two different ways.

Using factors: $3^2 \div 3^2 = \frac{{}^1\cancel{3 \times 3}}{\cancel{3 \times 3}_1} = 1$

Using the 'Division rule for indices': $3^2 \div 3^2 = 3^{2-2} = 3^0$

So we can see that $3^0 = 1$

Zero index rule

$a^0 = 1$ (for all numbers except $a = 0$)

Example

Simplify, giving a numerical answer:

a $\frac{t^7}{t^7}$ **b** $6^3k^4 \div 6^2k^4$

c $3(5^2x^4y)^0$

Answer

a $\frac{t^7}{t^7} = t^{7-7}$
$= t^0$
$= 1$

b $6^3k^4 \div 6^2k^4 = 6^{3-2}k^{4-4}$
$= 6^1k^0$
$= 6 \times 1$
$= 6$

c $3(5^2x^4y)^0 = 3 \times 1$
$= 3$

EXERCISE

1 Which of the following is the reciprocal of a^2?

A $-a^2$ **B** a^{-2} **C** $\frac{1}{a^2}$ **D** $\frac{1}{a^{-2}}$ **E** $\frac{2}{a}$

2 Write 'True' or 'False' for each of the following:

a $\frac{1}{h^{-2}} = h^2$ **b** $\frac{1}{m^7} = m^{-7}$ **c** $3p^{-2} = \frac{1}{3p^2}$ **d** $\frac{m^2}{m^{-4}} = \frac{m^4}{m^{-2}}$

e $\frac{1}{m^{-7}} = \frac{m^4}{m^{-2}}$ **f** $\frac{m^2}{m^{-4}} = m^2 \times m^4$ **g** $\frac{m^2}{m^{-4}} = m^2 \times m^{-4}$

3 Copy the following and fill in the correct numbers or pronumerals:

a $\frac{1}{a^4} = a^{\square}$ **b** $m^{-4} = \frac{1}{m^{\square}}$ **c** $\frac{12}{\square} = 12n^{-7}$ **d** $\frac{1}{p^{-4}q^{-3}} = p^{\triangle}q^{\square}$

4 Simplify, giving your answers with positive indices:

a m^{-5} **b** $3a^{-2}$ **c** $(ab)^{-2}$ **d** $\frac{1}{m^{-5}}$

5 Simplify, giving your answers with positive indices:

a $11a^{-4}b^2 \times a^{-2}$ **b** $\frac{a^6}{a^{-3}}$ **c** $\frac{y^{-2}}{y^3}$ **d** $\frac{2s^{-1}}{s^3}$

e $\frac{1}{(5w^3)^{-2}}$ **f** $\frac{a^4}{b^{-2}} \times b^3$ **g** $\frac{k^{-2}}{(k^3)^{-1}}$ **h** $\frac{c^3}{(c^2)^{-2}}$

6 Simplify:

a $y^6 \div y^6$ **b** $\frac{3g^2 \times g}{g^3}$ **c** $\frac{-2h \times h}{4h^2}$

d $\frac{a^7b^4}{a^6b^4}$ **e** $\frac{7w^4 \times 7w}{7^2w^4}$ **f** $\frac{2^4x^6 \times 2x}{2^3x^7}$

7 Evaluate:

a $6^1 + 5^0$ **b** $2^0 - 3^1 \times 4^0$ **c** $\frac{9^3}{9^3} + \frac{9^3}{9^2}$ **d** $5^0 - 5^{-1}$ **e** $\frac{3^{-3}}{6^{-2}}$

8 Here is a riddle with a mathematical twist. Evaluate each expression. Then write the letter identifying the expression above its answer in a copy of the code below to find the answer to the riddle: *What did the Grandmother say when she lost her parrot?*

P 100^0 **L** $3x^4 \div x^4$ **G** $5y^0 \times (5^3y^2)^0$

O $10^0 \times 9^0 + 2^0$ **N** $7^2(a^3b^4)^0 - 6^2$ **Y** $(2^0 + 3^0)^2 + 2^2$

Code:

___	___	___	___	___	___	___
1	2	3	8	5	2	13

Combined index operations

Simplifying an expression written in index notation sometimes requires using more than one of the index laws. A simplified expression is usually written with **positive indices**.

Example

Simplify, giving your answer in index notation with **positive indices:**

a $(g^4h^{-1} \times gh^3)^3$

b $\dfrac{3^2d^5e^2 \times 3d^3e^2}{3^3de \times (e^{-3})^2}$

c $\dfrac{2a^2b^3c \times 2^3(ab^4c^2)^2}{2^4a^3b^5c^6}$

Answer

a *Step 1*: Simplify inside the brackets, if possible. $(g^4h^{-1} \times gh^3)^3 = (g^5h^2)^3$

Step 2: Expand the brackets. $= g^{5 \times 3}h^{2 \times 3}$

Step 3: Simplify the remaining expression. $= g^{15}h^6$

b *Step 1*: Simplify the numerator and simplify the denominator, after removing the brackets.

$$\frac{3^2d^5e^2 \times 3d^3e^2}{3^3de \times (e^{-3})^2} = \frac{3^{2+1}d^{5+3}e^{2+2}}{3^3de \times e^{-3 \times 2}}$$

$$= \frac{3^3d^8e^4}{3^3de^{1+-6}}$$

Step 2: Simplify the remaining expression.

$$= \frac{3^3d^8e^4}{3^3de^{-5}}$$

Step 3: Express the answer with positive indices.

$$= 3^{3-3}d^{8-1}e^{4--5}$$

$$= 3^0d^7e^9$$

$$= d^7e^9$$

c *Step 1*: Expand brackets.

$$\frac{2a^2b^3c \times 2^3(ab^4c^2)^2}{2^4a^3b^5c^6} = \frac{2a^2b^3c \times 2^3 \times a^{1 \times 2}b^{4 \times 2}c^{2 \times 2}}{2^4a^3b^5c^6}$$

Step 2: Simplify the numerator.

$$= \frac{2a^2b^3c \times 2^3 \times a^2b^8c^4}{2^4a^3b^5c^6}$$

Step 3: Simplify the remaining expression.

$$= \frac{2^{1+3}a^{2+2}b^{3+8}c^{1+4}}{2^4a^3b^5c^6}$$

$$= \frac{2^4a^4b^{11}c^5}{2^4a^3b^5c^6}$$

Step 4: Express the answer with positive indices.

$$= 2^{4-4}a^{4-3}b^{11-5}c^{5-6}$$

$$= 2^0a^1b^6c^{-1}$$

$$= \frac{ab^6}{c}$$

EXERCISE

1 Write each of the following in simplest index form, expressing each answer with positive indices:

a $10b^5 \div 7b^3$

b $(4a^3)^2 \times 4^{-2}b^4$

c $(c^6d^8 \div cd^2)^3$

d $-2a^3 \times 2a^0 \times -3a^4$

e $2^2(dg^{-1} \times d^4g^{-2})^3$

f $\frac{(2m)^2 \times 3m}{(2m^0n)^4}$

g $a^4b^2 \times a^{-2}b^{-3} \times 3a^{-3}b^{-1}$

h $\frac{(-2g^3h^2)^3}{(-4gh^2)^2}$

i $\frac{j^3k^{-3} \times j^2k^2}{j^{-3}k^{-1}}$

j $(5^2h^3)^5 \times \left(\frac{h^6}{j^2}\right)^3$

k $\frac{k^{-2}m^4 \times k^{-3}m^{-2}}{km \times (k^{-2}m^3)^2}$

l $\frac{(5^{x+4})^2}{5^2}$

m $\frac{a^{x+2}}{a^{x+4}}$

n $\frac{1}{(a^3b^2)^4} \times a^{10}b^9$

o $\left(\frac{4a^2b^6}{2ab^8}\right)^3 \times a^3b^5$

2 Simplify, leaving your answer in index form:

a $\frac{2^{18}}{(2^5 \div 2^3)^2}$

b $(3^6 \times 3 \div 3^5)^4$

c $(6^3 \div 6^2 \times 6^4)^3$

d $\frac{(7^3 \times 5^2)^2}{7 \times 5^3}$

e $11(2^4)^0$

f $3(3 \times 5^4)^0$

Lesson 10

Scientific notation

Scientists often work with very small numbers or very large numbers. For instance:

- the distance to the farthest know star is approximately 95 000 light years, or 900 000 000 000 000 000 kilometres
- the size of a particular atom is 0.000 0005 millimetres.

It is convenient to write numbers like these using **scientific notation**. Another name for scientific notation is **standard form**.

A number is written in scientific notation like this.

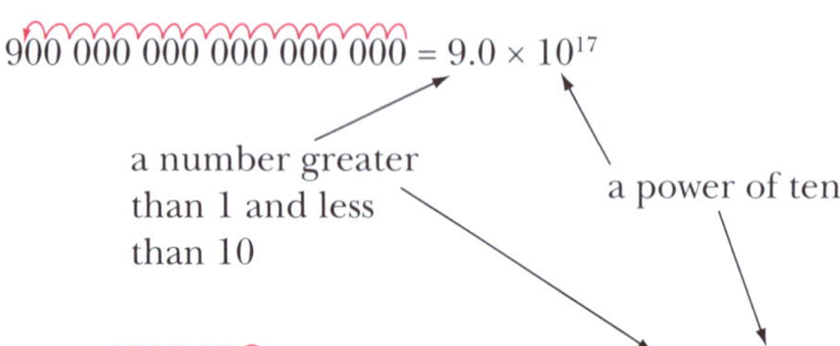

0.000 000 5 in scientific notation is 5.0×10^{-7}

Note that a large number (greater than 10) will have a positive power of 10, and a small number (less than 1) will have a negative power of 10.

Remember
- $10^0 = 1$
- A positive power of 10 gives a number equal to or greater than 10.
- A negative power of 10 gives a fraction or decimal.

In general, **scientific notation** or **standard form** can be represented as:
$a \times 10^n$ where $1 \leqslant a < 10$ and n can be any integer

Extremely large numbers and extremely small numbers often will automatically appear in standard form when entered on a calculator.

Example

a Write each of the following numbers in scientific notation:

i 34 000 000

ii 123 000

iii 0.004 69

iv 0.000 000 005

b Write each of these as ordinary numbers:

i 9.25×10^6

ii 3.7×10^{-4}

Answer

a i $34\,000\,000 = 3.4 \times 10^7$

ii $123\,000 = 1.23 \times 10^5$

iii $0.004\,69 = 4.69 \times 10^{-3}$

iv $0.000\,000\,005 = 5.0 \times 10^{-9}$

b i $9.25 \times 10^6 = 9.25 \times 1\,000\,000 = 9\,250\,000$

ii $3.7 \times 10^{-4} = 3.7 \times \frac{1}{10\,000} = 0.000\,37$

EXERCISE

1 Which of the following are written in scientific notation?

a 7×10.8 b 4.69×10^4 c 0.15×10^{-3}

d 10^6 e 2.1×10 f 7×10^{-2}

2 Express each of the following as a power of 10:

a $\frac{1}{1000}$ b 10 000 c 1

d 0.0001 e 1 000 000 f 0.1

3 Copy and complete this table:

Number	Calculations	Scientific notation
276	2.76×100	
8800	8.800×1000	
5166.66		
37		
	2.002×10	
	6.1755×100	
	$7.6 \times 10\,000$	7.6×10^5
	4×100	4×10^2
		1.66×10^0
		8.2×10^3
3 million		
		9.233×10^4
0.07		$7 \times 10^{\square}$
0.000 54		$5.4 \times 10^{\square}$
0.008 00		$8.00 \times 10^{\square}$
		1.0004×10^{-3}
		4.3333×10^{-5}
		1.5001×10^{-7}

4 Write each of the following numbers in standard form:

a 800 b 123 c 98 000 d 2 900 000

e 84.2×10^2 f 760 thousand g 773.2×10^3 h 98.765

i 17.6×10^4 j 20 million

5 Write each of the following as an ordinary number:
a 2×10^6 b 7.9×10^3 c 9×10^1 d 5.1×10^0

6 Write each of the following using scientific notation:
a 0.0023 b 0.004 c 0.000 000 078 d 0.000 000 49

7 Write each of the following as an ordinary number:
a 8.2×10^{-4} b 1.56×10^{-3} c 9×10^{-2}
d 4.03×10^{-1} e 7×10^{-7}

8 The sizes of four viruses are $\frac{7}{10\,000}$ mm, $\frac{65}{100\,000}$ mm, $\frac{712}{100\,000}$ mm, and $\frac{63}{10\,000}$. Write their sizes in scientific notation and arrange them from smallest to largest.

Did you know?

A googol is 10^{100}, that is the number one followed by 100 zeros. This is more than the number of atoms in the known universe! The name of this number is ten duotrigintillion.

A googolplex is the number one followed by one googol zeros, or 10 raised to the power of one googol:

$$10^{\text{googol}} = 10^{10^{100}}$$

$$10^{100} = 1 \text{ googol}$$

$$10^{10^{100}} = 1 \text{ googolplex}$$

Lesson 11

Applying scientific notation

We can carry out multiplication and division of numbers written in standard form using the index rules.

Example

a Find, in standard form: $(3.1 \times 10^5) \times (7.6 \times 10^7)$.

b Calculate: $(4.8 \times 10^7) \div (2.0 \times 10^3)$.
Give your answer in standard form.

Answer

a
$$\begin{aligned}(3.1 \times 10^5) \times (7.6 \times 10^7) &= 3.1 \times 10^5 \times 7.6 \times 10^7 \\ &= 3.1 \times 7.6 \times 10^5 \times 10^7 \\ &= 23.56 \times 10^{5+7} \\ &= 23.56 \times 10^{12} \\ &= (2.356 \times 10) \times 10^{12} \\ &= 2.356 \times 10^{13}\end{aligned}$$

b
$$\begin{aligned}(4.8 \times 10^7) \div (2.0 \times 10^3) &= \frac{4.8 \times 10^7}{2.0 \times 10^3} \\ &= \frac{4.8}{2.0} \times \frac{10^7}{10^3} \\ &= 2.4 \times 10^{7-3} \\ &= 2.4 \times 10^4\end{aligned}$$

EXERCISE

1 Write the answer to each of the following in scientific notation:
   a $2.0 \times 10^7 \times 1.4 \times 10^{-5}$
   b $(8.4 \times 10^{12}) \div (4 \times 10^4)$
   c $7.5 \times 10^{-3} \times 3 \times 10^4$
   d $(2.5 \times 10^{-6}) \div (5 \times 10^4)$

2 There are more than 600 million bacteria in a person's body.
   a Write this figure in scientific notation.
   b Work out and write in scientific notation the number of bacteria there would be in the people of a town with a population of 200 000.

3 Using scientific notation, write:
   a the number of seconds in a week
   b the number of seconds in 3×10^3 weeks.

4 In 2006, PNG exported 567 000 tonnes of gold.
   a How many kilograms of gold is this?
   b Write 567 000 in scientific notation.

Remember
1 tonne = 1000 kg

5 Light travels at 300 000 km/s.
   a Write 300 000 in scientific notation.
   b How far does light travel in:
      i 1 minute? ii 1 hour? iii 1 day?
   c What is the distance from Earth to Mars if light takes 41 minutes to reach Earth from Mars?

Remember
Distance = speed × time

6 The photograph of a plant cell magnified 12 000 times measures 36 millimetres. Work out the actual measurement of the cell, giving your answer in standard form.

7 a Every drop of blood contains about 25 million red blood cells. Write this figure in standard form.
   b Red blood cells only live for approximately 120 days. The body must replace them at a rate of about 3 million per second. How many blood cells would be replaced in 1 day? Write your answer in standard form.

8 The weight of a pin is approximately 0.0001 grams.
   a Express this weight in standard form.
   b How many pins would it take to give a weight of 1 kg?

Did you know?

A micrometre (μm) is one-millionth of a metre, that is 1×10^{-6} m. It is also known as a micron. Some examples of lengths measured in micrometres are given below:

- the width of a human hair is 80 μm
- the diameter of a fog or mist water droplet is 10 μm
- the diameter of a bacterium is 1 μm
- the width of a strand of spider's web is 7 μm.

Did you know?

If all your veins and arteries were laid out end to end, they would stretch 100 000 kilometres which is about two-and-a-half times around the world!

Evaluating powers using a calculator

Lesson 12

How can we find the value of, say, 6^3 using a calculator?

Of course, we could do the sum 6 × 6 × 6, using the multiplication key, to get 216.

On some calculators, there is a key for finding cubes, marked [x^3]. Using this, we would enter 6 [x^3] [=] to get 216. There is also usually a key for squares ([x^2]).

On scientific calculators, there is a special key for finding any powers.

It is marked [x^y].

So, we would enter 6 [x^y] 3 [=] to get 216.

Note that we enter the base (6), show that we want to calculate a power ([x^y]), then enter the index (3).

On some calculators, there is a [∧] key for finding 'the power of'.

Example

Use your calculator in work out the basic numeral for: **a** 8^4 **b** $(^-3)^3$

Answer

	Question	Calculator steps	Basic numeral
a	8^4	• For those with x^y on a key: 8 x^y 4 = 4096 • For those with x^y written **above** a key: 8 Shift x^y 4 = 4096 or 8 2nd F x^y 4 = 4096 • For those with a **graphics** calculator: 8 ^ 4 = 4096	4096
b	$(^-3)^5$	• For those with x^y on a key: 3 +/− x^y 5 = −243 • For those with x^y written **above** a key: 3 +/− Shift x^y 5 = −243 or 3 +/− 2nd F x^y 5 = −243 • For those with a **graphics** calculator. ((−) 3) ^ 5 = −243	$^-243$

Scientific notation on your calculator

You can also use a scientific calculator to work with numbers in standard form. To enter 3.5×10^7 the key sequence is:

3.5 EXP 7

EXERCISE

1 Use a calculator to evaluate:

a 12^3 **b** 5^4 **c** 7^3 **d** 8^5 **e** 2^9
f 3^7 **g** 9^5 **h** 4^6 **i** 11^4 **j** 5^{11}

2 Use your calculator to find the value of:

a 25^2 **b** $(^-11)^3$ **c** 0.5^6 **d** $(^-9)^4$
e 1.3^4 **f** $^-2.1^5$ **g** 10^5 **h** $(^-16)^3$
i $(16.5)^3$ **j** $(13)^4$ **k** $(36)^{0.5}$ **l** $(10\,000)^{0.25}$

3 What ordinary numeral would appear on the display of a calculator for each of these:

a 2.8 EXP 6 **b** 1.3 EXP −2 **c** 5.639 EXP 5
d 7.203 EXP 6 **e** 1.77 EXP −2

4 Use a calculator to evaluate each of these, writing your answer as an ordinary number:

a $7.0 \times 10^5 \times 1.4 \times 10^3$ **b** $(3.62 \times 10^{11}) \div (1.4 \times 10^5)$
c $8 \times 10^7 - 4 \times 10^6$ **d** $9.3 \times 10^4 + 5.21 \times 10^5$ **e** $(5.2 \times 10^3)^2$

EQUATIONS

Number patterns

A set of numbers arranged in a definite order and following a pattern is called a **sequence**. The numbers that make up a sequence are called the **terms** of the sequence. For example, 4, 7, 10, 13 … is a sequence. The first term of this sequence is 4, the second term is 7, and so on.

Another sequence might be: $^-5, ^-10, ^-15, \ldots$

We can often use logic to work out the hidden 'rule' that is being followed by the terms in the sequence. This might be very simple, as in the sequence 4, 7, 10, 13 …, for which we just add 3 to each term to get the next term. Or the rule might be more difficult to discover.

Starting with any number, a sequence can be generated by applying any of the operations +, –, × and ÷, either on their own or as a combination. The operations 'square' and 'square root' can also be used.

Example

a Write the first three numbers of a pattern, starting with 10 and doubling to find each new term.

b Write the next three terms of the sequence: 9, 14, 19 …

c What is the value of the pronumerals representing unknown numbers in each of these sequences?

i $20, 8, b, ^-16, ^-28, \ldots$ **ii** $12, 6, m, n, \frac{3}{4}, \frac{3}{8}, \ldots$

d Describe in words the rule for the pattern: 2, 5, 10, 17, …

Answer

a $10, 10 \times 2, (10 \times 2) \times 2 \ldots = 10, 20, 40 \ldots$

b We can see that each term in the pattern is increasing by 5, so the next three terms are 24, 29 and 34.

c **i** The second term is 12 less than the first term, and the fifth term is also 12 less than the fourth term, so we know that the pattern is decreasing by 12 each time. So $b = 8 - 12 = ^-4$

ii This pattern does not work using subtraction because the difference between the first and second terms is not the same as the difference between the fifth and sixth terms. We can check if it works using division: half of 12 is 6, and half of $\frac{3}{4}$ is $\frac{3}{8}$, so the pattern is halving (dividing by 2).
Therefore $m = 6 \div 2 = 3$, and $n = 3 \div 2 = \frac{3}{2}$

d There is more than one way to describe this pattern. One answer is 'Starting with 2, the numbers increase by 3, then 5, then 7' (that is, by two more each time). Another description is 'Starting with 1, square each counting number and add one'.

EXERCISE

1 Write the next three numbers in a pattern starting with 16 and:

a going up by 3s
b going down by 10s
c multiplying by $^-3$
d dividing by 2
e going up by 2, then up by 4, then up by 6
f going down by 1, then up by 3, then down by 5

2 State the value of each pronumeral and describe, in words, the rule for each of these patterns:

a 2 4 6 f 10 12
b 1 4 y 16 25 36
c 2 4 8 16 q 64
d w 9 13 17 21 25
e 4 1 $^-2$ p $^-8$ $^-11$
f 1 $\frac{1}{3}$ m $\frac{1}{27}$ $\frac{1}{81}$

3 Write the next three terms in each of these sequences and describe, in words, the rule for the pattern:

a 20, 25, 30, …
b 8, 1, $^-6$, …
c 9, 3, 1, …
d 3, $^-6$, 12, …
e $\frac{3}{4}$, 1, $1\frac{1}{4}$, …
f 4, 1.2, $^-1.6$, …
g 2, 5, 11, …
h 20, 9, $3\frac{1}{2}$, …
i $^-2$, $^-5$, $^-14$, …
j 5, $^-1$, $\frac{1}{5}$, …
k 4.5, 12, 27, …
l 1, 1.25, 1.5, …

4 Write the next three terms in each of these sequences and describe, in words, the rule for the pattern:

a 1, 2, 4, 8, …
b 1, 4, 9, 16, …
c $\frac{1}{2}$, $\frac{1}{3}$, $\frac{1}{4}$, …
d 1, 3, 6, 10 …
e 1, 2, 0, 3, $^-1$, …

5 How are the patterns in Question 4 different from the ones in Questions 2 and 3?

6 a Make up three patterns similar to those in the questions above. Try to make them hard to guess!

b Exchange your patterns with a partner and see who can work out the missing numbers the quickest.

c Explain the rule you used to create the pattern and compare this with the rule your partner used when working out the answers. (You might find that you described the same pattern differently.)

Challenge

A special sequence of numbers called Lucas numbers occurs in nature. The sequence is 1, 3, 4, 7, 11, 18, …

a Write the next three Lucas numbers.
b What positions are occupied by even numbers?
c Will the 129th Lucas number be odd or even?

Arithmetic sequences

A sequence of numbers formed by simply adding a fixed amount to each new term is called an **arithmetic sequence** or **arithmetic progression**.

For example, the sequence 9, 14, 19, 24 … starts with 9 and each of the numbers that follow is obtained by adding 5 to the previous number.

Another sequence might be 4, 1, $^-2$, $^-5$ … In this case, we are adding $^-3$ (that is, subtracting three), so we can see that an arithmetic progression might consist of a set of numbers that is increasing in size or decreasing in size.

A sequence is usually written t_1, t_2, t_3, and so on, where t_1 is the first term, t_2 is the second term … and t_n is the nth term (also called the **general term**). The '1' in t_1 is called a **subscript** (because it is written *below* the level of the text). The subscript of each term represents the term number.

The first number in the sequence is represented by the pronumeral 'a', while the fixed number being added is called the **common difference**, and is represented by d.

Example

A sign consists of a sequence of neon lights which form arrows when they are turned on.

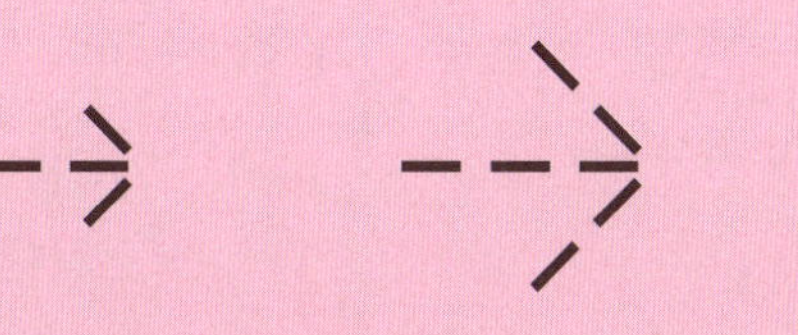

We can complete a table showing the term number and the number of neon lights making up that term.

Term (t)	1	2	3	4
Number of neon lights	4	7	10	

Can you state the number of lights which would be needed in the 4th term of this pattern?

Answer

In this example, $a = 4$, the first term, and $d = 3$, the common difference. So:

$t_1 = a = 4$

$t_2 = 7 = 4 + 3$ That is, $a + 1d$

$t_3 = 10 = 4 + 2 \times 3$ That is, $a + 2d$

$t_4 = 13 = 4 + 3 \times 3$ That is, $a + 3d$

The 4th term = 13

Looking at the pattern, we can see that each term consists of the starting number, a, and a multiple of the difference, d, one less than the term number. So:

$t_{30} = a + 29d$

and: $t_{45} = a + 44d$

and so on.

The general term of an arithmetic sequence is given by $t_n = a + (n - 1)d$

We can use this rule to quickly find any term of the sequence.

Remember

'Consecutive' means 'following one another'. Examples of **consecutive numbers** are 29 and 30, or 1046 and 1047. **Consecutive terms** are t_7 and t_8, or t_{78} and t_{79}.

Example

a Find t_{19} of the arithmetic sequence $^-3, 2, 7, \ldots$

b Find the difference between pairs of consecutive terms to determine which of these sequences are arithmetic progressions:

i $9, ^-2, 3, ^-8, \ldots$

ii $3, 4.5, 6, 7.5 \ldots$

iii $2, 7, 15, 31, \ldots$

iv $8, 4, 2, \ldots$

c Find a simplified expression for t_n for the sequence $7, 12, 17 \ldots$ and, use this to find the value of t_{200}.

d If $t_n = 5n + 11$, find the values of t_1, t_2 and t_3 and show that these form an arithmetic sequence.

Answer

a For the sequence $^-3, 2, 7, \ldots$ the first term is $^-3$ and the terms in the sequence are increasing by 5 each time. So $a = ^-3$ and $d = 5$.

$$t_{19} = a + 18d$$
$$= ^-3 + 18 \times 5$$
$$= ^-3 + 90$$
$$= 87$$

b i $t_2 - t_1 = ^-2 - 9 = ^-11$.

Checking this for $t_3 - t_2 = 3 - ^-2$
$= 3 + 2$
$= 5$

$t_2 - t_1 \neq t_3 - t_2$, so this is not an arithmetic progression.

ii $t_2 - t_1 = 4.5 - 3 = 1.5$

and: $t_3 - t_2 = 6 - 4.5 = 1.5$

and: $t_4 - t_3 = 7.5 - 6 = 1.5$

Because there is the same difference between consecutive terms, this is an arithmetic progression.

iii $t_2 - t_1 = 7 - 2 = 5$

and: $t_3 - t_2 = 15 - 7 = 8$

$t_2 - t_1 \neq t_3 - t_2$, so this is not an arithmetic progression.

iv $t_2 - t_1 = 4 - 8 = ^-4$

and: $t_3 - t_2 = 2 - 4 = ^-2$

$t_2 - t_1 \neq t_3 - t_2$, so this is not an arithmetic progression.

c For the sequence $7, 12, 17 \ldots$ $a = 7$ and $d = 5$.

$$t_n = a + (n - 1)d$$
$$= 7 + (n - 1) \times 5$$
$$= 7 + 5n - 5 \qquad \text{Using the 'Distributive rule'.}$$
$$= 2 + 5n$$

$\therefore$ A simplified expression for t_n is: $t_n = 2 + 5n$

So $t_{200} = 2 + 5 \times 200 = 1002$

d If $t_n = 5n + 11$:

$t_1 = 5 \times 1 + 11 = 16$

$t_2 = 5 \times 2 + 11 = 21$

$t_3 = 5 \times 3 + 11 = 26$

So the sequence is $16, 21, 26, \ldots$ This is an arithmetic sequence in which $a = 16$ and $d = 5$.

EXERCISE

1 i Identify which of the sequences below are arithmetic sequences and give reasons for your answer.

ii For those sequences which are arithmetic sequences, find the next three terms.

a $3, 5, 8, 12, \ldots$ b $12, 6, 0, \ldots$ c $\frac{1}{2}, \frac{1}{4}, \frac{1}{6}, \frac{1}{8}, \ldots$ d $1.2, 6.2, 11.2, \ldots$

e $^-6, ^-4, ^-2, \ldots$ f $5, 10, 20, \ldots$ g $^-10, ^-8, ^-5, ^-1, \ldots$ h $^-2, 0, ^-2, 0, \ldots$

2 i Identify a and d for each of the following arithmetic sequences.
ii Find t_{20} for each sequence.

a 4, 6, 8, 10, … b 0.5, 3.5, 6.5, … c $^-8, ^-3, 2, \ldots$
d $5, ^-3, ^-11, \ldots$ e $\frac{3}{4}, \frac{1}{2}, \frac{1}{4}, \ldots$ f 34, 43, 52, …
g 4, 4.6, 5.2, … h $30, ^-10, ^-50, \ldots$

3 i Find a simplified expression for t_n for each of the arithmetic progressions described below.
ii Use the t_n formula to find t_{50} each time.

a $a = 7, d = ^-1$ b $a = ^-9, d = 2$ c $a = 7.5, d = 0.5$

4 i Use each t_n formula given below to generate the first four terms of a sequence.
ii State the value of a and d for each sequence.
iii Find t_{50}.

a $t_n = 5n + 8$ b $t_n = 11 - 3d$ c $t_n = 15 + 0.5d$
d $t_n = 24 - 11d$ e $t_n = 200 - 20d$

Applications of arithmetic sequences and series

A sequence can be **finite** (having an end point) or **infinite** (going on indefinitely). For example the sequence of odd numbers 1, 3, 5, 7, … is infinite. This is indicated by the three dots at the end, showing that the sequence continues.

The sequence $9, 4, ^-1, ^-6, ^-11$. is finite because $^-11$ is the last term of the pattern.

A **series** is formed by adding the terms of a sequence.

For example 9, 14, 19, 24 … is a sequence, while 9 + 14 + 19 + 24 + … is a series.

A series can also be finite or infinite.

The numbers in some practical situations follow an arithmetic sequence. Below is one example.

Example

Karen is saving for a special trip. She deposits K50 in her bank account and plans to put in K15 every month from her pay. The account does not pay interest.

a Show that the savings at the end of months 1, 2 and 3 form an arithmetic sequence.
b State the value of a and d for this sequence.
c Find the value of t_{12}.
d Explain the significance of t_{12}.

Answer

a To start with, Karen's account has K50.
At the end of 1 month, the balance is:
K50 + K15 = K65
After 2 months, the balance is: K65 + K15 = K80
After 3 months the balance is: K80 + K15 = K95
So $t_1 = 65$, $t_2 = 80$ and $t_3 = 95$. The numbers form an arithmetic progression because they increase by the same amount each time.

b $a = 65$ and $d = 15$

c $t_{12} = a + 11d$
$= 65 + 11 \times 15$
$= 230$

d t_{12} is the balance in Karen's account at the end of 12 months and she has K230 in the bank.

EXERCISE

1 Correctly label each of the following patterns as a 'finite sequence', an 'infinite sequence', a 'finite series' or an 'infinite series':

a 11, 5, −1, −7, −13

b $9 + 3 + 1 + \frac{1}{3} + \ldots$

c 5 + 15 + 25 + 35 + 45

d 56, 44, 32, …

e 70 + 63 + 56 + 49

2 Frank is saving for a motorcycle. He deposits K30 in his bank account and then adds K3 every week from his pay. Assume there is no interest paid.

a Write the balance in Frank's account at the end of:
   i week 1
   ii week 2
   iii week 3.

b Explain why the numbers you wrote in part **a** form an arithmetic sequence.

c Write the value of:
   i a
   ii d

d Write the rule for t_n and simplify it.

e Use your t_n rule to work out how much is in Frank's account at the end of 52 weeks.

3 A population of ants is growing at the rate of 50 per week.

a If there are 2000 ants to start with, write the number of ants at the end of:
   i week 1
   ii week 2
   iii week 3.

b How many ants will there be at the end of 50 weeks? Use the rule for t_n of an arithmetic sequence to work out your answer.

4 The rungs of a ladder decrease in length by 0.9 cm. The first rung measures 50 cm.

a Write the length of the:
   i second rung
   ii third rung.

b For this arithmetic progression, state the value of:
   i a
   ii d

c Find the length of the eighth rung.

5 A new printing business offers customers the following prices for printing advertising booklets. The cost is 10 toea for the first page, 9.8 toea for the second page, 9.6 toea for the third page, and so on.

a For this arithmetic sequence, what are the values of:
   i a?
   ii d?

b Write a simplified rule for t_n.

c Use the rule to find the cost of the 30th page.

Enrichment: The sum of *n* terms

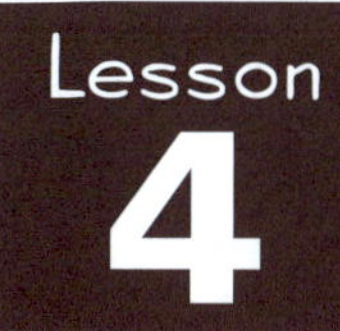

The formula for adding n terms of an arithmetic progression is:

$S_n = \frac{n}{2}[2a + (n - 1)d]$ where: $a = t_1$ and d = common difference

Example

Find the sum of 10 terms of the series 8 + 12 + 16 …

Answer

$a = 8$, $d = 4$ and $n = 10$.

Substituting into the S_n rule:

$$\begin{aligned} S_n &= \frac{n}{2}[2a + (n - 1)d] \\ &= \frac{10}{2}[2 \times 8 + (10 - 1) \times 4] \\ &= 5[16 + 9 \times 4] \\ &= 5[16 + 36] \\ &= 5 \times 52 \\ &= 260 \end{aligned}$$

Example

Some practical problems follow an arithmetic series. For example, a pattern is formed by coloured lights set out in a fan shape. The first row of the pattern has 1 light, the second row 2 lights, the third row 3 lights, and so on. What is the total number of lights in 20 rows?

Answer

In this case, we need to find the value of 1 + 2 + 3 + 4 … + 20.

$a = 1$, $d = 1$ and $n = 20$.

So:
$$\begin{aligned} S_n &= \frac{n}{2}[2a + (n - 1)d] \\ &= \frac{20}{2}[2 \times 1 + (20 - 1) \times 1] \\ &= 10[2 + 19] \\ &= 10 \times 21 \\ &= 210 \end{aligned}$$

EXERCISE

1 For each of the arithmetic series described below:

- **i** write the first three terms
- **ii** use the S_n formula to find the sum of the first 30 terms of the series.

a $a = 4$ and $d = 6$ **b** $a = {}^{-}10$ and $d = 1$ **c** $a = 100$ and $d = 10$

2 A pattern of stones is made by placing the stones in rows, starting with 1 stone in row 1, 3 stones in row 2, 5 stones in row 3, and so on, as shown on the right.

```
      •           Row 1
    • • •         Row 2
  • • • • •       Row 3
• • • • • • •     Row 4
                  etc.
```

a For this sequence, identify:

- **i** a
- **ii** d

b What would be the total number of stones in 20 rows?

3 Suppose that the ladder described in Question **4** on page 126 has 12 rungs. Find the total length of all the rungs.

Lesson 5

Writing algebraic expressions and equations

In Year 8, we used pronumerals to write algebraic expressions matching different situations. For example, if x represents Joseph's age in years:

- the statement '3 years older than Joseph' can be written as $x + 3$
- 'twice Joseph's age' can be written as $2 \times x$, or $2x$
- '5 years younger than Joseph' can be written as $x - 5$
- 'half Joseph's age' can be written as $x \div 2$, or $\frac{x}{2}$.

Using algebra to write statements is useful for solving equations and working out the answers to problems.

An **expression** is a collection of algebraic terms. For example $2a$, or $7m + 3n - 5$, or $6ab - b^2 + 4$ are all expressions. Expressions can be simplified by **collecting like terms**. Like terms are those terms which contain exactly the same pronumeral. If we are given a value for each pronumeral, we can find a value for the expression.

An **equation** is a statement containing 'equals' (=) and can be solved to find the value of a pronumeral.

Example

1 Simplify:

a $4h + 3h - h$

b $11b + 6c - 4b$

c $7 - 2g + 5g^2 - 6g$

2 Evaluate $5p - 6$ when:

a $p = 3$

b $p = {}^-1$

c $p = 2.5$

Answer

1 a $4h + 3h - h = 6h$

b $11b + 6c - 4b = 7b + 6c$

$11b$ and ${}^-4b$ are 'like terms' and can be added.

c $7 - 2g + 5g^2 - 6g = 7 - 8g + 5g^2$

2 a $5p - 6 = 5 \times 3 - 6$
$= 15 - 6$
$= 9$

b $5p - 6 = 5 \times {}^-1 - 6$
$= {}^-5 - 6$
$= {}^-11$

c $5p - 6 = 5 \times 2.5 - 6$
$= 12.5 - 6$
$= 6.5$

Remember

'$- h$' means '$- 1h$'.

EXERCISE

1 Write each of the following as an algebraic expression, using m for the 'certain number' each time.

a Six more than a certain number.
b Eight less than a certain number.
c Four times a certain number.
d A certain number divided by six.
e Twice a certain number is subtracted from nineteen.
f One is added to three times a certain number.
g A certain number is halved then subtracted from two.
h The sum of three and a certain number is multiplied by four.
i The product of a certain number and 11.
j The quotient of six and a certain number.
k Twice the difference between a certain number and four.
l Half the sum of a certain number and nine.

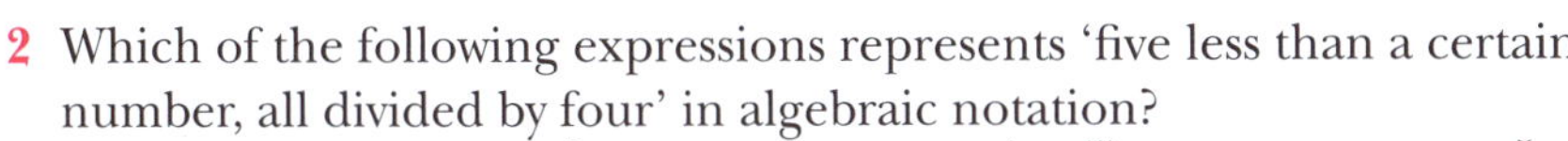

2 Which of the following expressions represents 'five less than a certain number, all divided by four' in algebraic notation?

A $5 - \frac{n}{4}$ **B** $\frac{5}{4} - 4n$ **C** $\frac{(n-5)}{4}$ **D** $n - \frac{5}{4}$

3 Simplify each of the following by collecting like terms:

a $9m + 2n - 4m$ **b** $3 + 7y - 2y + 1$

c $y^2 + 8y - 3y + 6$ **d** $4mn + 5n - 2nm - 8n$

e $5xy + 2x - 6y + 4x - y$ **f** $3a^2b - a^2b + 8ab - 11ba$

Remember

Pronumerals in algebraic terms are written in alphabetical order. We write $6ab$, not $6ba$.

4 Copy the table and evaluate each expression if:

i $m = 5$ **ii** $m = {}^-3$

Record your answers in the appropriate columns.

Help box

Evaluate means 'work out the value of'.

	Expression	i Value when $m = 5$	ii Value when $m = -3$
a	$2m + 3$		
b	$2(m + 3)$		
c	$m - 8$		
d	$2(m - 8)$		
e	$2m - 8$		
f	$3(m + 5) - 2$		
g	$\frac{3(m+5)}{2}$		

5 Copy the following table and complete it by:

i writing the algebraic equivalent, using n to represent the 'certain number'

ii writing the value of the number, n, which makes each statement true.

	Statement	Algebraic equivalent	Value of n
a	Half of a certain number equals six.	$n \div 2 = 6$ or $\frac{1}{2}n = 6$	12
b	Twice a certain number equals ten.		
c	Five times a certain number is twenty.		
d	When a certain number is subtracted from nineteen, the result is six.		
e	When ten is subtracted from a certain number, the result is six.		
f	One more than twice a certain number is nine.		
g	One less than three times a certain number is fourteen.		
h	When twice a certain number is added to seven, the result is thirteen.		
i	One-fifth of a certain number, then plus one, equals three.		
j	One is added to a certain number, and the result is halved to give nine.		
k	A number and the next consecutive number are added together to give thirty-five.		

Challenge

Colour mind reading puzzle

Carry out the following steps.

Step 1: Write the number representing the month you were born (for example April = 4, October = 10).

Step 2: Add 28.

Step 3: Now add the difference between 12 and the number of your birth month.

Step 4: Divide your answer by 2 and add 3 to find your special number.

Step 5: What letter of the alphabet does this number match? ($a = 1$, $b = 2$, and so on).

Step 6: Now think of a colour which starts with this letter.

a We bet we know what colour you have chosen! (See the 'Answers' in the back of the book).
Try this out on someone else.

b Write all the steps of this trick, using pronumerals, to show how it works.

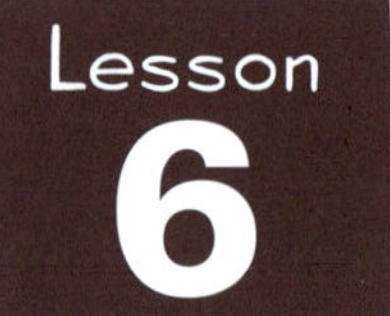

Grouping symbols and the 'Distributive rule'

Mathematical expressions and formulas often use the following grouping symbols:

( ) [ ] { }

They are called

- round brackets or parentheses, for example: $4(2w + 7)$
- square brackets, for example: $[(3a - 1) + (2a + 6)]$
- braces, for example: $\{6 + [(2q - 7) + q^2]\}$

To simplify such expressions as $4(2w + 7)$, in which there is a number in front of the bracket, we use the '**Distributive rule**' to **expand** (remove) the brackets.

The 'Distributive rule' is useful in many problems. See the example below.

Example

Find the area of a rectangle whose width is 5 and whose length is $(m + n)$.

Answer

We can draw a diagram showing this problem.

	m	n
5	A1	A2

Area of the rectangle

$= \text{length} \times \text{width}$

$= 5(m + n)$

The sum of the areas of the two smaller rectangles is:

$$A1 + A2 = 5 \times m + 5 \times n$$

But: Area of the rectangle = sum of the areas of the two smaller rectangles

So: $5(m + n) = 5 \times m + 5 \times n$

$= 5m + 5n$

Example

a Draw a diagram to show $3(a+b)$, and write a product statement to show the result.

b Expand the brackets to simplify $y(7+m)$.

c Expand $t(t+6)$.

d Expand $w(2w-9)$.

Answer

a

	a	b
3	A1	A2

$3(a+b) = 3 \times a + 3 \times b$
$= 3a + 3b$

b $y(7+m) = y \times 7 + y \times m$
$= 7y + my$

c $t(t+6) = t \times t + t \times 6$
$= t^2 + 6t$

d $w(2w-9) = w \times 2w - w \times 9$
$= 2w^2 - 9w$

Remember

We write numbers before pronumerals, and pronumerals in alphabetical order.

EXERCISE

1 Copy and complete these product boxes

a

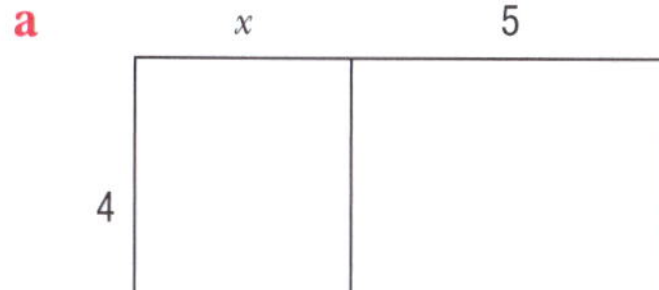

b

	m	n
7		

Remember

$t \times t = t^2$

c

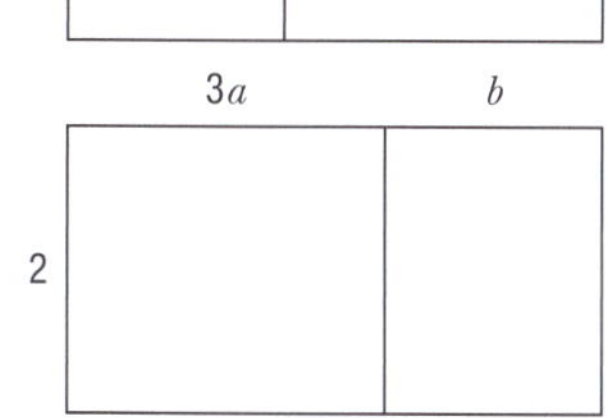

d

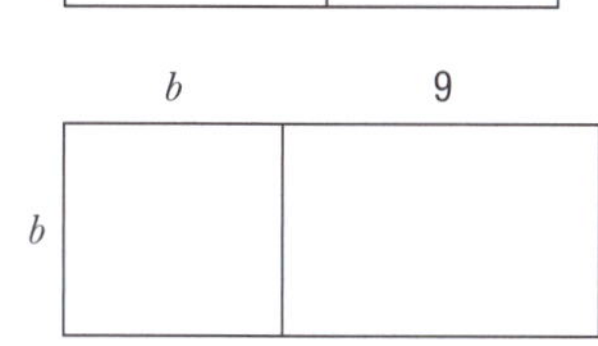

2 Draw product boxes to help you work out the following products:

a $5(w+3)$ **b** $2(t+z)$ **c** $3(4h+k)$ **d** $a(2b+c)$ **e** $2b(b+c)$

3 Expand the brackets:

a $4(m+11) = 4 \times$ __ $+ 4 \times$ __ $=$ **b** $3(5+2y)$ **c** $9(3+y)$

d $8(5+3t)$ **e** $6(2g+3)$ **f** $12(7a+b)$

4 Expand to remove the grouping symbols:

a $7(m-w) = 7 \times$ __ $- 7 \times w =$ __ **b** $6(9a-4)$ **c** $7(1-5x)$

d $4(8p-2q)$ **e** $9(2x-5y)$ **f** $10(6-4w)$

5 Expand:

a $a(p+4)$ **b** $g(c+11)$ **c** $5m(a+m)$ **d** $4r(r-5)$

e $3p(p-11)$ **f** $7a(a-7)$ **g** $2z(z+5y)$ **h** $5b(b-3c)$

6 Expand the brackets and simplify each of the following by collecting like terms:

a $7(m-4)+9m$ **b** $6a+2(a+5)$

c $10(w-1)+3w$ **d** $6(3+t)+2(t-4)$

e $5(a+1)+2(a-2)$ **f** $8(2+p)+3(p+4)$

g $4(t+v)+7$ **h** $5(2+m-n)+7(m+n)$

i $6(a-b+5)+3(a+7)$ **j** $2(j-3k)+4(5j+7k)+5j-12k$

k $a(a-2b)+3a(b-4a)$ **l** $m(3m-4n)+2n(m-5n)$

Remember

Like terms contain the same pronumeral, raised to the same power.

- $3b$ and ^-7b are like terms.
- $3b$ and $2b^2$ are not like terms.

Negative sign before grouping

When there is a negative number in front of brackets, this has the effect of **changing the sign inside the brackets**.

We can see this from the example below. With practice, we can leave out some of the steps, as in part **c** below.

We can apply the 'Distributive rule' to remove brackets and simplify expressions by collecting like terms, as in parts **d**, **e** and **f** below.

Example

Expand:

a ${}^-2(a+4)$

b ${}^-7(3m-1)$

c ${}^-w(3w+y)$

d $14c-3(2c+3)$

e $2(m+7)-5(m-2)$

f $g(g+7)-g(g-3)$

Answer

a ${}^-2(a+4) = {}^-2 \times a + {}^-2 \times ({}^+4)$
$= {}^-2a + ({}^-8)$
$= {}^-2a - 8$

b ${}^-7(3m-1) = {}^-7 \times 3m + {}^-7 \times {}^-1$
$= {}^-21m + 7$

c ${}^-w(3w+y) = {}^-3w^2 - wy$

d $14c - 3(2c+3) = 14c - 6c - 9$
$= 8c - 9$

e $2(m+7) - 5(m-2) = 2m + 14 - 5m + 10$
$= 7m + 24$

f $g(g+7) - g(g-3) = g^2 + 7g - g^2 + 3g$
$= 10g$

Remember

The rules for multiplying directed numbers are:

- **like** signs make a positive
  ${}^+ \times {}^+ = +$
  ${}^- \times {}^- = +$
- **unlike** signs make a negative
  ${}^+ \times {}^- = -$
  ${}^- \times {}^+ = -$

EXERCISE

1 Copy and complete each of the following;

a ${}^-5(x+3) = {}^-5 \times$ __ $+ {}^-5 \times$ __ $=$ __

b ${}^-2(m-7) = {}^-2 \times$ __ $+ ({}^-2) \times$ __ $=$ __

c ${}^-3(c-d) = {}^-3 \times$ __ $+ ({}^-3) \times$ __ $=$ __

d ${}^-8(5-3b) = {}^-8 \times$ __ $+ ({}^-8) \times$ __ $=$ __

e ${}^-2(3m+4n) = {}^-2 \times$ __ $+ ({}^-2) \times$ __ $=$ __

f ${}^-6(a^2-5a) = {}^-6 \times$ __ $+ ({}^-6) \times$ __ $=$ __

2 Remove the grouping symbols:

a ${}^-7(a-4)$ **b** ${}^-3(2m+9)$ **c** ${}^-5(3x+y)$

d ${}^-3(6-4b)$ **e** ${}^-2(m^2+6m)$ **f** ${}^-9(2x^2+3x-4)$

g ${}^-4(6p+q-2)$ **h** ${}^-6(3+7h-2h^2)$ **i** ${}^-7(4mn+3m^2-2n^2)$

3 Expand each of the following. (*Note*: A minus sign by itself in front of the brackets is the same as multiplying the contents of the brackets by ${}^-1$.)

a ${}^-(3m+4n)$ **b** ${}^-({}^-6+7y)$ **c** ${}^-(b-3c)$

d ${}^-(5t+11)$ **e** ${}^-(9+3m-4m^2)$ **f** ${}^-(xy-x^2)$

4 Remove the brackets from each of the following:

a ${}^-m(m+2n)$ **b** ${}^-h(2-3h)$ **c** ${}^-q(3q-9)$

d ${}^-b(4a+b)$ **e** ${}^-h(7h-5)$ **f** ${}^-w(7+5w)$

g ${}^-g(3g+h-6)$ **h** ${}^-t(5t+w-z)$

5 Remove the brackets and simplify:

a $^-4(a+7)-3a$ b $^-8(p-7)+11p$ c $^-2(4z+1)-z$
d $^-3(w-2)-5$ e $^-5(3t+8)-12t$ f $^-9(2m-1)+3m$

6 Remove the brackets and simplify:

a $8t-(t+v)$ b $7p-3(p+2)$
c $^-5a-2(a+3)$ d $4r-5(r-7)$
e $11h-9(h-1)$ f $5x-3(1-x)$
g $^-b(3b+7)-6b+1$ h $^-3p(2p+q)+7qp-5p^2$
i $14m+6mn-2n(m-3n)$ j $8y-6(y+z)-3z$
k $4m^2-3m(2m+7)+11m$ l $7y+6yz-y(3+2z)$

Using the 'Distributive rule'

Lesson **8**

Expressions can contain more than one set of grouping symbols, as in the example below.

Example

Simplify:
$3m(2m-5)+8m(m-1)$.

Answer

$3m(2m-5)+8m(m-1)$
$=6m^2-15m+8m^2-8m$ Using the distributive rule twice.
$=14m^2-23m$ Collecting like terms.

If brackets are placed **inside each other**, we use different bracket symbols to avoid confusion. An example is:

$3\{[5+7(4-1)]\div 2\}$

If we used only the ( ) symbols, the sum would look like this:

$3((5+7(4-1))\div 2)$

It would be hard to know where to begin!

Using the order of operations, and starting by working out the innermost brackets, we get:

$$\begin{aligned}3\{[5+7(4-1)]\div 2\} &= 3\{[5+7(3)]\div 2\}\\ &= 3\{[5+21]\div 2\}\\ &= 3\{[26]\div 2\}\\ &= 3\{13\}\\ &= 39\end{aligned}$$

Applying this to an algebraic expression, we simplify $7\{[a+5(a-1)]+9\}$ like this:

$$\begin{aligned}7\{[a+5(a-1)]+9\} &= 7\{[a+5a-5]+9\}\\ &= 7\{[6a-5]+9\}\\ &= 7\{6a+4\}\\ &= 42a+28\end{aligned}$$

EXERCISE

1 Expand each of the following and collect like terms:

a $5(m+7)+2(m-8)$ b $3a(a-1)+2a(7-a)$
c $9+5(2x-y)+6(2y+x)$ d $4m(m-3n)+2n(m+7n)$
e $a(2a+6)+5a(8-3a)$ f $d(7d-e)+e(3d+e)$
g $8(3w+4x-y)+7(3y-x+w)$ h $9(4w+2-z)+6w+3(2z-w+1)$

Weeks 6, 7, 8 and 9

2 Expand each of the following and collect like terms:

a $3(2m + 6n) - 4(m - 2n)$

b $5(x + 2y) - 3(y - 2z)$

c $9(w - 8) - 3(4w + 5)$

d $2(a + b) - 6(b - c)$

e $3t + 6(t - 4) - 2(5 - t)$

f $7(r - s) - 2(s + t)$

g $3(2a + 5) - 6(a - 11) - 3$

h $5(2m + n - 4) - m(1 - n)$

i $9(5 - 2t) - 6(3t + 1)$

j $^{-}2y(y + 3) - 5y - (y - 8)$

3 Expand and simplify:

a $3[2a + a(3 - a)]$

b $m[4(n + 5) + 2n]$

c $7[3(m - n) + 6n]$

d $2[5(w - y) - 11]$

e $^{-}[3(t + u) + 4(t - u)]$

f $^{-}2[a(3a + b) - b(2a + 3b)]$

4 Write an expression for the perimeter of each of the following:

i using $P = 2(l + b)$

ii in expanded form.

a $a + 3$; 2

b

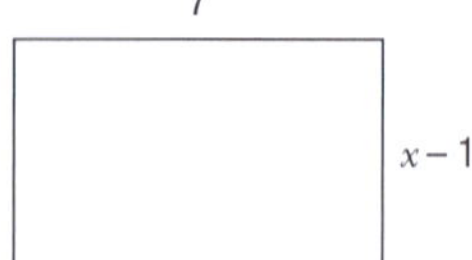

c

d

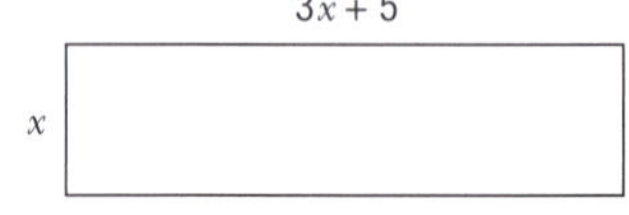

e

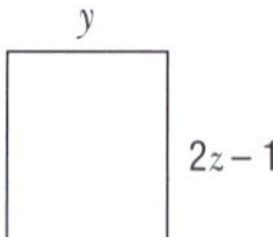

f

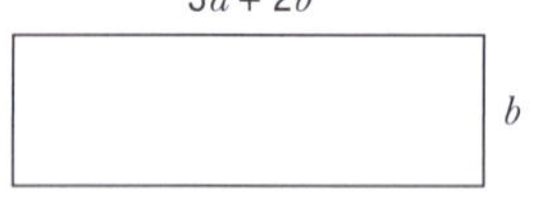

Remember

Area of a triangle = $\frac{1}{2} \times$ base $\times$ height

$A = \frac{bh}{2}$

5 Write an expression for the area of each of the following:

i as a product

ii in expanded form.

a 3; $x + 7$

b

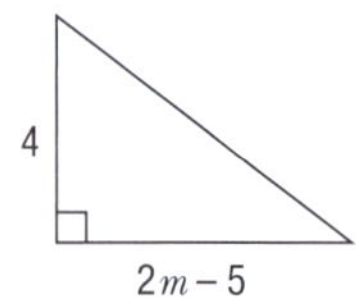

c

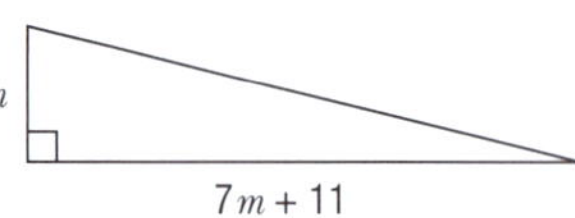

d

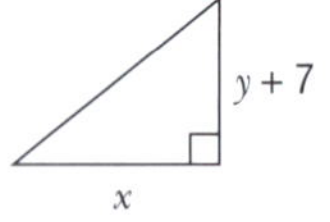

e

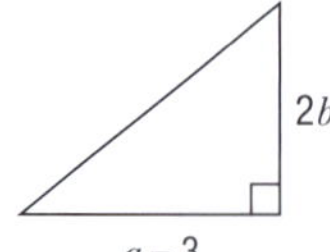

f

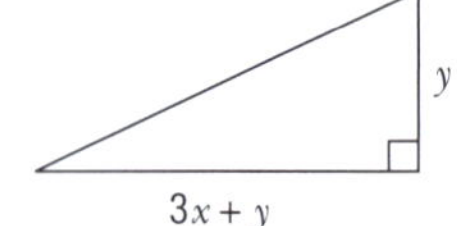

The 'Distributive rule' in reverse

We can use the 'Distributive rule' in reverse to **factorise** an expression. To **factorise** means to write an expression as a product of its factors.

The steps for factorising an algebraic expression are shown in the example below.

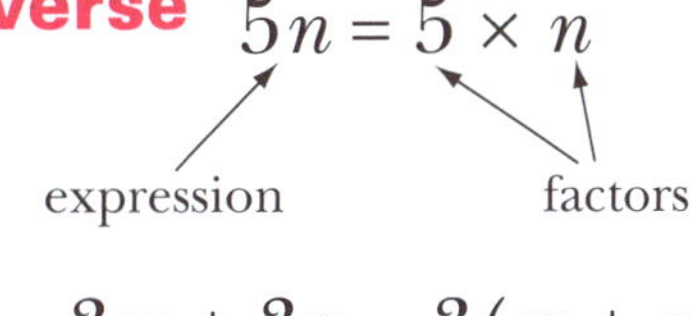

$$3m + 3n = 3(m + n)$$

expression factors

Example

Factorise $7b + 14c$.

Answer

Step 1: Write each term as a product of its factors. $7b + 14c = 7 \times b + 7 \times 2c$

Step 2: Find the HCF of the terms, which is 7. $= 7(\_\_+\_\_)$

Step 3: Write the HCF in front of the brackets and the remaining factors of each term inside the brackets. $= 7(b + 2c)$

Check

If you expand the brackets, you will obtain the original expression:

$7(b + 2c) = 7b + 14c$ **(Correct)**

We say $7(b + 2c)$ is in **factorised form** and $7b + 14c$ is in **expanded form**.

Example

Factorise:

a $ab + ad$

b $p^2 + 6p$

c $10w + 15wz$

d $^-4m - 8n$

e $a^2 - a$

Answer

a $ab + ad = a(\_\_\_\_\_\_)$
$= a(b + d)$
The highest common factor of ab and ad is a.

b $p^2 + 6p = p(\_\_\_\_\_\_)$
$= p(p + 6)$
p^2 is $p \times p$, so the HCF of p^2 and $6p$ is p.

c $10w + 15wz = 5w(\_\_\_\_\_\_)$
$= 5w(2 + 3z)$
The HCF is $5w$. The remaining factors are 2 and $3z$.

d $^-4m - 8n = {}^-4(\_\_\_\_\_\_)$
$= {}^-4(m + 2n)$
The HCF is $^-4$. *Note*: $^-8 = {}^-4 \times {}^+2$.

e $a^2 - a = a(a - 1)$
Note: $a = 1 \times a$, so removing (dividing by) a factor of 'a' leaves 1.

Remember

Taking out (dividing by) a **negative** factor gives the **opposite** sign inside the brackets.

EXERCISE

1 Write 'True' or 'False' for each of the following:

a $4m + 16 = 4(m + 12)$

b $4m + 16 = 4(m + 4)$

c $20 + 15x = 5(4 + 3x)$

d $20 + 15x = 5(4 + 15x)$

e $a(m + n) = am + n$

f $a(m + n) = am + an$

2 Match each expression below with its correct factorised form:

a $9m - 3n$ **A** $^-3(n + 3m)$
b $3n + 9m$ **B** $3(n + 3m)$
c $9m^2 - 3m$ **C** $3(3m - n)$
d $^-3n - 9m$ **D** $3m(3m - 1)$

3 Copy and complete each of the following:

a $8m + 16 = 8(\_\_ + \_\_)$ **b** $4a + 10b = \_\_(2a + 5b)$
c $7k - 14l = \_\_(k - 2l)$ **d** $5p - 10q = 5(\_\_ - \_\_)$
e $am + an = a(\_\_ + \_\_)$ **f** $20x + 15xy = 5x(\_\_ + \_\_)$
g $6p - 30q = \_\_(p - \_\_q)$ **h** $5am + 25cm = \_\_(a + 5c)$
i $cd - 8d = d(\_\_ - \_\_)$ **j** $8m + 8 = 8(\_\_ + \_\_)$
k $w^2 - 3w = w(\_\_ - 3)$ **l** $b - 3b^2 = \_\_(1 - 3b)$
m $p^2 + pq = p(\_\_ + \_\_)$ **n** $6y^2 + 9yz = 3y(\_\_ + \_\_)$
o $7t - 21t^2 = \_\_(1 - 3t)$ **p** $2a^2b + 4bc = 2b(\_\_ + \_\_)$

4 For each of the following:

i state the HCF **ii** factorise the expression.

a $3x + 9$ **b** $3s - 6$ **c** $3x + xc$ **d** $12w + 9ws$
e $16c - 16d$ **f** $10g + 5pg$ **g** $12 + 6x$ **h** $ab + bc$
i $12x - 6$ **j** $v + 2xv$ **k** $3x - 9$ **l** $3ca - bc$
m $4f - 4$ **n** $4dv - 2d$

5 Factorise:

a $2a + ac$ **b** $4p - 6pq$ **c** $x^2 + xy$ **d** $8h + 4h^2$
e $3w^2 - 6w$ **f** $v^2 + 2vz$ **g** $2d - 8d^2$ **h** $ab^2 + b^2c$
i $7x^2 - 6x$ **j** $f^2 + 2fg$ **k** $x^2 - 9xy$ **l** $ac - ac^2$
m $j^2 - 4jk$ **n** $2k^2 - 6k$

6 Copy each of the following and complete the factorisation:

a $^-6m + 12n = {^-6}(\_\_\_\_\_)$ **b** $^-2x - 12 = {^-2}(\_\_\_\_\_)$
c $^-4h^2 + 8h = {^-4h}(\_\_\_\_\_)$ **d** $^-3p + 3 = {^-3}(\_\_\_\_\_)$
e $^-3a - 6ab = {^-3a}(\_\_\_\_\_)$ **f** $^-6y^2 + 9y = {^-3y}(\_\_\_\_\_)$

Challenge

Reversing digits puzzle

Step 1: Choose and write a three-digit number.
Step 2: Now reverse the digits and write that number.
Step 3: Find the difference between the two numbers.
Step 4: Work out the prime factors of your answer.
Step 5: Select two other three-digit numbers and repeat Steps 1 to 4 above for each of them.

a What can you say about the factors in all three cases?
b Prove this general result, using *a*, *b* and *c* as the three digits. (*Hint*: The first three-digit number will have the value $100a + 10b + c$. What will the value be when the digits are reversed?)

Solving equations

Lesson **10**

Solving an equation is like 'undoing' a puzzle. Can you answer the following question?

I think of a number, double it and get 18. What number am I thinking of?

The answer to this is 9, and we can probably work that out in our heads, but we can also show the steps used for finding the answer in the way shown below.

Using the balance method

The two sides of an equation are like the two sides of a balance or scales. To keep the equation balanced, we must carry out the same operation **on both sides** of the equation.

If n represents the number, we can write the equation as:

$2n = 18$ To get n by itself on the left-hand side of the equation, we need to divide by 2. So we need to divide both sides by 2 to keep the equation balanced.

So: $\frac{2n}{2} = \frac{18}{2}$ Dividing both sides by 2.

$n = 9$

Check

We can verify (prove) that our answer above is correct by showing that it fits the original equation.

To do this, we substitute the answer we found for the pronumeral into the original equation, working out the left-hand side (LHS) and right-hand side (RHS) separately:

$2n = 18$

LHS $= 2n$ Substitute our answer of 9 for n.

$= 2 \times 9$

$= 18$ RHS = 18 (given)

LHS = RHS ∴ Our answer is correct.

Example

Solve $p + 4 = 1$.

Answer

$p + 4 = 1$

$p + 4 - 4 = 1 - 4$ Subtract 4 from both sides.

$p = {}^{-}3$

How do we know what operation to carry out? To 'undo' an equation we need to carry out the opposite operations to those used in the equation.

Operation	Opposite
Add (+)	Subtract (–)
Subtract (–)	Add (+)
Multiply (×)	Divide (÷)
Divide (÷)	Multiply (×)

EXERCISE

1 **i** Solve each of the following equations, showing your working.
ii Check that the answer is correct each time, setting out LHS and RHS.

a $n + 4 = 6$ **b** $n - 6 = 4$ **c** $n + 13 = 36$
d $n - 12 = 13$ **e** $n + 10 = 2.8$ **f** $n - \frac{3}{4} = -\frac{1}{6}$
g $n - 11 = -5.5$ **h** $d + 1.9 = -2$ **i** $8 + t = 9$
j $-\frac{3}{4} + r = 4$ **k** $2 + x = \frac{1}{2}$ **l** $-2.3 + c = -13.4$

2 Write an equation for each of the following statements, using g to represent the unknown number, and then solve the equation to find g.
- **a** 5 is added to a certain number and the result is 46.
- **b** If 23 is subtracted from a certain number, the result is 35.
- **c** 'In 24 years time I will be 99 years old,' said the old man. How old is he now?
- **d** Eight more than a certain number equals 5.
- **e** When a number is added to 12 the result is 9.
- **f** The sum of a certain number and 11 is 3.
- **g** Seven less than a certain number is −1.
- **h** Take 8 from a certain number and the result is $4\frac{1}{2}$.

3 **i** Solve each of the following equations, showing your working.
ii Check that the answer is correct each time, setting out LHS and RHS.

a $3n = 24$ **b** $4c = -1.2$ **c** $10r = 35$ **d** $-6w = 70$
e $-3x = -10$ **f** $7t = 2$ **g** $-5y = -17$ **h** $\frac{x}{3} = 15$
i $\frac{w}{4} = 2$ **j** $\frac{m}{9} = 0$ **k** $\frac{a}{7} = -5$ **l** $\frac{k}{7} = -1.2$
m $\frac{p}{5} = -3.6$

4 Set up equations for each of these and then solve them:
- **a** I think of a number, divide it by 5, and get an answer of 7. What is the number?
- **b** Six times a certain number is 21. What is the number?
- **c** Five boxes contain a total of 230 matches. How many matches are there in one box?
- **d** One-quarter of a number is −5. Find the number.
- **e** Double a number is 7. What is the number?
- **f** Half a number is $\frac{1}{4}$. What is the number?
- **g** Seven times a number is −4.2. What is the number?
- **h** The product of a number and 11 is 8. What is the number?

5 To make up your own 'mystery number' think of a number and carry out an operation (+, –, × or ÷) on it. Write six puzzles similar to those in Questions **2** and **4** about six 'mystery numbers'. Exchange your puzzles with a partner. See who can find the six correct answers the quickest.

Two-step and three-step equations

Lesson 11

Many problems involve several steps and more than one arithmetic operation. Consider the following problem:

I think of a number, double it and add 4, getting a result of 30. What is the number?

We can 'build' a written version of the problem:

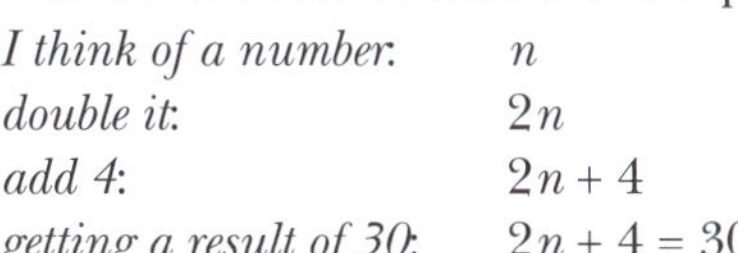

I think of a number:	n
double it:	$2n$
add 4:	$2n + 4$
getting a result of 30:	$2n + 4 = 30$

To solve the equation we use **inverse operations** to reverse the steps used to create the problem:

$2n + 4 - 4 = 30 - 4$ Subtract 4 from both sides.

$2n = 26$ By simplifying.

$\frac{2n}{2} = \frac{26}{2}$ Divide both sides by 2.

$n = 13$

Check

LHS $= 2n + 4$

$= 2 \times 13 + 4$

$= 30 =$ RHS Therefore, our answer is correct.

Example

Solve $\frac{x}{3} + 11 = 5$.

Answer

$\frac{x}{3} + 11 - 11 = 5 - 11$ Subtracting 11 from both sides.

$\frac{x}{3} = -6$

$\frac{x}{3} \times 3 = -6 \times 3$ Multiplying both sides by 3.

$x = -18$

EXERCISE

1 Choose the correct answer (**A**, **B**, **C** or **D**) each time:

a Which of the following is the value of m in the equation $4m + 7 = 11$?
A 4.5 **B** 16 **C** 1 **D** 0 **E** -1

b Which of the following is the solution to the equation $-3x + 5 = 11$?
A 2 **B** -2 **C** 9 **D** 19 **E** -0.5

c Which of the following is the solution to the equation $\frac{m}{2} + 1 = 7$?
A 16 **B** 13 **C** 3 **D** 15 **E** 12

d Which of the following is the solution to $15 - 4z = -19$?
A -1 **B** 1 **C** -8.5 **D** 8.5 **E** -8

e Which of the following is the answer to $15 = -3w + 1$?
A $5\frac{1}{3}$ **B** $-5\frac{1}{3}$ **C** -11 **D** $-4\frac{2}{3}$ **E** $4\frac{2}{3}$

2 Solve each of the following equations, showing your working neatly:

a $2n + 5 = 15$	**b** $4c + 3 = 27$	**c** $7 + 5t = 32$	**d** $2x + 10 = -34$
e $7 - d = 15$	**f** $-9x - 23 = 24$	**g** $15 + 3w = 15$	**h** $27 - 3x = 18$
i $45 + 3x = 21$	**j** $\frac{c}{4} + 13 = 2$	**k** $\frac{d}{5} + 2 = 7$	**l** $4 + \frac{x}{2} = \frac{3}{2}$
m $3 - 4x = 31$	**n** $15 - 7x = 8$	**o** $6 + 13c = 6$	**p** $7x - 14 = 0$
q $\frac{x}{5} - 3 = 9$	**r** $25 - 8s = -7$	**s** $\frac{d}{2} + 2.5 = 7.5$	**t** $36 - 6x = 0$

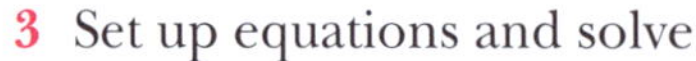

3 Set up equations and solve:

- **a** Twice a certain number plus 5 equals 13. What is the number?
- **b** The sum of 7 and four times a certain number is 2. What is the number?
- **c** Half a certain number is added to 8 to give 3. What is the number?
- **d** Two is added to three times a number, giving a result of 11. What is the number?
- **e** If a certain number is halved and then 5 is added, the result is 17. What is the number?
- **f** Three more than two times a number is $^{-}1$. What is the number?
- **g** One-half of a number plus 8 equals 4.5. What is the number?
- **h** When double a number is subtracted from 9 the result is 1.4. What is the number?
- **i** When 5 is added to one-eighth of a certain number, the result is 0. What is the number?
- **j** The sum of 14 and five times a certain number is 5. What is the number?

4 To make up your own 'mystery number' think of a number and carry out a combination of two operations (+, –, × or ÷) on that number. Write six puzzles similar to those in Question **3** about six 'mystery numbers'. Exchange your puzzles with a partner. See who can find the six correct answers the quickest.

Equations with grouping symbols

We have seen previously that brackets are expanded by multiplying **each** term inside the bracket by the term outside the bracket.

For example:

$$\begin{aligned} 3(2x+4) &= 3 \times 2x + 3 \times 4 \\ &= 6x + 12 \end{aligned}$$

We can use this rule when solving equations containing brackets.

Example

Solve $3(2x + 4) = 18$.

Answer

$3(2x+4) = 18$	
$3 \times 2x + 3 \times 4 = 18$	Expand the brackets on the left-hand side.
$6x + 12 = 18$	Simplify the result.
$6x + 12 - 12 = 18 - 12$	Subtract 12 from both sides.
$6x = 6$	Simplify.
$\frac{6x}{6} = \frac{6}{6}$	Divide both sides by 6.
$x = 1$	

It is possible to solve this equation using a different order of steps:

$3(2x+4) = 18$	
$\frac{3(2x+4)}{3} = \frac{18}{3}$	Divide both sides by 3.
$2x + 4 = 6$	
$2x + 4 - 4 = 6 - 4$	Subtract 4 from both sides.
$\frac{2x}{2} = \frac{2}{2}$	Divide both sides by 2.
$x = 1$	

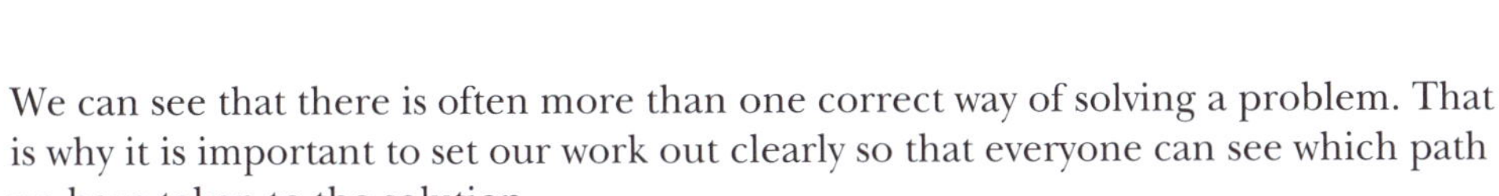

We can see that there is often more than one correct way of solving a problem. That is why it is important to set our work out clearly so that everyone can see which path we have taken to the solution.

Equations with fractions

Example

Solve $\frac{5y-4}{12} = 3$.

Answer

$\frac{5y-4}{12} \times 12 = 3 \times 12$ Multiply both sides by 12 to remove the denominator.

$5y - 4 = 36$

$5y - 4 + 4 = 36 + 4$ Add 4 to both sides.

$5y = 40$

$\frac{5y}{5} = \frac{40}{5}$ Divide both sides by 5.

$y = 8$

When there is more than one fraction in the equation, we multiply by the **lowest common denominator** to remove all denominators.

Example

Solve $\frac{x+1}{3} + \frac{1}{2} = \frac{7}{2}$.

Answer

$\frac{x+1}{3} \times 6 + \frac{1}{2} \times 6 = \frac{7}{2} \times 6$ Multiply both sides by 6, the LCD of 3 and 2.

$2(x+1) + 3 = 21$

$2x + 2 + 3 = 21$ Remove brackets.

$2x + 5 = 21$ Simplify.

$2x + 5 - 5 = 21 - 5$ Subtract 5 both sides.

$\frac{2x}{2} = \frac{16}{2}$ Divide both sides by 2.

$x = 8$

EXERCISE

1 Two students, Ivan and Imelda, solved the equation $2x + 1 = 3(x - 4) + 6$. Ivan's answer was $x = 4$, but Imelda's answer was $x = 7$. Check both answers, using LHS and RHS, to see who is correct.

2 Solve each of the following equations and check your answer showing LHS and RHS.

a $2(c + 3) = 8$ b $9(n + 7) = 18$ c $18(x - 1) = 36$

d $3(2t - 1) = {}^-6$ e ${}^-4(2d + 2) = 4$ f $5(w - 9) = {}^-5$

g $9(4x - 3) = 0$ h ${}^-5(2m + 1) = 10$ i $3(t - 2) = {}^-9$

j $5(2x - 3) + 7 = 2$ k $4 + 2(3m - 1) = {}^-4$ l $4(y - 2) + 2(3 - y) = 0$

3 Solve:

a $\frac{v+3}{2} = {}^-3$ b $\frac{5x-6}{3} = 5$ c $\frac{2c-3}{6} = 0$ d $\frac{m-1}{2} = {}^-1$

e $\frac{w+5}{3} = {}^-2$ f $\frac{7-x}{9} = 2$ g $\frac{6+v}{5} = 5$ h $\frac{t-11}{4} = {}^-2$

i $\frac{6-y}{2} = 5$ j $\frac{3(m-2)}{2} = 6$ k $\frac{4(2k-1)}{5} = 1$ l $\frac{3(d+2)}{7} = 2$

4 Write an equation to represent each of the following and solve the equation to find the unknown number:
   a I think of a number, add one, double the result and get an answer of 12. What is the number I thought of?
   b If I subtract five from a number and then triple the result I get 18. What is the unknown number?
   c If I add 7 to a number and then multiply by 8, I get 32. What is the number?
   d I subtracted 4 from a number, divided that result by 3 and got 11. What is the number?
   e I added 2 to a number, divided by 6 and got $^-5$. What is the number?
   f I think of a number, multiply by 7, subtract 2, divide the result by 3 and get 0. What is the number I thought of?
   g One-quarter of the difference between a number and 12 is 2.5. What is the number?
   h Twice the sum of 7 and a number is $^-3$. What is the number?
   i The sum of double a number and 7 is halved to give $^-3$. What is the number?
   j When I subtract 11 from a number, triple that result and add 2, I get 23. What is the number?

5 Write three 'mystery number' puzzles similar to those in Question 4 and exchange them with a partner. See who can find the correct answers the quickest.

Challenge

a *Try this*: Think of a number (between 1 and 20, say).
Square it.
Add this to the original number.
Divide by the number you started with.
Add 20.
Subtract the number you started with and divide by 3.
I bet the answer is 7.

b Use a pronumeral to represent the number you started with and write an expression for each step of the trick to show how this works.

c Make up your own 'mind reading' puzzle and try it on a partner.

Lesson 13

Equations with repeated pronumerals

Sometimes the pronumeral in an equation occurs more than once. For example, if we are given that the perimeter of this irregular pentagon is 20 units, we can write the equation:

$2x + (2x + 1) + 2 + (x + 1) + 3x = 20$

In this case, x occurs several times.

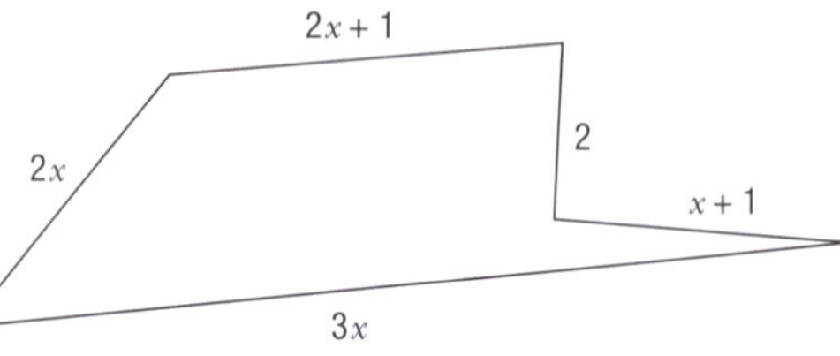

We can solve this equation as follows:

$$2x + (2x + 1) + 2 + (x + 1) + 3x = 20$$

$$2x + 2x + 1 + 2 + x + 1 + 3x = 20$$ Removing brackets.

$$8x + 4 = 20$$ Collecting like terms.

$$8x + 4 - 4 = 20 - 4$$ Subtracting 4 from both sides.

$$8x = 16$$

$$\frac{8x}{8} = \frac{16}{8}$$ Dividing both sides by 8.

$$x = 2$$

More difficult equations have pronumerals on both sides of the equation. To solve this type of equation, we need to remove the pronumeral on one side as shown in the example below.

Example

Find x if $9x + 4 = 2x - 3$.

Answer

We can remove $2x$ by subtracting it from both sides:

$$9x + 4 - 2x = 2x - 3 - 2x$$ Subtract $2x$ from both sides.

$$7x + 4 = {}^-3$$ Simplify.

$$7x + 4 - 4 = {}^-3 - 4$$ Subtract 4 from both sides.

$$7x = {}^-7$$ Simplify.

$$\frac{7x}{7} = \frac{{}^-7}{7}$$ Divide both sides by 7.

$$x = {}^-1$$

EXERCISE

1 Solve each of the following by finding the value of the pronumeral. Include a check of each answer using LHS and RHS.

a $3x + 4 + x = 20$	**b** $2n - 6 + 3n = 4$	**c** $5e - 8 - 4e = 0$
d $15 - 4r + r = 6$	**e** $10x + 5 + x = {}^-8$	**f** $3m + 5(m - 2) = 7$
g $7y + 3(2 + 4y) = 0$	**h** ${}^-2(p + 7) + 8p = {}^-20$	**i** $11 - 3(2y - 3) = {}^-10$
j $4p + 7 - 2(p - 1) = 10$	**k** $2x - 3(5 + 4x) = {}^-2$	

2 Solve each of the following.

a $19 - 3d = 4 - 4d$	**b** $3c + 9 = 7 - c$	**c** $2h + 11 = 5h - 7$
d $7x - 1 = 5x$	**e** $4 - 4x = 8 + 4x$	**f** $6b - 4 = 2b + 4$
g $10 + 4d - 5 = d + 2$	**h** $18 - 6t = 4 + t$	**i** $s + 4 + 5s = 22 - 3s$
j $2a + 3(a - 1) = 12$	**k** $3(3x - 4) + x = {}^-2$	**l** $3(x + 4) = 2(x - 3)$
m $5(3x - 12) = 7x - 12$	**n** $3(x + 1) + 2(x - 1) = x + 4$	
o ${}^-2(3 - x) + 3x = 2x + 3$	**p** $\frac{5(y - 3)}{2} = y$	**q** $\frac{3(4 - x)}{2} = 6x + 1$

Remember

The opposite sides of a rectangle are equal in length.

3 Find the value of x in each of the following rectangles:

a

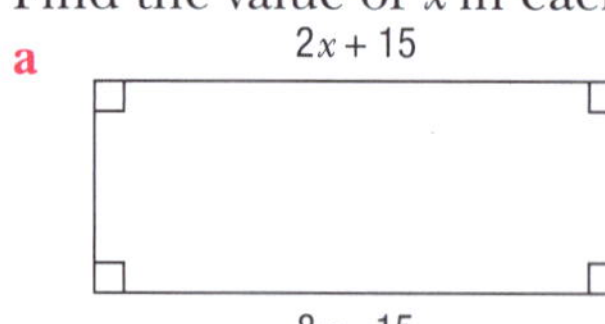

b

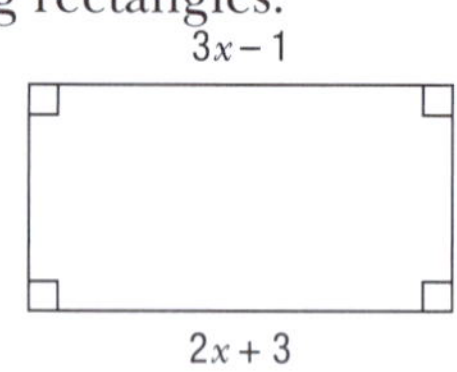

c

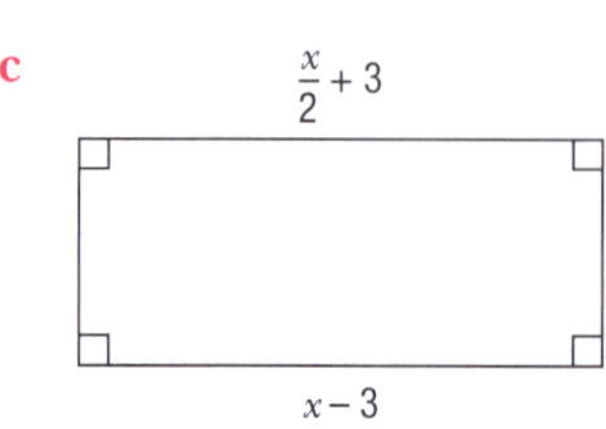

d

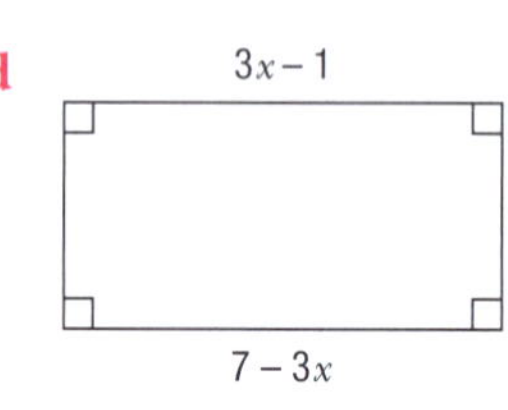

e

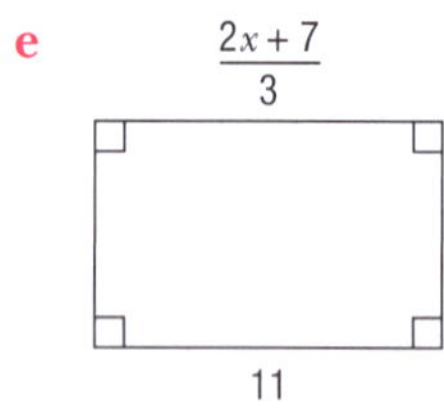

f

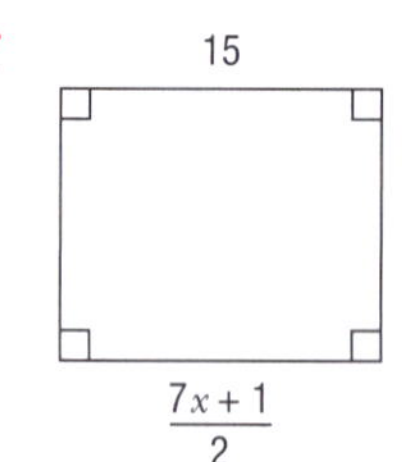

Lesson 14

Using equations to solve problems

Many problems described in words can be solved using equations. The steps we need to follow are listed below:

Step 1: Identify what we are being asked to find.

Step 2: Use a pronumeral to represent what we are being asked to find.

Step 3: Write an equation to match the problem statement.

Step 4: Solve the equation to find the value of the pronumeral.

Step 5: Give our answer(s) in words, including units.

Example

Joseph is twice as old as his sister, Leila. Their ages add to 15. How old is Leila?

Answer

Step 1: We are being asked to find Leila's age.

Step 2: Let Leila's age be x years.

Step 3: We have been told that Joseph is twice as old as Leila, so his age can be represented by $2x$ and our equation is $2x + x = 15$.

Step 4:

$$2x + x = 15$$

$$3x = 15 \quad \text{Simplify.}$$

$$\frac{3x}{3} = \frac{15}{3} \quad \text{Divide both sides by 3.}$$

$$x = 5$$

Step 5: Leila is 5 years old.

EXERCISE

1 If m = Mary's age in years:
  a write an expression for her age 5 years from now
  b write the equation corresponding with the statement, 'Four times Mary's age 5 years from now is 48'.
  c solve the equation to find Mary's age.

2 If I add K4 to 3 times the contents of my pocket, I would have K13.
  a Write an equation for this problem, using n to represent the number of kina in my pocket.
  b Solve the equation.
  c How much is in my pocket?

3 Polape said, 'If you triple my age today and add 10, you get one hundred.' Write an equation to solve this. How old is Polape?

4 Ari subtracted 3 from the number of fish he caught, divided that result by 5 and got 7. How many fish did Ari catch?

5 A flock of parrots was sitting in a tree. Five more joined them. Then half the total flew away, leaving 7 parrots. How many were in the flock?

6 A man is three times as old as his son and his son is five years older than the man's daughter.
  a If the daughter's age is represented by x, write an expression for:
    i the son's age   ii the father's age.
  b If their ages add to 55, how old is:
    i the daughter?   ii the son?   iii the father?

7 A triangle has angles of $(2m + 20)°$, $3m°$ and $(m + 70)°$. Find the value of m.

8 A crate of mangoes weighs 55 kilograms. If the crate weighs 7 kg more than the mangoes, how much do the mangoes weigh?

9 I think of a number, add 21 and divide the total by 7. If the answer is 5, what number am I thinking of?

10 The 20 students in a class are divided into three groups for an assignment. The first group has five more students than the second group, while the second group has twice as many as the third group. Find the number of students in each group.

11 Two brothers catch 35 fish. If one brother catches half as many again as the other brother, how many fish does each brother catch?

12 Mari bought 6 lollies and was given K1.40 change from her K5 note. How much was each lolly?

13 Three friends walked home from school. Yawie walked three times as far as Luania, while Luania walked one-quarter as far as Didya. If they walked 1 kilometre between them, how far did each person walk, in metres?

14 A box of 50 pencils contains only blue, red, green and yellow pencils. It has twice as many green pencils as blue pencils, four more red pencils than yellow pencils, and two more blue pencils than yellow pencils. How many are there of each colour?

Did you know?

The ancient Egyptians recorded this and similar problems in the Rhind papyrus, which was written around 1650 BC: A heap (that is a whole amount) and its seventh equals 19. What is the size of the heap?

1 a Write an equation for this problem, using h to represent 'a heap'.
  b Solve the problem.
  c Check your solution.

2 Here's another problem. (In this problem, 'Aha' represents the unknown, or x.) Aha, its heap and its fifth make 22. What is the value of Aha?

Remember

'Consecutive' means 'following one another'. Examples of **consecutive numbers** are 7 and 8, or 21 and 22.

15 Two consecutive numbers add to 51. Find the numbers.

16 When five consecutive numbers are added, their total is 140. What are the numbers?

17 Moses collects 7 more coconuts than Caleb, and Caleb collects twice as many as Paulius. If they collect 42 coconuts among them, how many does each collect?

18 Rose is 8 years older than Vera, who is twice as old as Keely. If their ages total 48, how old are they?

19 Four numbers, A, B, C and D, add to 50. Find their values given that B is 6 more than A, A is one-quarter of D, and D is three times C.

20 A group of inventors shared in a big prize. The first inventor won K1000. The second won one-third of what remained. The third won K5000 and the fourth won the same as the first. There was no prize money left after they had each taken their winnings. What was the value of the prize?

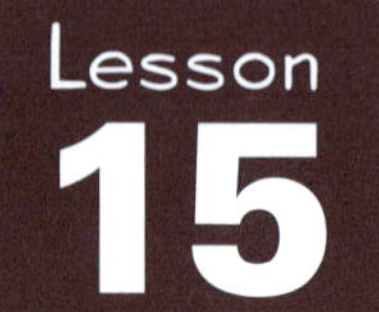

Lesson 15 Using formulas

We have already used formulas (or formulae) for many calculations, such as working out the area of a triangle using $A = \frac{1}{2}b \times h$.
A formula is a rule that shows the relationship between two or more variables, represented by pronumerals. Formulas are not just useful for mathematicians. They are constantly being used in everyday life. For example we can calculate interest at the bank, work out the dosage of medicine at the clinic, or navigate an aeroplane by substituting values into the right formula.

Example

The formula for calculating the perimeter of a rectangle is:

$P = 2(l + w)$ where P = perimeter, l = length and w = width

Find the perimeter of a rectangle with a length of 3.8 metres and a width of 1.2 metres.

Answer

$P = 2(l + w)$
$= 2(3.8 + 1.2)$
$= 2(5)$
$= 10$

So the perimeter is 10 metres.

Example

The approximate time taken for one complete swing of a pendulum can be calculated using the formula:

$T = \frac{1}{5}\sqrt{L}$ where T is the time in seconds and L is the length of the pendulum in centimetres.

How long will one swing of the pendulum take, if the length of the pendulum is 64 cm?

Answer

Substituting 64 for L into the formula gives:

$T = \frac{1}{5}\sqrt{64}$
$= \frac{1}{5} \times 8$
$= 1.6$

So it will take 1.6 seconds for one complete swing of the pendulum.

EXERCISE

1 Sae is travelling to the United States for a year. He knows that the weather where he is going is very hot in summer and very cold in winter, but the temperatures are measured in degrees Fahrenheit, not degrees Celsius which he uses at home. The formula relating the temperature in degrees Celsius to degrees Fahrenheit is:

$$F = 1.8C + 32$$

Use the formula to find the number of degrees Fahrenheit equal to each of these temperatures in degrees Celsius, giving your answer correct to one decimal place:

a 40°C **b** 22°C **c** −4°C **d** 0°C
e 32°C **f** −7°C **g** 45°C **h** 18°C

2 Kenime is using an old recipe book which says to set the oven temperature at 350°F. The formula for converting °F to °C is:

$$C = \frac{5}{9}(F - 32).$$

a At what temperature in °C (to the nearest degree) should Kenime set the oven?

b Kenime decides to write a list of some other cooking temperatures and paste it into her recipe book. Calculate the equivalent Celsius temperatures for each of the following, to the nearest degree:

i 250°F **ii** 300°F **iii** 400°F

3 Runners need good lung capacity so that they can get plenty of oxygen while running. The formula for lung capacity, *C*, in millilitres, is related to the runner's height, *h*, in centimetres, but is different for males and females.

a The formula for males is $C = h \times 23$. Find Leon's lung capacity if he is 150 cm tall.

b The formula for females is $C = h \times 20$. Find Didya's lung capacity if she is 160 cm tall.

c Who would have a greater lung capacity, and by how many millilitres: a man 1.6 m tall or a woman 1.75 m tall?

4 The stopping distance, *D*, of a car, in metres, is related to the speed, *V*, at which the car is travelling in kilometres per hour, by the formula:

$$D = \frac{V}{4} + \frac{V^2}{160}$$

a Calculate the stopping distances for cars travelling at:

i 40 km/h **ii** 60 km/h **iii** 80 km/h **iv** 100 km/h

b A driver in a 60 km/h zone is speeding, travelling at 70 km/h. How many more metres will he need to come to a stop than a driver who is travelling at 60 km/h?

c Testing shows that, when it is raining, a car's stopping distance is half as much again as the distance given by the formula. Find the stopping distance in the rain for a car travelling at 50 km/h, to the nearest metre.

5 The distance a person can see from a point above the Earth's surface is limited because of the Earth's curved surface. The formula for measuring the distance, D, **in kilometres** to the horizon, which is visible from a point h **metres** above the ground is given by:

$$D = 3.56 \times \sqrt{h}$$

a Jamie is in a helicopter 625 metres above the ground. Use the square root table on pages 391–2 to help you find how far he can see to the horizon.

b Use the square root table on pages 391–2 to find the distance to the horizon (correct to two decimal places) from each of these famous landmarks, given their heights:

i	the Eiffel Tower, Paris, France	300 metres
ii	the Empire State Building, New York, USA	381 metres
iii	Notre Dame Cathedral, Paris, France	142 metres
iv	Mount Everest, on the border of Nepal and Tibet	8848 metres

Mt Everest.

Lesson 16

Changing the subject of a formula

Sometimes we want to find the value of a part of the formula which is not the subject (or which is not the pronumeral that is by itself on one side of the equation). For instance, the formula connecting speed, distance and time is usually written:

$$\text{Speed} = \frac{\text{distance}}{\text{time}} \qquad \text{or} \qquad s = \frac{d}{t}$$

We can **change the subject** of the formula to d, using the same steps we use for solving equations:

$$s = \frac{d}{t}$$

$$s \times t = \frac{d}{t} \times t$$

$$s \times t = d$$

So: $\quad d = s \times t$

Suppose we are travelling at 70 km/h and want to calculate how far we have travelled in 4 hours.

Then: $d = 70 \times 4$

$= 280$

We have travelled 280 kilometres.

Example

Change the subject of the formula $A = \pi r^2$ to r.

Answer

$A = \pi r^2$

$\frac{A}{\pi} = \frac{\pi r^2}{\pi}$ Divide both sides by π.

$\frac{A}{\pi} = r^2$

$\sqrt{\frac{A}{\pi}} = \sqrt{r^2}$ Find the square root of both sides.

$\sqrt{\frac{A}{\pi}} = r$

EXERCISE

1 **i** Change the subject of each formula below to the pronumeral shown in brackets.

ii Evaluate the formula using the given values.

a	**i** $P = 2l + 2w$	(w)	**ii** $P = 36$ and $l = 12$
b	**i** $V = \frac{Ah}{3}$	(h)	**ii** $V = 5000$ and $A = 750$
c	**i** $A = s^2$	(s)	**ii** $A = 100$
d	**i** $A = \frac{bh}{2}$	(b)	**ii** $A = 48$ and $h = 12$
e	**i** $A = 2\pi rh$	(r)	**ii** $A = 880$ and $h = 35$
f	**i** $I = \frac{PRT}{100}$	(T)	**ii** $I = 780$, $P = 8000$ and $R = 6.5$
g	**i** $C = \pi d$	(d)	**ii** $C = 88$ (use $\pi \approx \frac{22}{7}$)
h	**i** $v = u + at$	(a)	**ii** $v = 460$, $u = 10$ and $t = 15$
i	**i** $A = 2\pi r(r + h)$	(h)	**ii** $A = 1188$, $r = 7$ (use $\pi \approx \frac{22}{7}$)
j	**i** $B = \frac{w}{h^2}$	(w)	**ii** $B = 20$, $h = 1.5$
k	**i** $B = \frac{w}{h^2}$	(h)	**ii** $B = 24$, $w = 61.44$
l	**i** $E = \frac{1}{2}mv^2$	(m)	**ii** $E = 11\,700$ and $v = 300$
m	**i** $E = \frac{1}{2}mv^2$	(v)	**ii** $E = 11\,250$ and $m = 25$

REVISION AND ASSESSMENT

Directed numbers

Choose the correct answer (**A**, **B**, **C** or **D**) for each of Questions **1** to **10**.

1 Which of the following statements is true?

A $^{-}7 > {}^{-}4$ **B** $^{-}4 < {}^{-}7$ **C** $^{-}4 > 7$ **D** $^{-}4 > {}^{-}7$

2 Which of the following is the largest number out of $^{-}5$, $^{-}2$, $^{-}7$ and $^{-}1$?

A $^{-}7$ **B** $^{-}2$ **C** $^{-}1$ **D** -5

3 Which of the following is the smallest number in the set $\{^{-}8, 0, 8, {}^{-}10\}$?

A $^{-}10$ **B** 0 **C** $^{-}8$ **D** 8

4 Which of the following is the number 6 more than $^{-}5$?

A 1 **B** $^{-}1$ **C** 11 **D** $^{-}11$

5 Which of the following equals $^{+}7 + {}^{-}3$?

A $^{+}10$ **B** $^{-}10$ **C** $^{-}4$ **D** $^{+}4$

6 Which of the following is the sum of $^{-}4$ and $^{-}7$?

A $^{+}11$ **B** $^{-}11$ **C** $^{+}3$ **D** $^{-}3$

7 Which of the following is the product of $^{-}3$ and $^{+}6$?

A $^{+}3$ **B** $^{-}9$ **C** $^{-}18$ **D** $^{-}2$

8 Which of the following is the quotient of $^{-}8$ and $^{-}2$?

A 16 **B** $^{-}4$ **C** 4 **D** $^{-}16$

9 Which of the following equals $^{-}7 + 2 \times {}^{-}3$?

A 13 **B** $^{-}13$ **C** $^{-}15$ **D** 15

10 Which one of these statements is false?

A The opposite of 8 is $^{-}4$.
B The opposite of 15 is $^{-}15$.
C 6 is farther from 0 than $^{-}3$ is.
D 8 is twice as far from 0 as $^{-}4$ is.

11 The scale on the thermometer shown on the right is marked off in intervals of 5 degrees (Celsius).

a Copy the thermometer and write the correct value beside each mark on the scale.

b Place the letters representing the following temperatures in the correct places on the thermometer you drew in part **a**.

A 22° **B** $^{-}4$° **C** 15° **D** $^{-}9$° **E** 12°

c List the temperatures given in part **b** in order, from warmest to coldest.

d If the temperature in a city is $^{-}2$°C at 4:00am and it rises 9°C by late afternoon, what temperature would it be then?

e What temperature is halfway between $^{-}4$°C and 12°C?

12 Copy the following statements and complete them using < or >.

a $^{+}6 \square {}^{-}4$ **b** $^{-}3 \square {}^{+}5$ **c** $^{-}9 \square {}^{-}10$

13 Find the sum of:

a $^{+}9$ and $^{+}5$ **b** $^{+}4$ and $^{-}3$ **c** $^{-}7$ and $^{+}2$

d $^{-}3$ and $^{-}12$ **e** $^{+}5$ and $^{-}6$ and $^{-}11$

14 Evaluate:

a $^{+}6 + {}^{+}3$ b $^{+}12 + {}^{-}8$ c $^{-}9 + {}^{-}5$ d $^{-}8 - {}^{+}4$

e $^{-}11 - {}^{+}7$ f $^{+}13 - {}^{-}9$ g $^{-}6 - {}^{-}7$ h $^{+}9 - {}^{+}8 - {}^{-}2$

15 Evaluate:

a $^{+}6 \times {}^{-}5$ b $^{-}7 \times {}^{-}4$ c $^{+}10 \times {}^{-}4 \times {}^{-}2$

d $^{-}24 \div {}^{-}8$ e $^{-}20 \div {}^{+}5$ f $40 \div {}^{-}5 \div {}^{-}2$

16 Find the value of:

a $\frac{50}{-5} + {}^{-}8$ b $35 \div {}^{-}7 \times {}^{-}4$ c $\frac{-36}{-4} + 10$

d $^{-}3(+7 + {}^{-}9)$ e $15 \div {}^{-}5 + {}^{-}75 \div 3$ f $^{-}18 \div {}^{-}3 + 7 \times {}^{-}5$

g $20 \div ({}^{-}4 - 1) \times 3$ h $3 \times {}^{-}4 \div {}^{-}2 + 11 - 5$ i $(8 - 11)({}^{-}1 + 5) \div {}^{-}2$

j $({}^{-}5 \times {}^{-}7 + 6 \times {}^{-}4) \div 2$ k $35 \div {}^{-}5 + 12 \div {}^{-}2 \times 3$

17 Five friends, Alina, Benny, Cara, Dillon and Ernie, work out how much money they each have (or owe) at the end of a holiday together. The amounts are marked on the number line below using dots. Use the number line and the clues to help you decide which dots the names of the five friends go beside.

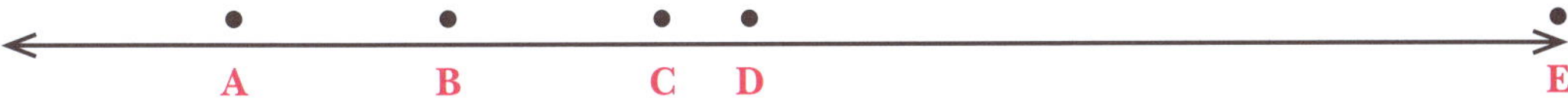

Clues:

- One person has no money.
- Cara has more money than Benny.
- Alina has the most money.
- One of the people owes as much money as another one has.
- Dillon has a little less money than Ernie.

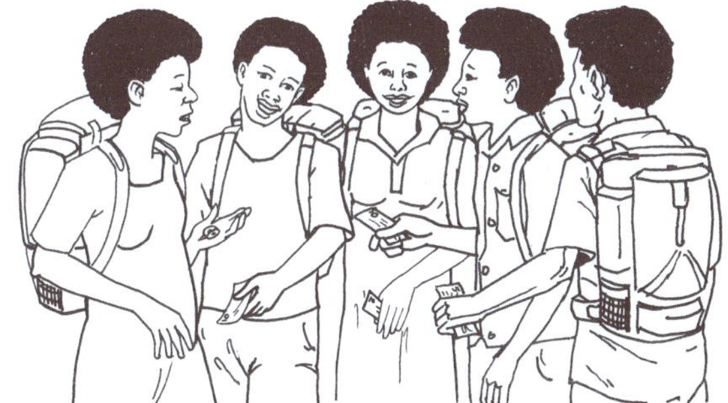

Indices

1 Match each of these with the correct answer:

a $\sqrt{121}$ A 625

b 25^2 B 12

c $\sqrt{25} \div 1^2$ C 1

d $3^2 + \sqrt{9}$ D 11

e $\sqrt{36} - \sqrt{25}$ E 5

2 Write 'True' or 'False' for each of the following:

a $3^2 \times 2^4 = 6^6$ b $2 \times 5^{-3} = \frac{2}{5^3}$ c $a^6 \div a^3 = a^2$

d $(3a^4b^2)^3 = 3a^{12}b^6$ e $27 \times 81 = 3^7$

3 Which of the following is an example of standard form?

A 10^3 B 6.29×10 C 21.5×10^5 D 9.7×10^4

4 Which of the following is equal to $5^0 - 5^1$?

A 5^{-1} B $^{-}4$ C $^{-}5$ D $\frac{1}{5}$

5 Which of the following is equal to 2^{2x+2}?
A $(2^{2x})^2$ **B** $(2^{2+2})^x$ **C** $2^{x+1} \times 2^2$ **D** $2^{2x} \times 2^2$

6 Which of the following is 864 expressed in index form, as a product of prime factors?
A $12^1 \times 6^1 \times 4^1 \times 3^1$ **B** $6^2 \times 4^1 \times 3^1 \times 2^1$ **C** $9^1 \times 3^1 \times 2^5$ **D** $3^3 \times 2^5$

7 Which of the following is equal to $3(5x^2y^2)^0$?
A 0 **B** 1 **C** 3 **D** 15

8 Write each of the following as a product of prime factors in index form:
a 98 **b** 1128

9 Write $5 \times 5 \times 5 \times 5$ in index notation.

10 State the power in the expression m^7.

11 Express x^4 in factor form.

12 If $m = {}^-2$ and $n = 1$, find the value of:
a m^2 **b** $3m^2$ **c** $-4n^3$ **d** m^3n **e** n^4
f $m^2 \div n$ **g** $n^3 \div m$ **h** m^3n^2 **i** m^4n^3 **j** $5mn^6$

13 Write each of the following in index form:
a $5 \times a \times a \times a \times a \times b \times b$ **b** $5 \times ab \times ab \times ab \times ab$ **c** $5ab \times 5ab \times 5ab \times 5ab$

14 Write each of the following in simplified index form:
a $x^4 \times x^2 \times x^5$ **b** $5a^3 \times {}^-4a^4$ **c** $x^3y^2 \times 6xy^5$ **d** $b^8 \div b^4$
e $14a^7 \div 2a^3$ **f** $\frac{z^{12}}{z^7}$ **g** $\frac{12a^2b^3}{3ab}$

15 Simplify each of the following, giving your answers with positive indices:
a $b^6 \times b \times b^5$ **b** $c^7 \times c^2 \div c^9$ **c** $(3^2de^4)^3$
d $5g^4 \times 3g^2$ **e** $\frac{e^7f^4}{e^5f}$ **f** $\frac{f^4}{f^3 \times f}$
g ${}^-3h^5j^2 \times 2h^3j$ **h** ${}^-5j^5 \times {}^-2j^0$ **i** $2n^2p^4 \times 3(n^3p^2)^2$
j $\frac{3m^5 \times 4n^7}{(2mn^2)^3}$ **k** $(a^2b^{-2})^2$ **l** $\frac{3km^5}{6k^3m^{-2}}$
m $\frac{k^4(m^5)^3}{(k^2)^2m^{10}}$ **n** $\frac{({}^-2q^4r^{-2})^3}{({}^-3q^3r^{-3})^2}$ **o** $\frac{(m^5)^2n^6p^3}{(m^2n^3p)^3}$
p $\left(\frac{g^2h^{-3}}{j}\right)^2$ **q** $\frac{2^5(s^3t^2)^2 \times 2^3st^2}{s^2t \times 2s^5t^6}$ **r** $\frac{(v^3w)^4 \times (v^3w^{-2})^3}{2(v^7w)^2}$

16 Evaluate:
a $4^1 - 12^0$ **b** $2^0 + 2^{-1}$ **c** $7^2 - 7^0 + 7^{-1}$ **d** $(\frac{1}{3})^{-1} + 2^{-1}$

17 Write each of these products as an ordinary number:
a 3.001×10^4 **b** 1.932×10^{-3} **c** 8.124×10^{-7} **d** 5.006×10^5 **e** 5.432×10^{-5}

18 Express each of the following in standard form:
a 121.04 **b** 0.0045 **c** 1000
d 70 thousand **e** 3 millionths **f** 3.75×10^{-5}

19 **a** The Mariana Trench, off the coast of Guam in the Pacific Ocean, is 11 034 m deep. Write 11 034 in standard form.
b Mt Everest is 8.85 km high. Write its height in metres, in standard form.

20 A goat's hair has a diameter of 21 micrometres. (1 micrometre = one millionth of a metre.)
  a Write this diameter in metres.
  b Convert the length to millimetres, giving your answer in standard form.
  c What distance would 500 of these hairs, placed side-by-side, take up in millimetres? Express your answer in standard form.

Equations

1 Which of the following is the missing number in the sequence 12, 6, __, $1\frac{1}{2}$, $\frac{3}{4}$?
  A 0 B 3 C 2 D $4\frac{1}{2}$

2 Which of the following is the 20th term of the arithmetic progression in which $t_1 = 7$ and $d = 2$?
  A 47 B 43 C 31 D 45

3 Which of the following is the expression representing 'seven more than twice a number'?
  A $m^2 - 7$ B $2m + 7$ C $m^2 + 7$ D $2m - 7$

4 Which of the following is equal to $a - 3b^2$ if $a = 2$ and $b = ^-3$?
  A $^-25$ B 29 C $^-9$ D 9

5 Which of the following is the simplest expanded form of $b(b + 3)$?
  A $b^2 + 3$ B $2b + 3$ C $b^2 + 3b$ D $3b^2$

6 Which of the following is the fully factorised form of $16a + 24ab$?
  A $4a(4 + 6b)$ B $8(a + 3b)$ C $2a(8 + 3b)$ D $8a(2 + 3b)$

7 Which of the following lists the pair of steps, in order, needed to solve the equation $2g - 5 = 15$?
  A *Step 1*: Add 5 to both sides. *Step 2*: Divide both sides by 2.
  B *Step 1*: Subtract 5 from both sides. *Step 2*: Multiply both sides by 2.
  C *Step 1*: Multiply both sides by 2. *Step 2*: Add 5 to both sides.
  D *Step 1*: Add 5 to both sides. *Step 2*: Multiply both sides by 2

8 Which of the following equations is the same as 'The sum of a certain number divided by 4, and added to twice that number is equal to 12'?
  A $\frac{x}{4} = x^2 + 12$ B $\frac{(x + 2x)}{4} = 12$ C $\frac{x}{4} + x^2 = 12$ D $\frac{x}{4} + 2x = 12$

9 Which of the following is the formula if s is made the subject of the formula $v^2 = 2gs$?
  A $s = \frac{v^2}{2g}$ B $s = \frac{2g}{v^2}$ C $s = v^2 \times 2g$ D $s = v^2 - 2g$

10 Which of the following is the value of S if $S = \frac{180(n - 2)}{n}$ and $n = 4$?
  A 45.5 B 352 C 180 D 90

11 Write the next three numbers in each of the following sequences:
  a 7, 13,19, … b 8, 24, 72, … c $^-3$, $^-8$, $^-13$, … d 16, 8, 4, … e 1, 5, 11, 19, …

12 Which of the sequences in Question 11 are arithmetic sequences?

13 Write the first three terms of an arithmetic sequence in which $a = 20$ and $d = ^-9$.

14 State the values of a and d for the arithmetic sequence: $^-11$, $^-5$, 1, …

15 What are the first three terms of the arithmetic progression for which $t_n = 7 + 8d$?

16 Find the value of t_{20} for the arithmetic sequence: 10, 11.5, 13, …

17 Simplify each of the following by collecting like terms:

a $15t + 9 - 6t - 4$ **b** $7p - p$ **c** $x^2 + 2x + 5 + 4x$

18 Expand and simplify:

a $9(3m - 11)$ **b** $^-4(p + 3)$ **c** $w(w + 7)$ **d** $2(x - 2) - 7$

e $4(m + 2) - 3(m + 1)$ **f** $b(b + 1) + b(b - 4)$ **g** $3(h - 7) - 4(9 - h)$ **h** $2(p + 4q) - 3(p - 5q)$

i $2(w - v) - 6(v - w)$ **j** $8(g + 2h) - 3(5h - 3g)$

19 Factorise:

a $6x + 14$ **b** $ab^2 + 6b$ **c** $7mn - 7n$

d $^-5t^2 + 10t$ **e** $5a^2 + 10a - 20ab$

20 **a** Show that $d = 3$ is the solution to the equation $2(4d - 3) = 18$.

b Show that $w = 1$ is **not** the correct solution to the equation $\frac{3w + 2}{4} = 7w - 6$.

21 Find the value of the pronumeral in each of the following:

a $x + 1 = {}^-7$ **b** $m + 2 = 23$ **c** $t - 4 = {}^-5$ **d** $v - 9 = 12$

e $3t = 12$ **f** $7y = {}^-25$ **g** $\frac{x}{5} = {}^-4$ **h** $\frac{m}{3} = 13$

22 Solve:

a $5m + 11 = 18$ **b** $4m + 5 = 9$ **c** $3t - 7 = {}^-5$

d $6 - 5x = 1$ **e** $3n - \frac{4}{5} = 5\frac{1}{5}$ **f** $5x = 33 - 6x$

g $5(3 - x) = 45$ **h** $2(3d - 3) = 18$ **i** $9(4 - 5y) = 81$

j $\frac{2p}{7} = 5$ **k** $\frac{3m}{4} = {}^-1$ **l** $\frac{3z + 7}{5} = 0$

m $\frac{3m + 1}{2} = m + 4$ **n** $\frac{6y}{7} + 2 = {}^-10$ **o** $\frac{2c + 1}{3} = 2c - 1$

23 Write an expression for each of the following, using m to represent 'a certain number'.

a Twelve more than a certain number.

b One-quarter of a certain number.

c Five less than double a certain number.

d Half the sum of a certain number and three.

e Triple a certain number added to one-half the number.

f The product of a certain number and one more than the number.

24 **a** Write the values of these points in ascending order:

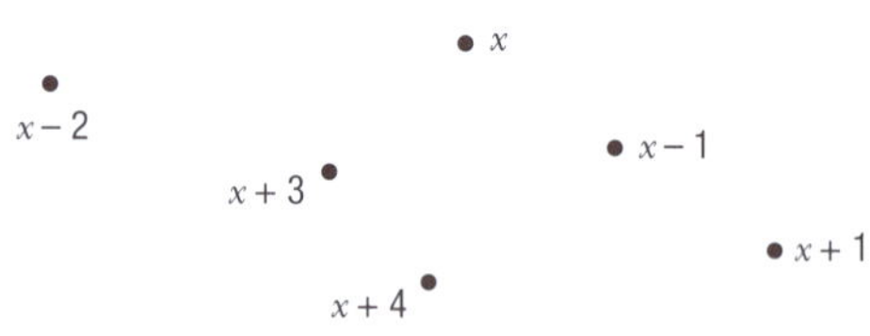

b Write the values of these points in descending order:

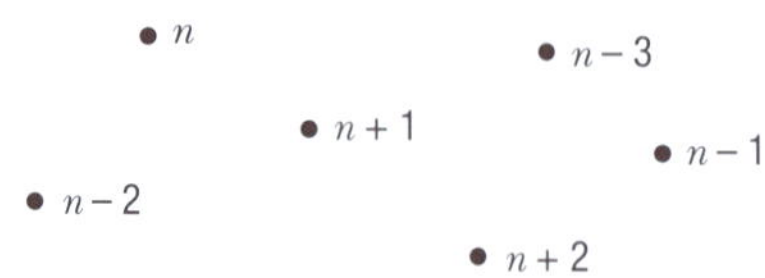

25 Using y to represent the 'certain number', write an equation for each of the following and solve.

- **a** Four times a certain number equals twenty.
- **b** When a certain number is divided by six the result is eight.
- **c** When twice a certain number is taken away from nineteen the result is nine.
- **d** When one is added to three times a certain number the result is twenty-two.
- **e** When the result of adding one to a certain number is multiplied by three, the answer is fifteen.
- **f** Adding one to a certain number and taking one third of the result gives the same value as the number. What is the number?

26 **a** This triangle is isosceles, so the two marked sides are equal in length. Find the value of x.

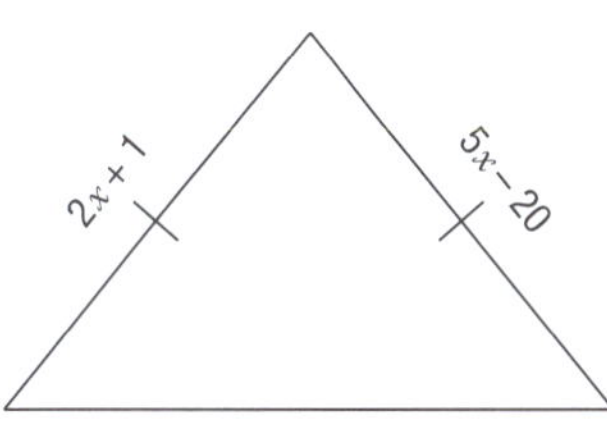

b Find the size of each angle in this triangle.

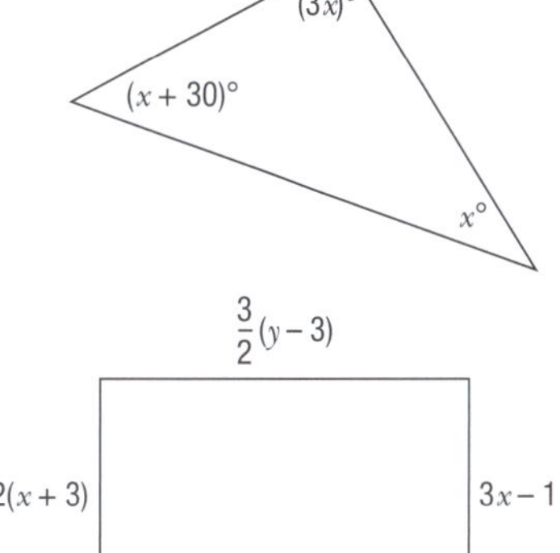

c What are the lengths of the sides of this rectangle?

$\frac{3}{2}(y-3)$

$2(x+3)$ $3x-1$

$y-1$

27 Flora has saved K4 less than twice the amount Jason has saved. Together they have K44. How much has each person saved?

28 Evaluate each of the following formulas, giving your answers correct to two decimal places:

- **a** $P = 2(l + b)$ given $l = 3.5$ and $b = 1.7$.
- **b** $A = 2\pi r(r + h)$ given $r = 4$ and $h = 6$. (Use $\pi = 3.14$.)
- **c** $C = \frac{5}{9}(F - 32)$ given $F = 80$.

29 **a** **i** Make n the subject of the formula $S = 2n - 4$.

ii Find n when $S = 12$.

b **i** Make s the subject of the formula $A = 6s^2$.

ii Find s when $A = 486$.

PROJECTS

Directed numbers: Practice with mental arithmetic

Here are some activities to help you become skilled at working out simple calculations with directed numbers, without using a pen and paper.

Activity 1: Stone target game

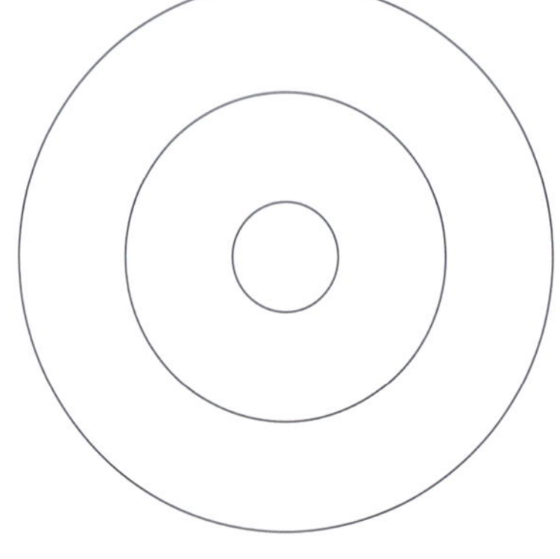

What you need:

On an area of ground outside the classroom, draw three concentric circles, as shown on the right. The diameter of the smallest circle should be about the length of your foot, the second circle should be a foot length away from the first circle, and the third circle should be a foot length away from the second circle.
This will be the target.

Write the following scores on the target:

- inside the inner circle, 9
- inside the next circle, 4
- inside the outer circle, $^{-}2$
- outside the target, $^{-}6$.

Measure out a distance of three paces from the centre of the circles and mark a line on the ground. This will be the line the players play from.
Each player will also need a stone about the size of a 20 toea coin.

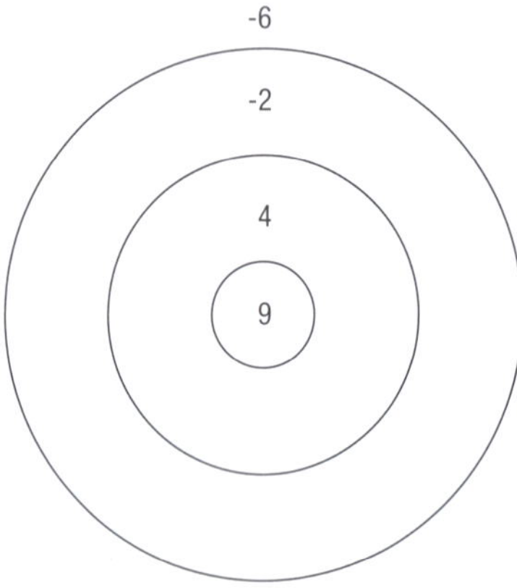

How the game works

Players take turns throwing their stone from the line into the circles.
They must add the number they land on to their score.
The first person to reach a score of 20 is the winner.

Activity 2: Number line game

What you need:

On a strip of ground outside the classroom (around 20 metres long), mark out a number line from $^{-}10$ to $^{+}10$, spacing the numbers approximately one metre apart.
Label the numbers so that they can be clearly seen when you stand beside any section of the number line.

How the game works

- Divide the class into two teams.
- In every turn, one person from each team moves to the number line and stands beside the number 0.
- The teacher (or a student) calls out a calculation which has an answer between $^{-}10$ and $^{+}10$.
  The contestants separately work out the answer and move to that position on the number line.
  No one else in the team should give any help with the calculation or the team will lose the point.
- The first contestant to reach the correct position on the number line scores a point for their team.
  If both contestants arrive at the correct position at the same time, they go back to 0 and are given a second question to answer.

The questions can be a simple addition, subtraction, multiplication or division sum using positive and negative numbers. Harder questions can use two or more operations.

Activity 3: Bingo

1 Students each draw a 3 × 3 grid and fill in their own 'lucky numbers', choosing numbers between $^{-}10$ and $^{+}10$.

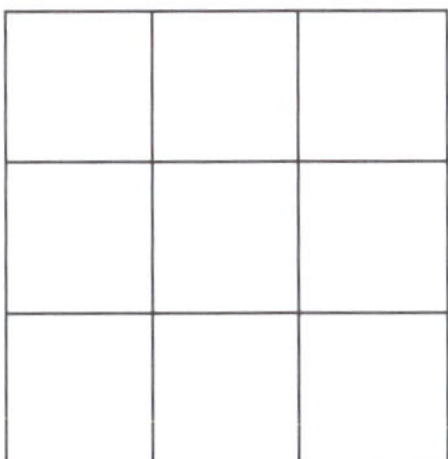

2 The teacher has a set of cards showing the numbers $^{-}10$ to $^{+}10$. The cards are shuffled.

3 The teacher (or a student) selects a card, calls the number and puts that card aside.

4 Students mark a cross on the number called if it is on their grid, then another card is selected and called.

5 The process is repeated until someone has all nine squares on their grid crossed and calls 'Bingo'.

The degree of difficulty of the game can be increased by the teacher calling a directed number calculation, instead of simply calling a number. For example, instead of the number 9, the teacher could call '$^{-}18 \div {}^{-}2$', or '$^{+}4 - {}^{-}5$', or '$^{-}3 \times {}^{-}3$'.

The Bingo game grid can also be increased in size, to a 5 × 5 grid, or larger, or the numbers to be chosen from can be increased ($^{-}30$ to 30, for example).

Students can make up their own pack of cards to be used in new games.

Indices: Everyday applications

Project 1: The chessboard problem

According to legend, the Shah (the ruler) of Persia was so impressed when he first saw the game of chess that he offered whatever reward the game's inventor chose.

The clever inventor asked for two grains of wheat for the first square of the chessboard, four grains for the second, eight grains for the third square, and so on, for all of the 64 squares on the chessboard.

1 How many grains would the inventor receive for:
   a the fourth square? b the fifth square? c the sixth square?

2 Construct a table, similar to the one below, showing the number of each chessboard square and the number of grains of wheat the inventor would be given for that square. Continue the table for each square of the chessboard.

S (Square number)	1	2	3	4	5	6	etc.
G (Number of grains)	2	4	8				
G as a power of 2							

3 What power of 2 does each value of G represent? Fill in the third row of the table.

4 Find the total number of grains of wheat the inventor would receive as a reward.

Project 2: Blood bank research

Human bodies contain about 8×10^4 microlitres (μL) of blood for each kilogram of body weight. One microlitre is one-millionth of a litre.

$$1\ \mu\text{L} = \frac{1}{1\,000\,000}\ \text{L} = 1 \times 10^{-6}\ \text{L}$$

Each μL of blood contains about 5×10^6 red blood cells. Red blood cells transport oxygen around the body.

1 Find the approximate number of red blood cells in a person who weighs 56 kg.

2 Find your weight in kilograms and estimate the number of red blood cells in your body.

3 Assuming the average blood donor gives about 470 mL, estimate the number of red blood cells in one donation.

4 Research the four main blood groups and draw a graph representing the percentage of the population belonging to each group.

5 Carry out the following research at a clinic where people donate blood.
   a How many litres of blood are donated:
      i each day? ii each week?
   b Show your answer for part a graphically, with the data broken into the four blood groups.
   c Convert the volume of blood you found in part a into the number of red blood cells and write this in standard form.
   d Find out the different uses of the blood donations.

Project 3: A forensic investigation

The investigation of a crime scene involves collecting evidence and identifying its source. In addition to fingerprints or blood samples, evidence collected might include hair and fibre samples, pollen, paints and chemicals.

A single human hair can reveal a person's race, what part of their body it came from, whether it was pulled out or dropped, and the state of the person's health.

Examination of fibres can identify the type of fibre (such as wool, cotton, or synthetic). These fibres can be compared to fibres from the clothing of victims and suspects, or to carpets, car upholstery and other textiles.

The problem

A forensic scientist was called to a crime scene where she found traces of blood and fibres from woollen cloth.

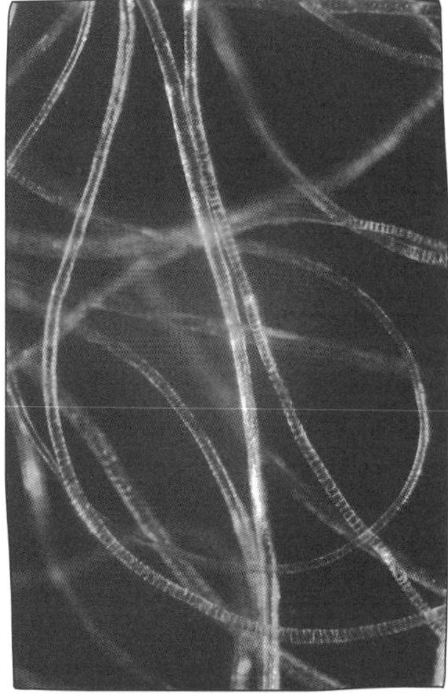

1 a Wool fibres are shown here as they appeared under the microscope. Each fibre is 18 µm wide. Express in standard form the width of a single fibre, in millimetres.

b How many fibres must be laid side by side to form 1 cm of cloth? (Give your answer in standard form.)

2 A drop of blood was found spattered on the carpet.

a There are approximately 5 million red blood cells in one cubic millimetre of the spattered blood. Express 5 million in standard form.

b In total there was 20.5 mm^3 of blood spattered on the carpet. Approximately how many red blood cells would this blood contain?

3 The forensic scientist examined the blood cells under a microscope (as shown on the right). The cells are shaped like biconcave discs. The diagram of the cross-section below shows the diameter and thickness of one cell. Write the diameter of the cell in metres, in standard form.

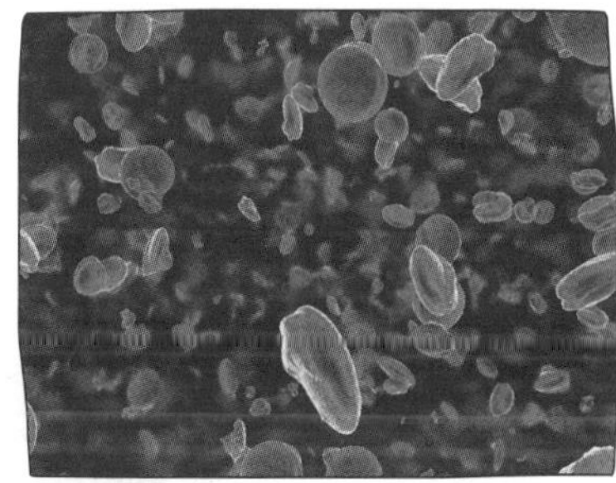

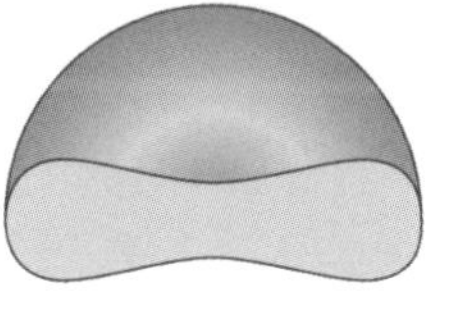

cut through centre

8 µm

2 µm

side view

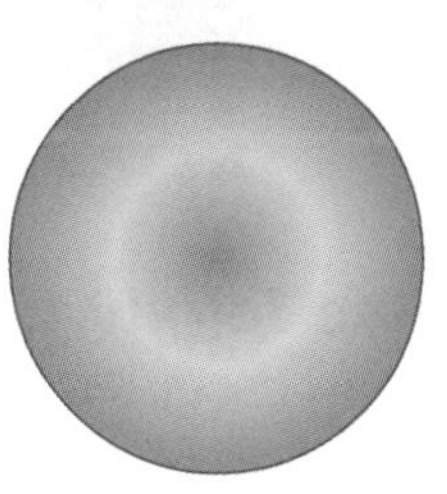

surface view

4 Red blood cells (RBC) contain pigment molecules called haemoglobin. These haemoglobin molecules carry oxygen around the body. A single RBC usually contains 280 million molecules of haemoglobin.

a Write, in standard form, the number of haemoglobin molecules in a single red blood cell.

b Estimate how many haemoglobin molecules would be in 1 mm^3 of the spattered blood. (Give your answer in standard form.)

Number patterns: The Fibonacci sequence

Fibonacci, whose real name was Leonardo da Pisa (around 1170–1250), discovered this special pattern: 0, 1, 1, 2, 3, 5, 8, 13, …

Can you see how the Fibonacci pattern continues?

Fibonacci numbers can often be found in nature (for example, in the patterns of the petals or seeds of flowers, or in the spiral pattern on pineapples, or pine cones).

Fibonacci also introduced the decimal (Hindu-Arabic) number system we use today, which replaced Roman numerals.

Project 1: The rabbit problem

Fibonacci explained his number pattern by referring to how quickly rabbit populations could grow. The problem is as follows:

> *Start with a pair of newborn rabbits, one male and one female. Assume they reach maturity at the age of 1 month, when they mate and, after another month, the female gives birth to a pair of rabbits, one male and one female. Continuing in this way, every female produces a new pair every month after they have reached 2 months of age.*

The number of pairs of rabbits can be seen on this 'tree'.

Start	1					1 pair of rabbits
Month 1:	1					1 pair of rabbits
Month 2:	1	1				2 pairs of rabbits
Month 3:	1	1	1			3 pairs of rabbits
Month 4:	1	1	1	1	1	5 pairs of rabbits

Your turn:

Copy the diagram and continue the tree to find the number of pairs of rabbits at the end of one year (assuming no rabbits die).

Project 2: Bee family trees

In a bee colony, only the queen bee produces eggs. If these eggs are fertilised, female bees (worker bees) are produced. Male bees (drones) are produced without fertilisation. This means that female bees have two parents but males have just one.

So, if we start with a male bee, his family tree looks like the diagram below (M represents a male and F represents a female).

M	1	current generation
F	1	1 generation before
M F	2	2 generations before
F M F	3	3 generations before
M F F M F	5	

As you can see, the number of bees in each generation follows the Fibonacci sequence.

Your turn:

1 Draw a family tree for a female bee, with up to four generations before.
2 How does this compare with your own family tree? Starting with M or F to represent you, draw a family tree up to six generations before you were born. Identify the number of people in each generation.
3 For your family tree, explain any pattern you can see for the number of people in each generation and use this to write the number of people in the 10th generation before you were born.

Project 3: A Fibonacci spiral

To construct this, start by drawing two one-unit squares side by side. Now draw a 2×2 square on top of those, then a 3×3 square beside, and a 5×5 square below that. The size of each subsequent square is the next number in the Fibonacci series. You can produce a type of spiral by joining quarter circles in each square, as shown below. This is similar to the spirals you can see in nature, such as in a nautilus shell.

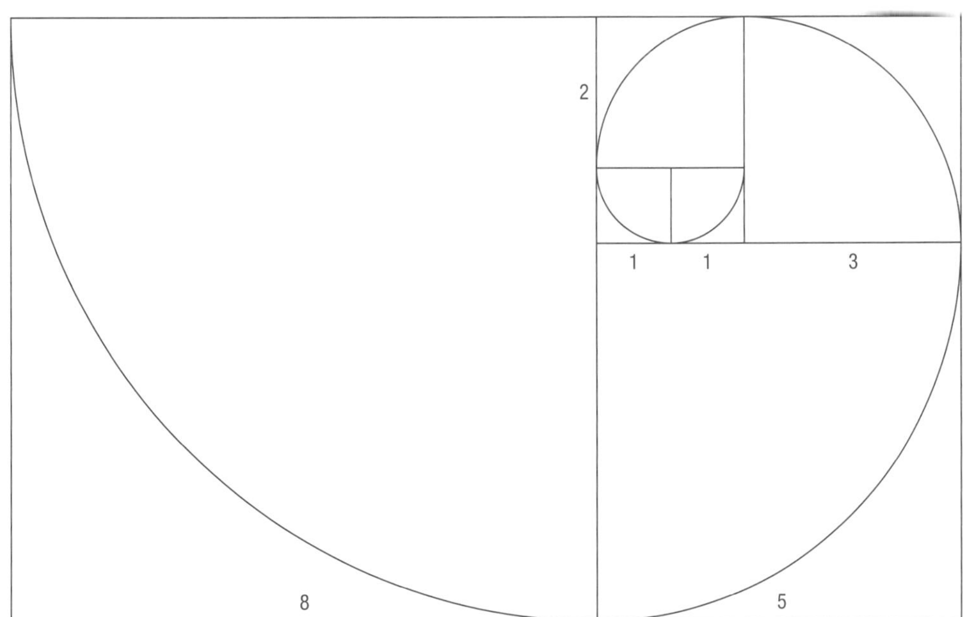

ENRICHMENT

Algebraic fractions

Algebraic fractions are fractions containing at least one pronumeral. The following are all algebraic fractions:

$$\frac{3w}{4} \quad \frac{9}{t} \quad \frac{6m^2}{12mn^3} \quad \frac{m-3}{2}$$

Simplifying algebraic fractions

Algebraic fractions can be simplified if the numerator and denominator have a common factor.

Step 1: Express both the numerator and the denominator as a product of factors.

Step 2: Cancel common factors in the numerator and the denominator.
(*Note*: Only **factors** can be cancelled. Terms joined by + or – cannot be cancelled.)

Step 3: Simplify both the numerator and the denominator.

Example

Simplify:

a $\frac{2m}{4}$

b $\frac{6ab}{3b}$

c $\frac{^-4x^2y}{12xy^3}$

d $\frac{9(m+1)}{3m}$

e $\frac{6a^2+3a}{12a}$

Answer

a $\frac{2m}{4} = \frac{2 \times m}{2 \times 2} = \frac{m}{2}$

b $\frac{6ab}{3b} = \frac{3 \times 2 \times a \times b}{3 \times b} = \frac{2 \times a}{1} = 2a$

c $\frac{^-4x^2y}{12xy^3} = \frac{^-4 \times x \times x \times y}{4 \times 3 \times x \times y \times y \times y} = \frac{^-1 \times x}{3 \times y \times y} = \frac{^-x}{3y^2}$

d $\frac{9(m+1)}{3m} = \frac{3 \times 3(m+1)}{3 \times m} = \frac{3(m+1)}{m}$

e $\frac{6a^2+3a}{12a} = \frac{3a(2a+1)}{12a} = \frac{(2a+1)}{4} = \frac{2a+1}{4}$

Adding and subtracting algebraic fractions

Algebraic fractions can be added and subtracted by changing them to fractions with a common denominator. This is the same method we use for all fractions.

Remember

When adding or subtracting fractions, 'simplify' means to write the answer as a single fraction.

Example

Simplify:

a $\frac{b}{3} + \frac{b}{4}$

b $\frac{d}{3} - \frac{2d}{5}$

c $\frac{x}{6} + 1$

d $\frac{4}{m} + \frac{9}{n}$

e $\frac{7}{x} - \frac{3}{x^2}$

Answer

a $\frac{b}{3} + \frac{b}{4} = \frac{4b}{12} + \frac{3b}{12} = \frac{7b}{12}$

b $\frac{d}{3} - \frac{2d}{5} = \frac{5d}{15} - \frac{6d}{15} = \frac{^-d}{15}$

c $\frac{x}{6} + 1 = \frac{x}{6} + \frac{6}{6} = \frac{x+6}{6}$

d $\frac{4}{m} + \frac{9}{n} = \frac{4n}{mn} + \frac{9m}{mn} = \frac{4n+9m}{mn}$

e $\frac{7}{x} - \frac{3}{x^2} = \frac{7x}{x^2} - \frac{3}{x^2} = \frac{7x-3}{x^2}$

Multiplying and dividing algebraic fractions

To multiply two fractions, we multiply the numerators together to form the new numerator, and multiply the denominators together to form the new denominator.
If there is a common factor in the numerator and denominator, we cancel to obtain the simplest fraction answer.
To divide two fractions, we invert the second fraction and then multiply, in the manner described above.
These are the same methods we use for multiplying and dividing all fractions.

Example

Find:

a $\frac{a}{2} \times \frac{a}{5}$

b $\frac{a}{2} \div \frac{d}{10}$

c $\frac{4}{n} \div \frac{8}{n^2}$

d $5 \div \frac{10}{m}$

Answer

a $\frac{a}{2} \times \frac{a}{5} = \frac{a \times a}{2 \times 5} = \frac{a^2}{10}$

b $\frac{a}{2} \div \frac{d}{10} = \frac{a}{{}_{1}\cancel{2}} \times \frac{\cancel{10}^{5}}{d} = \frac{a \times 5}{d} = \frac{5a}{d}$

c $\frac{4}{n} \div \frac{8}{n^2} = \frac{{}^{1}\cancel{4}}{{}_{1}\cancel{n}} \times \frac{\cancel{n^2}^{\,n}}{\cancel{8}_{2}} = \frac{n}{2}$

d $5 \div \frac{10}{m} = \frac{{}^{1}\cancel{5}}{1} \times \frac{m}{\cancel{10}_{2}} = \frac{m}{2}$

EXERCISE

1 Write equivalent fractions for each of the following by filling in the gaps:

a $\frac{2a}{4} = \frac{a}{\square} = \frac{\square a}{16} = \frac{2ab}{\square}$ **b** $\frac{x}{2} = \frac{3x}{\square} = \frac{\square x}{10} = \frac{\square x}{2a}$

2 Simplify:

a $\frac{3m}{12}$ **b** $\frac{5t^2}{10t}$ **c** $\frac{6mn}{3n}$ **d** $\frac{^-8y^2}{4y}$

e $\frac{5a^2}{a^3}$ **f** $\frac{4(t+5)}{2t}$ **g** $\frac{a(b-3)}{5a}$ **h** $\frac{7(m+1)}{m+1}$

3 Simplify:

a $\frac{p}{6} + \frac{p}{2}$ **b** $\frac{3m}{5} - \frac{m}{4}$ **c** $\frac{a}{4} + \frac{b}{7}$ **d** $\frac{2y}{7} - \frac{y}{3}$ **e** $\frac{^-2h}{3} + \frac{7j}{5}$

f $\frac{3}{a} + \frac{7}{a}$ **g** $\frac{y}{4} + 3$ **h** $4 - \frac{x}{9}$ **i** $\frac{7}{ab} + \frac{2}{bc}$ **j** $\frac{5}{d^2} + \frac{3}{d}$

4 Simplify:

a $\frac{x}{5} \times \frac{y}{3}$ **b** $\frac{2a}{3} \times \frac{b}{8}$ **c** $\frac{a}{7} \times \frac{14}{b}$ **d** $\frac{w}{12} \times \frac{24}{w^2}$

e $\frac{4h}{11} \times \frac{11}{4h}$ **f** $\frac{j^2}{6} \times \frac{^-12}{j}$ **g** $\frac{3a}{8} \times \frac{a}{6b}$ **h** $\frac{r}{2s} \times \frac{5}{rs}$

5 Simplify:

a $\frac{3}{y} \div \frac{1}{2y}$ **b** $\frac{m}{6} \div \frac{n}{12}$ **c** $\frac{3p}{7} \div \frac{5q}{14}$ **d** $\frac{w^2}{4} \div \frac{wz}{8}$

e $m \div \frac{mn}{3}$ **f** $\frac{t}{9} \div 2t$

Unit 3

Working with data

Unit summary

This unit focuses on everyday data and how it is collected, presented, analysed and interpreted.

People in many situations use data and statistics in order to make informed decisions:

- a manufacturer will analyse data when selecting the type of product that customers will buy
- a teacher will collect data and use statistics when grading students
- a farmer may collect data on crop yield from a particular type of plant so that a decision can be made about the most appropriate crop for a particular situation.

The PNG Government collects extensive data on the location, health and age distribution of the population. This enables the government to make decisions about the health, education, housing and transport services that it will provide and also enables it to explore trends that may affect Papua New Guinean society in the future.

To help build ideas of random sampling, the basic concepts of probability are also introduced in this unit.

The core section of this unit is assessed with tests that will enable a student to demonstrate the extent of his/her understanding of mathematical concepts and ability to correctly choose and apply mathematical techniques.

The options for this unit are described below.

Option A: Random events and simulation

This option builds on the work with probability and random events covered in the core of the unit, emphasising the effect that probability and randomness have on the outcome of games.

Assessment for this option involves a group project, designing and analysing a simple game of chance.

Option B: Statistical surveys

This option builds on the data collection and analysis part of the core, with particular attention to the collection of data and how this affects the reliability of conclusions made from a survey.

Assessment for this option involves a group project, designing and analysing a statistical survey conducted in the local community.

Urban, rural and national trends in the HIV epidemic in PNG

HIV prevalence (%): 0, 1, 2, 3, 4, 5, 6, 7

National — Urban - - Rural

Year of detection: 2000, 2001, 2002, 2003, 2004, 2005, 2006, 2007, 2008, 2009, 2010, 2011, 2012

HIV/AIDS infections reported in PNG, by province of detection (1987–2006)

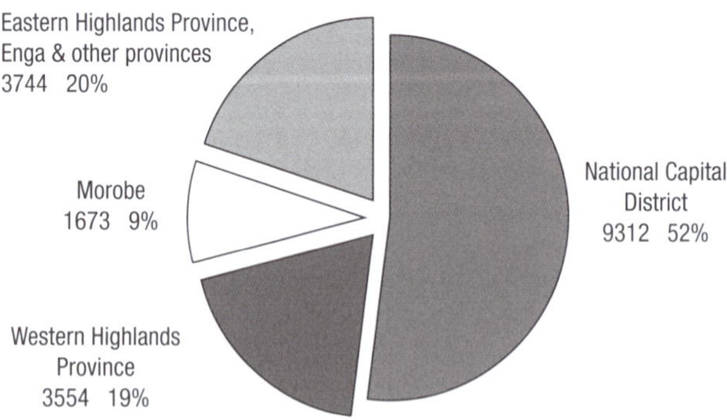

Syllabus references

9.3.1 Represent, interpret, analyse and solve problems using discrete and continuous data

9.3.2 Estimate and calculate probabilities

9.3.3 Communicate mathematical processes and results both orally and in writing

9.3.4 Undertake investigations individually and cooperatively in which mathematics can be applied to solve problems

Working with data: Topics

Week	Content	Page
1	**Recording data** **Sorting and organising data** • Types of data • Organising, displaying and describing categorical data • Organising, displaying and describing numerical data • Stem-and-leaf plots	166
2 and 3	**Measures of central tendency and spread** • Cumulative frequency graphs • Calculation of measures of centre and their interpretation • Calculation of measures of spread and their interpretation • Boxplots and comparing sets of data • Graphing and interpreting time-series data	182
4	**Misleading statistical graphs** • Identifying misleading graphs	197
5	**Experimental probabilities** **Probabilities based on symmetry** • Introduction to probability • Probability based on investigations • Probability based on experiments • Probability based on theory	200
6, 7, 8, 9 and 10	• **Option A or B** • Revision and assessment, and project work	215 242

RECORDING DATA and SORTING AND ORGANISING DATA

When information for a statistical investigation is collected and recorded, this information is referred to as data.

Before data is collected, there are decisions that must be made about the amount of data collected and the method of collection. This will be covered in detail in 'Option B: Statistical surveys' (see page 226).

The type of data and how the conclusions based on it are to be presented also need to be considered in the planning of an investigation.

Lesson 1

Types of data

There are two types of data that we commonly deal with and each of these types has appropriate ways to be **organised** (in tables, etc.), **displayed** (as graphs) and **interpreted** (drawing conclusions). The two types of data are:

a **categorical data** for which a **category** or **quality**, often a word, is recorded. For example the mode of transport to school for students could be investigated. The data recorded for that investigation would be from the categories: walk, bike, car or bus.

b **numerical data** for which a **number** is recorded. There are two types of numerical data:

  i **discrete numerical data**. The data recorded will be distinct values; often whole numbers. Discrete numerical data is often **counted** so that the data can be recorded. For example, if we investigated the number of people in the households in a village the data would have the values 1, 2, 3, …

  ii **continuous numerical data**. The data recorded can theoretically take any value on a number line. Continuous numerical data is often **measured** so that the data can be recorded. For example, if we investigated the speed of cars on a stretch of road, the data recorded would be from 0 km/h to the fastest speed that a car can travel, but is most likely to be in the range of 10 km/h to 120 km/h.

EXERCISE

1 For each of the following, write your response to the question and state whether the data being collected is **categorical** or **numerical**.

  a How many brothers and sisters do you have?
  b What is your favourite colour?
  c Which sport do you like to play?
  d How long does it take you to travel to school?
  e What is your height?
  f How many people live in your household?
  g Which town/village do you live in?
  h What is your favourite food?

2 For each of the following questions, write your response to the question and state whether your numerical answer is **discrete** or **continuous**.
  a How many children are in your family?
  b What is the length of your hand span?
  c How many pets does your family have?
  d How far is your vegetable garden from your house?
  e What is the length of your school desk/table?
  f How long is it until you leave school to go home?

3 Classify the following data as **categorical**, **discrete numerical** or **continuous numerical**:
  a the number of pages in daily newspapers
  b the maximum daily temperatures in Port Moresby
  c preferred football codes
  d positions taken by players on a football field
  e the times it takes for 12-year-olds to run one kilometre
  f the lengths of people's right feet
  g the number of goals shot by netballers

4 A selection of light poles in a big city was surveyed and the following data was collected. Classify the data as either **categorical**, **discrete numerical** or **continuous numerical**.
  a the diameter of each light pole (in centimetres) measured 1 metre from the base
  b the material from which each light pole is made
  c the location of each light pole (inner, outer, north, south, east or west)
  d the height of each light pole, in metres
  e the time since each light pole was last inspected
  f the number of inspections each light pole has had since installation
  g the condition of each light pole (very good, good, fair, unsatisfactory)

5 Give three examples of your own for each type of data (**categorical**, **discrete numerical** or **continuous numerical**). Discuss these with the class.

Organising, displaying and describing categorical data

Categorical data can be **organised** using a tally and frequency table and can be **displayed** using a column graph, a bar chart, a pie chart, or a side-by-side column graph when comparing similar categories.

The **tally and frequency table** below shows the results of a survey of the type of fuel the customers at a service station used to fill their tanks.

Type of fuel	Tally	Frequency
Unleaded	~~IIII~~ ~~IIII~~ ~~IIII~~ ~~IIII~~ ~~IIII~~ III	28
Leaded	~~IIII~~ ~~IIII~~ II	12
LPG	~~IIII~~ III	8
Diesel	II	2
	Total:	**50**

The data from a tally and frequency table can be displayed in each of the following ways.

Column graph:

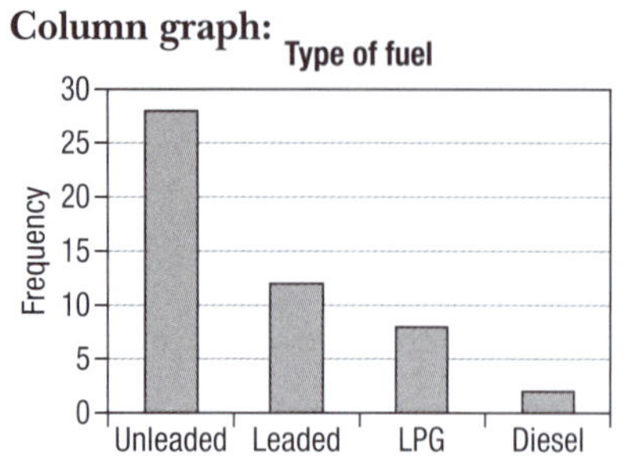

Bar chart:

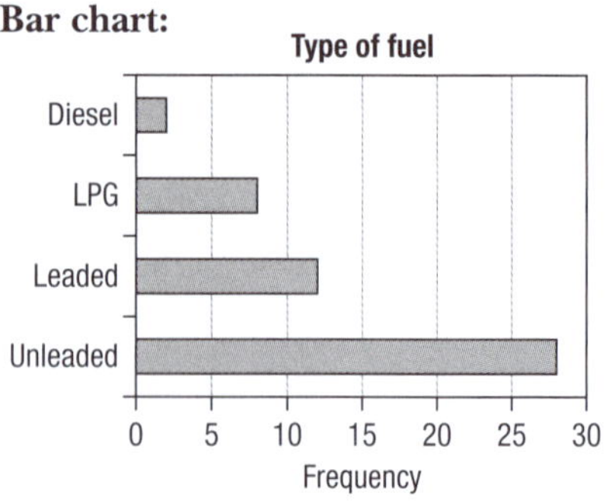

Pie chart:

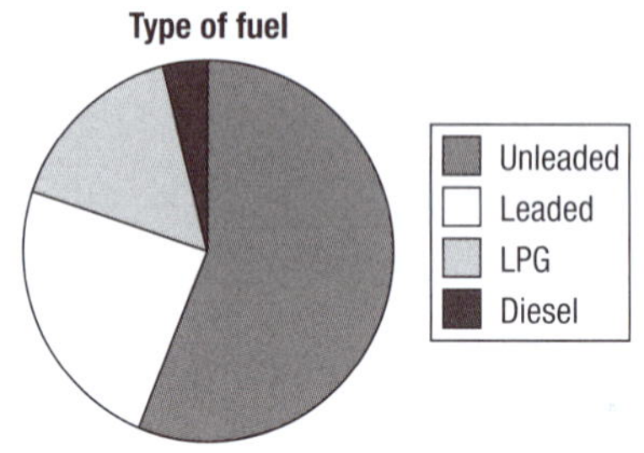

EXERCISE

1 An investigation was carried out into the type of school sport chosen by a group of 50 male students. The following data was collected:

rugby, rugby, basketball, swimming, volleyball, rugby, soccer, swimming, soccer, tennis, volleyball, rugby, rugby, swimming, basketball, basketball, tennis, rugby, volleyball, volleyball, soccer, soccer, rugby, basketball, swimming, basketball, tennis, rugby, rugby, swimming, soccer, volleyball, rugby, rugby, basketball, rugby, basketball, basketball, soccer, basketball, soccer, rugby, tennis, soccer, soccer, rugby, volleyball, soccer, swimming, volleyball

- a Construct a tally and frequency table for the collected categorical data.
- b Construct a bar chart for the data.
- c What is the most popular sport chosen by this group of students?
- d What percentage of the students chose volleyball?
- e Copy and complete the following statement about the data that was collected:

  'The most popular sport chosen by this group of male students was _______, with ____% of the students playing this sport. The least popular sport chosen was ________, with only ____% of students playing this sport. The second most popular sport chosen was _______ (____%), closely followed by _______ (____%).

  ____% of the students chose volleyball and ____% chose swimming.'

2 **a** Construct a pie chart for the categorical data in the table on the right. Label the sectors or include a legend to explain the colours you use.

b Copy and complete the following statement about the data that was collected:

Expenditure item	Weekly expenditure (K)
Food	35
Clothing	15
Rent	110
Travel	24
Power	16

'A total of ____% of the weekly expenditure was allocated to the essential items of food, rent and power. Rent was the most expensive weekly item, accounting for ____% of the weekly expenditure. Travel accounted for ____% of the weekly expenditure and ____% was allocated for clothing.'

Help box

To calculate the angle for each of the sectors in a pie chart, use:

$$\text{Angle} = \frac{\text{number in category}}{\text{total number}} \times 360°$$

Use a protractor to measure and mark the angles. Then colour and label the sectors.

3 **a** Construct a side-by-side column graph for the data in the table below. Label the horizontal axis 'Year' and the vertical axis 'Participation rate'.

	Labour force participation rates in PNG		
	1980	1990	2000
Male	65.5	76.7	68.4
Female	62	60.1	66.7

b Copy and complete the following statement about labour force participation rates:

'In each of the three years when data was collected, ________ had a greater participation rate than ________. The greatest difference between male and female rates was recorded in _____, when the difference in participation rate was _____%.

Participation rates decreased for males between _____ and _____. The decrease was _____%. In the same period the female participation rate increased by ____%.'

A **side-by-side column graph** compares two or more alternatives (male and female in this case) for each category.

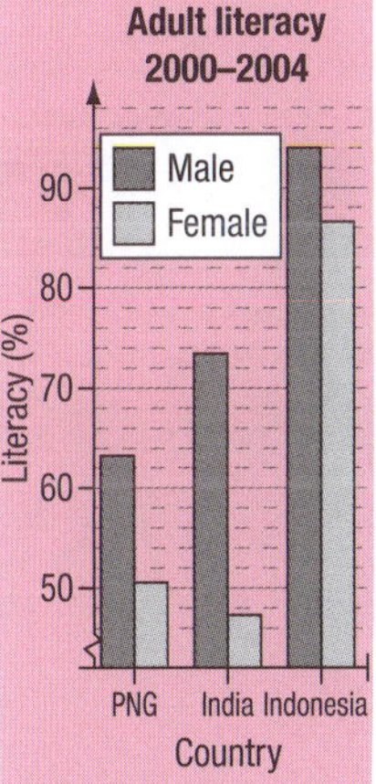

4 The table below shows the incidence of childhood malnutrition and low birth weight in various regions of PNG in 2006:

Region	Childhood malnutrition (%)	Low birth weight (%)
Gulf	39.73	10
Central	30.89	8
Enga	18.91	5
Chimbu	15.75	7
Morobe	38.1	6
Madang	43.96	14
Manus	21.63	5
North Solomons	27.88	7

a Construct a bar chart for the 'Childhood malnutrition' data in the table.

b Comment on the graph you constructed in part **a**, mentioning the regions with the highest incidence and lowest incidence of childhood malnutrition.

c Construct a side-by-side column graph for the 'Childhood malnutrition' and 'Low birth weight' data in the table. Label the horizontal axis 'Region' and the vertical axis 'Percentage'.

d Comment on your graph from part **c**, mentioning the 'Low birth weight' data. Compare the data for 'Childhood malnutrition' and 'Low birth weight'. Is there a link between the two sets of data?

Lesson 3

Organising, displaying and describing discrete numerical data

When discrete numerical data is collected, numbers are recorded. Discrete numerical data is often obtained by counting.

Tally and frequency tables can be used to organise the data, and **column graphs** are used to display the data.

Remember

A **tally and frequency table** is used to organise data.

Number in household	Tally	Frequency
1	||	2
2	||||	4
3	~~||||~~ ||	7
4	~~||||~~ ~~||||~~ ||	12
5	~~||||~~ ~~||||~~ ~~||||~~ |	16
6	~~||||~~ ~~||||~~ ||||	14
7	||||	4
8	|	1

Remember

A **column graph** is used to display the data.
The columns are not joined and are of equal width.

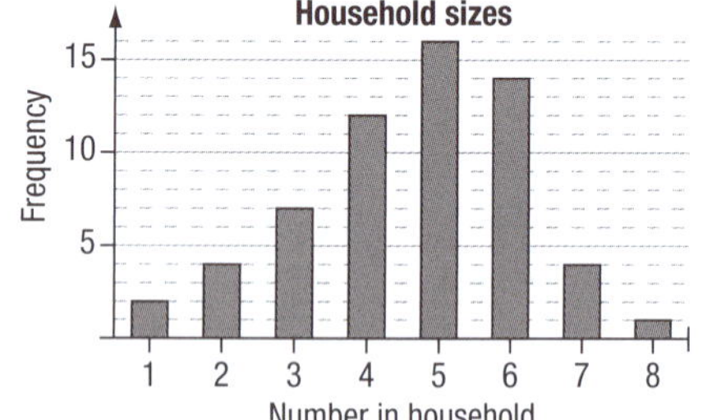

Alternatively, a **dot plot** can be used to organise and display the data at the same time. As we go through the data, we place a dot in the appropriate column to represent each piece of data. The dots must be evenly spaced so that the final plot is a frequency graph.

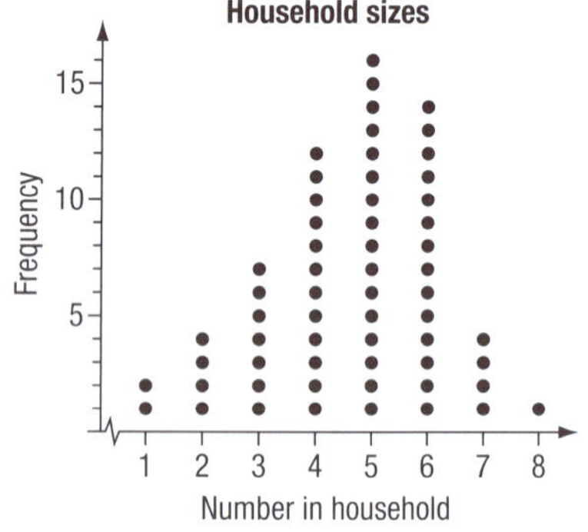

The distribution of a set of data

The distribution of a set of data that is equally spread on either side of a central value is called **symmetric**.

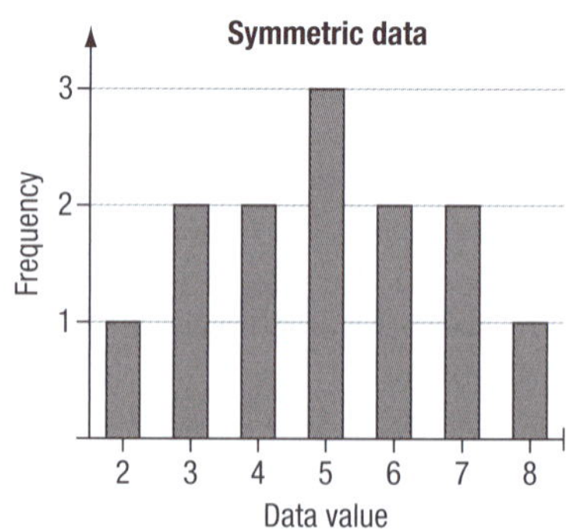

The distribution of a set of data that appears stretched to the right is said to be **positively skewed**.

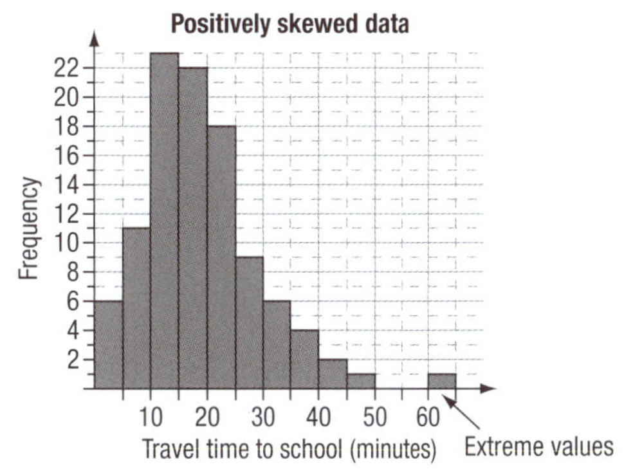

The distribution of a set of data that appears stretched to the left is said to be **negatively skewed**.

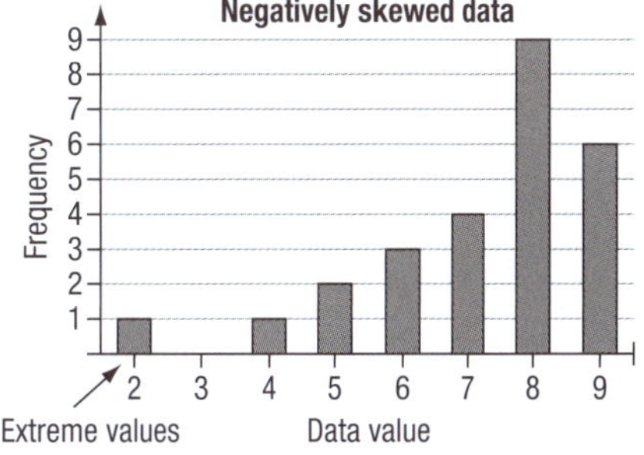

Extreme data values, often called **outliers**, are either much larger or much smaller than the body of data. They appear as isolated points on a graph.

EXERCISE

Number of toothpicks	Frequency
34	1
35	5
36	12
37	13
38	10
39	6
40	1

1 The number of toothpicks in a packet is stated to be 36 but the actual numbers vary. The number of toothpicks in each of 48 packets is counted and the data is recorded in this frequency table.

- a Construct a column graph for this discrete numerical data.
- b What was the most frequently occurring number of toothpicks in a packet?
- c What percentage of the packets contained fewer than 36 toothpicks?
- d Describe the distribution of the data. Is the distribution of the data symmetric, positively skewed or negatively skewed?

2 Leo, a bowler in a cricket team, recorded the number of wickets he has taken in each of the last 30 innings he has played. The data is listed below:

1 1 3 2 0 0 4 2 2 4 3 1 0 1 0 2 1 5 1 3 7 2 2 2 4 3 1 1 0 3

- a Construct a dot plot for the raw data.
- b Describe the distribution of the data. What is the extreme data value?
- c Copy and complete the following statement about the data:

  'Most often Leo took _____ wickets in an innings but, in ___% of the innings, he took two wickets. There were _____ innings when Leo did not take a wicket but, on his best performance, he took ____ wickets.'

3 A survey was conducted into the number of chickens kept by households. There were 200 households included in the survey and the following dotplot has been used to display the data:

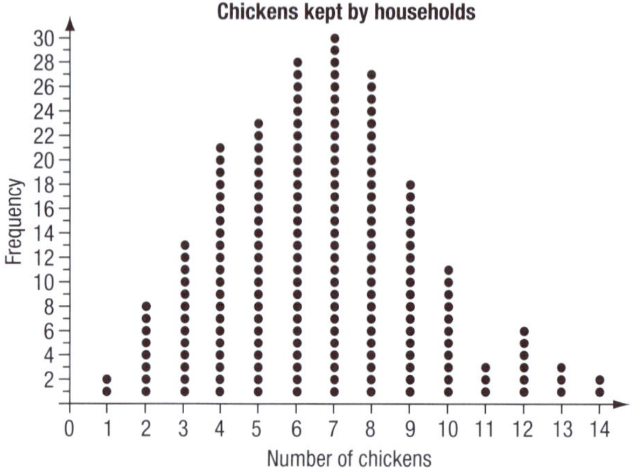

a Describe the distribution of the data as either **symmetric**, **positively skewed** or **negatively skewed**.

b Copy and complete the following statement about the data:

'The maximum number of chickens kept by a household in the survey was _____, and ____% of the households did not keep any chickens. The most frequently occurring number of chickens kept by a household was ____, and ____% of the households kept between 4 and 9 chickens inclusive. _____ of the households kept 10 or more chickens and ____% kept 3 or fewer chickens.'

Remember

'Inclusive' means that the numbers 4 and 9 are both included.

4 A class of students was asked to count and record the number of aunts (mother's or father's sisters) that they had.

The following data was recorded:

3 3 5 7 2 1 4 4 6 4 4 5 5 3 6 2 1 3 3 4 4 5 4 6 7 6 4 4 2 5 3 5

a Construct a graph to display this data.

b Describe the distribution of the data.

c Write a statement about the data mentioning the maximum/minimum values and the most frequently occurring value, including some percentages.

Class exercise

a Collect and analyse discrete numerical data from your class.
For example:
The number of pets/pigs/chickens or the number of siblings/aunts/uncles/cousins.

b Organise and display the data using a tally and frequency table and a column graph or, alternatively, a dotplot. Comment on the distribution of the data, mentioning the maximum/minimum values and the most frequently occurring value, including some percentages.

Stem-and-leaf plots (or stemplots)

Lesson 4

If there are many possibilities for the numerical values in a set of data, then a **stem-and-leaf plot** (or **stemplot**) can be used to organise and display the data. A stem-and-leaf plot groups the data and shows the relative frequencies but has the added advantage of retaining the actual data values. An example is given below.

The score, out of 50, on a test was recorded for 36 students:

25	36	38	49	23	46	47	15	28	38	34	9
30	24	27	27	42	16	28	31	24	46	25	31
37	35	32	39	43	40	50	47	29	36	35	33

The data is all two-digit numbers (the data value 9 will be recorded as 09) so the first digit will be the 'stem' and the last digit is the 'leaf' for each of the numbers. (The last digit is always the leaf.)

The stems will be 0, 1, 2, 3, 4, 5 to account for numbers from 0 to 50.

The stem-and-leaf plot is set-up with the stems in the left-hand column and, as we go through the data, the leaves are written in the right-hand column. For example, for the number 25, the leaf, 5, is put in the right-hand column beside the stem, 2.

Each number is recorded in a similar manner, making sure the leaves are evenly spaced.

Recording all the data as we read it from the list gives an **unordered stemplot**:

Stem	Leaf
0	9
1	5 6
2	5 3 8 4 7 7 8 4 5 9
3	6 8 8 4 0 1 1 7 5 2 9 6 5 3
4	6 7 2 6 3 0 7
5	0

Key: 2 | 5 means 25 marks.

Ordering the data, from smallest to largest, for each stem results in a table that is ordered for the whole set of data (an **ordered stem-and-leaf plot**).

Stem	Leaf
0	9
1	5 6
2	3 4 4 5 5 7 7 8 8 9
3	0 1 1 2 3 4 5 5 6 6 7 8 8 9
4	0 2 3 6 6 7 7
5	0

Remember

An **ordered set of numbers** is the numbers listed from smallest to largest (or from largest to smallest).

The data in the 'leaves' is evenly spaced (with no commas) so the shape of the graph can be seen as if it is viewed sideways:

Stem	Leaf
0	9
1	5 6
2	3 4 4 5 5 6 7 7 8 9
3	0 1 1 2 3 4 5 5 6 6 7 8 8
4	0 2 3 6 6 7 7 9
5	0

From the ordered stem-and-leaf plot, we can see the following important features:

i All the actual data values.

ii The minimum (smallest) data value (9).

iii The maximum (largest) data value (50).

iv The grouping of the data values and, in particular, the range of values that occurs most often in the data set (30–39).

v The general shape of the frequency graph (called the distribution of the data).

vi The descriptive statistics (the median, the range and the upper quartile and lower quartile can be found easily).

EXERCISE

Remember

Describing the distribution of a set of graphed data means commenting on the shape of the graph as either symmetric, positively skewed or negatively skewed. A positively skewed set of data is stretched towards the larger values.

1 The winning scores from 32 basketball games have been recorded and organised into the stem-and-leaf plot below.

Winning basketball scores

Stem	Leaf
3	5 9
4	7 5 9 6 2 6 7 8 9
5	4 8 4 0 1 3 4 4 3 4 5 6 8
6	6 2 1 2 4 6 6
7	1

Key: 4 | 6 means 46

a Rewrite this stem-and-leaf plot so that it is ordered.

b Describe the distribution of the data.

c Copy and complete the following statement about the data:

'Most frequently, the winning teams scored between ___ and ___ goals. The minimum winning score was ___ and the maximum was ___. ____% of the winning teams scored more than 60 goals, and ____% of the winning teams scored 45 or fewer goals.'

2 Data, from an investigation into the number of phonecalls made per week by 50 businesses, is given in the stem-and-leaf plot below.

Number of phonecalls per week

Stem	Leaf
0	2 4 5 5 7 9
1	0 0 1 1 2 6 7 8
2	1 1 2 3 5 6 6 6 8 9
3	0 0 1 1 2 2 4 6 7 7 7 9 9
4	0 2 2 3 4 4 7 9
5	2 4 4 5 5

Key: 3 | 4 means 34

a Is this stem-and-leaf plot ordered or unordered?

b What is the:

i minimum number of phonecalls per week made by a business in this investigation?

ii maximum number of phonecalls per week made by a business in this investigation?

c What percentage of the businesses in the survey made fewer than 30 phonecalls in the week?

d What percentage of the businesses in the survey made more than 40 phonecalls in the week?

e Describe the distribution of the data.

3 The following marks were scored on a test for which the maximum score was 100:

61	78	51	50	75	76	81	70	60	65
53	72	71	69	59	72	64	66	82	56
76	34	75	65	71	75	57	70	83	43

a Construct a stem-and-leaf plot of the data, using the stems: 3, 4, … 8.

b What percentage of the students scored 80 or more marks?

c If a score of 50 or more is a pass, what percentage of the students passed?

d Describe the distribution of the marks.

e Are there any data values that you would consider to be extreme values?

4 An investigation is being carried out into the number of fuel-powered vehicles (cars, trucks, or motorbikes) travelling along a particular road in a 24-hour period. The following data was recorded over 30 days:

121	133	94	94	129	130	145	127	88	96
145	143	138	136	122	136	110	119	106	111
132	147	107	103	154	140	135	131	148	128

a Construct a stem-and-leaf plot for the data, using 08, 09, 10, 11, 12, 13, 14 for the stems.

b Rewrite the stem-and-leaf plot so that it is ordered.

c What were the maximum number and minimum number of vehicles per day that travelled along the road in the 30 days?

d Copy and complete the following to describe the most popular grouping for this data:

'Most often there were between ___ and ___ vehicles per day along this road.'

e On what percentage of the days were there fewer than 100 vehicles travelling along the road?

f Describe the distribution of the data.

g Copy and complete the following statement:

'On half of the days there were ____ or more vehicles on the road.'

5 An investigation is being made into the number of mangoes produced by a particular type of mango tree grown in a district.

The number of ripening mangoes on 50 trees is counted, giving the following data.

85	102	75	74	94	100	105	94	84	82
77	96	95	93	83	96	88	83	106	80
96	83	98	89	95	99	81	94	86	101
77	68	94	87	103	92	79	87	96	104
87	87	99	101	90	91	84	78	114	69

a Construct a stem-and-leaf plot for the data.

b Comment on the data, mentioning the distribution, and the most frequently occurring class interval, and quoting some percentages.

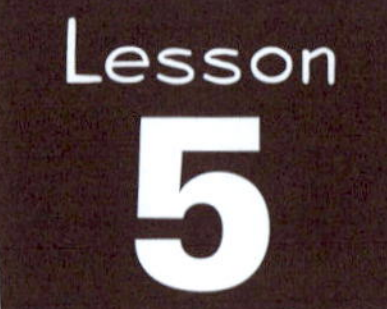

Organising, displaying and describing continuous numerical data

When continuous numerical data is recorded, there are likely to be many different values, so this data is organised by grouping it into **class intervals**.

A **tally and frequency table** can be used to organise the data and a **histogram** can be used to graph the data.

Alternatively, because the data is usually rounded to a set number of decimal places, a **stem-and-leaf plot** can be used to both organise and display the data. An example is given below.

The weights of the parcels sent on a particular day from a post office are recorded, in kilograms, rounded to one decimal place:

2.1 3.0 0.6 1.5 1.9 2.4 3.2 4.2 2.6 3.1
1.8 1.7 3.9 2.4 0.3 1.5 1.2

An ordered stem-and-leaf plot for this data would look like the one on the right.

For stemplots where the numbers are not whole numbers, it is essential to include a 'key' to explain the digits in the table.

For this stemplot, the key is:

Key: 1 | 2 represents 1.2 kg.

Stem	Leaf
0	3 6
1	2 5 5 7 8 9
2	1 4 4 6
3	0 1 2 9
4	2

EXERCISE

1 Construct a histogram for the continuous numerical data in the frequency table below.

Time to complete 100 m swim (seconds)	Number of swimmers
50–	3
55–	6
60–	16
65–	11
70–	2
75–	2

Remember

The class interval 50– would include all times from 50 seconds up to, but not including, 55 seconds.

Help box

On a histogram, the horizontal axis is a number line and the columns are joined together. The height of the columns represents the frequency.

The distribution of data in this example is positively skewed. There are no extreme values.

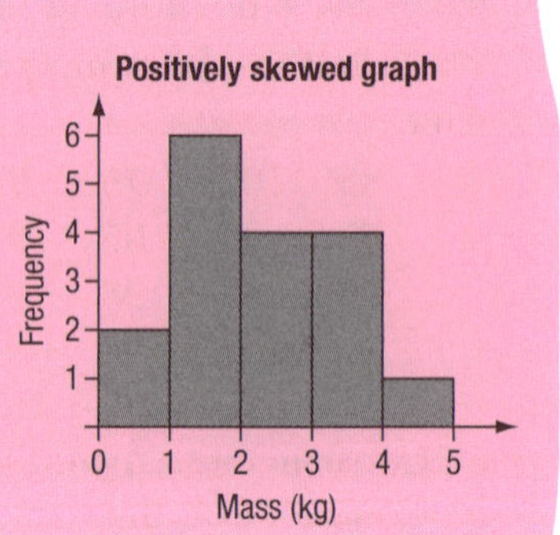

2 The speeds of vehicles travelling along a section of highway have been recorded and then displayed using the following histogram:

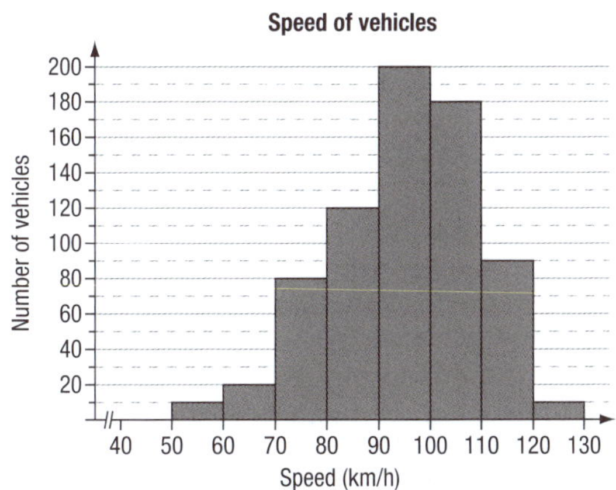

a How many vehicles were included in this survey?

b What percentage of the vehicles were travelling at speeds equal to or greater than 100 km/h?

c What percentage of the vehicles were travelling at speeds from 100 km/h up to 110 km/h?

d What percentage of the vehicles were travelling at speeds less than 80 km/h?

e If the owners of the vehicles travelling at 110 km/h or more were fined K180 each, what amount would be collected in fines?

3 The daily maximum temperatures (in °C) in Mount Hagen for one month are recorded below:

34 38 31 38 23 24 25 26 29 35 41 23 32 36 22 21
24 26 35 36 25 32 27 30 34 30 27 25 26 23 25

a Using class intervals of 5 degrees, construct a tally and frequency table for the data.

b Construct a histogram to display the data.

c Describe the distribution of daily maximum temperatures in Mount Hagen for the month.

d Construct a 'split-level' stem-and-leaf plot for the data using the stem '2' for the values 20°C to 24°C and the stem '2*' for the values 25°C to 29°C, and so on.

f What is the advantage of a stem-and-leaf plot, compared to a histogram?

Stem	Leaf
2	
2*	
3	
3*	
4	

4 Loaves of bread are intended to each weigh 900 g but the weight varies slightly from loaf to loaf. A baker has weighed a sample of sixty 900 g loaves of bread and has recorded their weights to the nearest gram:

901	904	913	924	921	893	894	895	878	885	896	910
901	903	907	907	904	892	888	905	907	901	915	901
909	917	889	891	894	894	898	895	904	908	913	924
927	885	898	903	903	913	916	931	882	893	894	903
900	906	910	928	901	896	886	897	899	908	904	889

a Copy and complete the stem-and-leaf plot on the right for the data about loaf weights from the previous page. The 'stems' have been split to give a better display of the distribution of weights. The stem 87 is used for weights from 870 g to 874 g and the stem 87* is the stem for weights from 875 g to 879 g, and so on.

b What percentage of the loaves weigh 900 grams or more?

c What percentage of the loaves weigh less than 890 grams?

d Describe the distribution of weights.

e The baker would like 95% of the loaves to weigh between 890 g and 920 g inclusive. Is this the case for this sample? Explain.

Stem	Leaf
87	
87*	
88	
88*	
89	
89*	
90	
90*	
91	
91*	
92	
92*	
93	

'Inclusive' means that 890 g and 920 g are both included.

5 The weight, in kilograms, of baggage going through the check-in counter of an airline for a particular flight is given below:

0.6	1.5	5.7	10.6	34.2	25.3	7.2	15.4	23.6	34.8
14.6	18.5	22.3	28.6	3.4	11.6	5.8	38.9	19.4	39.9
30.0	10.8	15.0	26.4	7.9	20.6	11.2	35.8	19.9	33.3

a Construct a tally and frequency table using the class intervals 0–, 5–, 10–, 15–, 20–, 25–, 30– and 35–.

b Draw a histogram to display the data.

c Why would it not be practical to construct a stem-and-leaf plot for this data?

Lesson 6

Cumulative frequency tables and graphs

When the data is **grouped** a **cumulative frequency graph** is used to estimate the number (or percentage) of the data that lies above or below a particular value. Cumulative frequency tables and graphs can be produced for both discrete and continuous numerical data.

Example

The times taken by each of 40 swimmers to complete a 100 m swim is recorded in the table of data on the right:

Time to complete 100 m swim (seconds)	Number of swimmers
50–	3
55–	6
60–	16
65–	11
70–	2
75–	2

From the table of data, a cumulative frequency table and graph can be drawn.

Time	Cumulative frequency
less than 50 (< 50)	**0**
< 55	0 + 3 = **3**
< 60	0 + 3 + 6 = **9**
< 65	0 + 3 + 6 + 16 = **25**
< 70	0 + 3 + 6 + 16 + 11 = **36**
< 75	0 + 3 + 6 + 16 + 11 + 2 = **38**
< 80	0 + 3 + 6 + 16 + 11 + 2 + 2 = **40**

The cumulative frequency graph below has been constructed by plotting and joining the cumulative frequency values represented by the coloured values in the table above: (50, 0), (55, 3), (60, 9) … (80, 40).

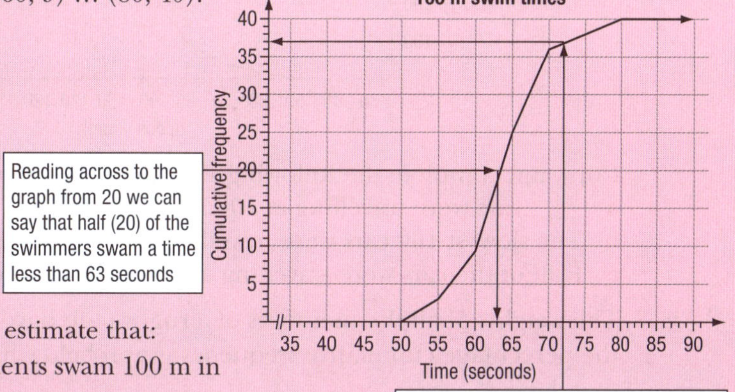

From the graph, we can estimate that:

i half (20) of the students swam 100 m in 63 seconds or less

ii 37 swimmers swam a time of 72 seconds or less (or 3 swimmers swam a time of more than 72 seconds).

A **percentage cumulative frequency** table and graph can be produced for the same data:

Time	Cumulative frequency	% cumulative frequency
less than 50 (< 50)	0	0
< 55	3	7.5
< 60	9	22.5
< 65	25	62.5
< 70	36	90
< 75	38	95
< 80	40	100

The percentage cumulative frequency is calculated as follows:

$\frac{9}{40} \times \frac{100}{1} = 22.5\%$

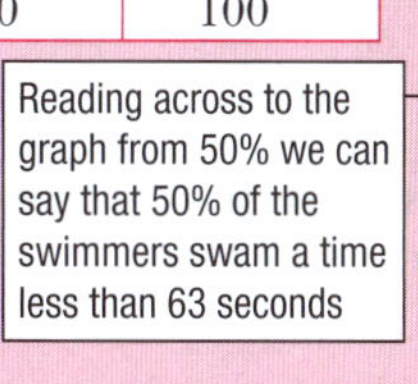

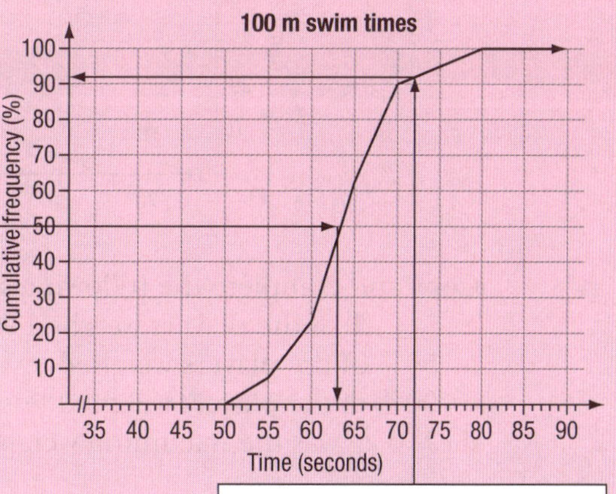

EXERCISE

1 The speeds of 710 vehicles travelling along a section of highway have been recorded (see Question 2 on page 177). A cumulative frequency graph has been produced below for this data.

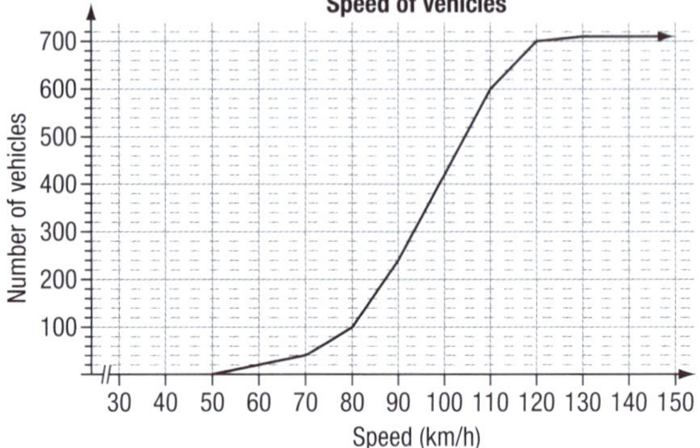

Copy and complete the following statements by referring to the graph:

a ____ cars were travelling at 100 km/h or less.

b The slowest 100 cars were travelling at ____ km/h or less.

c Half of the cars were travelling at ____ km/h or less.

2 The weights of all the members of a rugby club have been recorded and the percentage cumulative frequency graph below has been produced from that data.

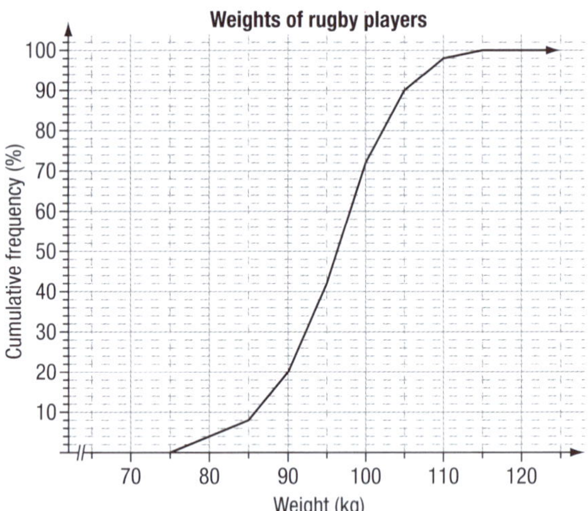

Copy and complete the following statements by referring to the graph:

a ____% of the players weighed 90 kg or less.

b 50% of the players weighed less than ____ kg.

c 90% of the players weighed less than ____ kg.

d 110 kg was the maximum weight for ____% of the players.

3 The cumulative frequency table below shows the heights of a sample of 5-year-old children.

Height (cm)	Cumulative frequency
< 95	0
< 100	4
< 105	19
< 110	38
< 115	47
< 120	50

a How many 5-year-old children were in this sample?

b Construct a cumulative frequency graph from this table, using 'Height (cm)' along the horizontal axis and 'Cumulative frequency' on the vertical axis.

c Copy and complete the following statements by referring to your graph:

i The 5-year-old children in this sample all had a height less than ____ cm.

ii Half of the children had a height less than ____ cm.

iii The shortest 10 children had a height of ____ cm or less.

iv The tallest 10 children were at least ____ cm tall, but not more than ____ cm tall.

4 The number of phonecalls made per week by a sample of 50 businesses has been recorded. The results are given in this frequency table.

Number of phonecalls	Frequency
0–	6
10–	8
20–	10
30–	13
40–	8
50–	5

a Copy and complete the following cumulative frequency and percentage cumulative frequency table.

Number of phonecalls	Cumulative frequency	Percentage cumulative frequency
< 10	6	
< 20	14	
< 30		48
< 40		
< 50		
< 60	50	100

b Construct a percentage cumulative frequency graph for this data.

c Copy and complete the following statements:

i 50% of the businesses in the sample made fewer than ____ phonecalls/week

ii 80% of the businesses made fewer than ____ phonecalls/week.

iii ____ % of the businesses made fewer than 35 phonecalls/week.

iv ____% of the businesses made fewer than 15 phonecalls/week.

Remember

< 10 means 'less than 10'.

MEASURES OF CENTRAL TENDENCY AND SPREAD

Measures of centre

A **measure of centre** provides a single value that indicates the centre of a set of data. There are three statistics that provide a measure of centre.

1 **Mean** is the statistical name for **'average'**.

$$\text{Mean} = \frac{\text{sum of data values}}{\text{number of data values}}$$

2 The **median** is the **middle value** of an ordered set of data. If there is an even number of data values, the two middle values are averaged to find the median.

3 The **mode** is the most frequently occurring value in the data set.

Comparing measures of centre

- The mean and the median do not need to be one of the data values.
- The mode is always one of the data values.
- If there are two modes then the set of data is described as bi-modal.
- The mean is affected by extreme values.
- The median is not affected by extreme values. If there are extreme values in a data set then the median is usually a better measure of centre.
- The mode is usually not affected by extreme values but it is not always a representative centre value for the data set.

Finding measures of centre for data given in frequency tables or graphs

Example

From this frequency table and dotplot (which show the same data), find the measures of centre:

a mean

b median

c mode.

Data value	Frequency	Data value × frequency
2	1	$2 \times 1 = 2$
4	1	$4 \times 1 = 4$
5	2	$5 \times 2 = 10$
6	3	$6 \times 3 = 18$
7	4	$7 \times 4 = 28$
8	9	$8 \times 9 = 72$
9	6	$9 \times 6 = 54$
Total:	**26**	**188**

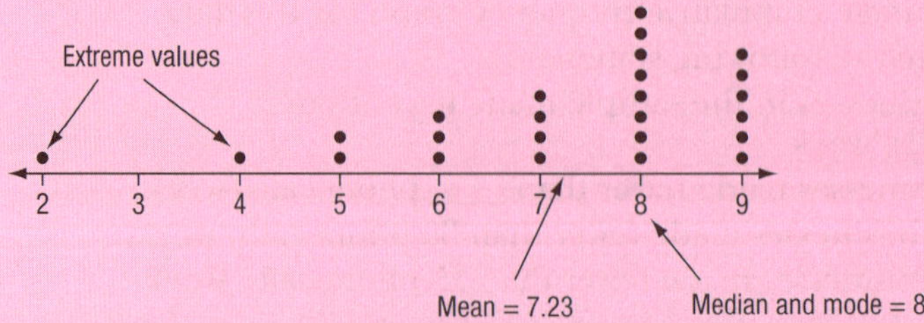

Answer

In this set of data there is one 2, one 4, two 5s, three 6s, four 7s, nine 8s and six 9s. So, to find the sum of the values in the table, we can multiply together the values in the first two columns (or, for the dotplot, we can multiply the number of dots by the data value each time). The total number of data values is the sum of the frequency column (or the number of dots).

a The **mean** $= \frac{188}{26} = 7.23$

b There are 26 pieces of data, so the **median** will be the average of the 13th and 14th values when the data is written in order;

2	4	5	5	6	6	6	7	7	7	7	8	**8**	**8**	8	8
8	8	8	8	9	9	9	9	9	9						

Both the 13th and the 14th values are '8'. The average of 8 and 8 is 8, so the **median is 8**.

c The **mode** is the value with the highest frequency (which is 8, because it has a frequency of 9). So the **mode is 8**.

Note: This data set is negatively skewed (stretched towards the left) so the **mean** has been dragged towards the extreme values.

Finding measures of centre from a stem-and-leaf plot

Example

From this ordered stem-and-leaf plot, find the:

a mean
b median
c mode.

Stem	Leaf
1	6 7 8 8
2	2 3 3 4 4 6 (7) 8 9
3	1 2 5 8
4	0 4 6
5	1

The median is 27, the 11th value.

Answer

a The **mean** is found by dividing the sum of all the data values by the number of data, making sure that the 'stem' is included with the 'leaf' each time:

Mean = $\frac{16 + 17 + 18 + 18 + 22 + 23 + 23 + 24 + 24 + 26 + 27 + 28 + 29 + 31 + 32 + 35 + 38 + 40 + 44 + 46 + 51}{21}$

$= 29.14$ (rounded to two decimal places)

b The **median** is the middle value of the 21 data values. That is the 11th value in this ordered data set. So, counting the leaves from the beginning to find the 11th value shows us the median is 27.

c The **mode** is the most frequently occurring value. there are two 18s, two 23s and two 24s in this set of data. So we can say that the mode is not distinct in this case and is not useful as a measure of centre.

Note: The mean (29.14) is larger than the median (27) indicating that the distribution is positively skewed.

For this **negatively skewed data set**, the mean is less than the median and mode because the mean is dragged towards the extreme small values.

In cases where there are extreme data values and the distribution of the data is skewed, the most suitable measure of centre to use is the **median** or the **mode**.

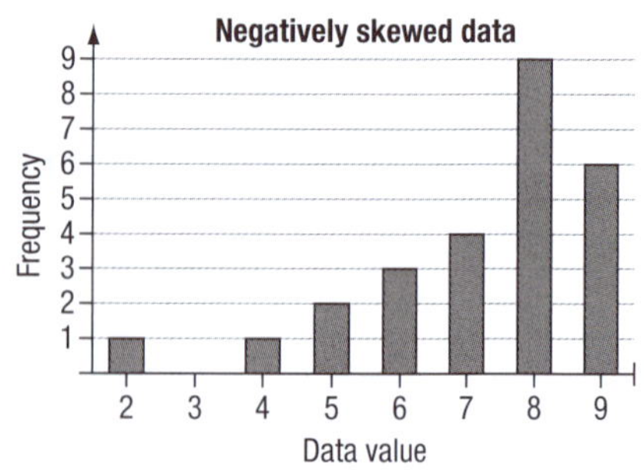

For this **symmetric data set** the mean, median and mode are all the same value (5). The graph is **symmetric** about the value 5.

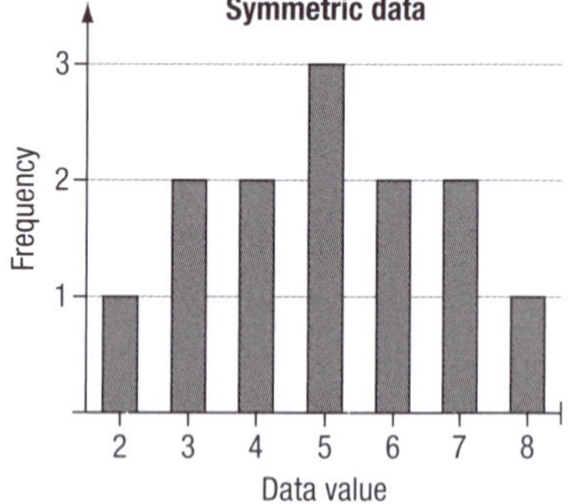

A **positively skewed set of data** would appear on a graph to be stretched to the right and the mean (19) would be greater than the median and mode (both in the 10–15 class interval). This set of data has an extreme value in the 60–65 class interval.

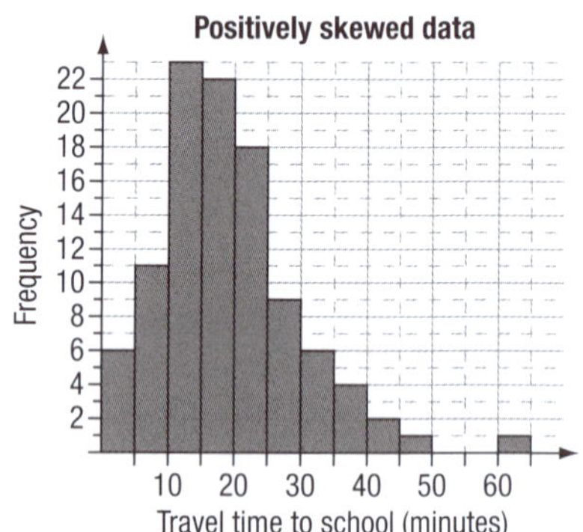

EXERCISE

1 Find the mean, median and mode for each of the following data sets.

a 3, 3, 4, 4, 4, 5, 5, 6, 6, 6, 6,7, 7, 8, 9

b 10, 11, 10, 15, 12, 15, 16, 10, 17, 12, 18, 14

c

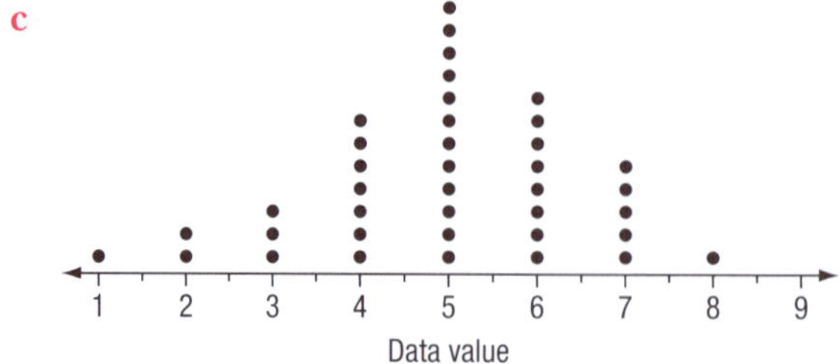

d

Data value	Frequency
5	10
6	12
7	15
8	11
9	9

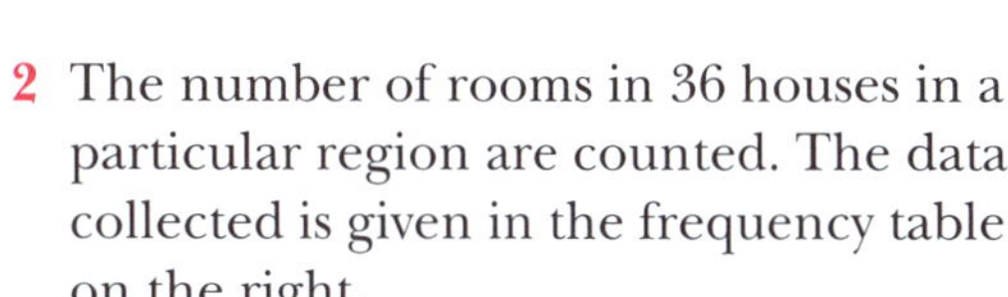

Number of rooms	Frequency
2	4
3	10
4	12
5	6
6	2
7	1
8	1

2 The number of rooms in 36 houses in a particular region are counted. The data collected is given in the frequency table on the right.

- **a** Construct a column graph for the data.
- **b** Describe the distribution of the data.
- **c** For this data, calculate:
  - **i** the mean
  - **ii** the median
  - **iii** the mode.
- **d** Copy and complete the following statement about the data:
  '___% of the houses in the survey had 4 rooms and the median number of rooms was ___. ___% of the houses had more than 5 rooms and ___% had 2 rooms. The maximum number of rooms for houses in this survey was ___.'

3 The test scores out of 30 marks for a class of 22 students are:

15	16	18	23	22	28	29	25	25	24	27
18	11	20	23	26	26	30	25	18	15	17

- **a** For this data, find:
  - **i** the mean
  - **ii** the median
  - **iii** the mode.
- **b** Give a reason why the mean is not the most suitable measure of centre for this set of data.
- **c** Give a reason why the mode is not the most suitable measure of centre for this set of data.

Number of messages	Frequency
0	5
1	8
2	11
3	10
4	6
5	3
6	3
7	2
8	1
9	0
10	0
11	1

4 The frequency table on the right records the number of text-messages made in a day by 50 fifteen-year-olds.

- **a** For this data, find:
  - **i** the mean
  - **ii** the median
  - **iii** the mode.
- **b** Construct a column graph for the data and show the position of the measures of centre (mean, median and mode) on the horizontal axis.
- **c** Describe the distribution of the data.
- **d** Why is the mean larger than the median for this data?
- **e** Which measure of centre would be the most suitable for this data set?

5 Estimate the mean, median and mode for each of the data sets graphed below.

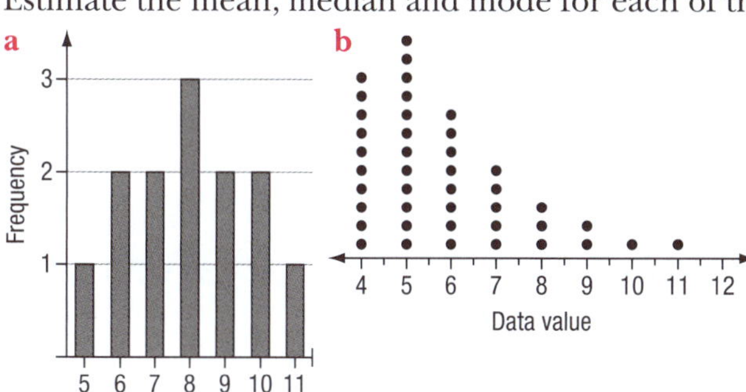

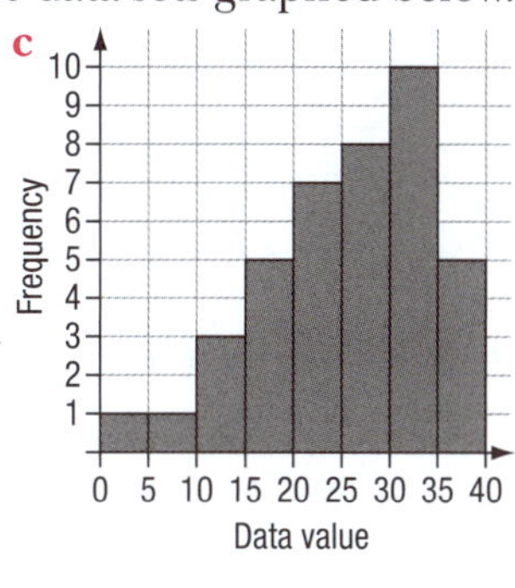

Measures of spread

A quick idea of a data set can be obtained if we have an indication of the centre of the data and the **spread** of the data.

Two commonly used statistics that indicate the spread of a set of data are:

i **Range** = (maximum data value – minimum data value)

ii **Interquartile range** (IQR) = (upper quartile – lower quartile)

The upper and lower quartiles

The median divides an ordered data set into two halves and these halves are divided in half again by the **quartiles**.

The middle value of the lower half of the data is called the **lower quartile**. One-quarter, or 25%, of the data values are less than the lower quartile. Three-quarters, or 75%, of the data values are greater than the lower quartile.

The middle value of the upper half of the data is called the **upper quartile**. One-quarter, or 25%, of the data values are greater than the upper quartile. Three-quarters, or 75%, of the data values are less than the upper quartile.

Example

For the following set of data, find:

a the median
b the range
c the upper quartile
d the lower quartile
e the interquartile range

2 3 3 4 4 5 5 5 6 6 7 7 8

Answer

a The median is the middle value when the data is placed in order, so the median is 5.

b The range is: **Maximum value – minimum value** $= 8 - 2$
$= 6$

c The median divides the data into two sets of six (an even number) so the upper quartile is the average of the middle values of the second six values (the ninth and tenth values, 6 and 7): Upper quartile $= \frac{6+7}{2}$
$= 6.5$

d The lower quartile is the average of the middle values of the first six values (the third and fourth values, 3 and 4): Lower quartile $= \frac{3+4}{2}$
$= 3.5$

Range = 8 – 2 = 6

2 3 3 | 4 4 5 5 5 6 6 | 7 7 8

3.5 Lower quartile
5 Median
6.5 Upper quartile

Interquartile range = 3

e Interquartile range = upper quartile – lower quartile
$= 6.5 - 3.5$
$= 3$

The interquartile range is the spread of the middle half (middle 50%) of the data. To describe this data set in words, we would say:

> 'The data is spread over 6 values (range = 6); centred around the value 5 (median = 5). The middle half of the data is spread over 3 values (interquartile range = 3).'

Consider an **ordered** data set with the 18 data values given here:

3 3 4 5 5 6 6 7 7 8 8 9 9 9 10 11 12 14

The median is not one of the data values. In this case, the median divides the data set into two sets of nine values:

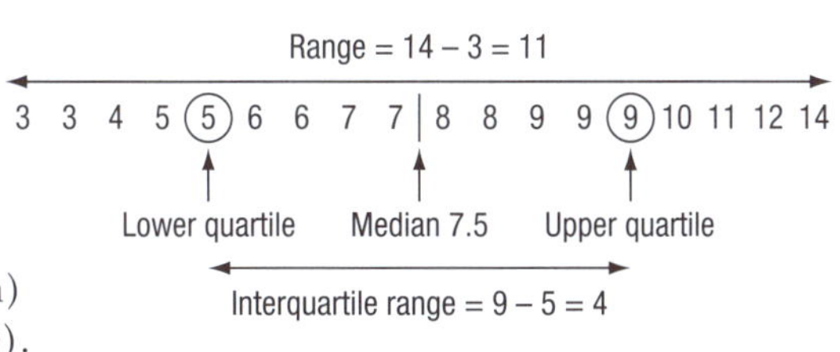

To describe the distribution of this data set in words, we would say:

'The data is centred at 7.5 (the median) and is spread over 11 values (the range). The middle half of the data is spread over 4 values (the interquartile range).'

EXERCISE

1 For each of the following data sets find:

- **i** the median (make sure the data is ordered)
- **ii** the upper and lower quartiles
- **iii** the range
- **iv** the interquartile range.

a 2, 3, 3, 3, 4, 4, 4, 5, 5, 5, 5, 6, 6, 6, 6, 6, 7, 7, 8, 8, 8, 9, 9

b 10, 12, 15, 12, 24, 18, 19, 18, 18, 15, 16, 20, 21, 17, 18, 16, 14

c

Data value	Frequency
3	2
4	7
5	10
6	8
7	3

d

0 1 2 3 4 5 6 7 8 9 10 11 12 13

Data value

2 The time, in minutes, spent by 20 people in a queue at a bank, waiting to be attended by a teller, has been recorded:

3.4	2.1	3.8	2.2	4.5	1.4	0	0	1.6	4.8
1.5	1.9	0	3.6	5.2	2.7	3.0	0.8	3.8	5.2

a Write the data in order.

b Find:

- **i** the median waiting time
- **ii** the upper quartile
- **iii** the lower quartile.

c Calculate:

- **i** the range of the waiting times
- **ii** the interquartile range of the waiting times.

d Copy and complete the following statements about the data:

'50% of the waiting times were greater than _____ minutes and 75% of the waiting times were less than _____ minutes. The minimum waiting time was ____ minutes and the maximum time was ____ minutes. The waiting times were spread over a range of ____ minutes.

3 A class of 32 students was asked 'How many aunts (mother's or father's sisters) do you have?' The following data was recorded:

3	3	5	7	2	1	4	4	6	4	4	5	5	3	6	2
1	3	3	4	4	5	4	6	7	6	4	4	2	5	3	5

a Write the data in order.

b Find:

i the median number of aunts ii the upper quartile

iii the lower quartile.

c Calculate:

i the range of the numbers of aunts

ii the interquartile range of the numbers of aunts.

d Copy and complete the following statements:

'50% of the students had more than ____ aunts and 75% of the students had fewer than ____ aunts. The minimum number of aunts for students in this class was ____ and the maximum number was ____ aunts. The number of aunts was spread over a range of ____.'

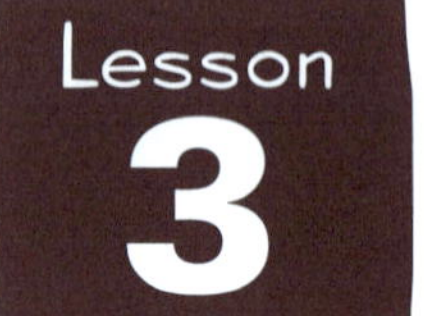

Lesson 3

Finding measures of centre and spread from stem-and-leaf plots and describing the distribution of the data set

The median, range, and interquartile range can be found easily from an ordered stem-and-leaf plot.

Example

For the data in this stem-and-leaf plot, find:

a the range

b the median

c the interquartile range.

Stem	Leaf
1	6 7 8 8
2	2 4 7 8 9
3	0 2 3 3 4 5 6 8
4	0 4 6
5	1

Answer

The data is ordered, so we can read the data from the smallest value to the largest value, combining the 'stem' with the 'leaf':

16, 17, 18, 18, 22, 24, 27, 28, 29, 30, 32, 33, 33, 34, 35, 36, 38, 40, 44, 46, 51

a The **minimum** is 16 and the **maximum** is 51, which means that: **Range** = 51 – 16 = 35

b The median is the middle value (the 11th data value in the list of 21 values) and, counting from the beginning, the **median** is 32 (circled in the stem-and-leaf plot shown).

Stem	Leaf
1	6 7 8 8
2	[2 4] 7 8 9
3	0 (2) 3 3 4 5 [6 8]
4	0 4 6
5	1

c The median divides the data into two groups of 10 data values. The average of the middle values of these two groups will be the lower quartile and the upper quartile respectively (boxed in the stem-and-leaf plot shown).

Lower quartile $= \frac{22 + 24}{2} = 23$

Upper quartile $= \frac{36 + 38}{2} = 37$

Interquartile range = upper quartile – lower quartile $= 37 - 23 = 14$

EXERCISE

1 For the data given in this stem-and-leaf plot, find:

a the median
b the upper quartile
c the lower quartile
d the range
e the interquartile range.

Stem	Leaf
0	3 4 7 9
1	0 3 4 6 7 8
2	0 0 3 5 6 9 9 9
3	1 3 7 8
4	2

2 The data in this stem-and-leaf plot is the result of counting the number of plants in the vegetable gardens of 29 households.

Stem	Leaf
2	0 2 5 5 8 9
3	0 1 3 5 6 6 8 9
4	2 2 4 7 7 8 9
5	0 0 1 2 6
6	2 5
7	2

a Find:
  i the median number of plants per household
  ii the upper quartile of the data
  iii the lower quartile of the data.

b Copy and complete the following statements:
'Half of the households have more than ____ plants.'
'75% of the households have more than ____ plants.'

c Find:
  i the range
  ii the interquartile range.

3 The height of 20 six-year-old children is recorded in the stem-and-leaf plot on the right.

Stem	Leaf
10	9
11	1 3 4 4 8 9
12	2 2 4 4 6 8 9 9
13	1 2 5 8 8

a For this data, find:
  i the median height
  ii the upper quartile
  iii the lower quartile.

b Copy and complete the following statements:
'Half of the children are more than ____ cm tall.'
'75% of the children are less than ____ cm tall.'

c For this data, calculate:
  i the range
  ii the interquartile range.

d Copy and complete:
'The children in the survey have heights spread over a range of ____ cm and the middle 50% of the children have heights spread over _____ cm.'

4 The number of PMV trips made per month by 50 people, is given in the stem-and-leaf plot on the right.

Number of PMV trips per month

Stem	Leaf
0	2 4 5 5 7 9
1	0 0 1 1 2 6 7 8
2	1 1 2 3 5 6 6 6 8 9
3	0 0 1 1 2 2 4 6 7 7 7 9 9
4	0 2 2 3 4 4 7 9
5	2 4 4 5 5

a For this data, find:
  i the median
  ii the upper quartile
  iii the lower quartile.

b For this data, find:
  i the range
  ii the interquartile range.

c Comment on the data, referring to the statistics found in parts a and b.

5 The minimum daily temperature (in °C) for Port Moresby in July has been recorded over a month (31 days):

21.4	24.6	19.8	19.4	23.2	24.3	25.1	23.1	21.3	20.9	20.1
23.4	23.2	22.9	21.2	23.5	22.0	21.2	25.3	20.6	23.5	21.3
23.9	22.2	23.2	23.9	20.8	23.2	21.7	25.0	18.4		

a Construct an ordered stem-and-leaf plot for this data.

b Find:

  i the median minimum daily temperature

  ii the upper quartile

  iii the lower quartile.

c Calculate:

  i the range **ii** the interquartile range.

d Copy and complete the following statements:

'75% of the minimum daily temperatures were greater than ______.
The minimum daily temperature, in Port Moresby, for July, was centred around ____°C and was spread over a range of _____°C.'

The boxplot (or box-and-whisker plot)

A boxplot is a visual display of the following descriptive statistics of a set of data:

- the minimum value
- the maximum value
- the median
- the upper quartile
- the lower quartile.

These five statistics form what is called the **five-number summary** of a data set.

Drawing a boxplot for a set of data

Example

Draw a boxplot for the following data:

2 3 5 4 3 6 5 7 3 8 1 7 5 5 9

Answer

We first need to order the data and find statistics that give us the five-number summary:

1 2 3 [3] 3 4 5 (5) 5 5 6 [7] 7 8 9 (15 pieces of data)

The **minimum** is 1.

The **maximum** is 9.

The **median** is the 8th value, which is 5 (circled above).

The **lower quartile** is the 4th value, which is 3 (boxed above).

The **upper quartile** is the 12th value which is 7 (boxed above).

A boxplot is constructed above a number line which is drawn to a scale that covers all the data values in the data set.

The boxplot is drawn with a rectangular 'box' representing the middle half of the data (the length of the rectangle reaches from the lower quartile to the upper quartile on the number line).

The two 'whiskers' extend from the ends of the rectangle to the maximum value and the minimum value on the number line.

A vertical line marks the position of the median in the rectangular 'box'.

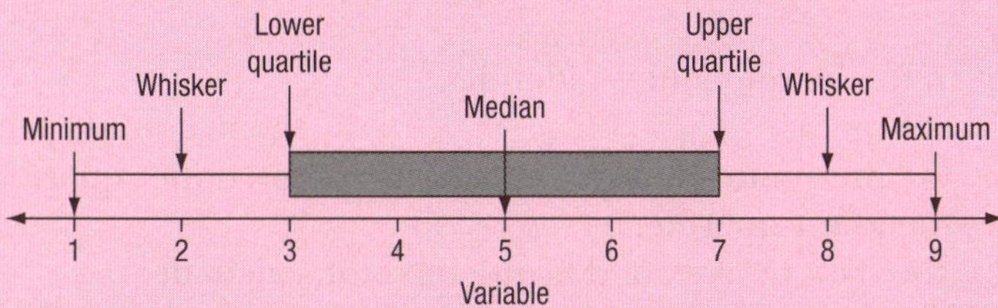

What does this boxplot tell us about the data?

The data is symmetrically distributed (the median, 5, is in the centre of the 'box' and the whiskers are the same length).

The data is spread over eight values (the range is 9 – 1 = 8) and the middle half of the data is spread over four values (the interquartile range is 7 – 3 = 4).

Note: Boxplots can be drawn for both discrete data and continuous data.

The boxplot for a negatively skewed set of data has a longer left whisker and generally appears stretched to the left:

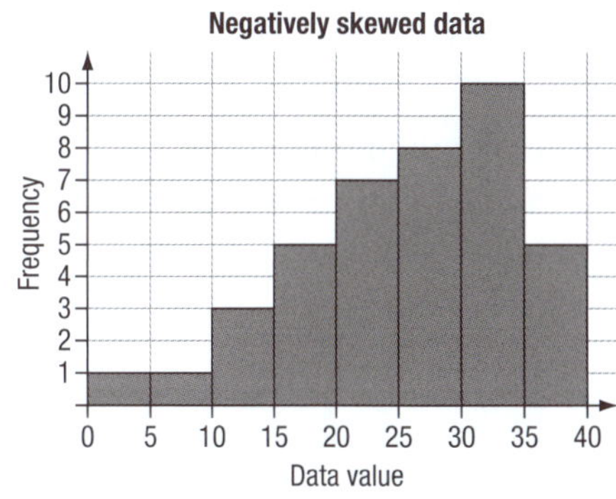

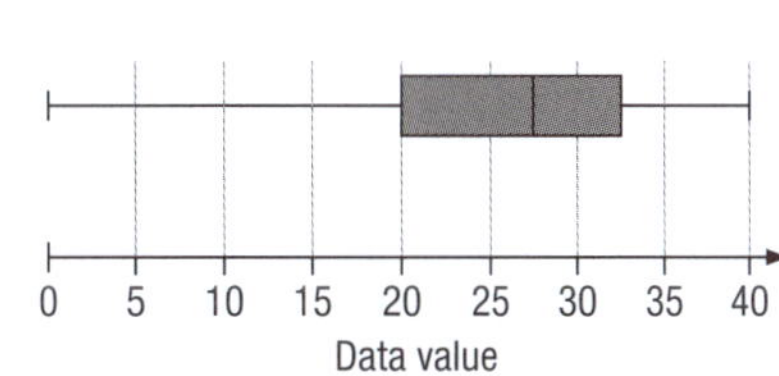

The boxplot for a positively skewed set of data appears stretched to the right:

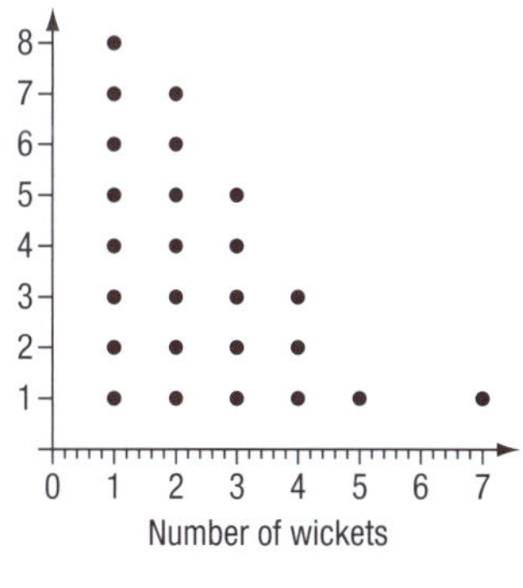

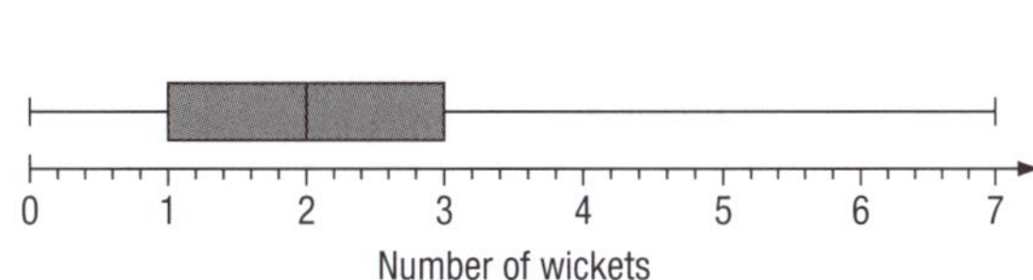

EXERCISE

1 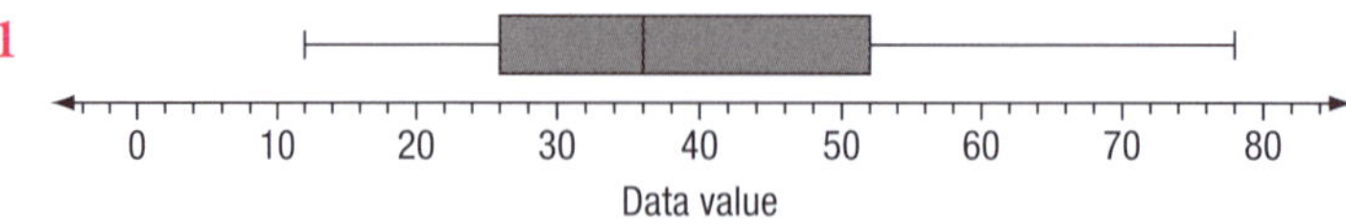

a From the boxplot above, identify:
   i the median
   ii the maximum value
   iii the minimum value
   iv the upper quartile
   v the lower quartile.

b For the data represented by the boxplot, calculate:
   i the range
   ii the interquartile range.

2 This boxplot summarises the results of a class for a test:

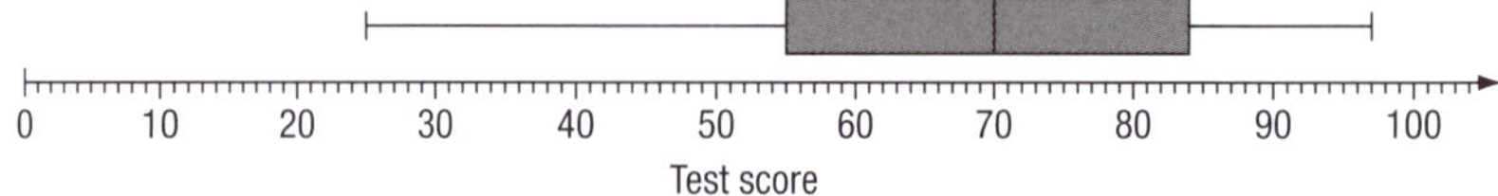

Copy and complete the following statements about the test results.

a The highest mark scored on the test was ___.

b The lowest mark scored on the test was ___.

c Half of the class scored a mark greater than ___.

d The top 25% of the class scored at least ___ marks on the test.

e The middle half of the class had scores between ___ and ___ on this test.

3 Draw a boxplot to display the following statistics for the number of cars travelling along a particular road. (Make sure you carefully scale and label the number line used.)

- Median = 32
- Lower quartile = 23
- Upper quartile = 37
- Minimum = 16
- Maximum = 51

4 The heights, in centimetres, of a sample of 16-year-olds have been measured and the following five-number summary of the data was calculated:

- Median = 162.5
- Lower quartile = 152
- Upper quartile = 168
- Minimum = 148
- Maximum = 185

a Draw, and carefully scale and label, the number line that covers all the values in the five-number summary.

b Draw the boxplot for the five-number summary above your number line.

c Calculate the range and interquartile range for this sample of 16-year-olds.

d Describe, in words, the distribution of the heights in the sample.

5 The weight, in grams, of a loaf of a particular brand of bread is stated to be 900 g. However, some loaves weigh more than this and some weigh less. A sample of loaves is carefully weighed and their weights are recorded in the following ordered stem-and-leaf plot.

Stem	Leaf
88	3 5 9
89	2 2 5 7 8 8 9
90	1 1 1 1 2 3 3 3 4 4 5 6 6
91	0 1 1 1 2 3 3 3 6 7 7 8 9
92	0 1 4 7

a For this set of data, identify:
   i the median
   ii the upper quartile
   iii the lower quartile
   iv the maximum weight
   v the minimum weight.

b Draw a boxplot for the data.

c For this data, find:
   i the interquartile range
   ii the range

d Copy and complete the following statements about the distribution of weights for the loaves of bread in this sample:
   i Half of the loaves of bread weighed ___ grams or more.
   ii ___% of the loaves weighed less than 898 grams.
   iii The weights of the middle 50% of the loaves in this sample were spread over ___ grams.
   iv The lightest 25% of the loaves weighed ___ grams or less.

e Is the distribution of weights in this sample symmetric, positively skewed or negatively skewed?

6 Two classes completed the same test. The boxplots drawn to summarise and display the results have been drawn on the same set of axes so that the results can be compared:

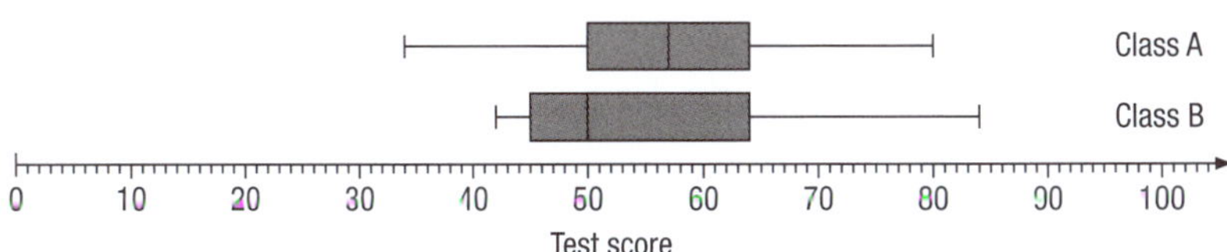

a In which class was:
   i the highest mark scored?
   ii the lowest mark scored ?
   iii there a larger spread of marks?

b Find:
   i the range of marks in class B
   ii the interquartile range for class A.

c If the pass mark was 50 for the test, what percentage of students passed the test in:
   i class A?
   ii class B?

d Describe the distribution of marks in:
   i class A
   ii class B.

e Copy and complete:

The students in class ___ generally scored higher marks.

The marks in class ___ were more varied.

Time series graphs

When numerical data is collected over **consecutive time intervals** then a line graph joining the data points is the appropriate type of graph. This is called a **time series graph**.

The consecutive time intervals can be seconds, minutes, hours, days, weeks, months, years, etc.

The time series graph on the right shows the percentage of live infants born with a low birth weight in Papua New Guinea for the years 2002 to 2006.

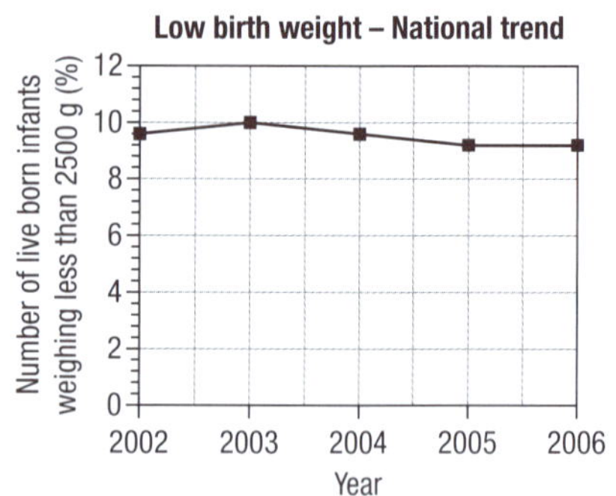

A continuous line is used to join the points on a time series graph so that any upward or downward pattern, called a **trend**, can be seen.

The graph above shows a slight downward trend from 2003 in the percentage of infants with a low birth weight, although the figures for 2005 and 2006 are almost the same.

If there is a **repeated pattern** apparent on a graph then the graphed data is said to be **seasonal**. Seasonal data is often seen on graphs for sales.

The graph below shows a seasonal pattern that is repeated every four quarters. There also appears to be an upward trend to the data.

Remember
A 'quarter' is three months.

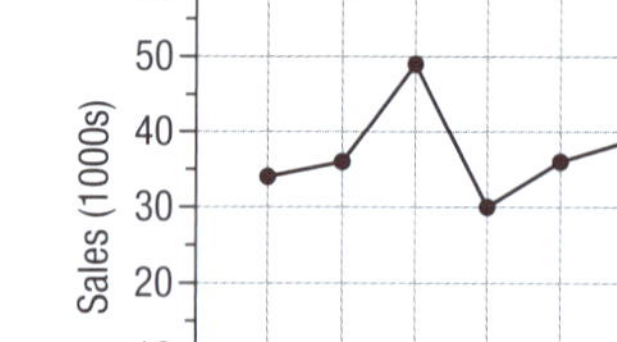

Time series graphs are used to **predict** future data. A trend line can be drawn to fit the data so that predictions can be made.

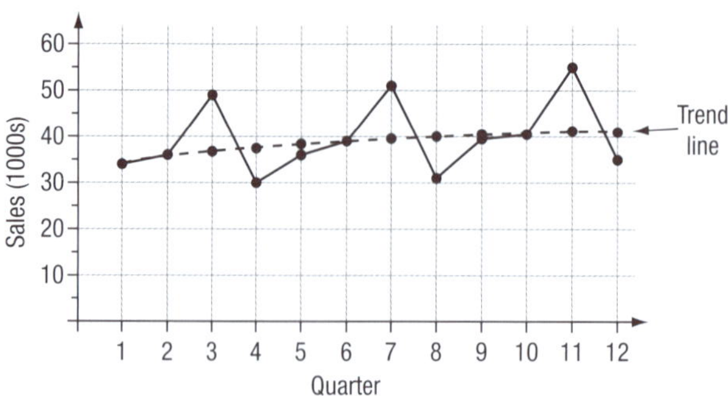

EXERCISE

1 The graph below shows the average daily rainfall in Madang for each month of 2003.

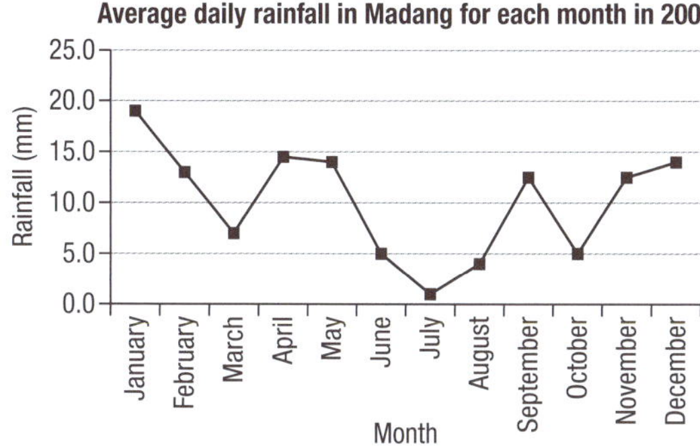

a What is the average daily rainfall in Madang in March 2003?
b Which month had the least average daily rainfall in 2003?
c In how many months was the average daily rainfall above 10 mm?

2 The graph below shows the incidence of HIV/AIDS in PNG for the years 1987 to 2005. Use the graph to answer the following questions.

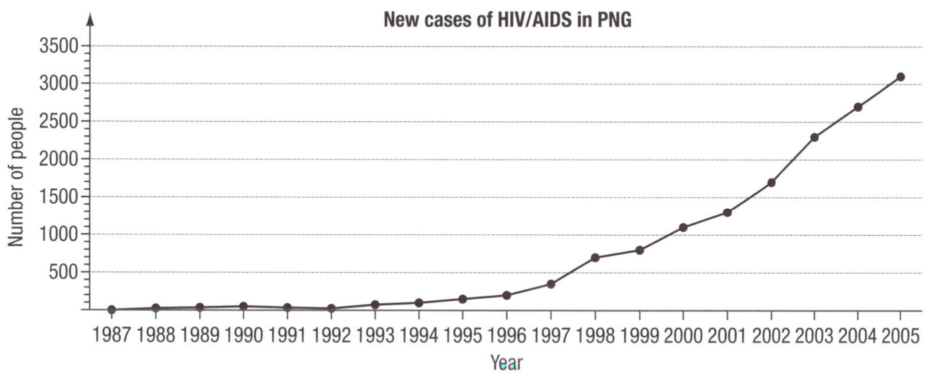

a How many new cases of HIV/AIDS were diagnosed in 2005?
b Describe the trend shown by the graph for the years 1994 to 2005.
c How many more new cases of HIV/AIDS were diagnosed in 2005 compared to 2001?
d Assuming the trend continued, use the graph to predict the number of new cases of HIV/AIDS that were diagnosed in 2006.

3 The table on the right records the total number of visitor (non-resident) arrivals in PNG for the years 1996 to 2005.

a Construct a time series graph for this data, starting the 'Visitors' scale on the vertical axis at 50 000.
b Comment on what happened to the number of visitors to PNG over this time period.
c Assuming the trend continues, use your graph to predict the number of visitors in 2007.

Year	Visitors
1996	61 392
1997	65 960
1998	67 816
1999	67 816
2000	58 448
2001	54 235
2002	53 670
2003	56 185
2004	59 013
2005	69 251
2006	77 730

4 The following graph shows the monthly rainfall in Daru for the years 2002 and 2003.

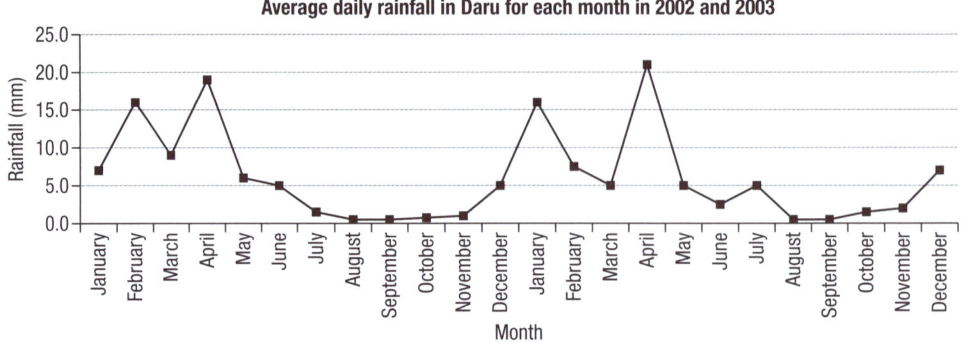

a In which month of the year did Daru have the highest average daily rainfall?

b In which month(s) of the year did Daru have the least average daily rainfall?

c Can you see a yearly pattern to the rainfall in Daru? What word do we use to describe this repeated pattern?

d Copy and complete the following statements about the rainfall in Daru.

'Rainfall in Daru is seasonal with the most rainfall falling in the month of _______ and the least amount of rain falling in the months of _______ and ________.

From the months of June to November the average daily rainfall is less than _______ mm.'

5 The graph on the right shows the amount, in millions of kina, of exports from PNG to Australia, and imports from Australia to PNG, for the years 2002 to 2006.

a What is the value of imports from Australia to PNG in 2003?

b What is the value of exports from PNG to Australia in 2006?

c Describe the trend of exports from PNG to Australia.

d Describe the trend of imports from Australia to PNG.

e What is happening to the **difference** between exports and imports for the period from 2002 to 2006?

f Use the graph to predict the value of the exports to Australia from PNG in 2007.

g Use the graph to predict the value of the imports from Australia to PNG in 2007.

MISLEADING STATISTICAL GRAPHS

Lesson 1

Misleading graphs

There are several ways in which graphs can be constructed so they are misleading.

i Using a 'cut-off' scale on the vertical axis

Consider the following graph:

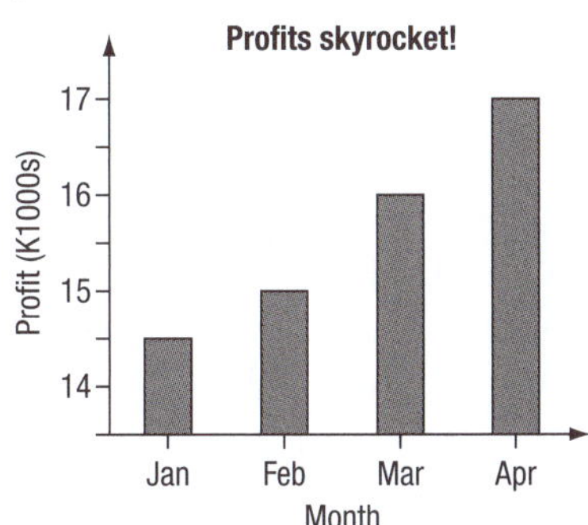

A close look at this graph reveals that the vertical scale does not start at zero and so exaggerates the increase in profits.

The graph should look like this:

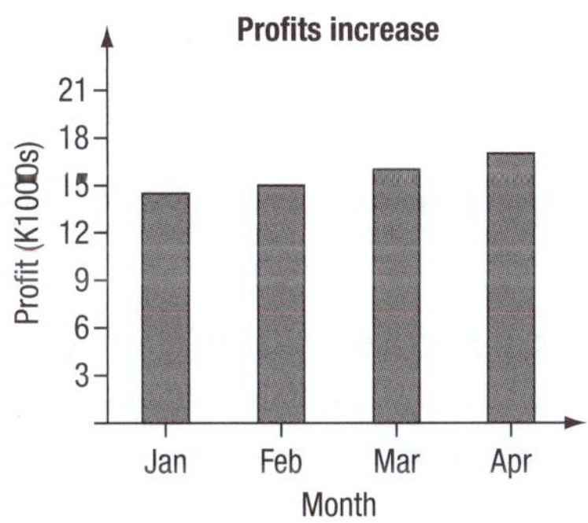

This graph shows the true picture of the profit increases and probably should be titled 'A modest but steady increase in profits'.

ii Giving the 'bars' on a bar chart (or column graph) a larger appearance

The bars on a graph can be made to look bigger by adding area or the appearance of volume when it is the **height** of the bar that is the dimension representing frequency. Consider the graph below comparing sales of different types of soft drink.

By giving the 'bars' the appearance of volume the sales of Kick drinks appear to be about eight times the sales of Fizz drinks.

On a bar chart, frequency (amount of sales in this case) is proportional to the height of the bar only and so the graph should look like this:

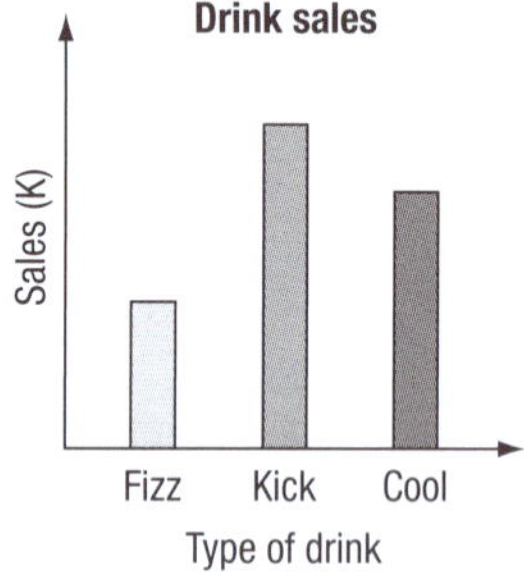

It can be seen from this bar chart that the sales of Kick are just over twice the sales of Fizz.

iii Scaling the axes in a misleading manner

In the example below, the vertical axis on the first graph has been stretched to accentuate the increase in readership of the *Messenger*. The graph with the correct scale is on the right.

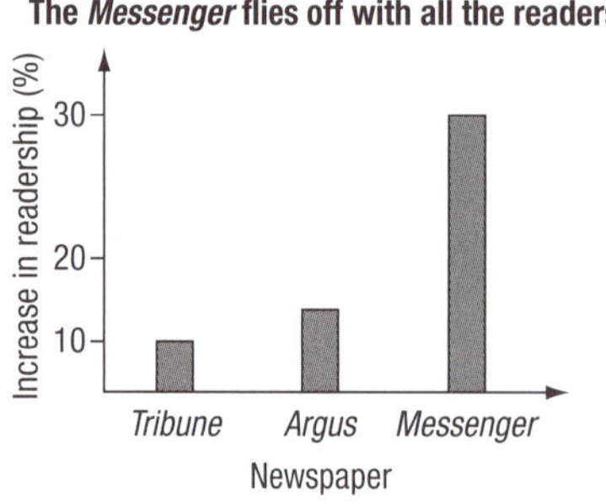

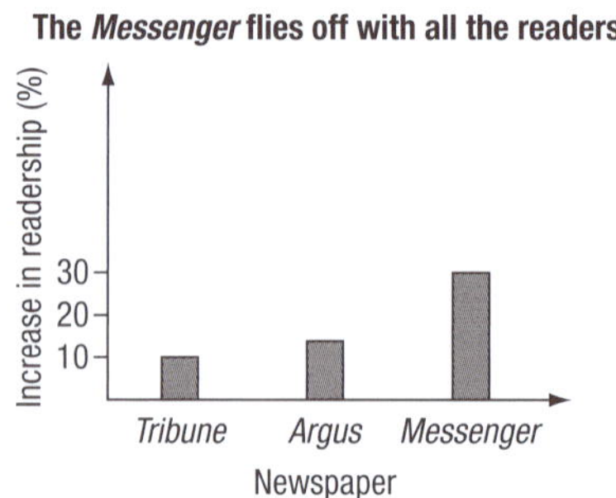

EXERCISE

1 Identify the misleading feature of each of the following graphs:

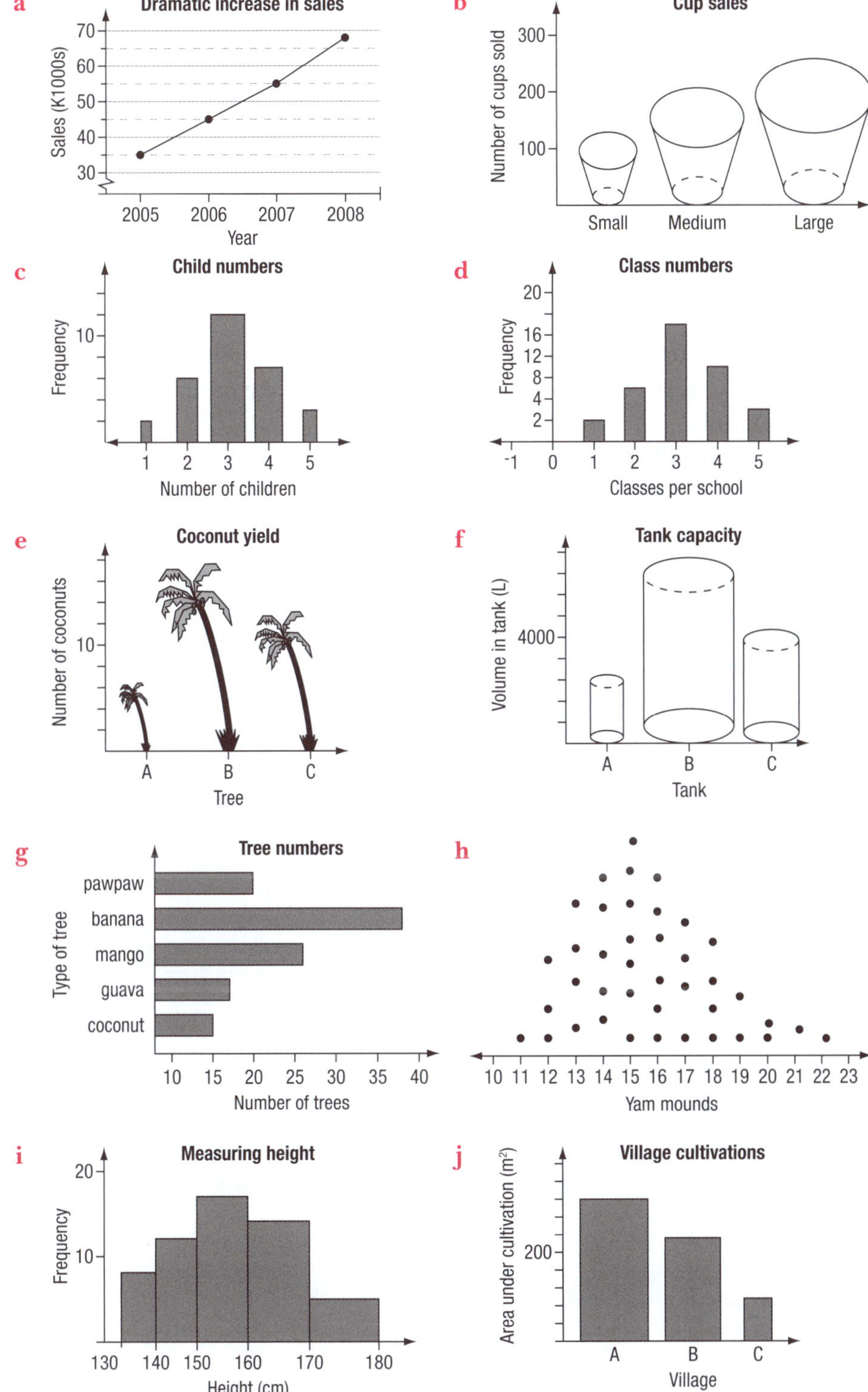

EXPERIMENTAL PROBABILITIES and PROBABILITIES BASED ON SYMMETRY

The probability of an event occurring is usually stated as a decimal or fraction between 0 and 1 inclusive. Sometimes it is stated as a percentage.
If the event is certain to occur it has a probability of 1. An impossible event has a probability of 0. All other events, that may or may not happen, have probabilities between 0 and 1.
Probabilities can be based on experiments or investigations or can be found using theoretical probability theory.
If probabilities are based on investigations then the quality and size of the sample for the investigation will have a large influence on the accuracy of the probabilities.
Similarly, the number of trials in an experiment will affect the accuracy of the probabilities obtained.

Lesson 1

Probability based on investigations

For probabilities based on investigations, the probability values are calculated as **proportions** or **relative frequencies**.
Consider the number of toothpicks in a packet (stated on the packet as 36). The number of toothpicks is counted for 48 packets and the results are given in the first two column of the frequency table shown here.
In the third column, the relative frequencies or proportions are calculated by dividing the frequency values by the total of the frequencies.
These proportions can be used as probabilities.
From the results in the table, we can say:

- 'The probability of getting 36 toothpicks in a packet is $\frac{13}{48}$, or approximately 0.271.'
- 'The probability of there being more than 36 toothpicks in a packet is $\frac{12}{48} + \frac{10}{48} + \frac{6}{48} + \frac{1}{48} = \frac{29}{48}$, or approximately 0.604.'

Number of toothpicks	Frequency	Proportion
34	1	$\frac{1}{48} = 0.021$
35	5	$\frac{5}{48} = 0.104$
36	13	$\frac{13}{48} = 0.271$
37	12	$\frac{12}{48} = 0.25$
38	10	$\frac{10}{48} = 0.208$
39	6	$\frac{6}{48} = 0.125$
40	1	$\frac{1}{48} = 0.021$
Total:	**48**	**1**

Discussion

How accurate do you think the probabilities from the table in the example above will be? How could you make these probabilities more accurate?

Consider the adult male and adult female literacy rate for several countries, shown by the graph on the right.

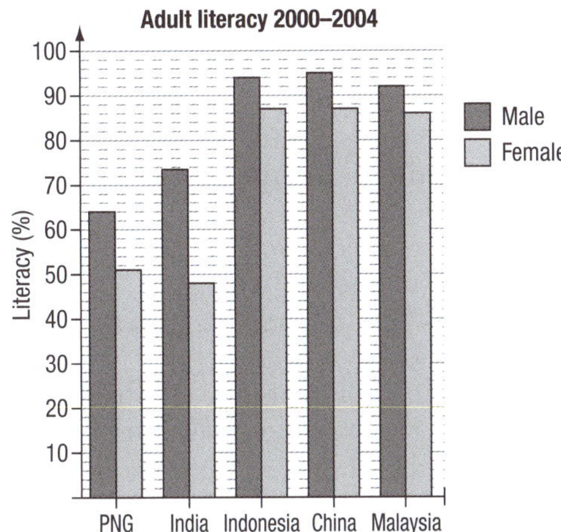

If you were asked, 'What is the probability that a male person from China is literate?' you could use the male literacy rate for China (95%) to say: 'The probability is 0.95'. The percentage is divided by 100 to give a probability number between 0 and 1 inclusive.

What is the probability that a female person from India is literate?

From the graph, the female literacy rate in India is 48%, so the probability is 0.48.

Remember

'Inclusive' means that 0 and 1 are both included.

Discussion

What do you think about the accuracy of these probabilities?

EXERCISE

Give your probability answers correct to two decimal places.

1 There are commonly used expressions associated with probability:

probably	likely	no chance	certain	very likely
unlikely	impossible	50-50	good chance	

Assign a probability, from 0 to 1, for each of these expressions and indicate their position on a probability number similar to the one shown below:

0   0.1   0.2   0.3   0.4   0.5   0.6   0.7   0.8   0.9   1

Impossible   **Certain**

2 Assign one of these probability descriptions to each of the events below:

probably	unlikely	likely	very likely
very unlikely	50-50	certain	impossible

a The Sun will rise tomorrow.
b On Thursday the next day will be Tuesday.
c It will rain this week.
d You will attend school every day for the rest of this year.
e You will catch a cold in the next three months.
f Your chickens will lay no eggs tomorrow.
g You will throw a six with a fair die.
h In 10 throws of a fair die, you will throw at least one six.

3 A graph showing the child malnutrition rates in several regions of PNG is shown on the right.

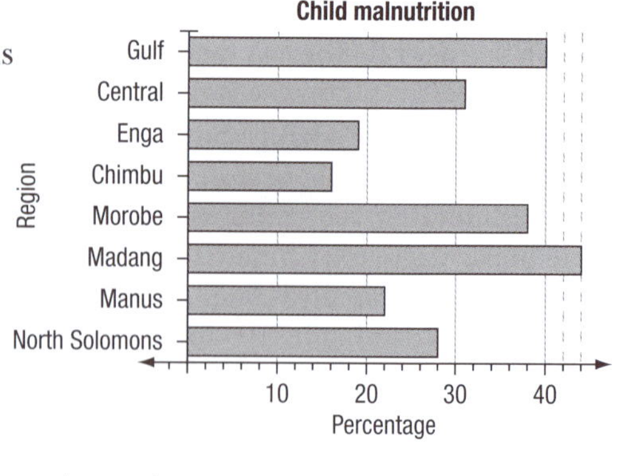

- **a** Use the percentages on the graph to find the probability that:
  - **i** a child from Madang is malnourished
  - **ii** a child from Chimbu is malnourished
  - **iii** a child from the Central region is **not** malnourished.
- **b** Explain how you found your answer to part **c**.

4 Two hundred households in PNG were asked the number of chickens that they keep. The data collected is shown on the graph on the right.

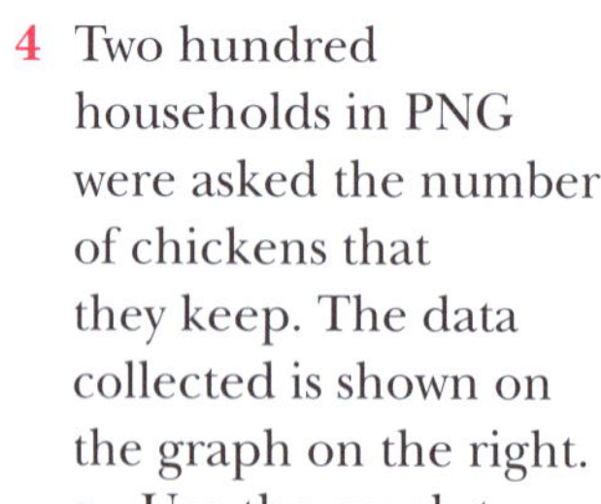

- **a** Use the graph to find:
  - **i** the probability that a household in PNG will keep seven chickens
  - **ii** the probability that a household in PNG will keep 14 chickens
  - **iii** the probability that a household in PNG will keep, at most, two chickens.
- **b** If the data for this investigation came from only one region of PNG, would the data represent all households in PNG?

Discussion

How could probabilities that apply to all PNG households be found?

5 The number of ripening mangoes on 50 trees has been counted and the results have been organised in the stem-and-leaf plot shown below.

Number of mangoes per tree

Stem	Leaf
6	8 9
7	4 5 7 7 8 9
8	0 1 2 3 3 4 4 4 5 6 7 7 7 7 8 9
9	0 1 2 3 4 4 4 4 5 5 6 6 6 6 8 9 9
10	0 1 1 2 3 4 5 6
11	4

Key: 9 | 8 means 98

- **a** Use these results to find the probability that a mango tree will produce:
  - **i** between 90 and 99 (inclusive) mangoes in a season
  - **ii** fewer than 70 mangoes in a season
  - **iii** 100 or more mangoes in a season
- **b** If the 50 trees in the sample were all of the same variety of mango tree would these figures apply to other varieties of mango tree?

6 The table on the right shows the age distribution of all people in PNG as recorded in the year 2000. If you have a calculator, find the answers correct to two decimal places, otherwise give your answers as fractions.

a From the table find the probability that:

 i a female is in the 10–14 age group
 ii a male is a child less than 10 years old
 iii a female is aged 60 or more
 iv a male is in the 30 to 49 age group
 v a female is in the 15 to 44 age group.

b i How accurate would these probabilities be in the year 2000?
 ii How accurate would these probabilities be now?

c From the table find the probability that a person from PNG is:

 i in the 0–4 age group
 ii aged 50 or more.

Age distribution in PNG		
Age	**Males**	**Females**
0–4	377 776	348 904
5–9	381 339	346 031
10–14	330 965	289 909
15–19	293 277	261 204
20–24	239 863	234 938
25–29	219 680	228 734
30–34	191 662	192 598
35–39	166 656	163 681
40–44	128 910	123 458
45–49	104 867	93 890
50–54	79 899	71 114
55–59	59 308	49 400
60–64	48 530	40 973
65–69	31 351	25 870
70–74	19 657	15 477
75–79	9782	7082
80–84	5041	3628
85–89	1984	1331
90+	1197	820
Total:	2 691 744	2 499 042

7 The manufacturers want to know the probability that a pane of glass, produced in their factory, contains more than one fault. They have examined fifty panes of glass for faults and have obtained the following results:

0 1 0 2 1 2 0 3 0 1 0 1 0 1 0 5 1 0 0 1 0 0 1 4 1
1 0 0 0 0 2 1 0 0 1 0 0 0 0 0 1 2 0 0 1 0 0 1 3 0

Use the results to estimate the probability that a pane of glass from this factory contains more than one fault.

Experimental probability

Lesson
2

Experimental probabilities are obtained from 'experiments' in which a task is repeated many times (the number of trials and all the possible outcomes are known.) The **experiment** could be tossing a coin, rolling a die, choosing a card from a pack, etc.

A **fair** coin will have an equal chance of landing 'heads' or 'tails'. Similarly a fair die will have an equal chance of landing on any one of the six numbers, and any card in a pack is equally likely to be selected.

An **estimate** of the **probability** of a particular outcome occurring is calculated as the number of favourable outcomes divided by the total number of trials:

$$\text{Probability} = \frac{\text{number of favourable outcomes}}{\text{number of trials}}$$

For example, a fair coin is tossed 10 times resulting in 3 'heads' and 7 'tails'.
In this case the **experiment** is 'tossing a coin', there are two **possible outcomes** ('heads' or 'tails') and the **number of trials** is 10.
If we wanted to find the probability of 'tossing a head' then, in this experiment, the number of **favourable outcomes** is 3, so an estimate of the probability of tossing a head is $\frac{3}{10}$ or 0.3.
We can say: **Pr(head) = 0.3**.
Suppose the coin is tossed another 10 times, this time resulting in 5 heads and 5 tails.
In the combined 20 trials we have 8 heads and 12 tails, so we can now estimate Pr(heads) $= \frac{8}{20} = 0.4$

Discussion

Will everybody get the same results from tossing a coin 10 times?
How can we improve the accuracy of the probability of 'tossing a head'?

In general, the greater the number of trials the better the estimate of the probability becomes.

EXERCISE

1 Toss a fair coin 10, 20, 30, 40 … 100 times so that you can estimate the probability of a coin landing tails when tossed. Find an estimate of the probability after each set of 10 tosses. (Copy and complete a table like the one below, as you go.)

Trial	Outcome		Total number of tails	Estimate of Pr(tails)
	Head	Tail		
1–10	|||	𝍸 ||	7	$\frac{7}{10} = 0.7$
11–20	𝍸	𝍸	12	$\frac{12}{20} = 0.6$
21–30				
31–40				
41–50				
51–60				
61–70				
71–80				
81–90				
91–100				

a What do you estimate Pr(tails) should be when tossing a coin?
b After how many trials did your estimate match your answer from part **a**?

Remember
Pr(tails) $= \frac{\text{number of tails}}{\text{number of trials}}$

2 The task is to toss two coins together 100 times and record the results in a table so that you can estimate the probability of tossing 'a head and a tail'.

a Before you start, answer the following questions:

i What is the experiment?

ii How many trials will there be?

iii What are the possible outcomes?

iv What is the favourable outcome in this experiment?

v How will you estimate Pr(a head and a tail)?

vi What do you think Pr(a head and a tail) will be?

b i After you have found your estimate of Pr(a head and a tail), combine the results of your experiment with those of another student, so that you have a better estimate of Pr(a head and a tail). For this calculation you need to combine the two sets of results for 'a head and a tail' to get the total number of outcomes of 'a head and a tail', and the total number of trials will be 200.

ii Is this estimate, using combined results, closer to the answer you had in part **a vi** than your estimate in part **a v** was?

Challenge

You will need about 20 items, of identical shape but of two different colours, and a container to hold the items so that they are not visible. Dried beans would be a good choice (some can be coloured easily). Playing cards could also be used.

a Working in pairs, one person sets up the experiment by choosing 10 items, being sure to have at least one item of each colour, and placing them in the container without showing his/her partner.

b The other person (the sampler) selects an item from the container, records the colour, then returns the item to the container. This selection process is repeated for a total of 10 trials. If you are using cards, they need to be shuffled between selections.

c The sampler uses the results to help him/her estimate the number of items of each colour that are in the container.

d The sampler completes another 10 trials and combines these results with the previous results from part **b**.

e The sampler uses the combined results from parts **b** and **d** to estimate the number of items of each colour in the container. The sampler asks the partner who set up the experiment if the estimate is correct. If the estimate is not correct, the sampler continues sampling until the estimate describes the correct proportion.

f Change roles and repeat the experiment.

g Write a paragraph to report your experience as the sampler.

More experimental probability

Experiments using everyday items can be used to obtain estimates of probabilities.

EXERCISE

1 a Design your own probability experiment. (Do this in pairs and possibly outside the classroom). Here are some ideas:
- Tossing a can. (What is the probability that the can lands on its side or on its end?)
- Make your own spinner for which not all outcomes are equally likely.
- Tossing a drawing-pin (thumb-tack). (Estimate the probability that it will land with its flat top down or on its side.)
- Tossing a matchbox. Balance the matchbox on the edge of a table and flip it onto the floor. (Estimate the chance that the matchbox will land on its top, on its bottom, on an edge, or on a side.)
- Using cards. You can work with a smaller pack than normal. For example, you could use a pack with no court cards (that means no Jacks, Queens and Kings) or you could use only the court cards.
- Construct your own 'die'. Use cardboard or a wooden block to make a die which has four rectangular sides of the same size (numbered 1 to 4) and two other square sides (numbered 5 and 6), as shown in the diagram. (Use experimental probabilities to find the bias of this die.)

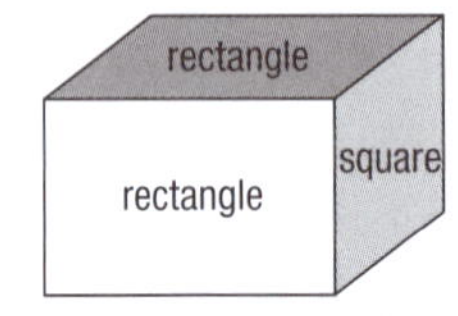

b Carry out your experiment and record all your results. Write a paragraph of your findings, mentioning:
- the experiment
- a list of all possible outcomes
- the probabilities that you are estimating
- the number of trials
- equally likely outcomes
- bias
- any conclusions you can make based on your results.

Theoretical probability: Sample space and equally likely outcomes

When tossing a die the possible outcomes are 1, 2, 3, 4, 5, or 6.

The set of all possible outcomes of a probability experiment is called the sample space.

The sample space for tossing a die can be written as {1, 2, 3, 4, 5, 6}. The curly brackets indicate a set.

Remember
Curly brackets are called braces.

For a fair die each of the possible outcomes is equally likely and the probability of throwing any one of the numbers will be the same.

If all outcomes of an experiment are equally likely, the probability of any one of the outcomes occurring is:

$$\frac{1}{\text{number of possible outcomes}}$$

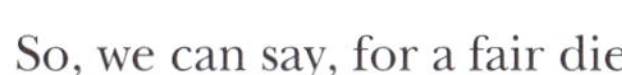

So, we can say, for a fair die:

- $Pr(\text{throwing a 1}) = \frac{1}{6}$
- $Pr(\text{throwing a 2}) = \frac{1}{6}$
- $Pr(\text{throwing a 3}) = \frac{1}{6}$
- $Pr(\text{throwing a 4}) = \frac{1}{6}$
- $Pr(\text{throwing a 5}) = \frac{1}{6}$
- $Pr(\text{throwing a 6}) = \frac{1}{6}$.

Note that:

$$\begin{aligned}\text{The sum of all the probabilities} &= \frac{1}{6} + \frac{1}{6} + \frac{1}{6} + \frac{1}{6} + \frac{1}{6} + \frac{1}{6} \\ &= 1\end{aligned}$$

EXERCISE

1 Write the sample space for each of the following probability experiments and state whether the outcomes are equally likely.

a Flipping a fair coin and noting whether it lands 'heads' or 'tails'.

b A spinner, divided into four equal sectors, numbered 1 to 4, is spun and the number in the sector on which it comes to rest is recorded.

c The spinner on the right is spun and the number in the sector on which it comes to rest is recorded.

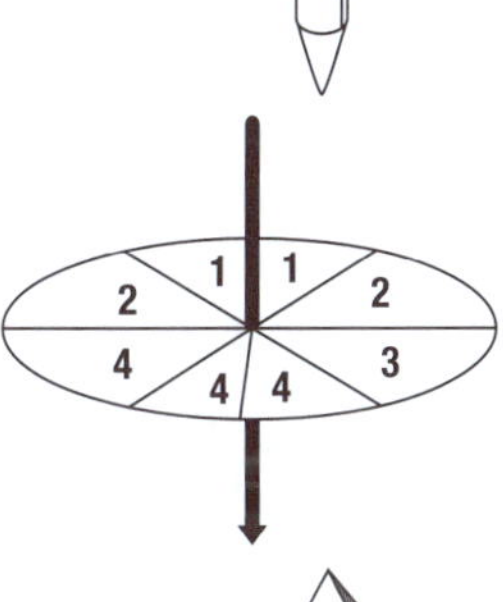

d A regular tetrahedron, with sides coloured red, blue, yellow and green, is tossed and the colour of the side on which it lands is recorded.

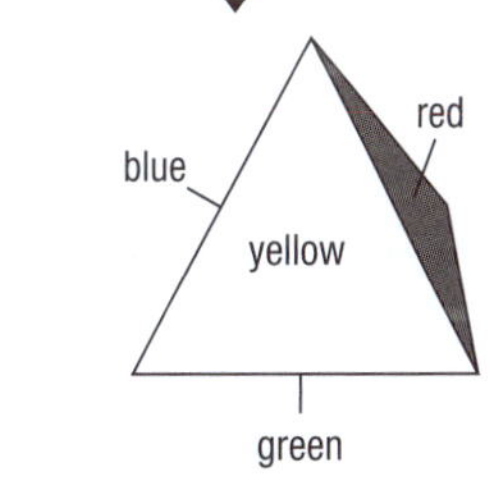

e An octahedron, with one side coloured yellow, two sides coloured red, two sides coloured blue and three sides coloured green, is tossed and the colour of the side on which it lands is noted.

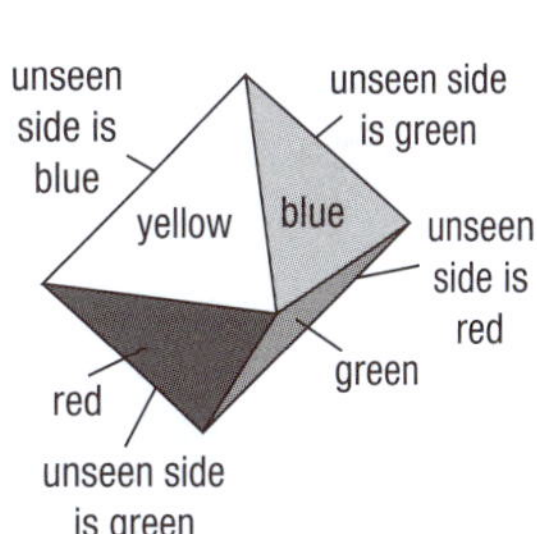

f One ball, from a bag containing four blue balls and six red balls, is selected. The colour is recorded and then it is returned to the bag.

2 Find the probability of the following outcomes of experiments in which each of the outcomes is equally likely:
  a selecting the Ace of hearts from a pack of 52 playing cards
  b throwing a tail when tossing a fair coin
  c that a randomly selected person in your maths class will be you
  d that the day that you were born on was Sunday
  e that the day on which a randomly selected person was born was 21 April
  f that 'red' is the colour not showing when a regular tetrahedron, with sides coloured red, blue, yellow and green, is tossed

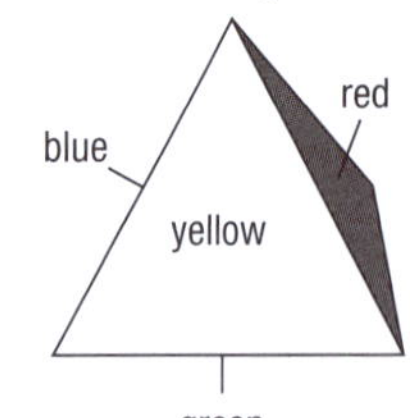

  g that a newborn baby is a boy
  h if you were to randomly choose a school table to sit at in your classroom then it would be the one closest to the door.

Challenge

a i Carefully construct a spinner using a 3 cm × 4 cm rectangle of cardboard and a pin on which to spin it, taking care to place the pin in the 'centre' of the rectangle where the diagonals cross.

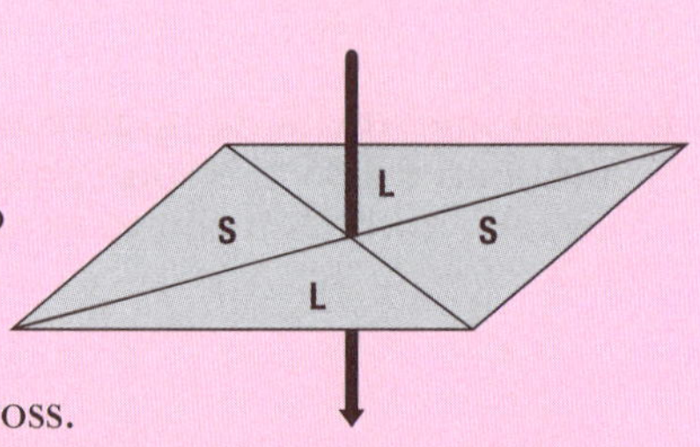

  ii Mark the edges of the rectangle L (for long) and S (for short).
  iii Guess the probability that the spinner will land on one of the longer sides.
  iv Conduct 50 spins, recording in a tally table whether the spinner lands on a long side or a short side each time.
  v Use your results to calculate the probability that your spinner lands on a long side. Do your results match your guess from part iii? Compare your results with those of others in your class.
  vi What factors in the construction of the spinner and in conducting the trials could influence the results?
  vii Write a report on your findings.
  viii Do you think your results would be the same if the spinner was, say, 3.5 cm × 4 cm?

b Make another spinner whose size is different from 3 cm × 4 cm and repeat the experiment from part a.

Event space

Consider the situation in which the probability experiment is to roll a die and each of the possible outcomes {1, 2, 3, 4, 5, 6} is equally likely.

If you were asked, 'What is the probability that you roll an even number?', then 'rolling an even number' is referred to as an **event**.

In general: The probability of an event occurring = $\dfrac{\text{the number of outcomes favourable to the event}}{\text{number of possible outcomes}}$

Of the six outcomes that are possible when throwing a die, three of these are favourable to 'throwing an even number'. These are {2, 4, 6}.

So we can say:

$$\text{Pr(even number)} = \frac{3}{6} = \frac{1}{2}$$

EXERCISE

Write all your probabilities as fractions in their simplest form.

1 A fair die is thrown.

  i List the favourable outcomes for each of the events listed below.

  ii Find the probability of throwing each of the following.

a an odd number **b** a number greater than 4

c a multiple of 3 **d** a factor of 6

e a prime number

2 One card is drawn at random from a pack of 52 playing cards.

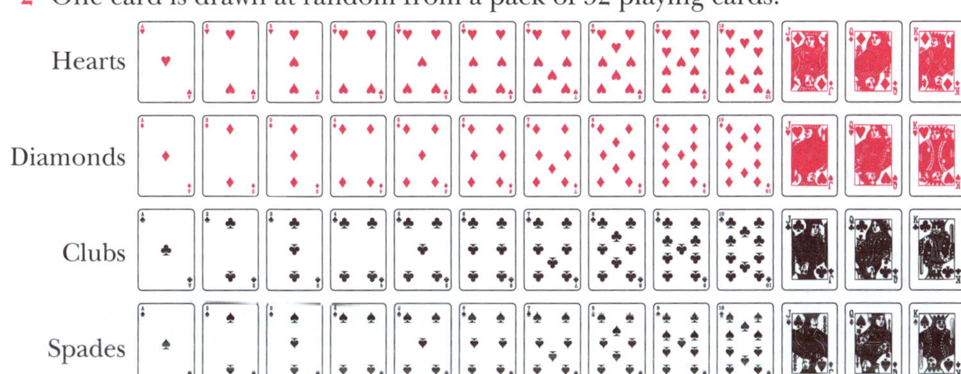

Find the probability of choosing:

a an Ace **b** a spade

c a 7, 8, 9 or 10 **d** a red card

e a court card (that is a Jack, Queen or King)

f the Ace of hearts

g a red card between the numbers 3 and 10 (not inclusive).

3 A spinner is made from a circle divided into eight equal sectors and numbered as shown.

What is the probability that, when spun, the spinner will land on:

a a one?

b a two?

c a three?

d a four?

e an even number?

f a factor of four?

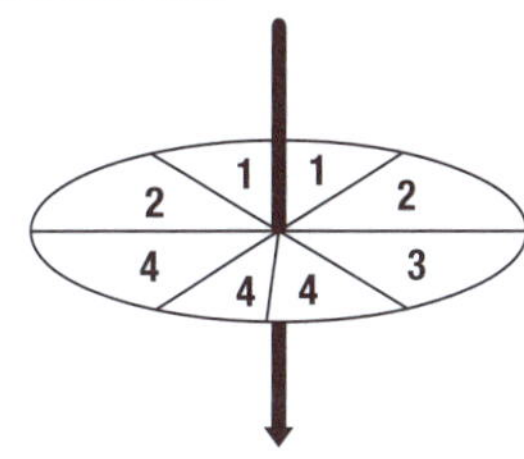

4 A carton of 12 eggs contains three brown eggs, four speckled eggs and five white eggs. If an egg is chosen at random what is the probability that the egg is:

a white?
b speckled?
c not white?
d brown or white?

5 A spinner is made from a regular decagon with the sectors numbered 1 to 10, so that the outcomes 1 to 10 are equally likely.

a A is the event 'spinning an even number'.
B is the event 'spinning a number greater than 5'.
C is the event 'spinning a multiple of 3'.
Find:

i Pr(A) ii Pr(B) iii Pr(C)

b When the spinner is spun once, are you more likely to get:

i a prime number or an odd number?
ii a number greater than 5 or a number less than 5?
iii a multiple of 3 or a factor of 8?

6 A roulette wheel is circular and is divided into 37 equal sectors numbered 0 to 36. The sector numbered 0 is coloured green. Half the numbers 1 to 36 are coloured black and the other half are coloured red.

A small ball is dropped onto the spinning roulette wheel and eventually lands on one of the numbers when the wheel stops spinning.

A gambler can bet on:

- any one of the numbers 0 to 36
- evens or odds
- reds or blacks
- groups of 2, 3, 4 … 18 numbers.

What is the probability that the ball will land on:

a a red number?
b a black number?
c the green zero?
d an even number?
e a 1, 2 or 3?
f one of the numbers 1 to 12?
g one of the numbers 19 to 36?

7 If one letter is chosen at random from the word 'PROBABILITY' what is the probability that the letter is:

a P?
b I?
c Q?
d B?
e a consonant?
f a vowel?

Illustrating the sample space

Suppose two coins are tossed together. We can illustrate the **sample space** (all the possible outcomes) using a **tree diagram** or a **grid diagram**:

Tree diagram

1st coin | 2nd coin | Outcome

H — H — HH
H — T — HT
T — H — TH
T — T — TT

Grid diagram

		Coin 1	
		H	**T**
Coin 2	**H**	HH	TH
	T	HT	TT

The diagrams above both show that there are four possible outcomes (HH, TH, HT, TT).

What is the probability of throwing two tails? Only one of the four outcomes, **TT**, is favourable to two tails, so:

$$\text{Pr(two tails)} = \frac{1}{4}$$

What is the probability of throwing a head and a tail? Two of the outcomes, **HT** and **TH**, are favourable to throwing a head and a tail, so:

$$\begin{aligned}\text{Pr(a head and a tail)} &= \frac{2}{4}\\ &= \frac{1}{2}\end{aligned}$$

What is the probability of throwing two heads? Only one of the four outcomes, **HH**, is favourable to two heads, so:

$$\text{Pr(two heads)} = \frac{1}{4}$$

EXERCISE

1 A tree diagram illustrating the sample space for the experiment 'throwing a coin and a die together' is shown on the right.

- **a** How many possible outcomes are there?
- **b** What is the probability of throwing:
  - **i** a two and a head?
  - **ii** a six and a tail?
  - **iii** an even number and a head?
  - **iv** a number greater than 2 and a tail?

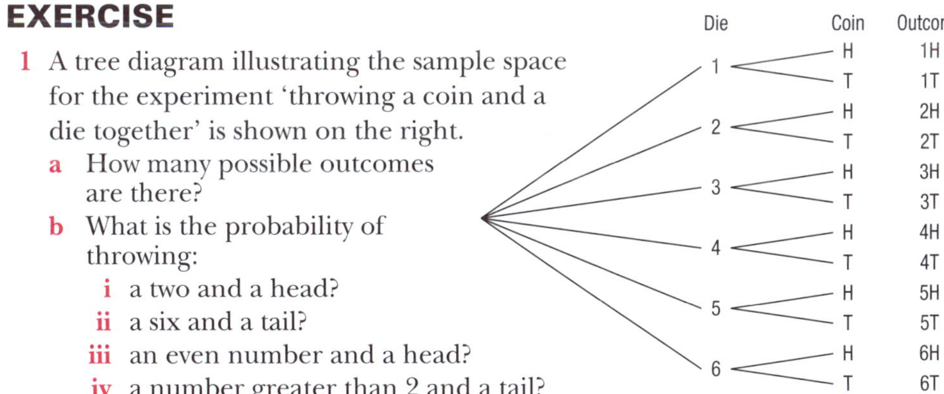

2 Three coins are tossed together.

a Copy and complete this tree diagram to illustrate the sample space for this experiment.

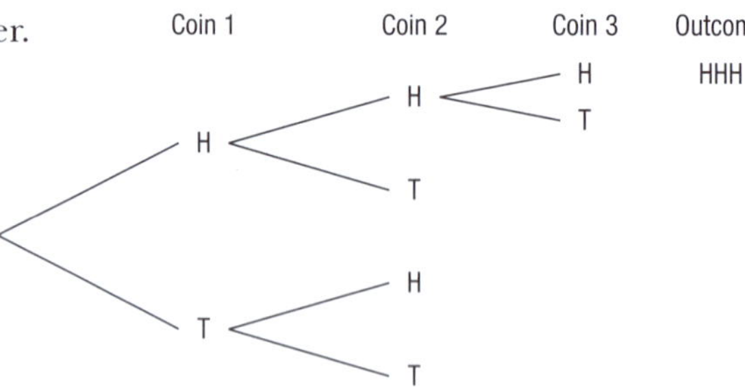

b State the four possible combinations of heads and tails that you can get from this experiment.

c Write the favourable outcomes and the probability of throwing:

i three tails **ii** two heads and a tail

iii two tails and a head **iv** three heads.

d Calculate the sum of all the probabilities from part **c**.

3 Two identical spinners are numbered 1 to 4, as shown on the right. On each spinner each outcome 1, 2, 3 or 4 is equally likely to occur. Both spinners are spun and the numbers they land on are recorded.

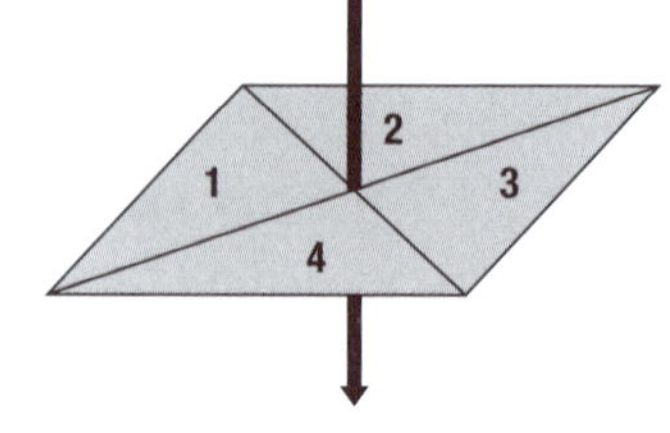

a Copy and complete the grid diagram below to show the sample space for this experiment.

		Spinner 1			
		1	2	3	4
Spinner 2	1	1 and 1	2 and 1		
	2	1 and 2			
	3	1 and 3		3 and 3	
	4				

b How many possible outcomes are there for this experiment if '1 and 2' is not the same outcome as '2 and 1'? Are all the outcomes equally likely?

c Find the probability of throwing

i two 4s **ii** a 1 and a 3, in no particular order

iii two of the same number **iv** two different numbers

v numbers with a difference of two

vi two numbers greater than 2

vii one number greater than 2 and the other less than 2.

d **i** Copy and complete the table below for the two spinners if the outcomes are the sums of the two numbers.

		Spinner 1			
		1	2	3	4
Spinner 2	1	2	3		
	2	3			
	3	4		6	
	4				

ii Copy and complete the table on the right of events and probabilities for the results recorded in the table in part **d i** above.

Event: 'sum of'	Probability
2	
3	$\frac{2}{16} = \frac{1}{8}$
4	
5	
6	
7	
8	

Comparing experimental and theoretical probabilities

Lesson 7

There are some situations in which we can conduct a number of trials of an experiment to obtain probability estimates and also use theoretical probability to check these experimental results.

EXERCISE

You will be investigating the outcomes and the associated probabilities when two dice are thrown together and the numbers that appear uppermost on the dice are **added**.

1 The first task is to record some data. Work in pairs, one person rolling the dice and the other recording the sum of the two numbers rolled. Complete 50 rolls of the dice each, so that you have a total of 100 data values recorded. Take care to count the number of rolls of the dice as you go, so that you do not go over 100.

Decide on a method of rolling the dice so that the outcomes are not biased (perhaps use a cup or container to shake and then roll the dice).

The sum of the two numbers uppermost is 9

- a What is the smallest sum that can be recorded for two dice?
- b What is the largest sum that can be recorded for two dice?
- c Is there a sum for every number between the smallest number and the largest number that can be recorded for two dice?
- d Set up a tally and frequency table to record the data. When you have finished the 100 rolls of the dice, add a 'frequency' column to your table and then construct a graph to display the results. (**Alternatively**, you could use a **dotplot** that combines the frequency table and graph.)
- e Use your results to estimate the probability of throwing:
  - i a total of 12
  - ii a total of 7
  - iii a total of 10
  - iv a total of 3.
- f From your table of results, can you determine which total you are most likely to roll when rolling two dice?

2 We are going to use theoretical probability to find the answers for this question, and Questions 3 and 4 following.

- a Set-up a grid diagram, similar to the one below, to show the **sample space** when two dice are rolled:

		Die 1					
		1	2	3	4	5	6
Die 2	1	1 and 1	2 and 1				
	2						
	3						
	4			3 and 4			
	5						
	6						

- b How many possible outcomes are there?

c Are all the outcomes equally likely to happen?

d Copy and complete the following table

Event 'sum of'	Number of favourable outcomes	Probability (fraction)
2		
3		
4		
5		
6		
7		
8		
9		
10		
11		
12		
Total:	36	

Remember

$$\text{Probability of an event} = \frac{\text{the number of outcomes favourable to the event}}{\text{the number of possible outcomes}}$$

e i From the table above which sum are you more likely to obtain when rolling two dice?

ii Is your answer for part **e i** the same answer you found using the experimental trials? If not, why do you think the results differ?

3 a From the theoretical probability table in Question 2, find:

i Pr(rolling two 6s) ii Pr(not rolling two 6s)

b i Add the probabilities from parts **a i** and **a ii**.

ii The two events 'rolling two 6s' and 'not rolling two 6s' are called **complementary events**. What is the special property of complementary events?

4 From the theoretical probability table in Question 2, find:

a Pr(rolling two of the same number)

b Pr(rolling two numbers with a sum greater than 9)

c Pr(rolling two numbers with an even sum)

d Pr(rolling two numbers with an odd sum)

e Pr(rolling two even numbers)

f Pr(rolling two numbers that are both factors of 6)

5 Write a paragraph reporting on the comparison between theoretical probability and experimental probability as you saw it in Questions 1 to 4 above. Include in your report:

- the steps that you took to ensure that the experimental results were unbiased
- the results from the experiment and the sum that you found most likely to be rolled when rolling two dice
- the results from the theoretical investigation and the sum that you found most likely to be rolled when rolling two dice
- some explanation as to why your results are not exactly the same for the experimental and theoretical probabilities.

OPTION A: RANDOM EVENTS AND SIMULATION

A random event is one that occurs without having a plan, a pattern, or a pre-arranged order.

- **Many events in everyday life occur randomly. The gender of a child is a random event. The winning number drawn in a raffle is a random event.**
- **Randomly selected samples are used for statistical investigations. The random nature of the sample means that there is no pattern or plan (no bias) in choosing one element of the population over another and this means that reliable conclusions about the population can be made based on the sample.**
- **Many games also rely on randomness to provide the element of chance and fairness. The roll of a fair die will randomly produce one of the numbers from 1 to 6; the toss of a coin will randomly fall 'heads' or 'tails'; and in most card games the cards are shuffled so that they can be randomly dealt.**

Simulation is when the conditions of a situation are reproduced. For example we can simulate the randomness of the gender of a child by allocating 'heads' to a boy and 'tails' to a girl and then tossing a coin.

Using physical objects to simulate random events

Physical objects can be used to generate random outcomes. Some are listed below.

- **A carefully constructed spinner:** A spinner can be made specifically for a situation. For example, if we needed to randomly choose three objects from a possible eight objects, spinning a spinner with eight equally likely outcomes three times would be appropriate for this random selection. However, the physical limitations of making a spinner mean that a spinner with 10 outcomes is probably the maximum size that would give reliably random outcomes.
- **Selecting marked objects from a bag:** Care needs to be taken to make sure that the objects are selected without bias. The objects should all be the same size and shape, should not be visible to the selector and should be returned to the bag and mixed well after each selection. There are also physical limits to the use of this method (depending on the size and strength of the bag and the size, weight and number of the objects).
- **A coin:** This can be used if there are only two possible outcomes. Each of the outcomes is assigned either 'heads' or 'tails' and the coin is tossed.
- **A die:** A die can be used if the number of possible outcomes is a factor of 6.
- **A set of marked cards:** Use as many cards as there are possible outcomes. Physically, this method is limited to about 100 cards. Playing cards can sometimes be used if we take care to allocate each card to an outcome.

EXERCISE

1 a Evelyn is expecting a baby. What are the two possible gender outcomes for her baby?

b Explain how each of the following could be used to simulate the outcomes:

i a coin
ii a die.

2 a David and Therese plan to have two children. Their two children will either be two boys, two girls, or a boy and a girl. You can use two coins to simulate the gender outcomes for two children. What will represent:

i a boy?
ii a girl?

b Toss two coins to simulate the gender outcomes of 50 families with two children. Record your results.

c From your results for part **b**, estimate the probabilities for each of the three outcomes (two boys, two girls, or a boy and a girl). Are they the same? You might need to do some more trials to test this.

d Comment on your results.

Class discussion

a How could you simulate the number of boys and girls in a four-child family using each of the following?

i **A bag of coloured balls.** How many colours are necessary? What proportion should there be of each colour? If you choose one ball do you need to put it back before you choose another ball? Why?

ii **A pack of playing cards.** How will you allocate 'girl' or 'boy' to a chosen card. How many cards will you need to select for one four-child family. If you choose one card do you need to put it back before you choose another card?

b If you were asked, 'What is the probability of having **at least one girl** (in a four-child family)?' how would you estimate this probability?

3 Leo thinks he is a good tennis player. He has found that he wins two out of three matches in the tournaments that he plays.

a Name two physical random number generators that you could use to simulate the outcome of Leo's tennis matches? Explain how each of these would be used.

b Leo is to play in an elimination tournament that starts with 8 players. If Leo wins all three matches then he will have won the tournament. How will you simulate Leo's performance in this tournament?

c Choose one of the physical random number generators from part **a** and conduct at least 30 trials(30 tournaments).

i Record your results, noting the number of times Leo wins the tournament.

ii Use the results to estimate the probability of Leo winning a three-match tournament.

d Comment on your results.

Using a random number table

Lesson **2**

Random number tables appear as a list of the digits from 0 to 9. The digits are generated by a computer and are grouped in fives so that the table is easier to read. You can start anywhere in the table and move across or down to select your random numbers.

Random number table:

39634	62349	74088	65564	16379	19713	39153	69459	17986	24537
14595	35050	40469	27478	44526	67331	93365	54526	22356	93208
30734	71571	83722	79712	25775	65178	07763	82928	31131	30196
64628	89126	91254	24090	25752	03091	39411	73146	06089	15630
42831	95113	43511	42082	15140	34733	68076	18292	69486	80468
80583	70361	41047	26792	78466	03395	17635	09697	82447	31405
00209	90404	99457	72570	42194	49043	24330	14939	09865	45906
05409	20830	01911	60767	55248	79253	12317	84120	77772	50103
95836	22530	91785	80210	34361	52228	33869	94332	83868	61672
65358	70469	87149	89509	72176	18103	55169	79954	72002	20582
72249	04037	36192	40221	14918	53437	60571	40995	55006	10694
41692	40581	93050	48734	34652	41577	04631	49184	39295	81776
61885	50796	96822	82002	07973	52925	75467	86013	98072	91942
48917	48129	48624	48248	91465	54898	61220	18721	67387	66575

Suppose we want to randomly choose 10 different numbers from 1 to 50 inclusive. We need to look at two digits at a time, because there are two digits in the largest number (50). The numbers 1 to 9 would appear as 01 to 09 in the random number table. Any two-digit numbers larger than 50 are ignored.

Starting in the top left-hand corner and working down, we would have the numbers **39**, **14**, **30**, (64 is ignored because it is larger than 50), **42**, (80 ignored), (00 ignored), **5 which appears as 05**, (95, 65, 72 ignored), **41**, (61 ignored) and **48**. Then, returning to the top and using the next pair of digits in each row, we have (63, 59, 73, 62, 83, 58 ignored), **20**, **40**, (83 ignored) and **35**.

The ten random numbers between 1 and 50 inclusive are 39, 14, 30, 42, 5, 41, 48, 20, 40 and 35.

Suppose we want to randomly choose one number between 400 and 500 inclusive. This time, we need to look at three digits at a time because there are three digits in all the numbers from 400 to 500.

Starting in the top left-hand corner and working down, we would have the numbers 396, 145, 307, 646, **428** ... The first four numbers are ignored because they are outside the range of 400 to 500. So the randomly chosen number is 428.

Some rules for using a random number table to select a set of numbers

- Choose a starting point and a direction and continue with this for the entire selection. Don't jump around the table!
- Ignore numbers that are larger or smaller than you need.
- Ignore numbers that are repeated.

Remember

'Inclusive' means that the numbers 1 and 50 are both included.

Discussion

Why do you think it is necessary to use random number tables? Why can't we just write a list of ten numbers between 1 and 50?

EXERCISE

1 To find each of the following sets of random numbers, start at the top left-hand corner of the random number table on the right and work downwards.

37968	33329
35845	62037
97097	20740
52650	19348
90510	26807
09017	41054
47841	99014
17274	04202
97792	69054
92894	85699

Find:

a a random list of five different numbers between 1 and 9 inclusive.

b a random list of ten different numbers between 50 and 99 inclusive.

c two randomly selected numbers between 500 and 600 inclusive.

Remember

'Inclusive' means that the numbers mentioned are both included.

2 Starting at the top left-hand corner of the random number table below, and working downward, find a random list of six different numbers between 100 and 200.

64628	89126	91254	24090	25752	03091	39411	73146	06089	15630
42831	95113	43511	42082	15140	34733	68076	18292	69486	80468
80583	70361	41047	26792	78466	03395	17635	09697	82447	31405
00209	90404	99457	72570	42194	49043	24330	14939	09865	45906
05409	20830	01911	60767	55248	79253	12317	84120	77772	50103
95836	22530	91785	80210	34361	52228	33869	94332	83868	61672

3 Starting at the top left-hand corner, of the random number table from Question 2, and going across from left to right, work out how to use the random number table to help you choose:

a one day from the days of the week (Sunday is day one)

b one day from the month of March

c one day from a year with 365 days. Work out the date (day and month) of this day?

4 Explain how you would use random numbers to randomly choose one student from your mathematics class.

5 To win a lottery you need to match the five chosen numbers in the range 1 to 45 inclusive. Using the random number table from Question 2, start at the top left-hand corner, and going across from left to right, choose five different numbers in the range 1 to 45 inclusive.

Challenge

Win the lottery

To win the prize money in a lottery, you need to match four numbers from the numbers 1 to 9 inclusive. It costs K1 for each entry and you win the prize money of K100 if you match the numbers.

a Use a random number table to make your selections of four numbers from 1 to 9 inclusive. How will you do this? Are there any digits that you are going to ignore? What will you do if a number is repeated in your chosen numbers?

b Find a starting point on the random number table and choose a direction in which to work. Do not change this direction for the whole exercise. Write the four numbers that you randomly choose for each entry.

c How much would you have needed to spend so that you won the prize money if the winning numbers were 2, 4, 6 and 8?

Simulation using a random number table

Random number tables can be used in the place of other physical objects that produce random outcomes.

For example, a coin has two possible outcomes, both equally likely to occur.
A random number table can be used to simulate the outcomes for tossing a coin if we allocate half the digits 0 to 9 to represent 'heads' and the other half to represent 'tails' (we could use 0, 1, 2, 3 and 4 as 'heads' and 5, 6, 7, 8 and 9 as 'tails').
Alternatively, we could use the odd numbers for 'heads' and the even numbers, including 0, for 'tails'.

EXERCISE

1 You are asked to use a coin for a probability experiment but you do not have a coin.

a Explain how you could use a random number table to simulate the outcomes for tossing a coin.

b Write the number(s) that will represent a 'head' and the number(s) that will represent a 'tail'.

c Use the list of random numbers below to simulate the tossing of a coin 50 times, recording the number of heads and the number of tails obtained:

39634 62349 74088 65564 16379 19713 39153 69459 17986 24537

2 You are asked to use a die for a probability experiment but you do not have a die.

a Explain how you could use a random number table to simulate the outcomes for rolling a die.

b Start at the top left-hand corner of the random number table below and, moving down, simulate the rolling of a die 50 times, recording each number obtained.

64628	89126	91254	24090	25752	03091	39411	73146	06089	15630
42831	95113	43511	42082	15140	34733	68076	18292	69486	80468
80583	70361	41047	26792	78466	03395	17635	09697	82447	31405
00209	90404	99457	72570	42194	49043	24330	14939	09865	45906
05409	20830	01911	60767	55248	79253	12317	84120	77772	50103
95836	22530	91785	80210	34361	52228	33869	94332	83868	61672

3 For this question, you will use the random number table to simulate the outcomes for tossing a coin. Even numbers, including 0, will represent a 'head' and odd numbers will represent a 'tail'.

a Start at the top left-hand corner of the random number table below and, moving across from left to right, use the table to simulate **the number of throws necessary to get a 'tail'**. What are the possible outcomes?

You may find it helpful to write out the table and mark when you get a 'tail'. For example (8 0 [5]) (8 [3]) ([7]) (0 [3]) (6 [1]) etc.

The brackets indicate the number of throws necessary to get a tail:

- three for the first tail
- two for the second tail
- one for the third tail
- two for the fourth tail
- two for the fifth tail.

80583	70361	41047	26792	78466	03395	17635	09697	82447	31405
00209	90404	99457	72570	42194	49043	24330	14939	09865	45906
05409	20830	01911	60767	55248	79253	12317	84120	77772	50103

b Construct a tally and frequency table for your results and graph the results after you have completed at least 50 trials. (Alternatively, construct a dotplot for the results.)

c Use your results to find an estimate of the probabilities of it taking 1, 2, 3, … throws to toss a 'tail'.

d Would the probabilities be very different for the number of throws necessary to toss a 'head'? Explain.

e Was it easier to use the random number table rather than tossing actual coins?

A research activity

The birthday problem

What do you think the chances would be that, if you have a random group of 30 people, at least two of them will have their birthday on the same day of the year? Most people would say that the chance would be very small but, in fact, it is about 70%. If you have 30 people in your classroom or school, test this theory.

EXERCISE

1 You can test whether two people share the same birthday using a random number table.

a How many possible birthdays are there in a non-leap year?

b How will you use the random number table to simulate these birthdays?

c Simulate the birthdates of 30 people and see if you have any matches. Complete several simulations.

d Combine the results for the whole class, counting the total number of trials and the number of times that there was a match. From these results, estimate the probability that two people in a group of 30 would have their birthday on the same day.

e If you have access to the Internet, investigate the theoretical probability of this 'problem'. Carry out a search using the words 'The probability birthday problem'.

Other random number generators

Lists of random numbers can also be obtained using a calculator or computer that has a random number generator.

i A scientific calculator

Most random number generators on scientific calculators produce random numbers between 0 (included) and 1, with three decimal places.

The numbers generated can be used in a similar manner to those in a random number table. Ignore the decimal point and use as many digits as needed.

ii A graphics calculator

A graphics calculator will generate a list of random numbers within a given range, one at a time or in groups.

Using a TI83 calculator, we key (MATH) ◄ to get to the probability menu (PRB), choose 5:randInt((ENTER), type in the range of numbers, 1 to 25 inclusive, and a random number in the range 1 to 25 will be generated each time we press (ENTER).

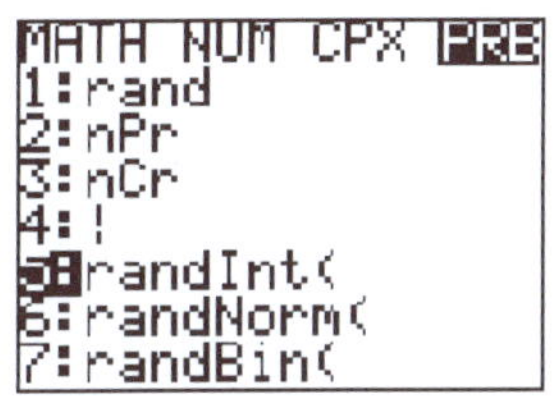

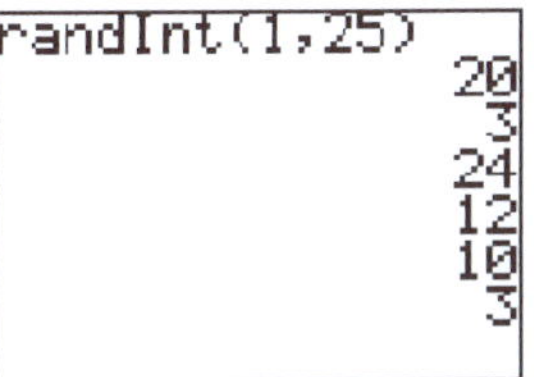

Alternatively, if we want to generate a group of numbers in the range 1 to 25, we type 1,25 for the range and then ,5) for the number in the group (5 in this case). Groups of five random numbers in the range 1 to 25 inclusive will be generated each time we press (ENTER).

```
randInt(1,25,5)
 {19 14 3 18 24}
  {13 8 16 18 7}
   {15 5 1 25 2}
    {3 6 18 12 9}
  {8 17 15 16 3}
```

iii A computer

A computer will generate a table of random numbers within a given range.

The following table was generated using a spreadsheet on the computer, typing =RANDBETWEEN(1,25) in the first cell and copying this into the other cells.

10	5	1	5	25	1	18	6	14	2
7	21	9	7	3	10	24	17	21	25
14	11	22	7	13	19	16	2	1	25
2	15	7	21	15	1	7	15	8	13
9	22	2	2	11	13	9	20	2	16
22	10	1	6	17	4	22	5	6	6
7	20	8	15	6	25	25	8	16	10

EXERCISE

Class exercise

1 Sam leaves for school at the same time each day and, walking by himself, he arrives at school with 10 minutes to spare.
On his route to school, he passes the houses of four of his friends, Kema, Sean, Vincent and Lucas. If one of these friends joins Sam on the journey to school, he takes 3 minutes longer to get there. If two friends join him, his journey takes 6 minutes longer than if he is by himself. If three friends join him, the journey is 9 minutes longer than when he is alone. If all four friends join him, the journey is 12 minutes longer than when he is alone. The chance that Sam will meet any one of his friends on the journey to school is $\frac{3}{5}$.

a Work in pairs and use simulation to estimate the probability of Sam being:

i 10 minutes early for school
ii 7 minutes early for school
iii 4 minutes early for school
iv 1 minute early for school
v 2 minutes late for school.

You can use any available simulator for this activity but you must explain how you are using it. You must also give an example of one trial journey and what it means in terms of the time that Sam gets to school.

b Record your results from the trials in a suitable table, graph them and give the probability estimates for parts **a i** to **a v** above.

c Compare your results with those of other people in your class. Comment on the accuracy of your results.

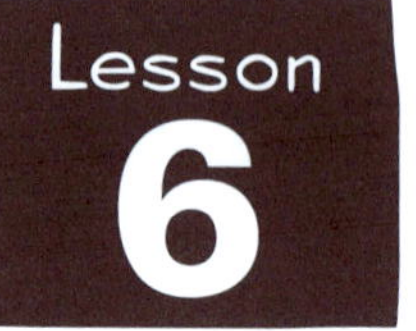

Activity: Devise your own simulation

In this activity, you will use a random number generator to simulate the outcome of something that relates to you.

ACTIVITY

Use a random number generator to simulate the situations below.

You will need to decide which random number generator you are going to use each time and how the numbers it generates relate to the outcomes.

Use your chosen simulator to conduct a number of trials and estimate the probability.

a You have found that you get to school on time on seven days out of eight. What is the probability that you will be on time on every day of a five-day school week?

You might construct a spinner for this exercise, with seven sectors marked 'on time' and one marked 'late'. You would need to spin the spinner five times for one trial and count the number of 'on times' you spin. You would need to do this for at least 30 trials.

b Your sporting team, on average, wins 5 out of 6 matches. What is the probability that you go through a six-match season undefeated?

A die could be used for the simulation, with the numbers 1 to 5 representing 'wins' and the number 6 representing a 'loss'. For one trial of a six-match season, you would need to toss the die 6 times (or use 6 dice). You would need to do at least 30 trials.

The number of trials to get the tokens

Lesson 7

Many manufacturers run campaigns that include an incentive so that customers will buy the product again.

For example a sweet manufacturer may include sports cards with their product so people keep buying the product because they want to collect the whole set of sports cards.

Some manufacturers include different tokens in their product and, when all the tokens are collected, the customer can claim a prize.

The cards or tokens in these products are usually randomly distributed and are not visible to the purchaser when the product is bought.

EXERCISE

1 A gum manufacturer includes a token inside each packet of gum and, when the customer has collected all six different tokens, he/she can exchange the tokens for a free packet of gum.

How many packets of gum would we need to buy to collect all the tokens, assuming the tokens are randomly included in the gum packets?

- a Guess the number of packets of gum that we would need to buy so that we are 90% sure that we will have all six different tokens?
- b Describe how you could simulate collecting the six tokens. State the type of random number generator you are going to use in the simulation and describe how the numbers are assigned to each of the tokens.
- c Do at least 50 simulations of 'the number of packets of gum needed to collect all six tokens', recording your results in a table. Use a graph to display the results.
- d Comment on your results.
- e Compare your results with those of other students in your class.
  - i How similar are your results to those of the other students?
  - ii If you combine your results with those of the other students, would the estimate be more accurate?

2 a Repeat the simulation from Question 1 for products that contain:

i 2 different tokens ii 3 different tokens iii 4 different tokens.

b Can you see a pattern in the number of packets that we need to buy to get all the tokens?

Games of chance

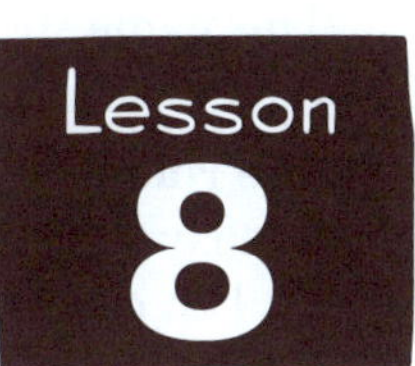

People from all cultures play games.

Games of chance are usually based on a physical object that gives a random outcome, for example dice.

Dice give the **chance factor** to the game because the outcome from throwing a die cannot be precisely predicted.

Some games rely on both chance and skill. Skill becomes part of the game when, for instance, a player is required to make a decision after considering a previous outcome.

Discussion

What games of chance do you play?

Are these games based on chance only or is there some degree of skill involved?

Weeks 6, 7, 8, 9 and 10

When playing the card game of Blackjack (or 21) the card-counting skill of a player can affect the outcome of the game. In this game, a player can see all the cards dealt in the first round and can calculate the likelihood of being dealt a favourable card in the second round.

Many games of chance are also associated with gambling. Gambling can be just for fun, involving tokens, or it can involve money. Gambling for money can be addictive for some people and could lead to serious problems for the gambler and his/her family.

EXERCISE

This exercise contains some games for you to play and analyse. Some of the games depend on chance only and some depend on chance and skill. (It may take several lessons to complete this Exercise.)

1 This game can be played in groups (of 2 to 6).

You will need:
- a die
- a shaker, to use for shaking and rolling the die
- grid paper, with a width of 20 (to 40) lines and a length of 20 (to 40) lines (see page 392)
- a token (of a size that fits on the grid) for each player.

The rules:
- Locate a centre square on the grid paper, to be where every player starts, and position the tokens on this square.
- Players take turns to throw the die and move their token.
- If a player throws an **even number**, he/she can move the token either to the **right or up** the number of squares corresponding to the even number thrown.
- If a player throws an **odd number**, he/she can move the token either to the **left or down** the number of squares corresponding to the odd number thrown.
- Players must move on each of their own throws.
- If a player gets stuck on an edge, he/she moves to the opposite end of the row/column and continues counting. (You may decide to modify this rule.)
- The first player to reach a corner square of the grid is the winner.

After you have played a few games (start on a smaller grid first, then play on larger grids) answer the following questions:

a Is this game a game of chance only, or can you use skill to influence the outcome of the game?

b If you think that you can use skill to influence the game how could you do this? State some strategies that will help you to win a game.

2 Modify the game in Question 1, so that you use a pack of cards (or part of a pack of cards) to determine the moves. Decide which cards correspond to left, right, up or down and the number of moves that correspond to the card that is drawn. Write a list of rules for this game and play the game several times.

3 This game should be played in groups of four or more and is a version of Blackjack or 21.

You will need:
- a pack of playing cards.

The rules:
- Players take it in turns to be the dealer. Shuffle the cards well before each round.
- The cards are each given a value: The cards 2 to 10 have their face value and all the court cards (Jack, Queen and King) have the value 10.
- The Ace has either the value 1 or 11 depending on the circumstances. If a value of 11 is assigned to the Ace and this puts the total over 21, the value of the Ace can be reduced to 1.
- It costs 5 points to play a game. The points are collected from the players at the beginning of each game. Everyone starts with 50 'points' and wagers 5 'points' per game. The game is easier to play if the 'points' are represented by objects, such as beans, counters or matches, that are easily collected.

- The dealer deals two cards to each person, one at a time. These cards should be visible to everyone.
- The winner of the round is the person whose score is closest to 21 and this person takes all the points collected at the beginning of the game.
- If someone has a score of 21 (for example a Jack and an Ace) after the first two cards are dealt, they are the winner.
- If there are two or more winners then the points are shared between those winning players.
- If the first two cards are dealt (and no-one has won) any player can buy another card. Buying a card costs another 5 points.
- If the total score of a player's cards is higher than 21, the player is out of the round and loses his/her points.
- If no-one wins the game, the points from that game are kept over for the next game and are added to the points collected at the beginning of the next game.
- The overall winner is the person who accumulates 100 points first.

After you have played a few games answer the following questions:

a Is this game a game of chance only or can you use skill to influence the outcome of the game?

b If you think that you can use skill to influence the game how would you do this?

c Does the number of players influence the outcome of this game?

d Can you think of a way that you could change this game so that more skill is involved?

4 Sevens

This game is best played by 3 to 5 people.

You will need:
- a die
- a shaker
- 150–200 counters (or beans). There needs to be a bank of counters (beans).

The rules

- Each player starts with 20 counters.
- The first player rolls the die, taking note of the number uppermost.
- Before rolling the die again, the player 'bets' any number of counters on either 'a total of 7', 'a total less than 7' or 'a total greater than 7'. This bet must be declared so that everyone knows what it is.
- The player rolls the die again and calculates the total of the two throws.
- If the 'declared bet' is 'less than 7' or 'greater than 7' and this matches the total rolled, the player receives from the bank twice the number of counters bet. If the total is 7 and this was the declared bet, the player receives three times the counters bet.
- If the total rolled does not match the declared bet, the player loses his/her counters.
- Players take turns to roll the die and declare bets.
- The first person to accumulate 50 (or 100) counters is the winner.

Play this game several times and then answer the following questions:

a Is this a game of chance only, or can you use skill to influence the outcome of the game?

b If you think that you can use skill to influence the game how would you do this?

c State what your declared bet would be, if your first throw of the die was:

i 1 **ii** 2 **iii** 3 **iv** 4 **v** 5 **vi** 6

d What is the probability of getting your declared bet for each case in part **c**?

e What is the probability of getting 'a total of 7' for the two throws of the die?

OPTION B: STATISTICAL SURVEYS

Populations and samples

A **survey** is a method of collecting information for a statistical investigation. The information collected is commonly called **data**. Governments, businesses and individuals collect data from surveys so that they can make informed decisions about things that affect us in everyday life and so that they can plan for the future.

Before data is collected for a statistical investigation, a decision needs to be made as to whether the data will be collected from the **whole population** or from a **sample** of the population.

In some statistical investigations it is possible to collect data for the whole population. Every 10 years, the **National Statistical Office of Papua New Guinea** conducts a survey of the whole population of PNG, called a **census**. This census collects data relating to population distribution and housing, and provides valuable information for future planning. Because the whole population of PNG is surveyed, the census provides accurate information but it can be a very time consuming and costly exercise.

The word '**population**' does not always mean 'all the people in PNG'. Generally, the word means all the people or things to which the conclusions of a statistical investigation would apply (in other words, all members of the particular group in question).

For example, if we want to investigate a theory related only to our school, the population would be all the students who attend the school.

If, however, we want to investigate something that applies to all school children in PNG then the population will be all the school children in PNG, and the students in our school would be a **sample** of the population.

For most statistical investigations, surveying the whole population is either too time consuming or too costly so, in these cases, a sample is chosen to represent the population. If data is only to be collected from a sample, then a sample must be chosen that **represents the population** so that reliable conclusions about the population can be made. Conclusions from a sample will still never be as accurate as conclusions made by surveying the whole population.

When planning a statistical investigation, we need to also consider the **type of data** that will be collected and how the conclusions are going to be presented. There are two types of data that we commonly deal with (categorical data and numerical data) and each of these types of data has appropriate ways to be organised (tables, etc.), displayed (graphs) and interpreted (conclusions). This was covered in the core part of this unit.

Discussion

Discuss the reasons you think it is important that the PNG government collects data about:

i the region in which people live
ii the age distribution of people in a particular region.

EXERCISE

1 At the last census, it was found that there was an average of 5.2 people in each household in PNG.
  - a What is a census?
  - b Who conducts a census?
  - c How often is a census conducted in PNG?
  - d When was the last census conducted in PNG?
  - e Why isn't a census conducted every year?
  - f What happens if you do not want to answer the questions in a census?
  - g State some difficulties that may have been encountered when collecting the data for 'the average number of people in a household in PNG'.
  - h Why do you think it is important that such information as 'the average number of people in a household in PNG' is known? Suggest some ways this information might be used.

2 From a school with 120 students, 25 students are randomly selected to complete a survey on their school uniform. In this situation:
  - a what is the population?
  - b what is the size of the population?
  - c what is the sample?
  - d what is the size of the sample?

3 i Discuss how you would investigate each of the following.
  ii Give reasons why you would use a representative sample or the whole population for each investigation.
  - a the age distribution of students in your school
  - b the average number of matches in a box
  - c the number of 15 year olds who possess a mobile phone
  - d the heights of 15-year-old girls in PNG.

4 A research company is investigating people's opinions about whether smoking should be banned in all public buildings in PNG. The researchers ask the opinion of people who are smoking in the street of a large town.
  - a What is the population for this survey?
  - b Will the data that the researchers collect be from a sample or from the whole population?
  - c Do you think a sample or the whole population should be used for this investigation? Give reasons.

5 A television station is using a telephone poll to investigate the question: 'Should all Papuan adults be tested for HIV/AIDS?' Viewers are asked to phone a given number to vote either 'Yes' or 'No'.
  - a What is the population in this case?
  - b Will the data that the television station collects be from a sample or from the whole population?
  - c Do you think a sample or the whole population should be used for this investigation? Give reasons.

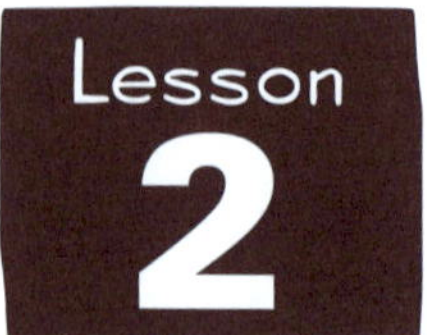

Biased samples

In an investigation in which a sample is chosen to represent the population, the sample must be chosen so that the results will not show **bias** towards a particular outcome. If the sample is biased towards a particular outcome, any conclusions that are made from the investigation will not apply to the whole population and much time and money will have been wasted.

For example, if the purpose of a survey is to obtain an accurate indication of how the population of PNG is going to vote at the next election, surveying a sample of Papuans from only one village would probably not provide information that represents all of PNG. It is possible that the sample would be unbiased, however it is unlikely.

Discussion

Discuss how/why the sample in the example above could be biased.

EXERCISE

1 Give reasons why the selected sample in each of the following could be a biased sample:
   a A researcher is investigating the incidence of obesity in PNG. She asks her friends, relatives and work colleagues what they weigh.
   b To investigate the level of 'left-handedness' in PNG, a researcher calculates the proportion of left-handed children in a particular school.
   c The principal of a school wishes to know the proportion of students who are late for school. On the next Monday, she stands at the school gates for 10 minutes and counts the number of students who are late.

2 A research company is investigating people's opinion on whether smoking should be banned in all public buildings in PNG. The researchers ask the opinion of people who are smoking in the street of a large town. Explain why the data collected is likely to be biased.

3 A television station is using a telephone poll to investigate the question: 'Should PNG adults all be tested for HIV/AIDS?' Viewers are asked to phone a given number to vote either 'Yes' or 'No'. How is the data biased if it is used to represent the views of all Papua New Guineans?

4 A customer complains to a manufacturer that the fish, in a tin of the manufacturer's fish have a strange taste. The manager of the manufacturing company tastes a sample of the fish that are being processed that day and says, 'It tastes fine to me.'
   a What is wrong with the manager's sampling method?
   b Why do manufacturers include batch numbers and dates of manufacture on their products?

Discussion

Suggest another way in which each of the samples in Questions 1 to 4 of this exercise could be selected.

Unbiased samples

Lesson **3**

There are many ways of choosing an unbiased sample.

One of these is a **simple random sample** which is a sample chosen so that **every person or object of the population has an equal chance of being chosen**.

'**Random**' means that there is no pattern or plan (no bias) in choosing one element of the population over another. If the sampling is done at random, then it is possible to draw valid conclusions about the whole population from the results for the sample.

There are various simple means of randomly selecting a sample. The first method we will look at involves the use of random numbers.

Using random number tables to choose a sample

Random number tables appear as a list of the digits from 0 to 9. The digits are generated by a computer and are grouped in fives so that the table is easier to read. You can start anywhere in the table and move across or down to select your random numbers.

39634	62349	74088	65564	16379	19713	39153	69459	17986	24537
14595	35050	40469	27478	44526	67331	93365	54526	22356	93208
30734	71571	83722	79712	25775	65178	07763	82928	31131	30196
64628	89126	91254	24090	25752	03091	39411	73146	06089	15630
42831	95113	43511	42082	15140	34733	68076	18292	69486	80468
80583	70361	41047	26792	78466	03395	17635	09697	82447	31405
00209	90404	99457	72570	42194	49043	24330	14939	09865	45906
05409	20830	01911	60767	55248	79253	12317	84120	77772	50103
95836	22530	91785	80210	34361	52228	33869	94332	83868	61672
65358	70469	87149	89509	72176	18103	55169	79954	72002	20582
72249	04037	36192	40221	14918	53437	60571	40995	55006	10694
41692	40581	93050	48734	34652	41577	04631	49184	39295	81776
61885	50796	96822	82002	07973	52925	75467	86013	98072	91942
48917	48129	48624	48248	91465	54898	61220	18721	67387	66575

Suppose we want to randomly choose 10 different numbers from 1 to 50 inclusive. We need to look at two digits at a time, because there are two digits in the largest number (50). The numbers 1 to 9 would appear as 01 to 09 in the random number table. Any two-digit numbers larger than 50 are ignored.

Starting in the top left-hand corner and working down, we would have the numbers **39**, **14**, **30**, (64 is ignored because it is larger than 50), **42**, (80 ignored), (00 ignored), **5 which appears as 05**, (95, 65, 72 ignored), **41**, (61 ignored) and **48**. Then, returning to the top and using the next pair of digits in each row, we have (63, 59, 73, 62, 83, 58 ignored), **20**, **40**, (83 ignored) and **35**.

The ten random numbers between 1 and 50 inclusive are 39, 14, 30, 42, 5, 41, 48, 20, 40 and 35.

Remember

'Inclusive' means that the numbers 1 and 50 are both included.

Discussion

Why do you think it is necessary to use random number tables? Why can't we just write a list of ten numbers between 1 and 50?

Some rules for using a random number table to select a set of numbers

- Choose a starting point and a direction and continue with this for the entire selection. Don't jump around the table!
- Ignore numbers that are larger or smaller than you need.
- Ignore numbers that are repeated.

Example

A teacher wants to choose a simple random sample of three students from a class of 32 students. She has a class list of the students in alphabetical order.

How can she obtain this simple random sample of three students using a random number table?

Answer

The population size in this case is 32 and the sample size is 3. (The teacher could assign a number from 1 to 32 to the students on the class list.) Then she would need to choose, from the random number table, three numbers in the range 1 to 32 inclusive.

She will look at the digits in the random number table, two at a time, ignoring any numbers that are outside the range 1 to 32 and any repeated numbers.

Starting in the second row of the table:

14595 35050 40469 27478 44526 67331 93365 54526 22356 93208

and moving across, the numbers in the random sample would be:

14, (59, 53, 50, 50, 40, 46, 92, 74, 78, 44, 52, 66, 73 ignored), **31**, (93, 36, 55, 45 ignored), and **26**.

The numbers in brackets, described as ignored, are not part of the random sample because they are outside the range 1–32.

Based on this sample of random numbers, the teacher would choose the 14th, 26th and 31st students from the class list as the random sample of three students.

EXERCISE

1 Start at the top left-hand corner of the random number table on the right and working downward, find:

a a random list of five different numbers between 1 and 9.

b a random list of ten different numbers between 50 and 99.

37968	33329
35845	62037
97097	20740
52650	19348
90510	26807
09017	41054
47841	99014
17274	04202
97792	69054
92894	85699

2 Starting at the top left-hand corner of the random number table below, and working downward, find a random list of six different numbers between 100 and 200.

64628	89126	91254	24090	25752	03091	39411	73146	06089	15630
42831	95113	43511	42082	15140	34733	68076	18292	69486	80468
80583	70361	41047	26792	78466	03395	17635	09697	82447	31405
00209	90404	99457	72570	42194	49043	24330	14939	09865	45906
05409	20830	01911	60767	55248	79253	12317	84120	77772	50103
95836	22530	91785	80210	34361	52228	33869	94332	83868	61672

3 Describe how you would use a random number table to randomly select one of the 20 provinces of PNG.

4 The following calendar is for June 2008:

Sunday	Monday	Tuesday	Wednesday	Thursday	Friday	Saturday
1	2	3	4	5	6	7
8	9	10	11	12	13	14
15	16	17	18	19	20	21
22	23	24	25	26	27	28
29	30					

For each of the samples described below:

- **i** state the population and the size of the population
- **ii** state the size of the sample
- **iii** use the random number table, from Question **1**, starting from the bottom left-hand corner and moving horizontally, to find the sample.

a three days in the month **b** a Thursday in the month

c four consecutive days in the month (*Hint*: How many groups of four consecutive days are there in this month? This will be your population size.)

5 **a** Describe how you would use a random number table to choose 30 dates, repeats allowed, from the year (365 days).

b Use a random number table to select 30 numbers between 1 and 365 inclusive. Write the numbers including any repeats.

c Do any of the numbers appear twice? (See 'The birthday problem' on page 220.)

6 To choose one household in a large town, an investigator randomly chose a street in the town and then, after finding out how many houses were in the street, randomly chose a house. Would every house in this town have an equal chance of being chosen? Explain.

Other random number generators

Physical objects can be used to generate random outcomes. Some are listed below.

- **A coin:** This can be used if there are only two possible outcomes or the size of the population is 2. Each of the outcomes, or members of the population, is assigned either 'heads' or 'tails' and the coin is tossed.
- **A die:** A die can be used if the number of possible outcomes or the size of the population is a factor of 6.
- **A carefully constructed spinner:** A spinner can be made specifically for a situation. For example, if we needed to randomly choose three objects from a possible eight objects, spinning a spinner with eight equally likely outcomes three times would be appropriate for this random selection. However, the physical limitations of making a spinner mean that a spinner with 10 outcomes is probably the maximum size that would give reliably random outcomes.
- **Selecting numbered objects from a bag:** Care needs to be taken to make sure that the sample is random. The objects should all be the same size and shape, should not be visible to the selector and should be returned to the bag and mixed well after each selection. There are also physical limits to the use of this method (depending on the size and strength of the bag and the size, weight and number of the objects).

- **A set of numbered cards:** We will need a card for every object in the population, so this method is limited to a population of about 100. Playing cards can sometimes be used if we take care to allocate the cards to members of the population.
- **Lists of random numbers:** These can be obtained from a calculator or computer that has a random number generator.

Ways of generating lists of random numbers

i A scientific calculator

Most random number generators on scientific calculators produce random number between 0 (included) and 1, with three decimal places.

The numbers generated can be used in a similar manner to those in a random number table. Ignore the decimal point and use as many digits as needed.

ii A graphics calculator

A graphics calculator will generate a list of random numbers within a given range, one at a time or in groups.

Using a TI83 calculator, we key (MATH) ◀ to get to the probability menu (PRB), choose 5:randInt ((ENTER), type in the range of numbers, 1 to 25 inclusive, and a random number in the range 1 to 25 will be generated each time we press (ENTER).

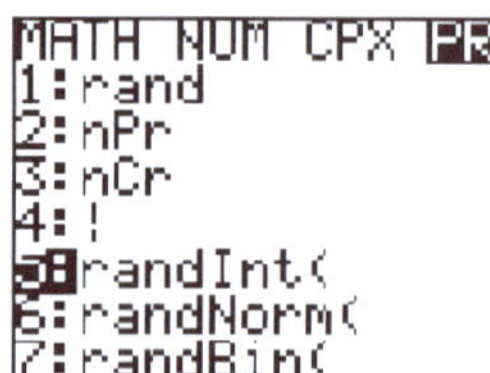

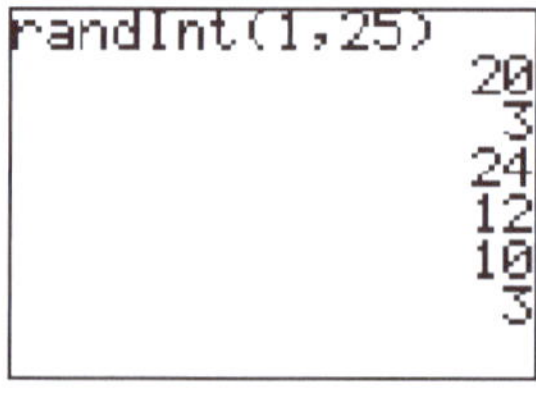


```
randInt(1,25,5)
 {19 14 3 18 24}
 {13 8 16 18 7}
 {15 5 1 25 2}
 {3 6 18 12 9}
 {8 17 15 16 3}
```

Alternatively, if we want to generate a group of number in the range 1 to 25, we type 1,25 for the range and then ,5) for the number in the group (5 in this case). Groups of five random numbers in the range 1 to 25 inclusive will be generated each time we press (ENTER).

iii A computer

A computer will generate a table of random numbers within a given range.

The table on the right was generated using a spreadsheet on the computer, typing =RANDBETWEEN(1,25) in the first cell and copying this into the other cells.

10	5	1	5	25	1	18	6	14	2
7	21	9	7	3	10	24	17	21	25
14	11	22	7	13	19	16	2	1	25
2	15	7	21	15	1	7	15	8	13
9	22	2	2	11	13	9	20	2	16
22	10	1	6	17	4	22	5	6	6
7	20	8	15	6	25	25	8	16	10

EXERCISE

1 Describe how you would use each of the following to choose a sample of one from a population of two:

a a coin

b a die.

2 Describe how you would use a die to choose a sample of two different objects from a population of:

a three different objects

b six different objects.

3 a Make a spinner for a population of eight. How will you make sure that each number, 1 to 8, has an equal chance of being spun?

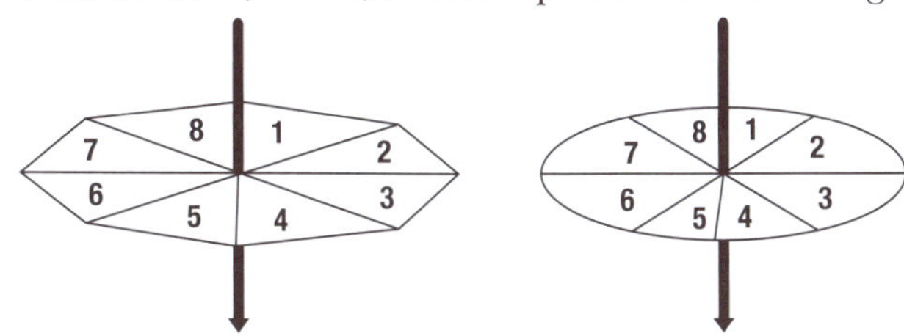

b Check that your spinner is unbiased by conducting a number of spins (at least 50) and recording your results in a dotplot.

c Describe how you would use the spinner to select a sample of three different numbers. Then use the method you described to choose your sample of three numbers.

d Continue choosing and recording samples of three numbers until you have 20 samples. How many of your samples are the same?

4 You have a population of 40 from which you want to randomly choose a sample of five.

a Describe how you could use a pack of playing cards to choose this random sample.

b If you have a pack of playing cards, use the method you described in part a to choose a sample of five from a population of 40.

5 You task is to randomly select eight students from your school. How could you do this so that every student in the school has an equal chance of being chosen? In your explanation, mention:

- the population size
- the sample size
- your method of selecting the sample
- why your sample is a random sample
- the difficulties you might encounter using your chosen method of selection.

Using a sample to estimate a characteristic of a population

In some situations we are required to know the proportion of a population that has a particular attribute.

For example, the directors of a company that manufactures scissors may wish to know the proportion of the population that is left-handed, so they can manufacture enough scissors to suit left-handed people.

Usually it is not possible to survey the whole population, so a random sample is used. The proportion of people in the sample with the attribute of being left-handed is used to make an estimate of the proportion of the people in the whole population having that attribute.

Example

A particular province contains 32 000 households. People in a random sample of 200 households in the province were asked: 'Do you grow tomatoes in your garden?' Twenty-four of the households answered 'Yes' to this question. From this data, how many households in the whole province would you estimate to grow tomatoes?

Answer

To estimate the number of households, x, of the 32 000 households in the province, that grow tomatoes we use:

Proportion of sample that grow tomatoes = proportion of province that grow tomatoes

$$\frac{24}{200} = \frac{x}{32\,000}$$

$$\frac{24 \times 32\,000}{200} = x$$

$$x = 3840$$

Using the sample, it is estimated that 3840 households in the province grow tomatoes.

Sometimes several random samples are used to estimate a single attribute and the results are analysed to give a better picture of the population proportion.

EXERCISE

Class exercise

In this exercise, you will work in pairs to demonstrate how a sample can be used to estimate the proportion of an attribute in the population.

You will know the proportion of the population with the attribute (10%) and will see how well the random samples that you select demonstrate this proportion.

You will need:

- a population of 100 similar objects.
  These could be:
  - Objects that are all the same size, for example dried beans or pebbles. (You will need to put these in a bag so that random selections can be made).
  - Cards. You could use two packs of playing cards discarding four of the cards. (You will need to shuffle the pack regularly.)
  - Numbers from a random number table, using 00 to 99.

Step 1: Mark 10 of the objects with a colour. These will represent the individuals with the attribute in the population. This is easy to do with a pencil or marking pen if you are using beans. If you are using cards, you can decide that, for instance, cards 1 to 5 of hearts are the marked cards. if you are using random numbers, you could decide that 00 to 09 are the marked numbers.

You will now have a population of 100 objects of which 10% (10) have a particular attribute.

If you select a random sample of 10 from the population, how many objects with the attribute would you expect to get?

Step 2: Select a random sample of 10 from the population of objects. Count the number of marked objects in your sample.
Start a dotplot showing the number in the sample that have the attribute.

Step 3: Return the previous sample to the population and make sure that the objects are well mixed. Select another sample of 10. Count the number of marked objects in this sample and record this number on your dotplot.

Help box

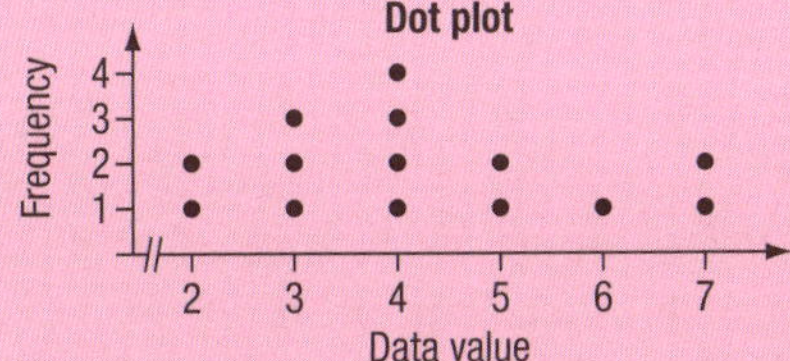

The mean for the dotplot above is:

$$\frac{2 \times 2 + 3 \times 3 + 4 \times 4 + 2 \times 5 + 1 \times 6 + 2 \times 7}{14} = 4.21 \quad \textbf{To two decimal places}$$

Step 4: Repeat Step 3 above, so that you have taken and recorded at least 20 samples and recorded the result on your dotplot.
When you have finished selecting your samples, answer the questions below.

1 Are the results of the sampling what you expected? Explain.
2 Looking at your dotplot, do the results indicate that the population proportion is 10%?
3 From your dotplot, calculate the mean number of objects with the attribute that were found in your samples. Does this figure indicate that the proportion having the attribute in the population is 10%.
4 Compare your results with those of some other groups. Are their results similar to your results? If they are different, how are they different?
5 If you use a larger sample size, of say 20, do you think your results would be more accurate?

Using a sample to estimate a statistic of the population

Lesson 6

Sometimes statistics (mean, median, mode, maximum, minimum, range, etc.) calculated from samples are used to estimate the statistics for a population.
If the sample size is too small, any statistics that are based on this sample may not provide a good estimate of the population statistics.

For example, measuring the heights of three 15-year-olds would not give a very reliable estimate of 'the heights of 15-year-olds'.

Statistics calculated from a large unbiased sample should give a reliable estimate of the population statistics but, often, there are time and cost restrictions preventing the use of large samples.

Determining the size of a sample that will reliably represent a population is a complex exercise so, often, a sample of a convenient size is chosen. This means, however, that the conclusions based on the sample may not be relevant to the whole population.

EXERCISE

Class exercise

For this exercise, you are going to use a sample, and statistics based on that sample, to estimate the size of a population.

For this exercise, every object in the population is numbered, using 1, 2, 3, … N, so the population size will be N.

Your teacher will know the value of N, but you will be required to use random samples to find this number.

Your first sample will contain five numbers (in the range of 1 to N inclusive) produced on a random number generator by your teacher.

Subsequent samples can be obtained from your teacher and combined with your other samples to obtain a 'better' estimate of N.

You will be required to keep records of all your sample numbers and calculations.

The **median** and the **mean** are measures of the centre of a list of data. To find an estimate of the maximum value of the data we multiply them by 2.

Discussion

- How could you use a sample of five numbers to estimate N?
- If you use the largest of your five numbers, how good an estimate of N do you think this will be?
- If you find the mean (average) of the numbers, what is this statistic estimating? How could you estimate N from the mean?
- If you find the median of the numbers what is this statistic estimating? How could you estimate N from the median?

1 Obtain a sample of five random numbers, in the range 1 to N, from your teacher.
   - a Find the **median** of these five numbers and use this to find an estimate of N.
   - b Graph all the estimates from the class on a dot plot. Comment on the graph and how it can be used to estimate N.
   - c Is there a number in your sample that is larger than the largest estimate? Would this be a better estimate of N? Why?

2 Using your sample from Question 1 again:
  a find the **mean** of your five numbers and use this to find an estimate of N
  b graph all the class estimates (rounded to the nearest whole number) on a dot plot
  c how do the estimates obtained using the mean compare with the estimates obtained using the median?

3 From your teacher, obtain another sample of five random numbers, in the range 1 to N.
  a Combine this new sample with your original sample and:
    i find the **median** of the ten numbers, and another estimate of N
    ii find the **mean** of the ten numbers, and another estimate of N.
  b Are your estimates in part **a** smaller than the largest value in your sample? If your estimates are smaller than the largest value, use the largest value from your sample as your estimate.
  c Graph all the new estimates on class dotplots.

4 Ask the teacher to reveal the value of N.
  a How close were your estimates?
  b Which of your estimates (from the median, mean or maximum value) was your best estimate of N?

Discussion

How well did the work you did in the class exercise indicate the value of N? Look at each of the class dotplots of the estimates.

Capture-recapture experiments: Using a sample to estimate the size of a population

Capture-recapture experiments are used to estimate the size of a closed population of animals/insects/fish (for example finding the number of fish in a pond).

A sample of animals is captured and all of these animals are marked or tagged then released back into the population.

At a later date, having given the tagged animals time to randomly distribute themselves back into the population, another sample is captured. The number of tagged animals in this new sample is counted.

If n is the size of the original sample (all tagged) and N is the size of the population (that we need to estimate), then the proportion of the tagged animals in the population is $\frac{n}{N}$.

If S is the size of the second sample, and there are s tagged animals in this sample, then the proportion of tagged animals in the second sample is $\frac{s}{S}$.

In capture-recapture experiments, we use the sample proportion to estimate the population proportion so:

$$\frac{s}{S} = \frac{n}{N}$$

Rearranging this equation gives an equation for estimating the population size, N:

$$\frac{s}{S} = \frac{n}{N}$$

$$s = \frac{nS}{N}$$ Multiplying both sides by S.

$$sN = nS$$ Multiplying both sides by N.

$$N = \frac{nS}{s}$$ Dividing both sides by s.

Example

Lillian wants to estimate the number of cockroaches that she has in her kitchen area. She traps 12 of the cockroaches, marks them with a white dot and sets them free. After two days she traps another group of 10 cockroaches and three of these have white dots.

For this experiment, the original sample size is $n = 12$. (These are all marked with a white dot.) The second sample size is $S = 10$ and three of these are marked, so $s = 3$.

Answer

An estimate of the size of the population, N, of cockroaches in Lillian's kitchen is given by:

$$N = \frac{nS}{s} = \frac{12 \times 10}{3} = 40$$

Lillian has estimated that there are 40 cockroaches in her kitchen area.

EXERCISE

Class exercise

Work in pairs to complete this exercise.

A capture-recapture experiment can be demonstrated using rice (or small beans).

You will need:

- a jar of rice or beans (the number of grains of rice/beans in this jar is the population size, N, that you are going to estimate)
- a smaller container (for collecting a sample of the rice/beans)
- food dye or marker pens

Step 1: Start with the jar of rice/beans. Use the smaller container to collect a sample of the rice/beans. Count the number of grains of rice or beans in the sample. This is the value of n. Dye these grains of rice or beans with the food colouring or marker pens. Then return the dyed objects to the original population jar.

Step 2: The dyed objects need to be distributed evenly throughout the population jar, so spend some time shaking the jar until you are satisfied that this is done. (If a glass jar is used, you will see when this has been done.)

Step 3: Take another sample from the population jar of rice grains/beans. Count the number altogether in this sample, S, and the number of grains in the sample that are dyed, s.
Use the formula $N = \frac{nS}{s}$ to estimate the size of the population in the jar.

Step 4: Return your sample to the jar, mix again, and repeat Step 3 to find another estimate of the population size, *N*. Did you get the same value for *N* as you did for your first calculation? How close were your two estimates?

Step 5: If you have time, take more samples to make other estimates. These could be graphed to give a better idea of the population size.

Note: In this case, it is possible (if you have the time and patience), to count all the rice/beans in the jar and check the accuracy of your results and this method of estimation.

Step 6: Write a paragraph reporting on this experiment, giving your estimates of the population size and your conclusions about the accuracy of these estimates.

Discussion

Are there any local insects/fish/animals whose population size you could estimate using this capture-recapture method?

Designing a questionnaire for a statistical investigation

Lesson 8

A **questionnaire** is a set of questions designed to obtain information from people. Questionnaires can be conducted in person, by mail or over the phone. Questionnaires conducted in person are more effective, especially if the questioner is recording the answers as they are given. There is often a large non-response rate for mail or phone questionnaires.

A questionnaire relies on the person who is being surveyed giving truthful or accurate answers to the questions and this can be influenced by:

- the directness of the questions. (The person being surveyed should know immediately what form the expected answer should take. They should not need to ask the questioner such questions as, 'What do you mean by … ?')
- the convenience of the survey. (A few direct questions, in a logical order, will gain a more cooperative and accurate response than a long questionnaire in which the relevance of some questions is not obvious.)
- the relevance of the questions. (The purpose of the survey should be stated to the person being surveyed, so they can judge the relevance of what is being asked.)
- the questions not being leading questions. (Avoid questions with value judgements built in, such as: 'Do you agree that criminals should be put in gaol?'. It would be difficult to find many people who would disagree with the question expressed in this form.)
- the questions not being of a private nature, where responding with truthful answers could embarrass the person being questioned or the questioner. (Some people might not want to tell you their age, their weight, their HIV/AIDS status, their income, etc.)

If you have designed a questionnaire to collect information it is a good idea to test it on a few friends first to see if it provides the required information.

EXERCISE

1 The purpose of the questionnaire on the right is to obtain people's opinion on whether smoking should be allowed in public areas and to see if there is any difference between the responses of men and women.

a Consider the directness of each of the questions in the questionnaire. (It is helpful, when considering the questions, to think about the answers that may be given to them.) Are any of the questions not direct enough?

b Which of the questions in the questionnaire do you consider to be irrelevant to the purpose of the investigation?

c Which questions do you consider to be leading or biased questions?

d Which questions may have made the person being surveyed uncomfortable about giving a truthful answer?

e Which of the questions in the questionnaire would you include in your own questionnaire for this investigation?

Questionnaire

i 'Male' or 'Female'? (Answer filled in by the questioner.) ______

ii Do you smoke cigarettes? Yes/No

If 'Yes', how many cigarettes do you smoke daily?
• 0–10 • 11–20 • 21–30 • more than 30

iii Do you work in a public building? Yes/No

iv Do you smoke at work? Yes/No

v Do you agree that smoking is bad for your health? Yes/No

vi Do you think smokers consider the effect of their smoking on other people? Yes/No

vii Should people be prevented from smoking in public areas? Yes/No

viii Do your parents smoke? Yes/No

ix Should people be prevented from smoking in each of the following places:

- schools? Yes/No/Undecided
- child care centres? Yes/No/Undecided
- hospitals? Yes/No/Undecided
- medical centres? Yes/No/Undecided
- government buildings? Yes/No/Undecided
- sporting grounds? Yes/No/Undecided
- hotels and clubs? Yes/No/Undecided
- church? Yes/No/Undecided
- sports clubs attended by children? Yes/No/Undecided
- in the street? Yes/No/Undecided
- on public transport? Yes/No/Undecided

2 The purpose of this investigation is to find out what students do in their out-of-school time and whether there is any difference in what boys and girls do in this time. Working in pairs, ask each other the questions below and discuss their effectiveness.

i What do you do after school?
ii Do you play sport after school?
iii Do you do anything interesting after school?
iv Do you need to work around home when you are not in school?
v 'Male' or 'Female'?
vi Do you do any school work in your out-of-school time?

a Comment on the relevance, and the directness of the questions, and the possible answers to the questions.

b Design a questionnaire that will provide some data about what students do in their out-of-school time. Test the questionnaire on several students to see if it provides the data that you want.

c What type of data (categorical or numerical) will you collect with this questionnaire?

d How will you organise (in tables), and display (as graphs) this data?

Collecting numerical data for a statistical investigation

Lesson

9

Numerical data can be collected by a questionnaire if it involves some characteristic of the local community that can be counted or measured.

Numerical data is either **discrete** or **continuous** data and each of these types has their appropriate method for organisation (tables) and display (graphs).

Statistics such as the mean, median, mode, range, and interquartile range can be found for numerical data and boxplots can be used to display these statistics.

EXERCISE

1 The purpose of this investigation is to find the distribution of the heights of 10-year-old boys in PNG.

- a Comment on the following methods of data collection.
  - i You have a 10-year-old brother and you measure him and five of his friends.
  - ii You obtain a list of all 10-year-old boys from a school and ask each of them their height.
  - iii You obtain a list of all 10-year-old boys from a school and measure their height.
  - iv You ask the height of any 10-year-old boy that you come across.
  - v There are 50 ten-year-old boys at a school and you randomly choose a sample of 30 of these boys and measure the height of each one of them.
- b Which of the data collection methods in part **a** is preferable. Why? Would the sample used in your choice of data collection be a good representation of the population?
- c What type of numerical data would you be collecting in this case?
- d How would you organise and display the data?
- e Which statistics can you calculate for this data?
- f How well will the conclusions from your sample represent the whole population of 10-year-old boys in PNG?

2 The purpose of this investigation is to find the difference, if any, between the number of letters in the words used in a book and the words used in the local newspaper.

- a Randomly choose a page and a paragraph on that page from both the book and the newspaper. Describe how you did this, including the random number generator that you used.
- b What type of numerical data would you be collecting in this case?
- c Collect the data for both the book and newspaper by counting the letters in the words of the chosen paragraph. What will you use to organise the data?
- d Graph the data for the book and the newspaper, using a suitable graph. Can you use a single graph for both sets of data? (Refer to the core part of this topic, page 168, for the appropriate types of graph.)
- e Find statistics for each set of data and construct a boxplot to display both sets of statistics on one set of axes.
- f Comment on what you have found, mentioning statistics to support your conclusions.

REVISION AND ASSESSMENT

Recording data and Sorting and organising data

1 Which of the following describes the diagram on the right?
- **A** bar graph
- **B** histogram
- **C** column graph
- **D** boxplot

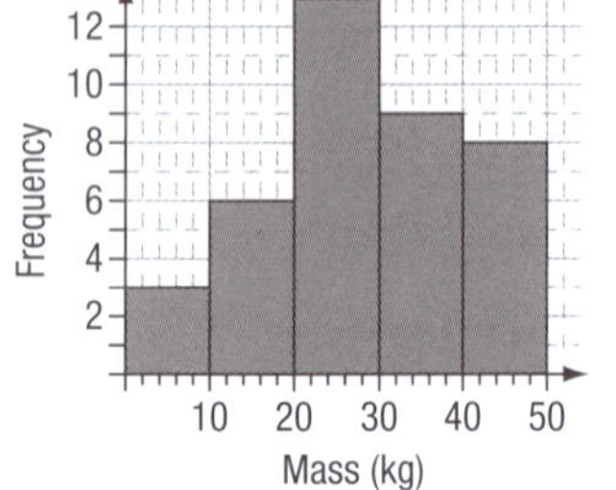

2 Which of the following describes the distribution of the data graphed on the right?
- **A** symmetric
- **B** positively skewed
- **C** negatively skewed
- **D** having an upward trend

3 In a particular village, the number of each type of fruit tree was counted. The results are displayed in the graph on the right.
- **a** What type of data is 'Type of fruit tree'?
- **b** How many trees were grown in this village?
- **c** What is the most frequently grown type of fruit tree in this village?
- **d** What percentage of the trees are mango trees?

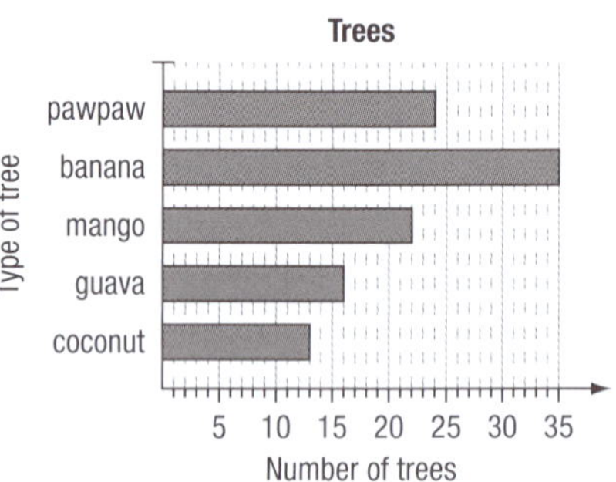

4 State whether each of the following numerical variables is discrete or continuous:
- **a** the times taken to run 400 m
- **b** the numbers of passenger in cars
- **c** the weights of year 9 students.

5 The speed of cars on a road is an example of which of the following:
- **A** discrete numerical data
- **B** categorical data
- **C** symmetric data
- **D** continuous numerical data

6 The following data represents the number of sweets in a packet.

22 24 24 25 26 22 23 24 28 26 27 26 27 25 25 27 27 24 24 23
25 24 27 26 26 26 25 25 22 23 23 25 25 25 23 27 25 24 26 24

- **a** Construct a dotplot for this data.
- **b** What percentage of the packets contained 25 or more sweets?
- **c** Find the median number of sweets in a packet.
- **d** Describe the distribution of the data.

Measures of central tendency and spread

1 Which of the following describes the middle value of a set of ordered data:

A mean B median C mode D average

2 The range of a set of data is the difference between which of the following:

A the mean and the median
B the upper quartile and the lower quartile
C the upper quartile and the median
D the maximum and minimum data values.

3 For the data given in the frequency table on the right:

a construct a column graph

b find:

i the mean

ii the median

iii the mode.

Number of points	Frequency
2	4
3	7
4	8
5	9
6	3

4 a For the following data, draw a stem-and-leaf plot, using the stems 8, 9, 10, 11, and 12:

87 103 95 85 125 113 122 95 115 128 119
83 110 126 120 105 104 99 108 121 117 106

b Redraw the stem-and-leaf plot so that it is ordered.

c For the given data find:

i the median ii the range.

5 A fast food outlet boasts that the time taken to fill an order is 4 minutes. To support this claim, a survey was carried out to investigate the time taken to fill the orders of 40 customers on the busiest day at the fast food outlet. The following results for the times taken to fill the orders, in minutes, were recorded:

3.1 2.2 2.9 2.6 4.6 5.3 4.5 2.2 4.2 3.3 5.2 1.9 2.2 3.1 3.8 1.8
4.9 6.5 1.4 6.7 5.5 4.1 3.8 3.7 4.1 2.9 5.2 7.3 6.4 2.4 4.4 6.2
2.5 4.4 4.6 2.3 1.9 2.2 5.5 2.8

a Construct a stemplot for this data, using the stems 1, 2, 3, … 7.

b Rewrite the stemplot so that is ordered.

c Find the median time taken to fill an order.

d Find the modal time taken to fill an order.

e Give a reason why the mode is not the most suitable measure of centre for this data.

f What percentage of the customers needed to wait more than 4 minutes to have their order filled?

6 For the data set on the right, find the:

a mean

b median

c mode.

Data value	Frequency
3	3
4	6
5	7
6	8
7	2

7 The weights of bags of fifty apples, in kilograms, has been measured and provided the following data:

6.1 6.3 6.3 6.4 6.4 6.5 6.5 6.5 6.6
6.6 6.6 6.6 6.7 6.7 6.8 6.8 6.9 6.9

a For this data find:

i the median ii the upper and lower quartiles

iii the range iv the interquartile range.

b Construct a boxplot to display the statistics from part a, making sure you carefully scale and label the axis (the number line below the boxplot).

c Describe the distribution of weights of these bags of apples, mentioning each of the statistics from part a.

8 A basketball team scored an average of 35 points in each of its first four games of the season. How many points did the team score in its fifth game if the average score for each of the five games, after the fifth game, was 40?

9 The heights of thirty 15-year-olds were measured and the data is given in the frequency table on the right.

a Draw a histogram to display the data from the frequency table.

b What percentage of the students were less than 160 cm tall?

Height (cm)	Frequency
150–	2
155–	4
160–	7
165–	10
170–	4
175–	2
180–	1

c Copy and complete the cumulative frequency table on the right.

d Construct a cumulative frequency graph to display the data from the table.

e Half of the students were shorter than a particular height. Find this height from the cumulative frequency graph.

Height (cm)	Cumulative frequency
< 155	2
< 160	
< 165	
< 170	
< 175	
< 180	
< 185	30

10 The cumulative frequency graph below has been constructed for data that recorded the time, in seconds, that the members of a group of athletes each took to run 100 m.

a How many athletes were in the group?

b How many athletes ran a time less than 11.6 seconds?

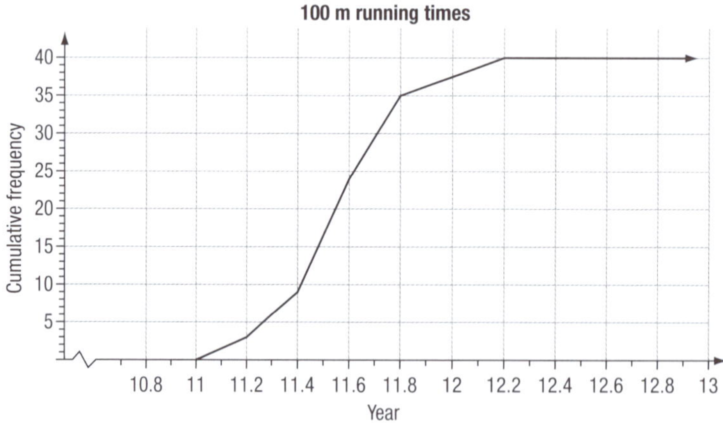

c Copy and complete:

The slowest five athletes ran a time that was ____ seconds or more.

11 The height of each child in a group of 13-year-old girls and 13-year-old boys has been measured. The statistics for the two groups are displayed in the following boxplots.

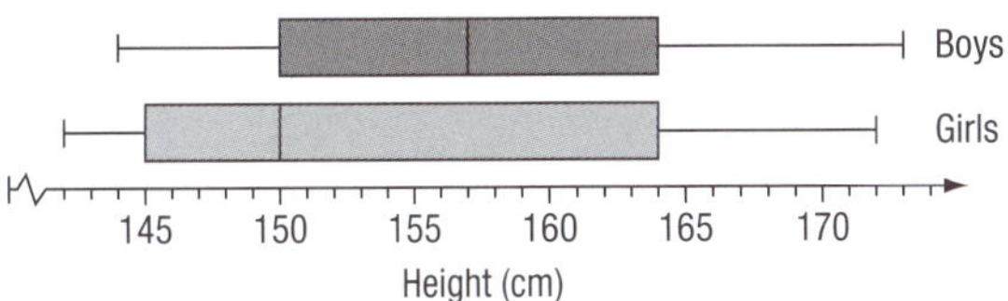

a From the boxplots find:
   i the median height of the group of boys
   ii the maximum height for the group of girls
   iii the range of the girls' heights
   iv the interquartile range of the boys' heights.

b Describe the distribution of the heights of the girls.

c Copy and complete the following:
   i 75% of the boys were ____ cm or taller.
   ii Half of the girls measured ____ cm or more.
   iii The tallest 25% of both groups were ____ cm or more.

12 A shop has collected data regarding the number of customers it has had each day over two weeks. The data is shown on the graph below.

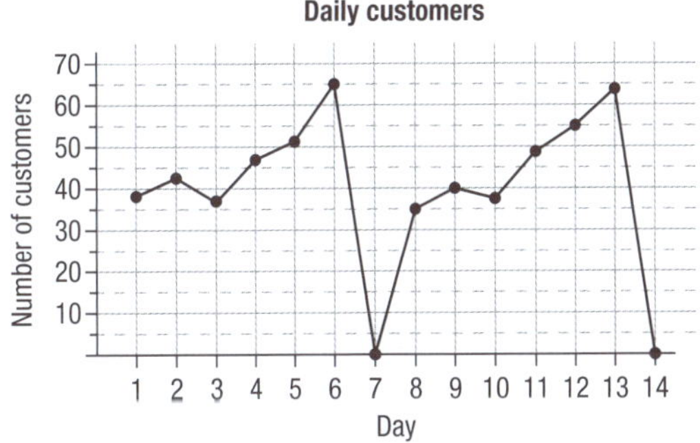

a On which of the 14 days was the shop closed?

b On how many days in the two weeks were there more than 50 customers?

c If 'Day 1' is Monday, what day is the busiest day of the week?

d What was the maximum number of customers in one day during the two-week period?

e Describe the pattern of the data.

13 The height of a tree was measured at the beginning of each year for ten years and the data has been graphed on the right.

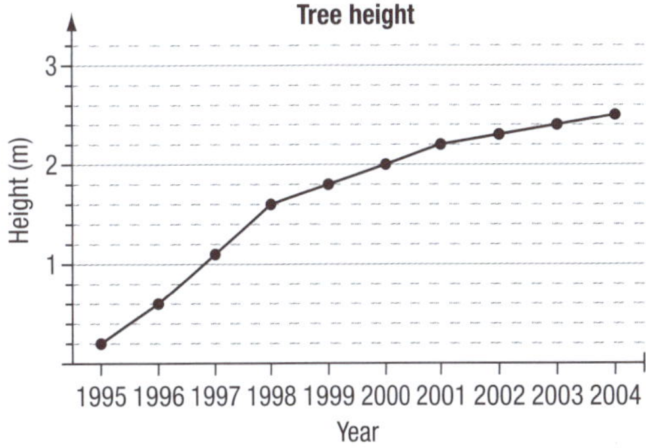

a What is the name of the type of graph shown?

b What was the height of the tree at the beginning of 2001?

c In which year did the height of the tree reach 2 metres?

d Describe the change in height of the tree over the ten years.

e Use the graph to predict the height of the tree in 2005.

Misleading statistical graphs

1 **a** Identify the misleading feature of this graph.
b Redraw the graph without the misleading feature.

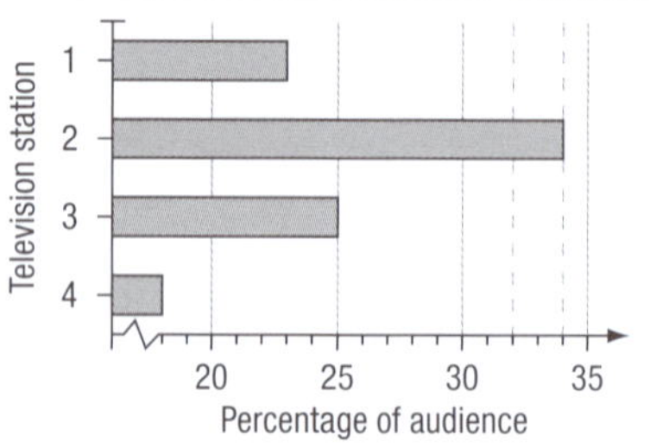

Experimental probabilities and Probabilities based on symmetry

1 Two hundred people were surveyed and asked the question: 'In which country was your car manufactured?' The results are shown in the table on the right.

Using these results, what is the probability that the car of a randomly selected person is:

a manufactured in Australia?
b manufactured in Japan or Korea?
c not manufactured in Japan?

Country	Frequency
Australia	83
Japan	76
Korea	18
Germany	13
Others	10
Total:	**200**

2 In a class of 40 students, 18 are girls and 22 are boys. Eight of the girls and 12 of the boys play basketball. A student from this class is selected at random. Find the probability that this student is:

a a boy
b a girl who plays basketball
c a boy who does not play basketball
d a student who does not play basketball.

3 The frequency table on the right records the data obtained when 24 households were asked: 'How many dogs do you own?'

Based on the figures in the table, find the probability that a household, selected at random, will own:

a two dogs
b no dogs or one dog
c more than two dogs
d at least one dog.

Number of dogs	Frequency
0	4
1	11
2	6
3	2
4	1

4 A school has students from two villages: village A and village B. There are 28 boys in the school, and 15 of them are from village A. There are 22 girls in the school, and 12 of them are from village B. A student from this school is selected at random. Find the probability that the student selected is:

a a girl
b a boy from village B
c a girl from village A
d from village A.

5 A cylindrical can is tossed 120 times and it is observed that it lands 38 times on its end.

a If 'tossing a can' is a probability experiment how many trials were conducted?

b Use the results of the experiment to estimate the probability that the can will land on its end next time it is tossed.

c How could you improve the accuracy of your experimental probability?

d Use a descriptive probability word (such as 'certain', 'unlikely', etc.) to describe the probability that the can will land on its side.

e If the experiment is repeated, are you likely to get the same results?

6 Which of the following is the probability of selecting one black Ace from a pack of 52 playing cards?

A $\frac{1}{13}$ B $\frac{1}{2}$ C $\frac{1}{26}$ D $\frac{1}{52}$

7 One ball is drawn out of a bag containing 3 black, 4 red and 5 blue balls. Which of the following is the probability that the ball is coloured red?

A $\frac{1}{4}$ B $\frac{1}{3}$ C $\frac{1}{2}$ D $\frac{4}{11}$

8 In a class of 45 students, there are 25 boys and 20 girls. Four of the boys and five of the girls are in the school swimming squad. Which of the following is the probability that a student from this class, chosen at random, is in the school swimming squad?

A $\frac{1}{5}$ B $\frac{1}{4}$ C $\frac{1}{3}$ D $\frac{1}{2}$

9 If one letter is chosen randomly from the letters of the word 'CONTINUOUS', which of the following is the probability that the letter is a vowel?

A $\frac{1}{2}$ B $\frac{4}{9}$ C $\frac{3}{8}$ D $\frac{5}{9}$

10 A fair coin is tossed twice and on both occasions it has landed 'heads'. If the coin is tossed a third time, which of the following is the probability that it lands 'heads' on this third toss?

A $\frac{1}{3}$ B $\frac{1}{8}$ C $\frac{1}{2}$ D $\frac{1}{4}$

11 A card is drawn from a pack of 52 playing cards. Find the probability that it is:

a black

b red or black

c a red King

d the Ace of diamonds

e not a court card

f a number between 3 and 8 inclusive.

Remember

In a pack of cards, the 'court cards' are the Jacks, Queens and Kings.

12 A letter is randomly chosen from the letters of the word 'MATHEMATICS'.

a List the sample space.

b Are all the outcomes equally likely?

c Calculate the probability that a randomly selected letter is:

i an 'M'

ii an 'E'

iii an 'O'

iv a vowel

v a consonant

vi an 'A' or a 'T'.

13 One green, one red, one blue and one yellow disc are placed in a tin. A disc is removed at random, its colour is noted and it is then replaced. A second disc is drawn and its colour is also noted.

- **a** Construct a probability tree diagram to represent this information.
- **b** Use your probability tree diagram from part **a** to find the probability that:
  - **i** the first disc is green and the second disc is red
  - **ii** both discs are green
  - **iii** the discs are different colours
  - **iv** one disc is yellow and one disc is blue.

14 The spinner shown on the right is spun.

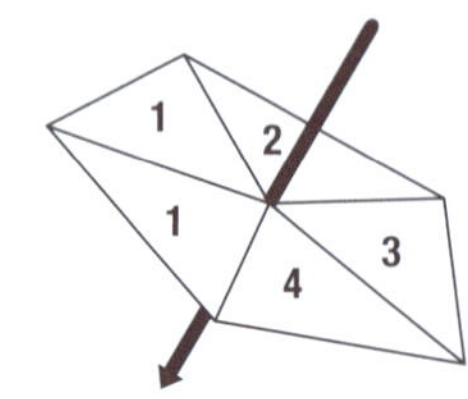

- **a** List the sample space.
- **b** Are all outcomes equally likely?
- **c** Find the probability that the spinner will land on '1'.
- **d** Find the probability that the spinner will not land on '1'.

15 A twelve-sided die, with sides numbered 1 to 12, is rolled. Find the probability that the number uppermost is:

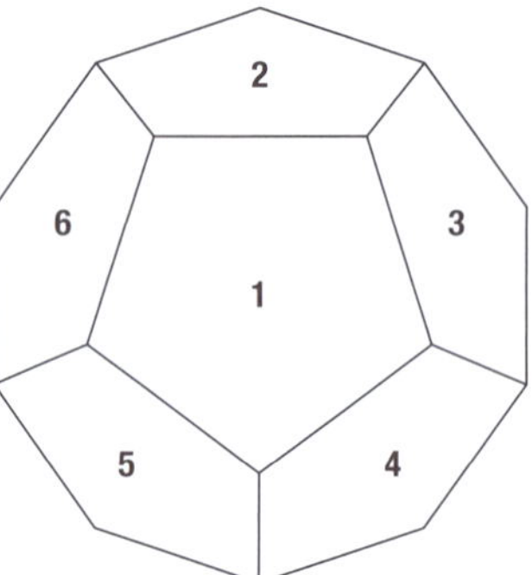

- **a** a '2'
- **b** a '5' or a '10'
- **c** a multiple of 3
- **d** a factor of 12
- **e** a prime number.

16 A fair three-sided spinner is made from an equilateral triangle and is marked as shown.

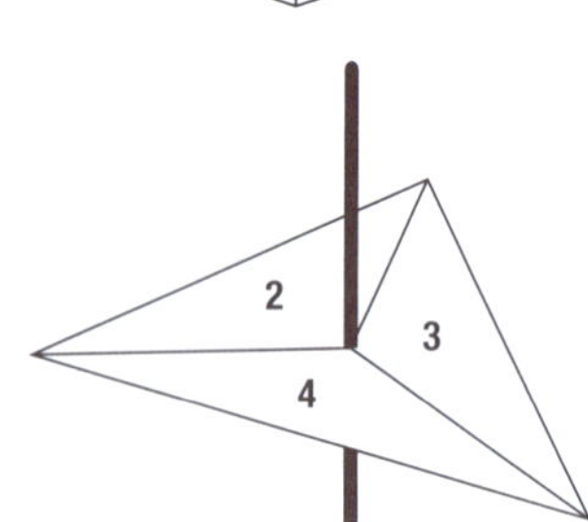

- **a** Write the sample space.
- **b** Are all three outcomes equally likely?
- **c** The spinner is spun twice. Construct a grid diagram to show the outcomes.
- **d** Use your grid diagram from part **c** to find the probability that:
  - **i** both numbers are the same
  - **ii** the numbers are different
  - **iii** the sum of the numbers is 4
  - **iv** the sum of the numbers is 6
  - **v** the sum of the numbers is more than 6.

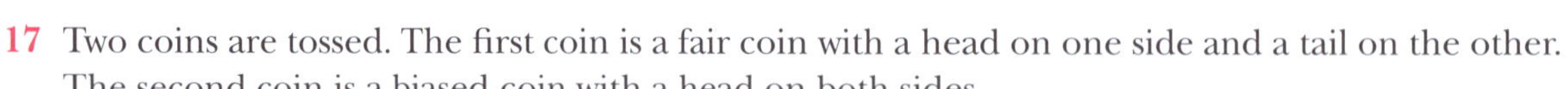

17 Two coins are tossed. The first coin is a fair coin with a head on one side and a tail on the other. The second coin is a biased coin with a head on both sides.

- **a** Draw a tree diagram to illustrate the possible outcomes when these two coins are tossed.
- **b** Use your tree diagram from part **a** to find the probability that you will get:
  - **i** two heads   **ii** a head and a tail   **iii** two tails.

18 A six-sided die has three sides marked with 5 dots and the other sides are marked with 1, 2 and 3 dots respectively. The die is rolled and the number of dots uppermost when the die lands is noted.

a State the sample space.

b Are all outcomes equally likely?

c What is the probability that the number of dots uppermost is five?

d What is the probability that the number of dots uppermost is a factor of 6?

e What is the probability that the number uppermost is a prime number?

19 The court cards (Jacks, Queens, and Kings) are removed from a pack of 52 playing cards. The remaining cards, with face values of 1 to 10 inclusive, are shuffled and one card is selected at random. Find the probability that the card selected is:

a a 4

b a red 10

c a black card and a number less than 10

d a red card and an even number

e the 5 of hearts or the 6 of clubs.

20 A target for a game of darts is constructed from three concentric circles, as shown in the diagram on the right. The score is 10, 5 or 1, depending on where the dart lands on the board.

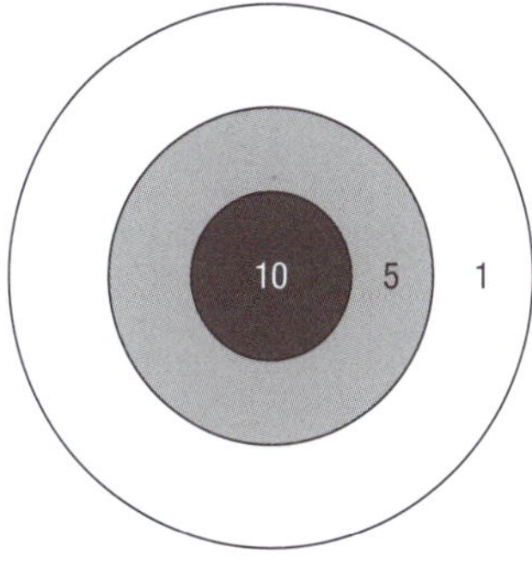

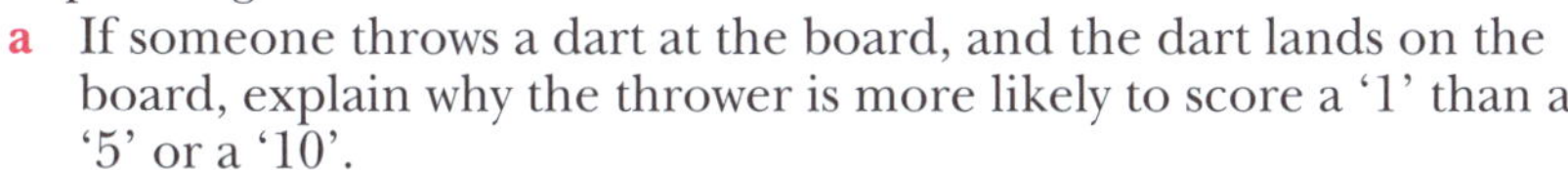

a If someone throws a dart at the board, and the dart lands on the board, explain why the thrower is more likely to score a '1' than a '5' or a '10'.

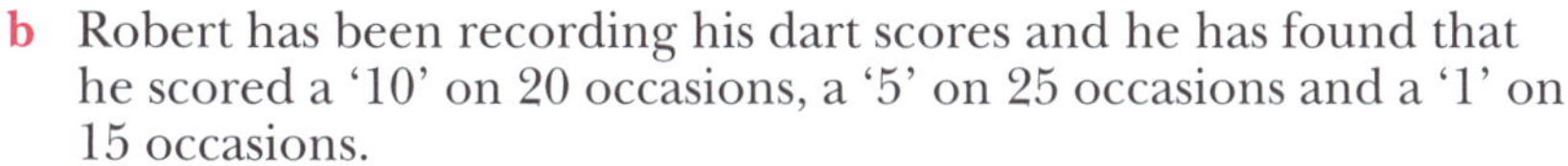

b Robert has been recording his dart scores and he has found that he scored a '10' on 20 occasions, a '5' on 25 occasions and a '1' on 15 occasions.

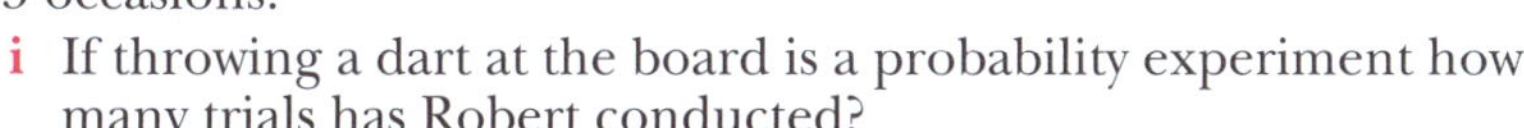

i If throwing a dart at the board is a probability experiment how many trials has Robert conducted?

ii Based on Robert's trials what is the probability that he will score a '5' when he throws a dart.

c Robert throws two darts at the board.

i Copy and complete this tree diagram to illustrate each total score that Robert could possibly get.

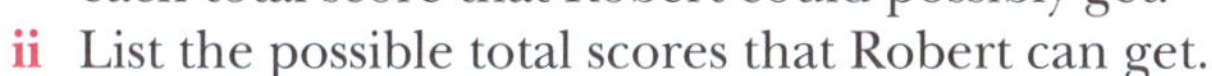

ii List the possible total scores that Robert can get.

iii Are the total scores you listed in part ii equally likely to occur?

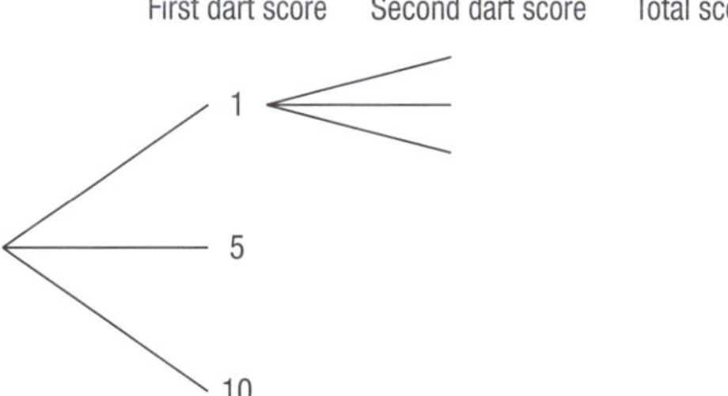

PROJECTS ASSESSMENT TASK TWO

Option A: Assessment

For assessment of this option, you will be required to work in groups to design and analyse **a simple game of chance**. Your teacher will give you more details of what you are required to do and how you should go about presenting your project but, in general, you will need to do the following.

- Establish your group. This may be done by your teacher, or you may be able to form your own groups.
- Spend some time discussing and trying some games. What equipment (dice, coins, cards, spinners, etc.) are you going to use for your game? If you have done most of the exercises in this section you should have some good ideas.
- Once you have decided on a possible game, you should write the list of rules, keeping it as short as possible, and play the game within your group.
- If you are happy with your choice of game, you can plan the different tasks for the analysis and presentation of the game and decide which group members are going to complete each of these tasks. It is a good idea to elect someone to write all your discussions in a bound exercise book, so a written record is kept.
- Ask another group to play the game using your list of rules.
- Analyse the probability of winning the game and how this is going to be included in your final report.
- Decide how you are going to present the report about your game to the class and/or teacher.

Option B: Assessment

For assessment of this option, you are required to work in groups to **conduct a statistical investigation**. Your teacher will give you the details of what you are required to do, but there are some things that you need to be aware of before you start.

Think about what you would like to investigate

This could be some characteristic of the local population, such as:

- the number per household of a particular type of livestock kept in your community
- cash crops grown locally
- vegetables grown locally
- mobile phone ownership or usage
- type of vehicles on a road
- time spent playing sport per day
- activities of schoolchildren out of school time
- the number of siblings/cousins.
- the type of snack food eaten (and how often)
- languages spoken in the community
- something that was investigated in the last census (for example the number of people per household, or the age distribution of people in your community).

Tasks involved in the investigation

For your investigation you will need to:

- state the purpose of your investigation
- decide who/what will be surveyed (Will you be using a sample of a larger population or are you surveying the whole population?)
- decide how you are going to choose a random sample if you are using a sample of the population
- decide how the information will be obtained (Will it be by questionnaire, observation, measuring or counting?)
- collect the data
- identify the type of data that you will be collecting (for example categorical, discrete numerical or continuous numerical)
- use the work from the core part of this unit to help you decide the appropriate tables, graphs and statistics for the analysis of the data that you have collected
- write a report on your investigation mentioning all of the items above. (Also include any problems that you encountered, how you dealt with the problems and how these may have affected the data collected and the final conclusions.)
- present a report of your investigation to the class.

Planning for the investigation

Establish groups for completing this project.

Within your group, discuss what you are interested in investigating. For each suggestion, go through the list above and discuss whether it would be practical to conduct an investigation into that suggestion. If you think that you have decided on a suitable subject to investigate, explain it to your teacher and ask for his/her opinion.

Once you have decided on what you are investigating, you can start to plan the different tasks for the investigation and decide who is going to do each of these tasks. It is a good idea for one group member to be appointed to write all your discussions in a bound exercise book, so a written record is kept.

Unit 4 Design in 2D and 3D geometry

Unit summary

This unit focuses on mathematics relating to the properties of two-dimensional (2D) plane figures and three-dimensional (3D) solids. Applications of 2D and 3D geometry can be found in the design and construction of many traditional items such as baskets, billums and other artefacts.

Students will be examining the properties of different shapes and applying these in classifying polygons and solving related problems. Patterns and designs which can be made from polygons will also be explored.

Students will estimate and calculate the area of two-dimensional shapes and the total surface area and volume of three-dimensional prisms and other commonly occurring solids. The relationship between surface area and volume will be explored.

The core of this unit will be assessed using tests focusing on the properties of shapes and solid figures; estimating, measuring and deducing angle sizes; and estimating and calculating the surface area and volume of prisms.

Students will also be required to complete one of the two options described below.

Option A: Construction

In this option, students will examine the symmetry properties of 2D figures and the accurate construction of angles and parallel lines, polygons and nets of solids. Students will also identify similar shapes and make scale drawings.

Option B: Deductive reasoning

This option explores the origins of many commonly used properties of angles, parallel lines, polygons and circles. Students will consider how proofs of these properties are logically demonstrated and applied.

Assessment for the option selected will be based on an individual directed investigation involving either the construction of angles, lines, regular polygons, nets and solids (Option A) or the explanation of the deductive steps given in proving a particular result (Option B). Students will present their findings in writing by way of, for example, a report, an illustrated chart, a poster, a flow chart or an article.

Syllabus references

- **9.4.1** Identify situations where patterns and measurement are applied traditionally
- **9.4.2** Classify shapes into families and their subgroups and justify reasoning
- **9.4.3** Interpret, analyse and solve measurement problems and justify selections and applications of formulae
- **9.4.4** Communicate mathematical processes and results in writing
- **9.4.5** Undertake investigations individually in which mathematics can be applied to solve problems

Design in 2D and 3D geometry: Topics

PROPERTIES OF PLANE FIGURES

Lesson 1

Polygons and plane figures

Plane figures are two-dimensional shapes. In other words, they are shapes that can be drawn on a flat surface.

Polygons are plane figures formed from a number of straight lines that make a closed shape. Polygons can be classified according to:

i their number of sides
ii whether they are **concave** or **convex**
iii whether their sides are equal or unequal in length.

Copy the table below into your work book.

Number of sides	Name of polygon
3	Triangle
4	Quadrilateral
5	Pentagon
6	Hexagon
7	Heptagon
8	Octagon
9	Nonagon
10	Decagon
12	Dodecagon

Did you know?

The Greek mathematician **Euclid**, who lived around 300 BC, is often referred to as the 'Father of geometry' because of the important work he did on the subject. His book, *Elements*, is the most complete study of geometry ever written.

Discussion

What polygons can be seen in your classroom or outside? Discuss in class any examples of these polygons in everyday life. Do you know other words with a similar prefix (tri-, quad-, and so on) and their meaning?

A **regular polygon** is one with all its sides equal and all its angles equal.

Figure A, below, is a **regular pentagon**. Figure B is an **irregular pentagon**.

A **concave** polygon has at least one of its internal angles greater than 180°. It looks as if a side has caved in. A **convex** polygon has no angles greater than 180°.

Figures A and B below are both convex pentagons. Figure C, below, is a concave pentagon.

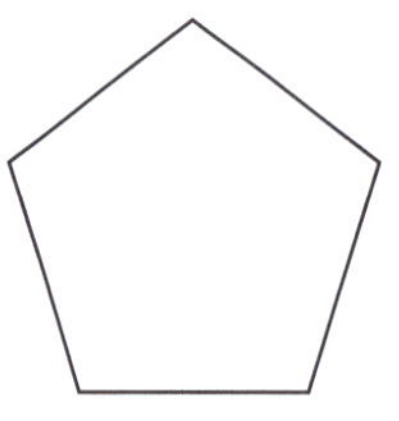

Figure A

Figure B

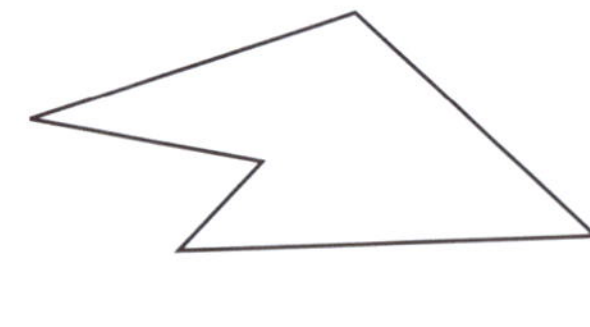

Figure C

EXERCISE

1 What is the name commonly used for a:

a a regular triangle?

b a regular quadrilateral?

2 Identify each of the following shapes as either concave or convex:

a 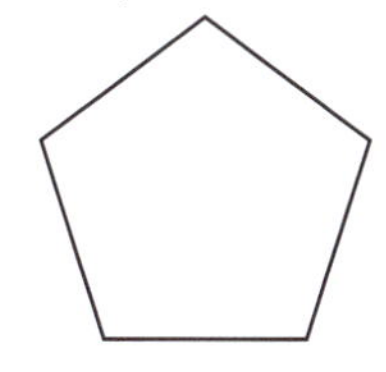

b 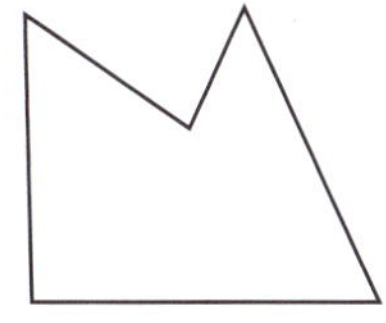

c 

d 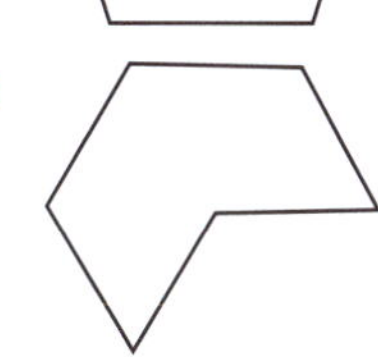

e 

f 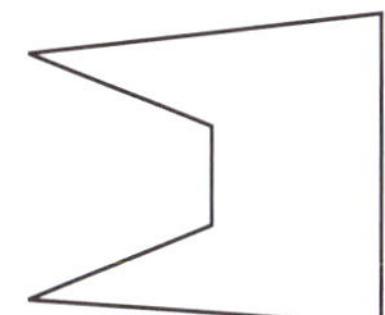

3 Sketch each of the following shapes:

a a regular hexagon

b a convex hexagon

c a concave hexagon

d a concave irregular pentagon

e a convex regular octagon

4 Why do we use some polygons in our buildings instead of others? Test the strength of different plane shapes by using straws and string to make the shapes. Thread the string through the straws and tie the ends. Which is the strongest (most rigid) shape?

5 **Paper knots**

You can make a number of regular polygons using one or two strips of paper. Here are the instructions for making a pentagon and a hexagon.

Pentagon:

Step 1: Take a strip of paper about 10 cm long and 1 cm wide.

Step 2: Form the paper into a loop, as shown on the right.

Step 3: Tuck one end of the strip over the other and up into the centre space.

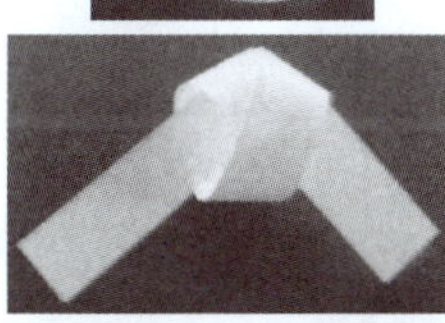

Step 4: Carefully pull both ends of the strip to close the knot, as shown on the right, pressing the knot flat with your fingers.

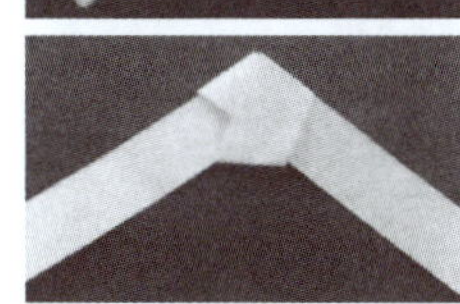

Step 5: Tuck the ends of the strip behind the knot and cut off the extra lengths of paper.

Hexagon:

Step 1: Take two strips of paper with the same dimensions.

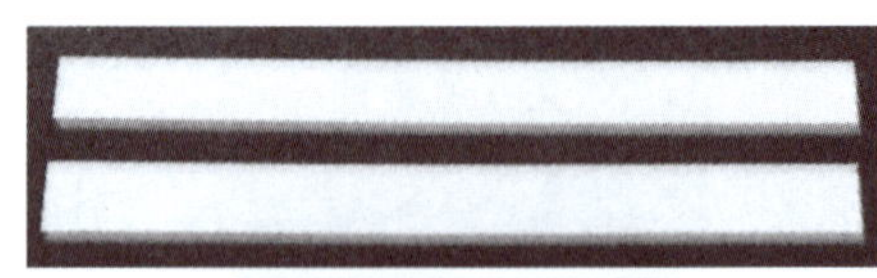

Step 2: Arrange the strips into two linked loops, as shown on the right.

Step 3: Carefully pull the knot closed, easing both of the strips and pressing the hexagon flat as it forms.

Step 4: Fold back the ends of the paper strip and cut away the extra lengths of paper.

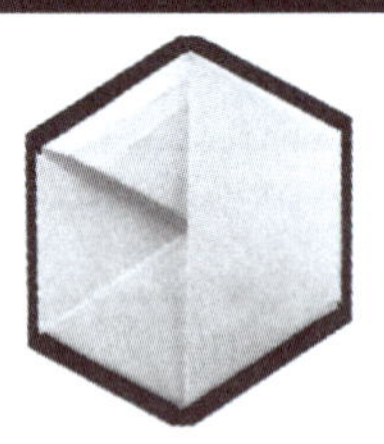

6 Each of the following polygons can be turned into a square by making one straight cut and rearranging the two pieces.

- **i** Name each polygon.
- **ii** Copy each polygon, mark where the straight cut goes and show with a dotted line where the shapes fit together, as shown in the example on the right.

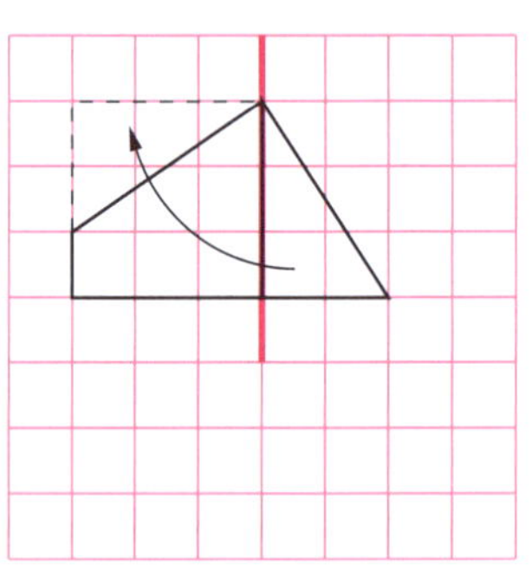

a 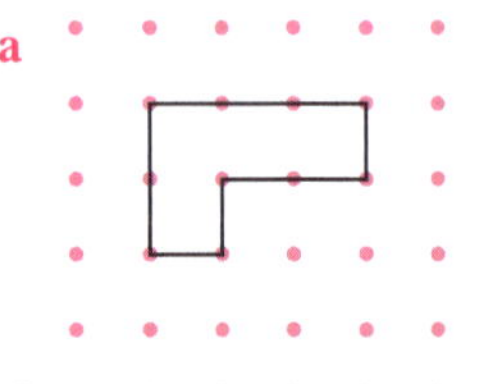

b 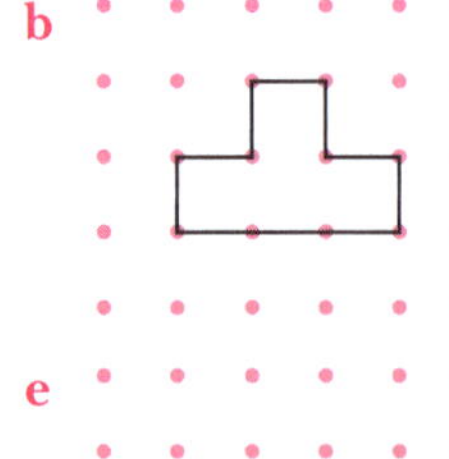

c

d

e

f 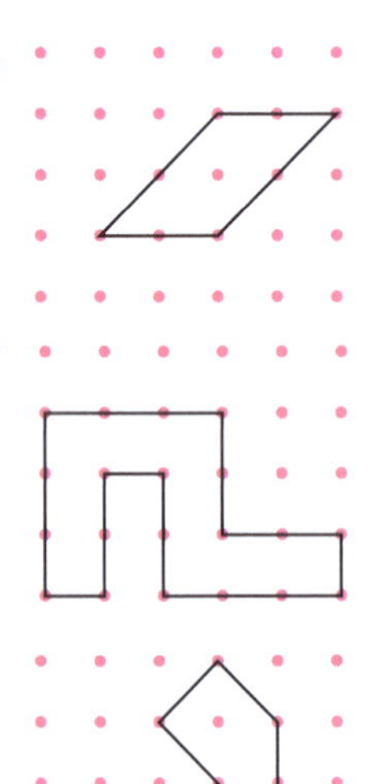

g

h 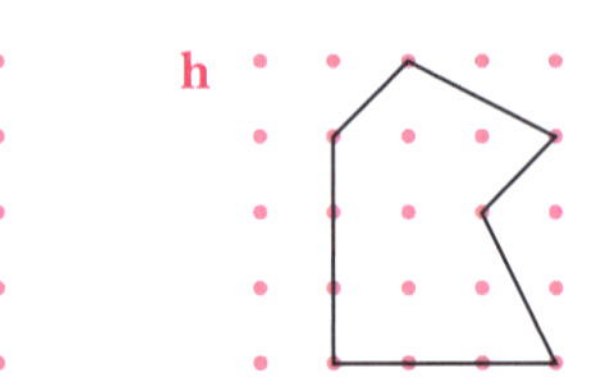

i

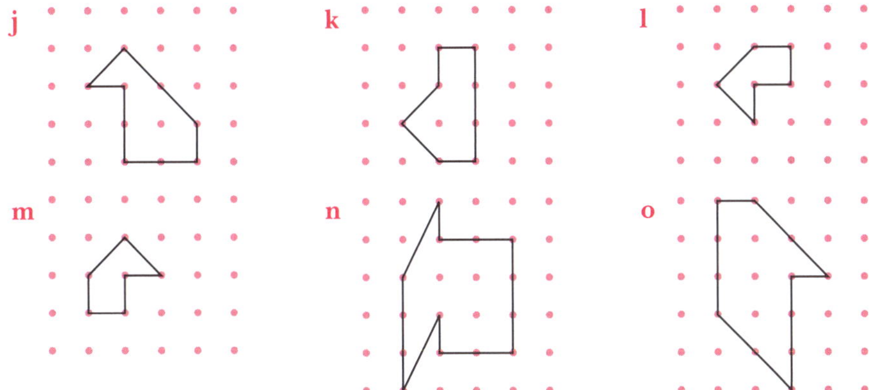

7 On coloured paper, draw a rectangle measuring10 cm by 8 cm. Lightly mark in 1 cm grid points and rule the lines shown on the diagram below. Cut out the seven pieces and rearrange them to form the six alien spacecraft shown.

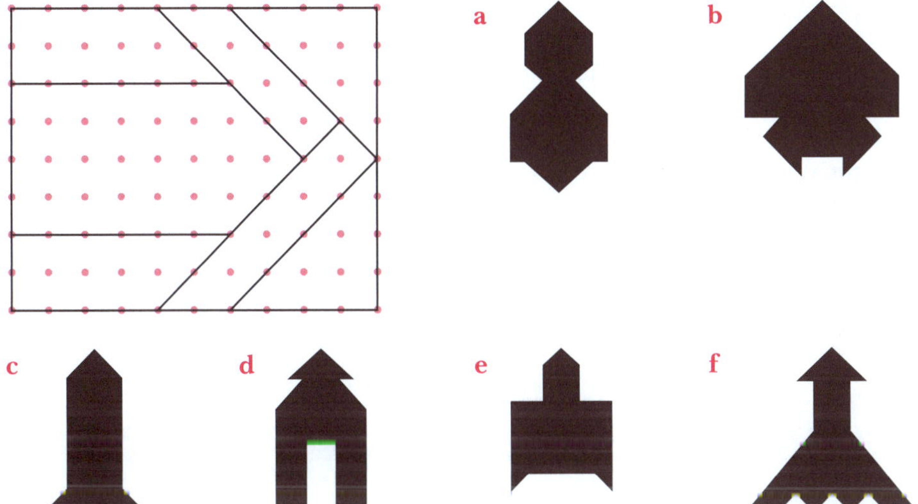

Triangles

Triangles are the simplest polygons, with three sides and three interior angles. They can be described in different ways according to their sides or their angles.

Some triangles are identified by their sides:

Scalene triangle	Isosceles triangle	Equilateral triangle
The longest side is opposite the largest angle. All sides are of different lengths.	Two sides are equal. Angles opposite those sides are equal.	All sides are equal. All angles are equal to 60°.

Some triangles are identified by their angles:

Acute-angled triangle	Right-angled triangle	Obtuse-angled triangle
All three angles are less than 90°.	One of the angles is equal to 90°.	One of the angles is greater than 90° but less than 180°.

Last year, we learned that the **angle sum of a triangle is 180°**. This can easily be shown by tearing the three angles from a triangle and placing them side by side, where they form a straight angle (a straight line).

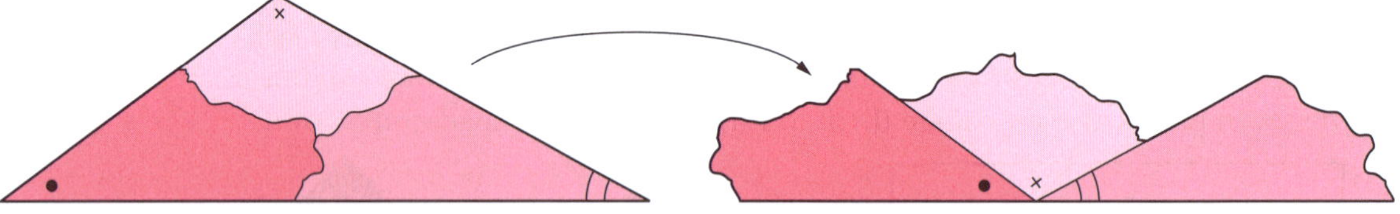

We can use this property to find the size of an unknown angle in a triangle.

Example

Find the value of the pronumeral in the triangle below.

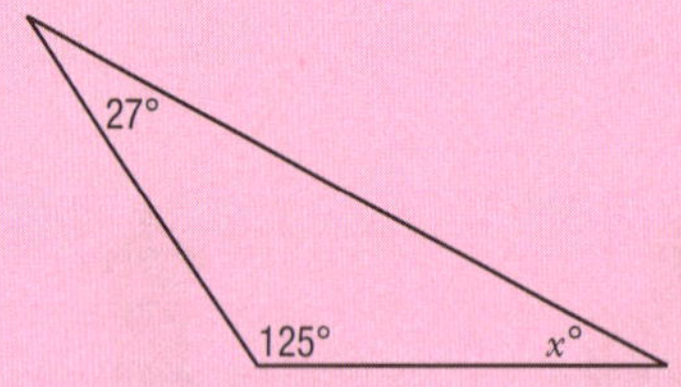

Answer

$x = 180 - (125 + 27)$
$= 180 - 152$
$= 28$

So the missing angle is 28°.

Exterior angles

If the sides of a triangle are extended, they form **exterior angles** (that is, angles on the outside of the figure).

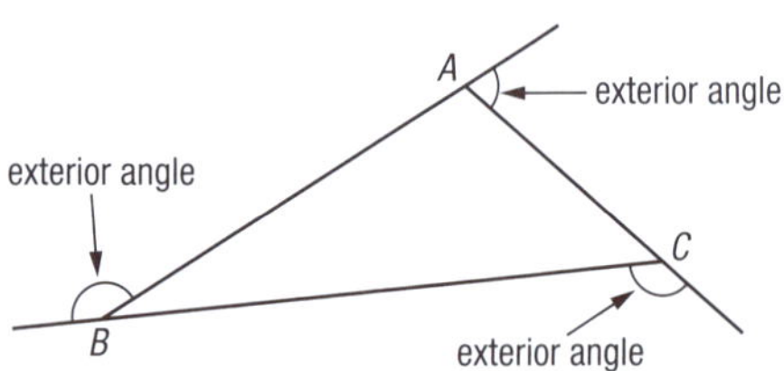

In each case, the size of the exterior angle is equal to the sum of the two interior opposite angles.

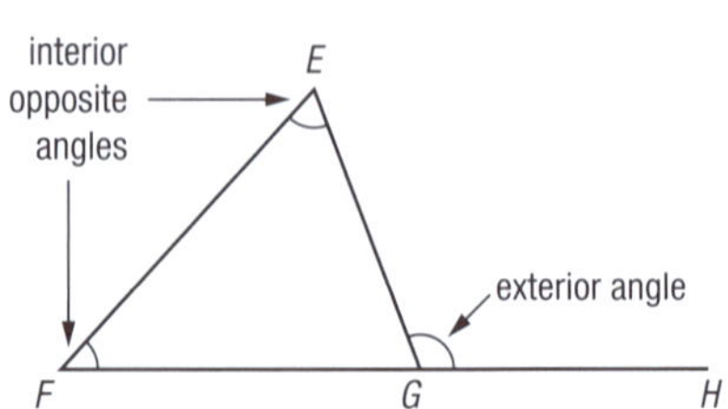

We can use this property to find the size of an unknown angle in a triangle.

Example

Find the value of the pronumeral in the triangle below.

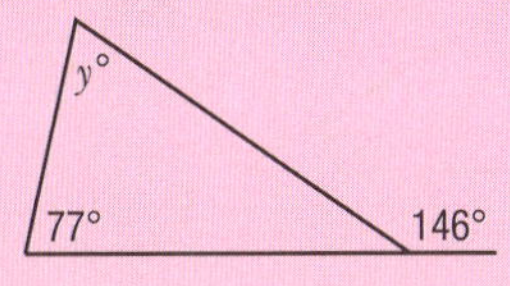

Answer

$77 + y = 146$

$y = 146 - 77$

$= 69$

So the missing angle is 69°.

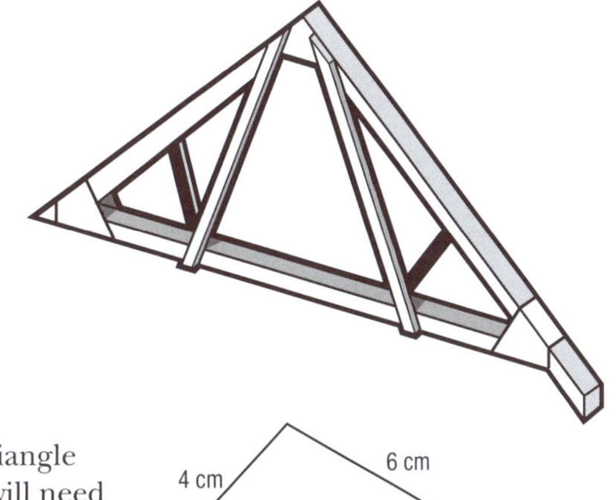

Triangle construction

Sometimes we need to construct triangles accurately, for instance when making a roof truss.

Here are the instructions for drawing a triangle given the lengths of the three sides. You will need a sharp pencil, a pair of compasses and a ruler.

4 cm
6 cm
8 cm

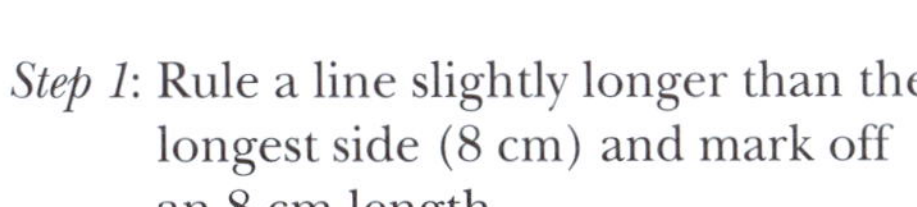

Step 1: Rule a line slightly longer than the longest side (8 cm) and mark off an 8 cm length.

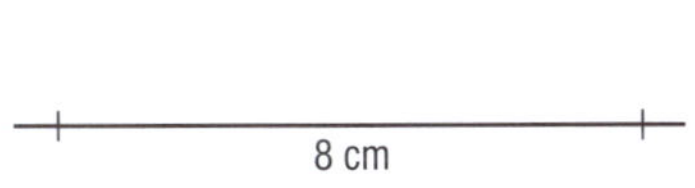

Step 2: Measure a radius of 6 cm on the compasses and draw an arc at one end of the line.

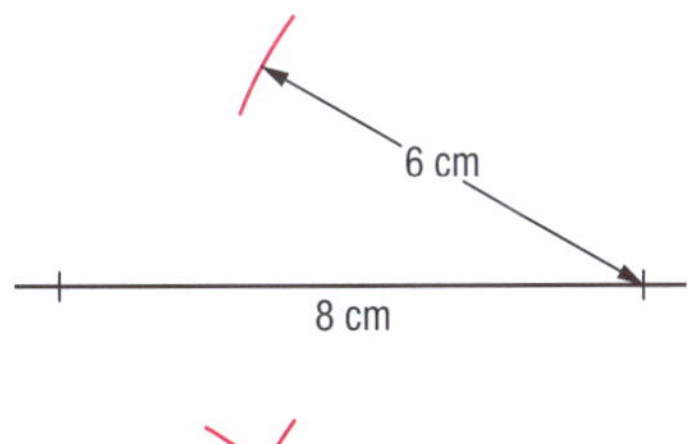

Step 3: Measure a radius of 4 cm on the compasses and draw an arc at the other end of the line.

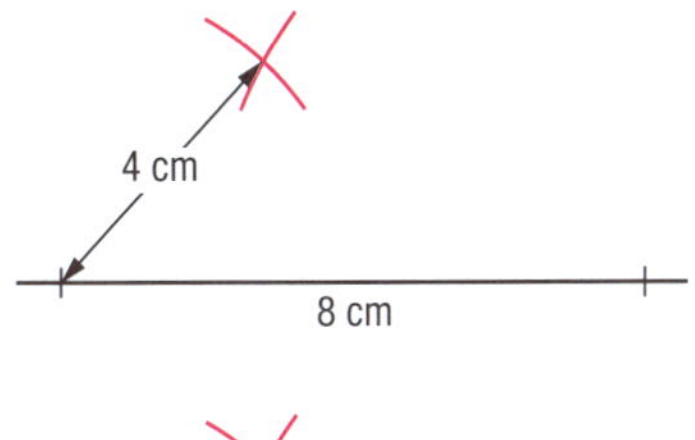

Step 4: Join the ends of the line to the point where the arcs meet.

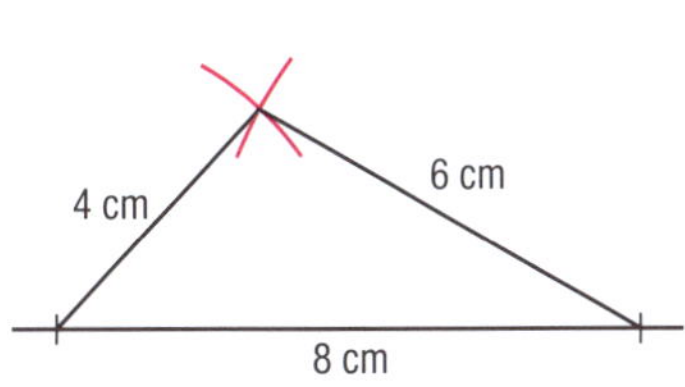

EXERCISE

1 Using your knowledge of triangle properties, find the value of the pronumeral in each of the following:

a 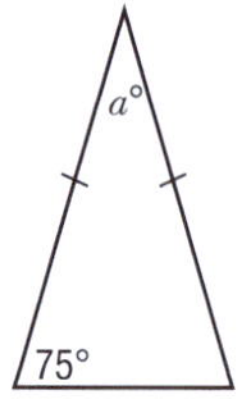

b 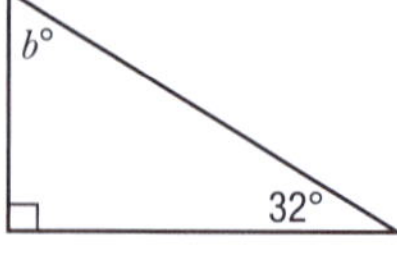

c 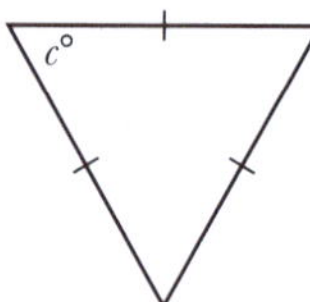

d 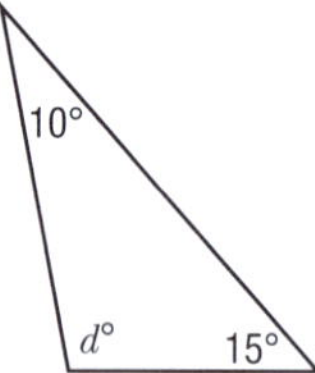

e 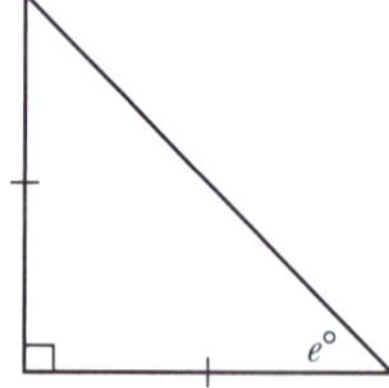

f 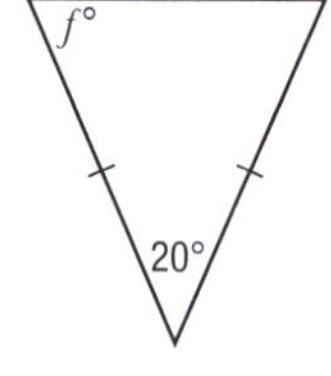

2 Use your knowledge of the relationship between the exterior and interior angles of a triangle to find the value of the pronumeral in each of the following:

a 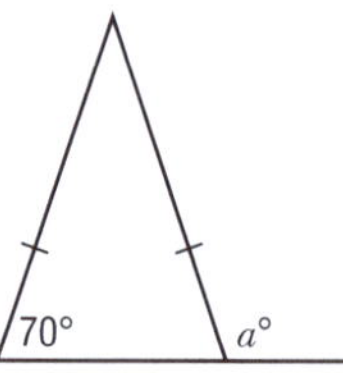

b 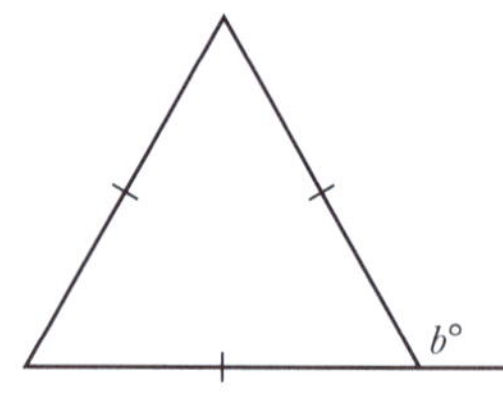

c 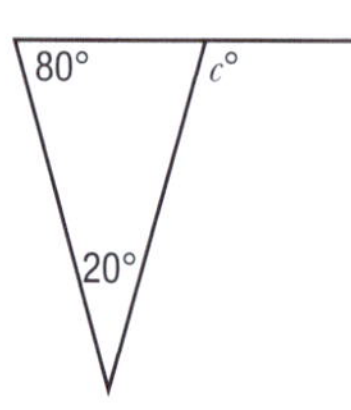

d 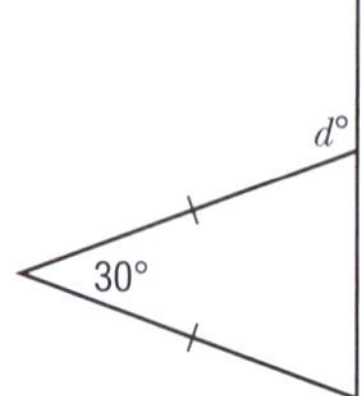

e 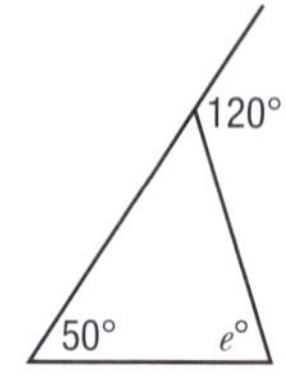

f 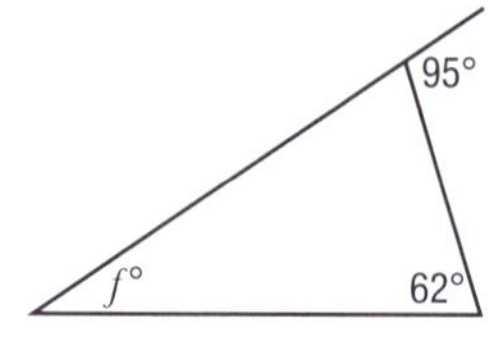

3 Name each of the triangles in Question 1 using:
 i the name according to its sides ii the name according to its angles.

4 Using a pencil, a ruler and compasses, construct triangles with the following sides:
 a 3 cm, 4 cm and 5 cm
 b 10 cm, 10 cm and 6 cm
 c 8 cm, 8 cm and 8 cm
 d 3 cm, 5 cm and 7 cm
 e 4 cm, 6 cm and 12 cm

5 Explain your answer to Question 4e?

Quadrilaterals

The 'family' of four-sided polygons has seven members:

- quadrilateral
- trapezium
- rhombus
- kite
- parallelogram
- rectangle
- square.

Their relationship is shown in this diagram:

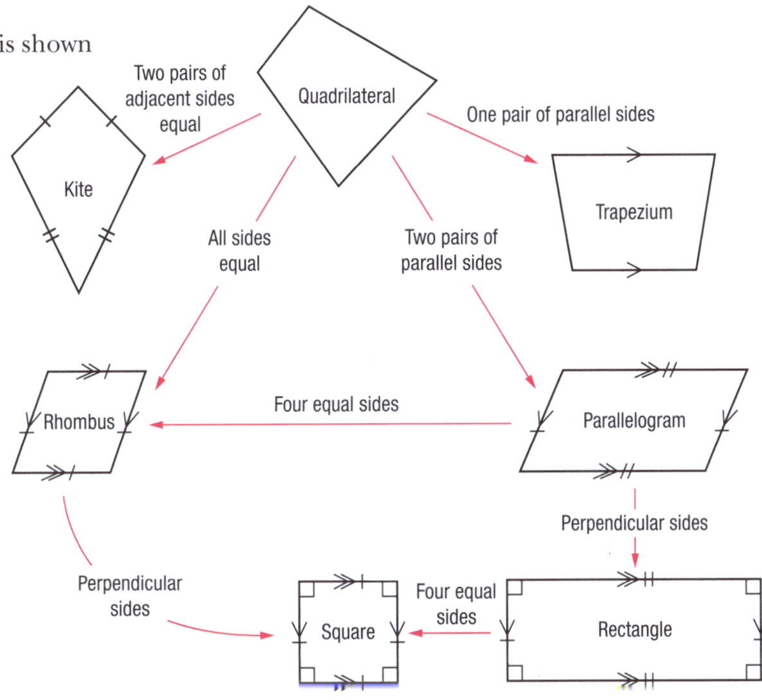

Challenge

Indoor investigation

Copy each shape above into your workbook. Measure the sides and angles, and correctly tick the boxes to complete the following table.

Property	Parallelogram	Rectangle	Rhombus	Square	Kite	Trapezium
Opposite sides equal						
Opposite angles equal						
All sides equal						
All angles equal (90°)						
Adjacent sides equal						
Diagonals equal						
Diagonals bisect each other						
Diagonals intersect at right angles						
One pair of parallel sides						
Two pairs of parallel sides						

Outdoor investigation

For this investigation, you will need three straight sticks (lengths of bamboo or similar), two of which are equal in length (about the length of your arm) and the third shorter.

- **i** Choose two of the sticks and place them on the ground so that they match each of the following descriptions.
- **ii** Mark on the ground the four sides of the quadrilateral which has the sticks as diagonals.
- **iii** Identify which quadrilateral is formed each time.
  - **a** Diagonals are equal and bisect each other at right angles
  - **b** Diagonals are equal and bisect each other, not at right angles
  - **c** Diagonals are unequal, and bisect each other at right angles
  - **d** Diagonals are unequal, and bisect each other, not at right angles
  - **e** Diagonals are unequal, cross at right angles, but do not bisect
  - **f** Diagonals are unequal, do not cross at right angles, and do not bisect.

Angle sum

You may have investigated the sum of the angles of a quadrilateral in Year 8. The angle sum of 360° can be demonstrated in the following way:

Step 1: Draw a quadrilateral.

Step 2: Mark each of the four angles.

Step 3: Tear off the angles and place them next to each other as shown.

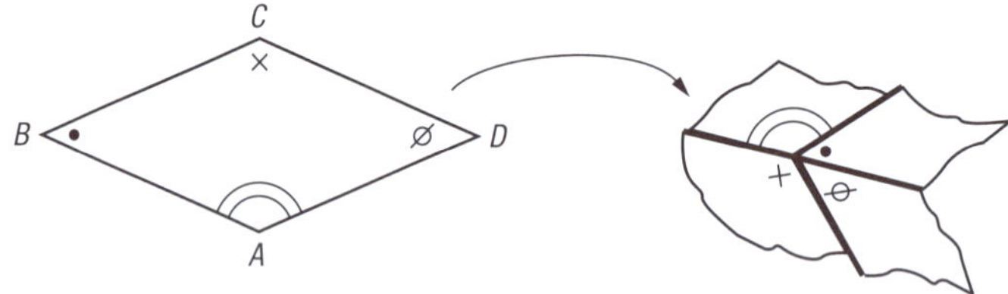

Because the angles make a complete revolution, we know that:

The angle sum of any quadrilateral is 360°.

We can use this information to find the unknown angle in a quadrilateral.

Example

Find the value of the pronumeral in each of the following quadrilaterals.

a

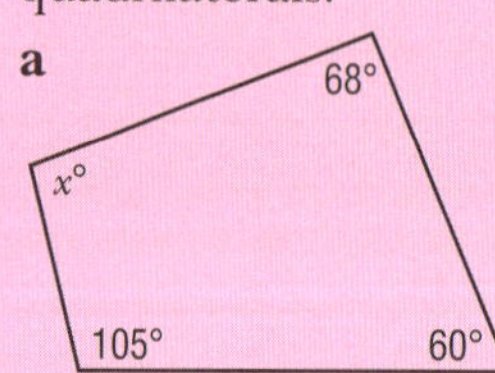

b

120°, 74°, $y°$ (with one right angle)

Answer

a $60 + 68 + 105 + x = 360$

$233 + x = 360$

$x = 127$

b $90 + 120 + 74 + y = 360$

$284 + y = 360$

$y = 76$

Constructing squares and rectangles

Constructing two right angles and measuring the sides shown will enable us to construct accurate squares and rectangles.

Here are the steps for constructing a 90° angle (as shown below).

Step 1: Draw a line *AB*.

Step 2: With centre *A*, draw a circle to cut *AB* at *L*.

Step 3: Keeping the same radius, use *L* as the centre and draw an arc to cut the circle at *M*.

Step 4: Again, keeping the same radius, use *M* as the centre and draw an arc to cut the circle at *N*.

Step 5: With the same radius, and using *M* and *N* as centres, draw arcs intersecting at *O*.

Step 6: Join *A* to *O*.

Angle *OAB* is 90°.

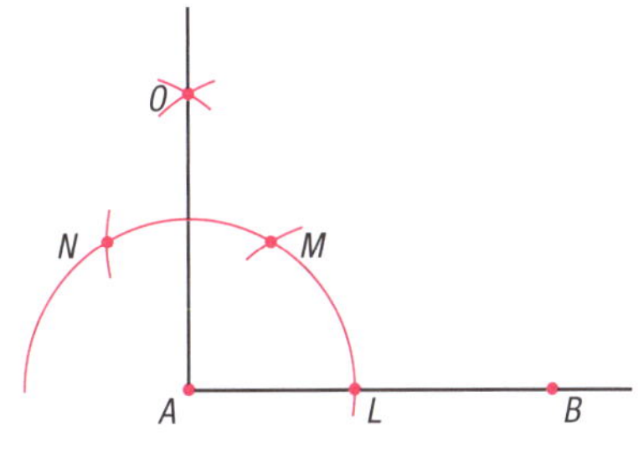

Example

Construct rectangle *ABCD* with sides 6 cm and 3 cm.

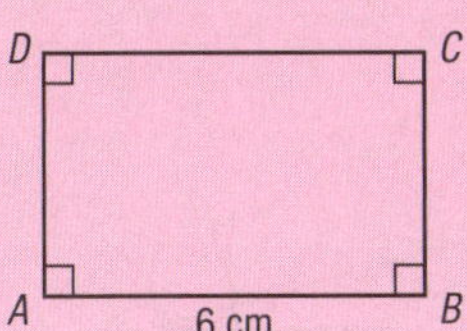

Answer

Rule a line 6 cm long for *AB*. Construct 90° angles at *A* and *B*. Measure 3 cm lengths for the sides *AD* and *BC*. Join *CD*.

EXERCISE

1 Name the quadrilateral (or quadrilaterals) with these properties:

- **a** four equal sides
- **b** one pair of equal opposite sides
- **c** two pairs of parallel sides
- **d** two pairs of equal opposite angles
- **e** four equal angles
- **f** no equal angles
- **g** one pair of parallel sides.

2 Sketch a quadrilateral:

- **a** with two pairs of equal opposite angles
- **b** with two right angles and no equal sides
- **c** with no equal angles
- **d** with one pair of parallel non-equal sides and one pair of equal sides
- **e** that is convex, and has two pairs of adjacent equal sides.

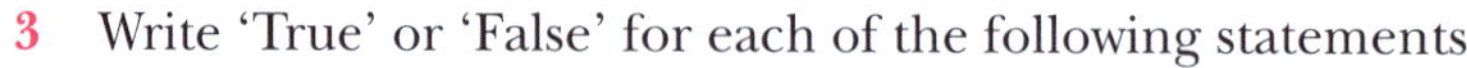

3 Write 'True' or 'False' for each of the following statements:

a all squares are rectangles

b all rectangles are squares

c all rectangles are parallelograms

d all rhombuses are parallelograms

e all kites are rhombuses

4 Sketch a quadrilateral, $LMNO$, so that two adjacent sides are equal and form a right angle. What kinds of special quadrilateral could $LMNO$ be?

5 Sketch a quadrilateral, $ABCD$, so that $AB \parallel DC$ and $BC \perp AB$. Must $ABCD$ be a rectangle?

6 Find the unknown angle in each of these quadrilaterals:

a

$a°$

b

$b°$

c

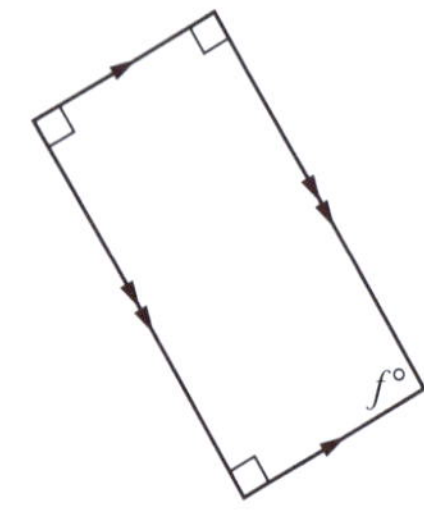

d

d°
220°
14°

e

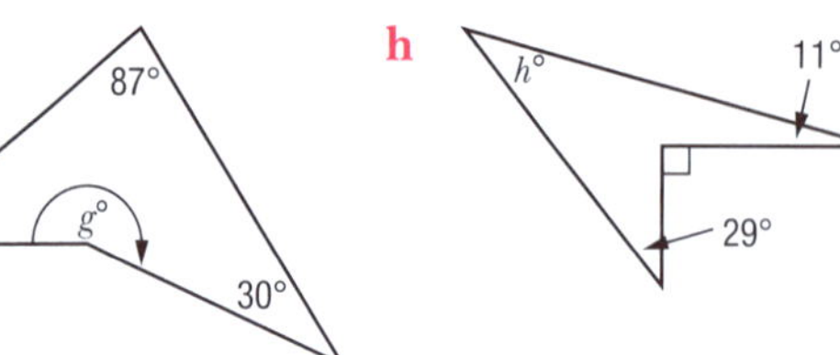

f

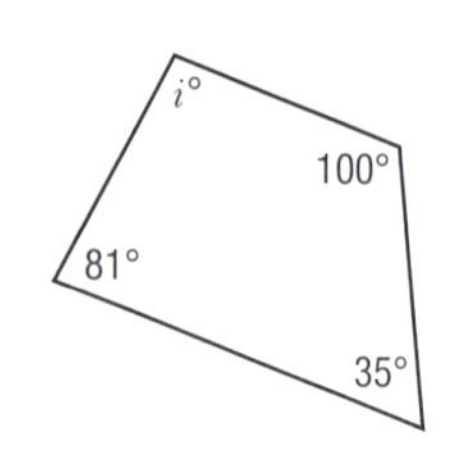

g

87°
21°
g°
30°

h

h°
11°
29°

i

i°
100°
81°
35°

7 Which figures in Question 6 are concave?

8 Construct the quadrilaterals sketched below, using a pencil, a ruler and a pair of compasses.

a square, with sides 8 cm

b rectangle

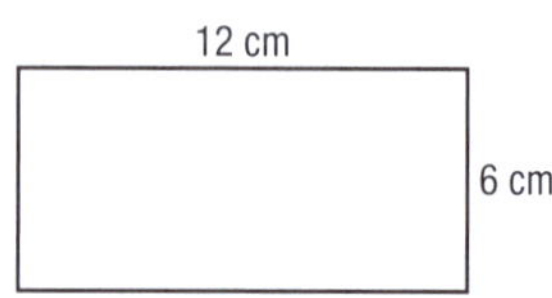

Interior and exterior angles of a polygon

Interior angles

The **interior angle** of any polygon is the angle between two adjacent sides. For example, in the pentagon below, $\angle ABC$ is an interior angle. For a regular polygon, the interior angles are all the same size.

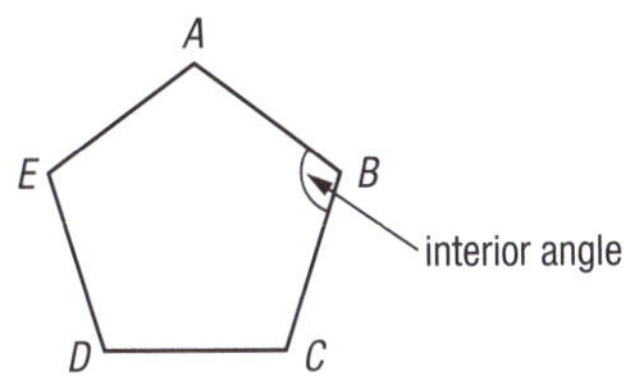

Challenge

Investigation 1: Sum of the interior angles

a To discover the sum of the interior angles of any polygon, copy the polygons drawn below. Join one vertex to each of the other vertices.

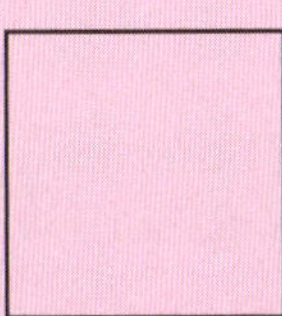
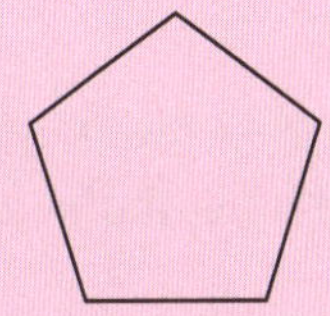
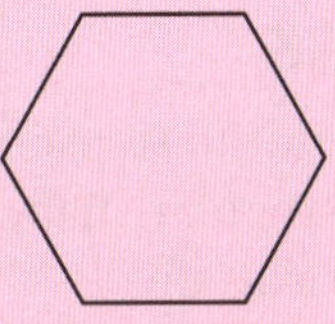
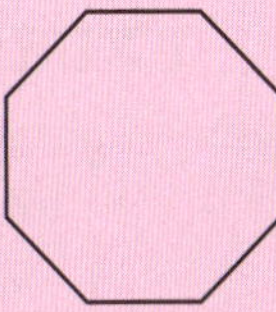

b For each polygon:

 i state how many triangles are formed

 ii explain how you would work out the angle sum of each polygon

 iii complete the corresponding row on a copy of the table below.

Polygon	Number of sides, *n*	Angle sum
Triangle		
Quadrilateral		
Pentagon		
Hexagon		
Octagon		

Exterior angles

The **exterior angle** of any polygon is the angle between the extension of one side and the adjacent side. For example, in the pentagon on the right, $\angle MBC$ is an exterior angle.

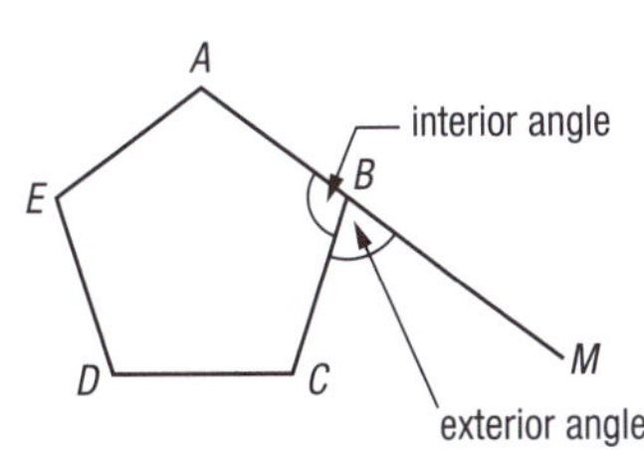

Challenge

Investigation 2: Sum of the exterior angles

Earlier, we looked at the properties of the exterior angles of triangles. Here we will investigate the sum of the exterior angles of a polygon.

Exterior angle of a pentagon

a Copy the pentagon on the right (or sketch your own).

b Extend each of the sides to form the exterior angles of the pentagon. Mark each of the exterior angles with an x, taking care to continue in the same direction (clockwise or anticlockwise).

c Carefully cut out each exterior angle and place the angles adjacent to each other. You will find they make a complete revolution. This is true for any polygon.

The sum of the exterior angles of a polygon is 360°.

For a regular polygon, the size of each exterior angle is $360° \div n$.

EXERCISE

1 a Repeat the steps in Investigation 2, using other polygons and including both regular shapes and irregular shapes. Record your results in a table similar to this.

Polygon	Number of sides	Size of each exterior angle	Sum of exterior angles
Triangle			
Quadrilateral			
Pentagon			
Hexagon			
Octagon			

b Copy and complete this statement:

To find the size of the exterior angle of any polygon ________________ ________________.

2 a Referring back to Investigation 1, copy and complete the following table:

Number of sides, n	3	4	5	6	8
Number of right angles, S					

b Work out a rule connecting S and n and write this as an equation: $S =$ ____.

c Use your rule to work out the sum of the interior angles of:

i a heptagon   ii a nonagon   iii a decagon.

d Sketch an example of each figure referred to in part c and prove your answer is correct by working out how many triangles are formed when you join one vertex to all of the others.

3 Find the value of each pronumeral in this figure:

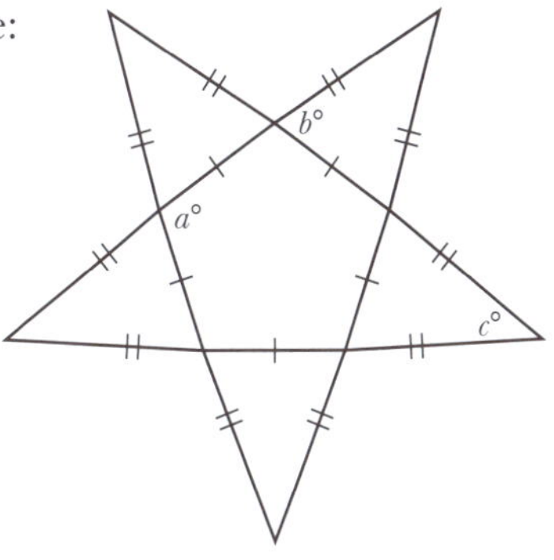

What are tessellations?

Tessellations are patterns made by covering a space with one or more shapes so that there are no gaps. Tessellation patterns traditionally appear in tiling patterns on floors or walls, but can also be used in handcrafts.

Simple tessellations can be made using one shape, while more complex patterns use a combination of polygons.

EXERCISE

1 Here are examples of designs for floor tiles. Name the shapes used in each design.

a 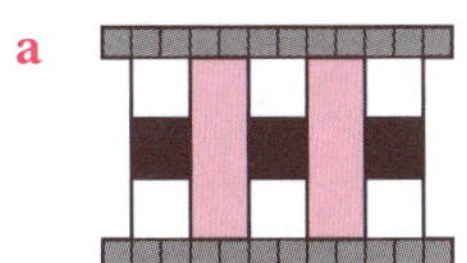

b 

c 

d 

e 

Weeks 1 and 2

2 Choose one of these designs and draw it on grid paper or dot paper:

a  b c d e 

3 These tiling patterns use only one shape. Using the same type of grid, copy the tile shape and continue the pattern to fill the space completely.

a

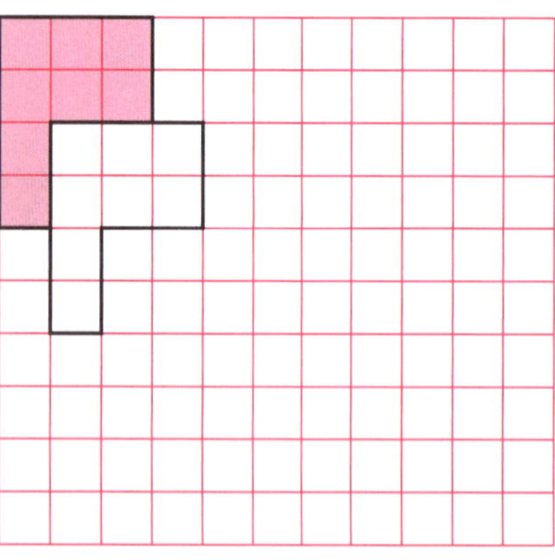

b

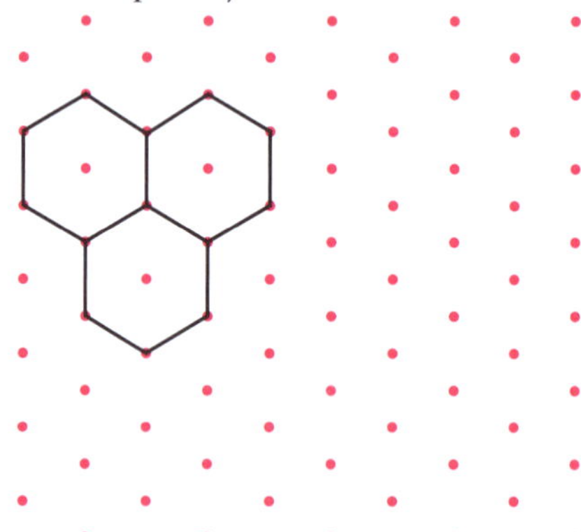

c

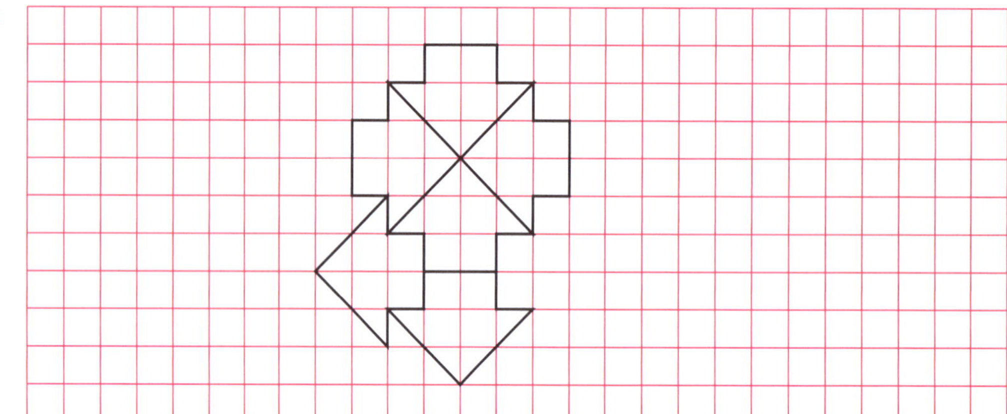

4 Copy and complete these tessellations, using the two shapes already used each time:

a

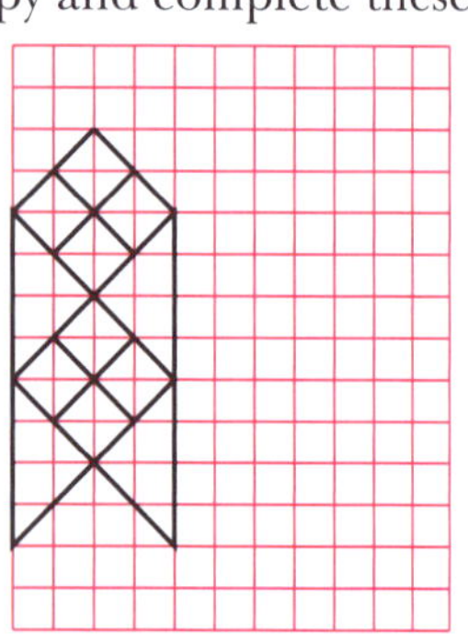

b

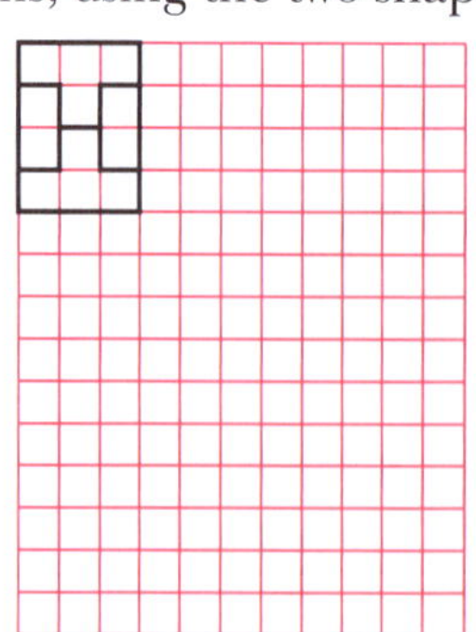

c

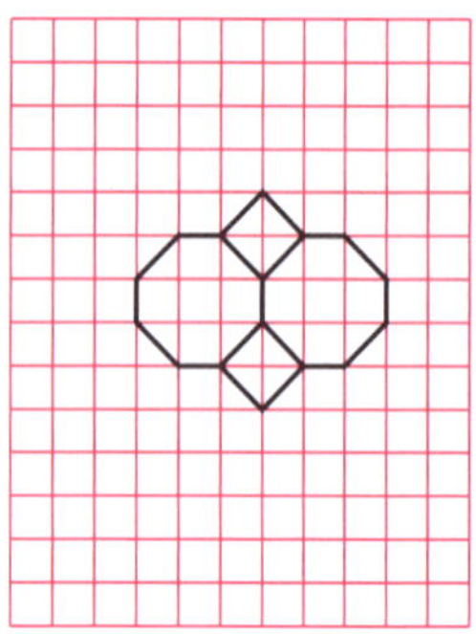

Challenge

Reproduce these traditional tiling patterns on grid paper as shown and continue the pattern?

a

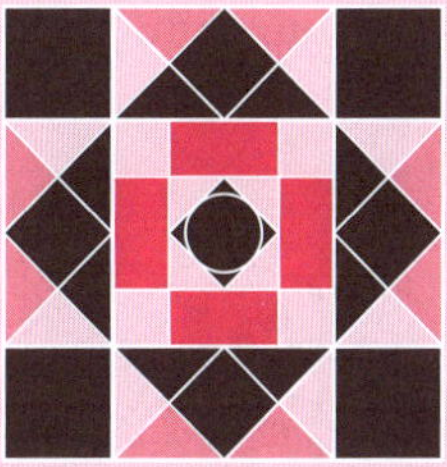

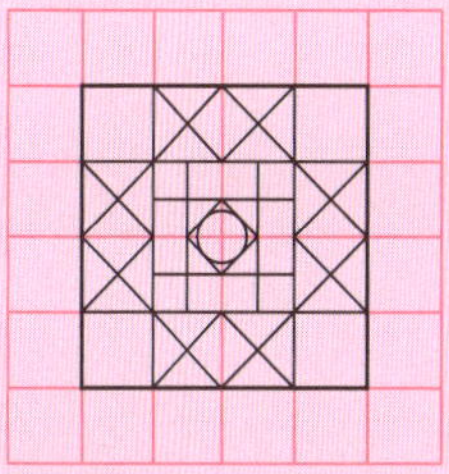

b

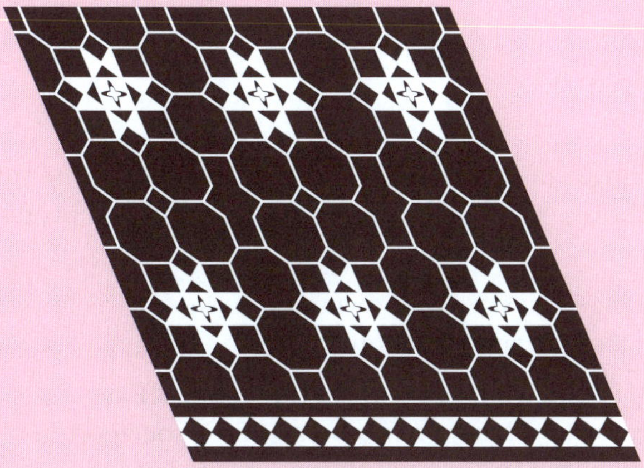

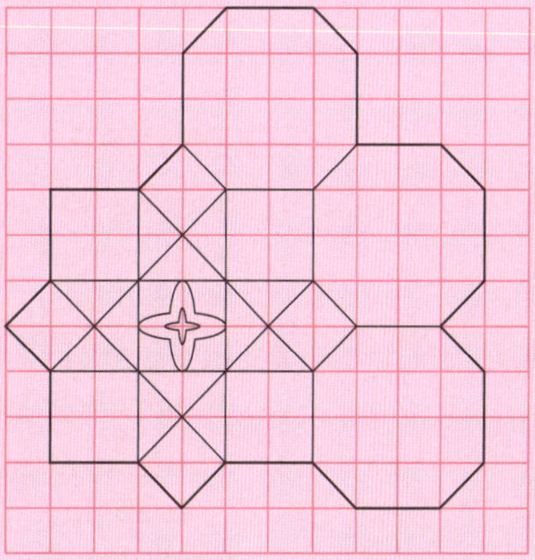

Making a shape that will tessellate

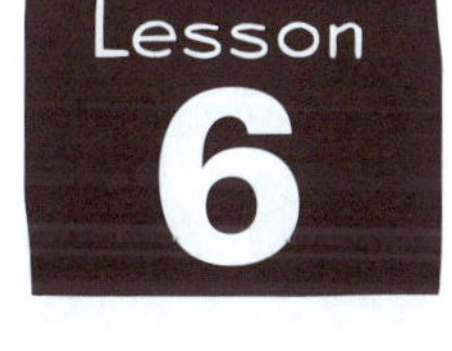

The Dutch artist Maurits Escher is famous for his graphic art. He made sketches of tiling patterns in old buildings, and from that idea developed tessellations like the onc shown below.

Two of the simplest shapes that will make a tiling pattern are a rectangle and a rhombus. Using these, you can try Escher's technique of changing the sides of a basic shape to make tiling patterns of animals (or anything you like).

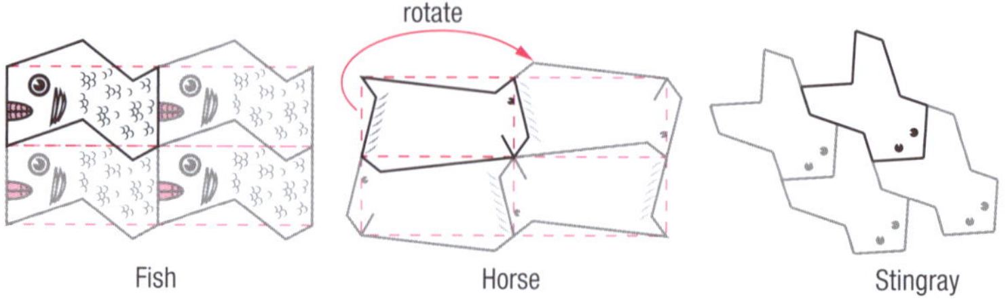

EXERCISE

1 Starting with a rectangle of cardboard, alter one pair of opposite sides (as in the fish design above). Cut out your template and use it to produce a tessellation.

2 In a similar way, change two sides of a rectangle or square (as in the horse design above) and produce a tessellation from your design.

3 a Investigate which regular polygons from this list tessellate:
triangle square pentagon hexagon octagon
b Write a statement about your conclusion.

Patterns in nature: Fractals

In nature, most objects are not made up of simple polygons such as triangles and squares. Many are shaped like **fractals**. These are complex geometric patterns, examples of which can be seen in the shape of snowflakes, or the branching of ferns or river systems.

Fractals are formed from the iteration (repeated application) of a geometric rule. The fractal pattern below looks like something we would find in nature.

This fractal image resembles a fern.

The Sierpinski Triangle (above) is a fractal formed from iteration.

EXERCISE

1 Here are the instructions for drawing the Sierpinski Triangle, starting with a smaller version of the figure.

- **a** Draw an equilateral triangle (isometric grid paper will make this easy).
- **b** Join the midpoints of the three sides to form a smaller equilateral triangle. Imagine this has been removed.
- **c** In each of the three remaining equilateral triangles, join the midpoint of the sides, as in part **c**.
- **d** Continue in this way, ignoring the middle triangle and joining the midpoints of the sides of the smaller triangles, until the triangles are too small for you to continue.

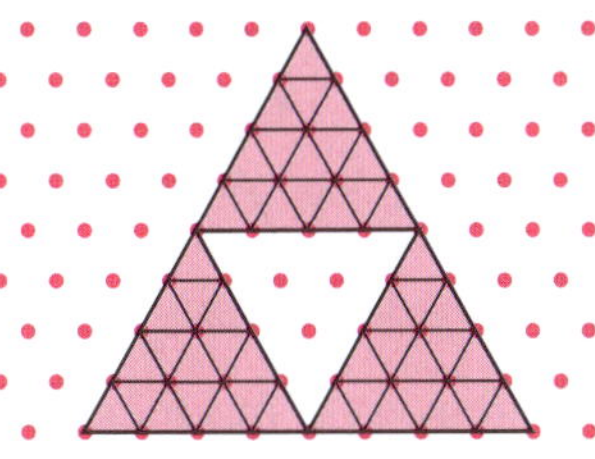

2 On isometric paper, reproduce this variation of the Sierpinski Triangle.

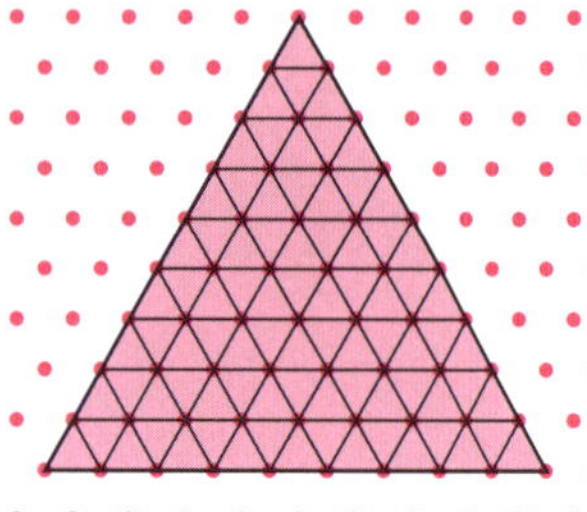
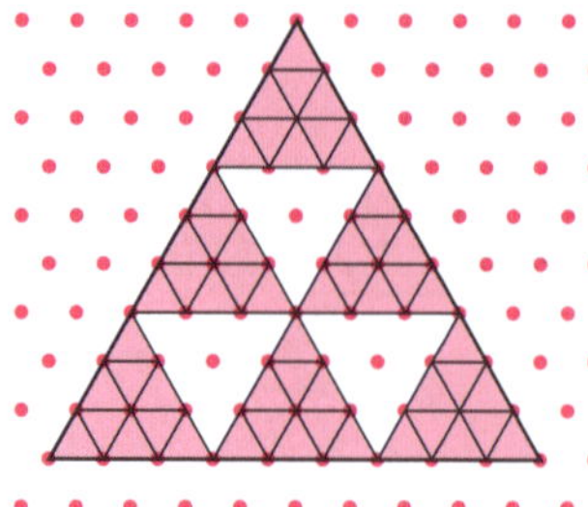
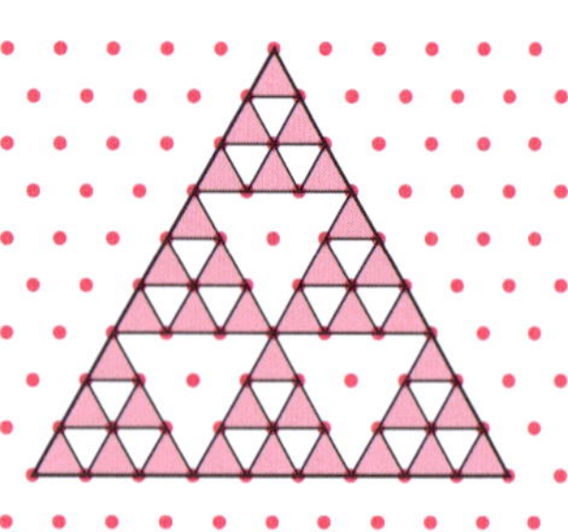

3 Below is a fractal known as the Von Koch curve, created in 1904 by the Swedish mathematician Niels von Koch. Starting with a horizontal line (27 units is a good length) on isometric paper, divide the line segment into thirds and replace the centre third with an equilateral triangle as shown. Continue in the same way as far as you can.

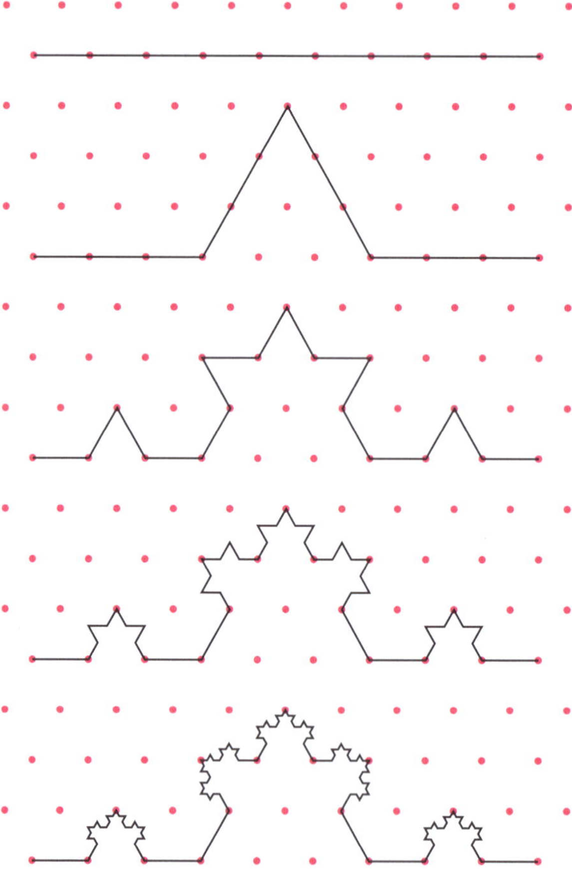

4 Use the same method as used in Question 3, but start with a closed polygon, such as a triangle or square.

ANGLES AND LINES

Lesson 1

About angles

Angles are important in many practical activities, such as constructing buildings and roads, or navigating boats or planes.

Copy and complete the following table, which summarises angle names according to their size.

Angle	Name	Description
	acute	
	right	90° exactly (a right angle is usually marked using a small square)
		between 90° and 180°
	straight	
		between 180° and 360°
	rotation or revolution	

Did you know?

The measurement of angles dates back more than 2500 years when it was believed the Earth rotated around the Sun in a circular path. The time for one complete rotation or revolution was measured as 360 days, and the portion of rotation in one day was called a **degree** (°).

You will remember how to use a protractor to measure angle sizes, as shown below.

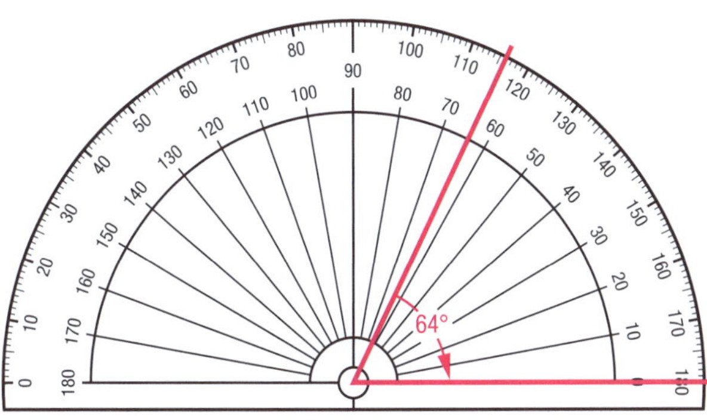

This angle measures 64°.

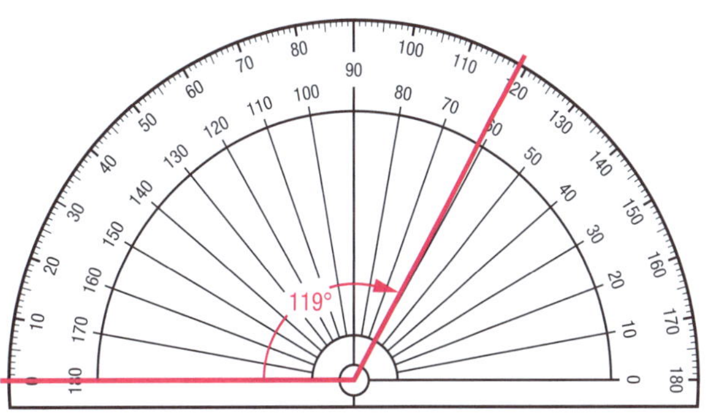

This angle measures 119°.

EXERCISE

1 Use a protractor to measure each of the following angles to the nearest degree.

a

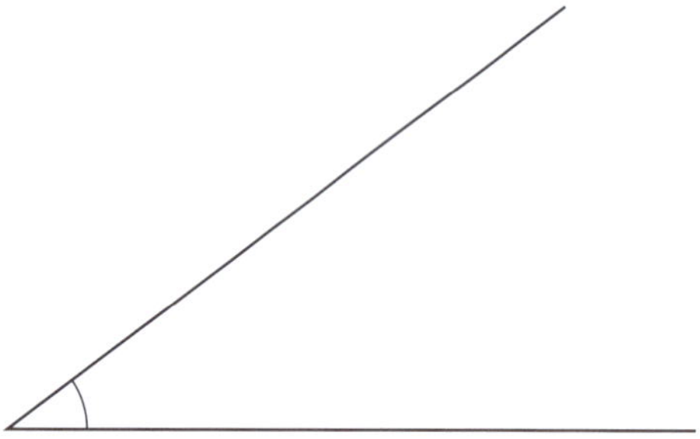

b

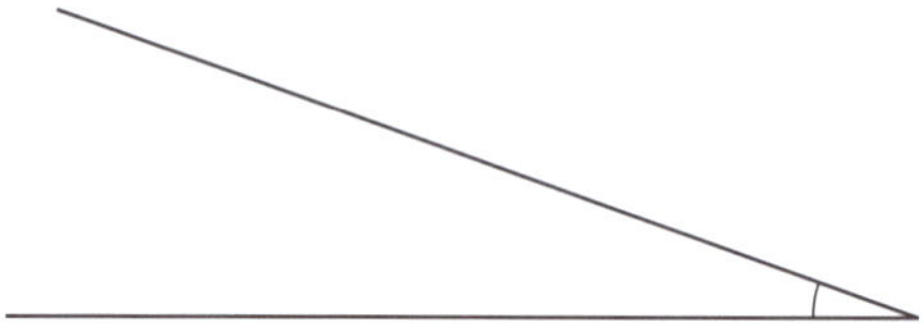

c

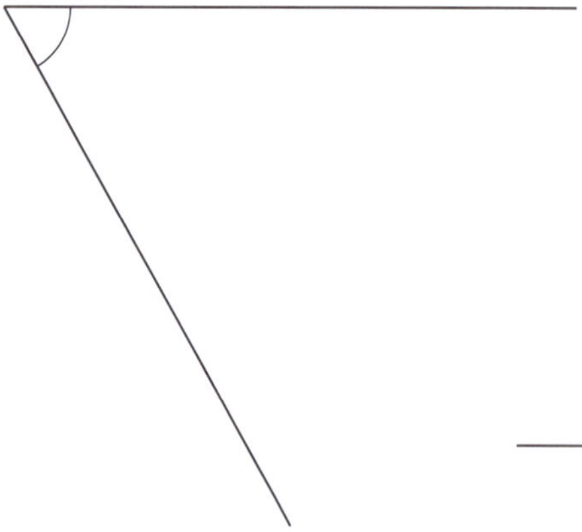

d

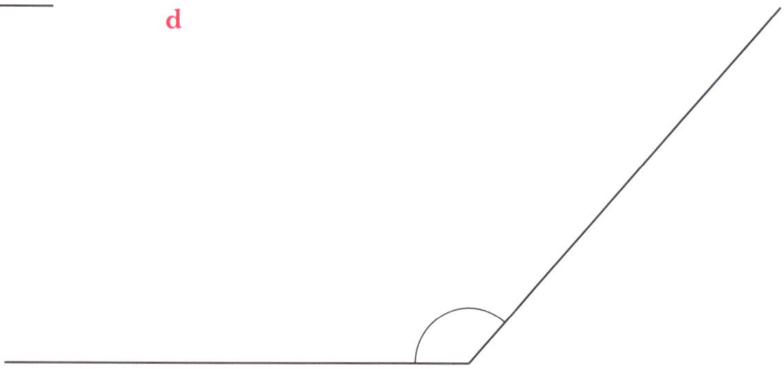

e

f

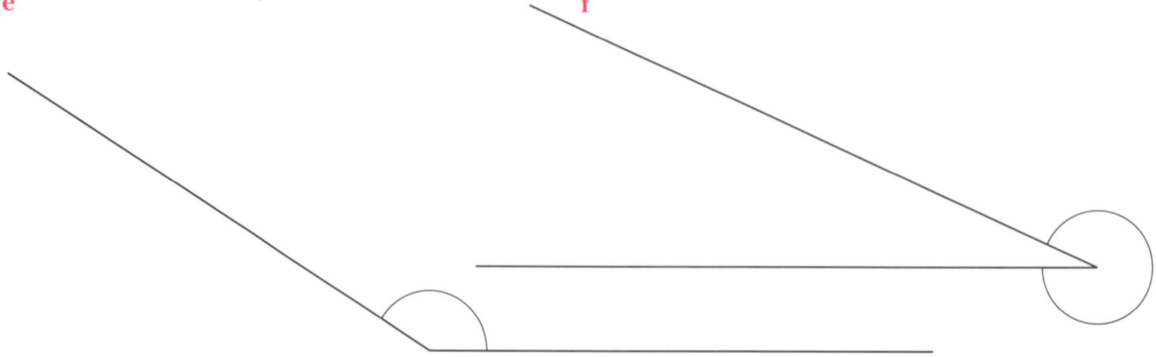

2 It is also useful to be able to estimate angle sizes. Think about your answers for Question 1, and then use that information to estimate the size of the following angles. Record your estimates in a table similar to the one on the right, and then measure to see how accurate you were.

Question part	Estimate	Actual measurement
a		
b		
. . .		

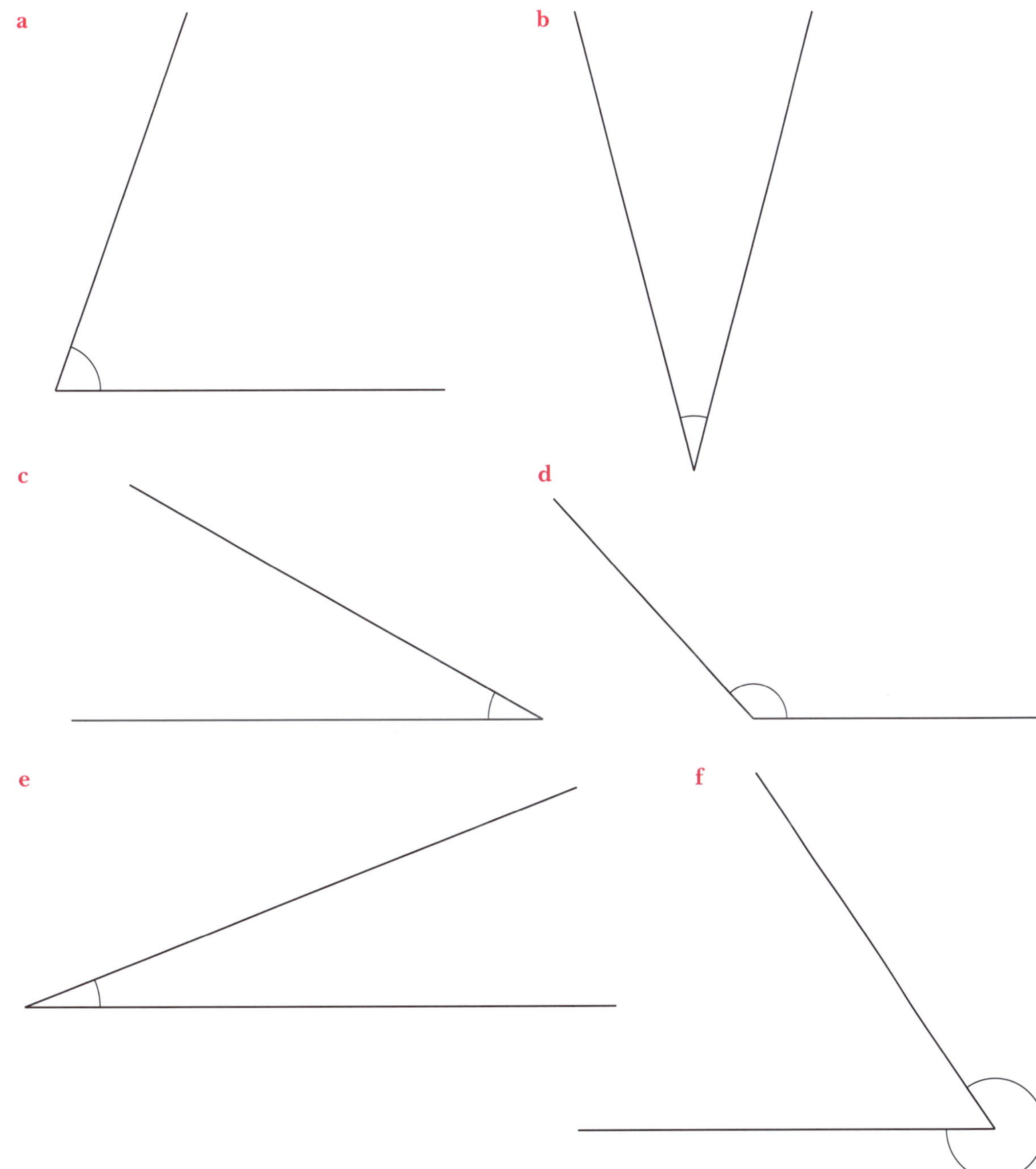

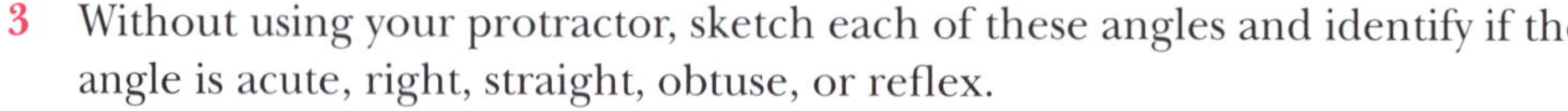

> **Remember**
>
> Angles can be named with three capital letters, with the vertex being the middle letter. They can also be named with a single letter, as shown below.
>
> L M N
>
> ∠*LMN* or ∠*M*.

3 Without using your protractor, sketch each of these angles and identify if the angle is acute, right, straight, obtuse, or reflex.

a 78° **b** 150° **c** 90° **d** 280° **e** 10°

4 Sketch and label each of these angles:

a an obtuse angle, ∠*V*
b a reflex angle, ∠*ABC*
c a right angle, ∠*PQR*
d an obtuse angle, ∠*W*

Lesson 2

Angle pairs

Aaron is constructing a framework for a chicken coop. The end view looks like the diagram below.

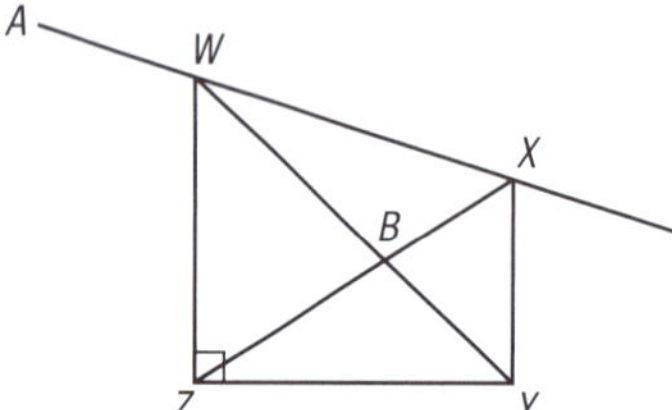

The angles between the lengths of wood can be described by their relationships to each other:

- ∠*WXZ* and ∠*ZXY* are **adjacent angles** (they share a common arm)
- ∠*WZX* and ∠*XZY* are **complementary angles** (they add to 90°)
- ∠*AWZ* and ∠*ZWX* are **supplementary angles** (they add to 180°)
- ∠*WBZ* and ∠*XBY* are **vertically opposite** angles
- ∠*WBZ*, ∠*ZBY*, ∠*YBX* and ∠*XBW* are **angles in a revolution** (they add to 360°).

EXERCISE

1 From the chicken coop described above, name a different pair of:

a adjacent angles **b** complementary angles
c supplementary angles **d** vertically opposite angles.

2 Which of the following pairs of angles are complementary?

a 30° and 50° **b** 50° and 40° **c** 58° and 42°
d 19° and 71° **e** 44° and 66°

3 Write the angle which is complementary to:

a 45° **b** 16° **c** 68° **d** 13° **e** 25°

4 Which of the following pairs of angles are supplementary?

a 70° and 90° **b** 93° and 87° **c** 45° and 135°
d 65° and 125° **e** 103° and 77°

5 Write the angle which is supplementary to:

a 100° **b** 134° **c** 38° **d** 43° **e** 89°

6 Find the value of the pronumerals in the following diagrams:

a

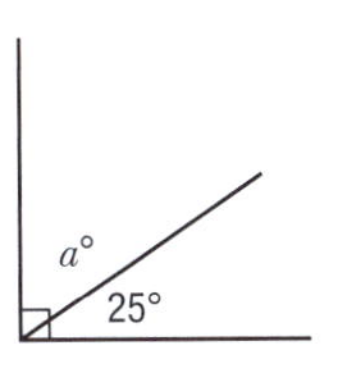

b

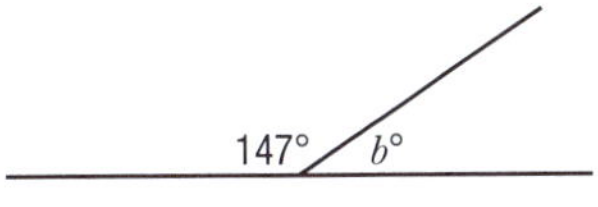

c

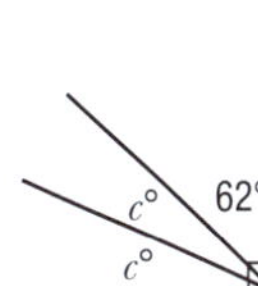

d

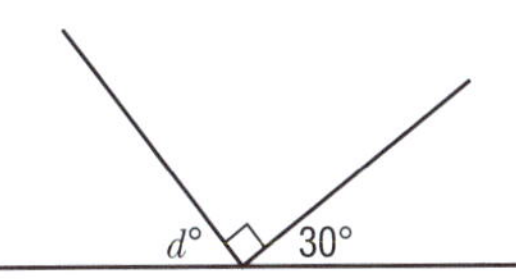

e

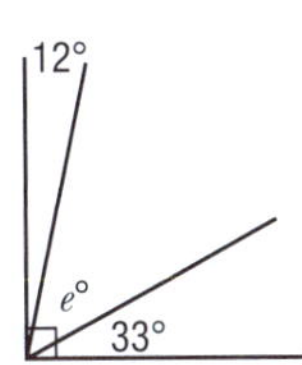

f

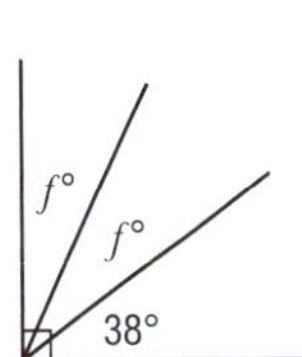

g

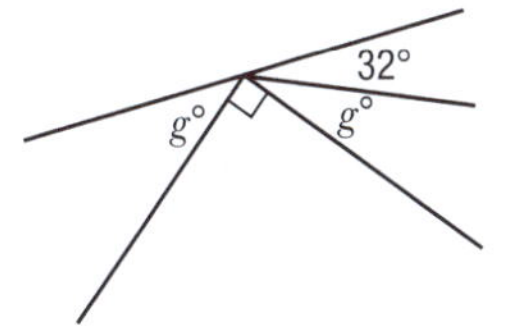

h

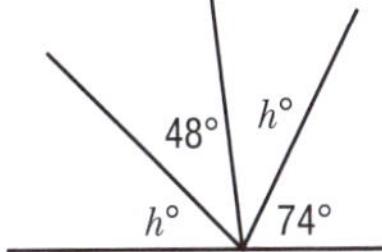

i

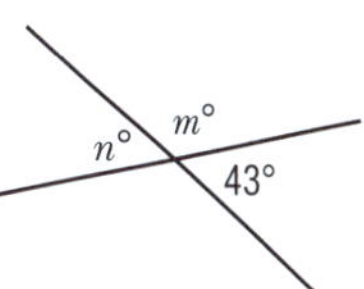

j

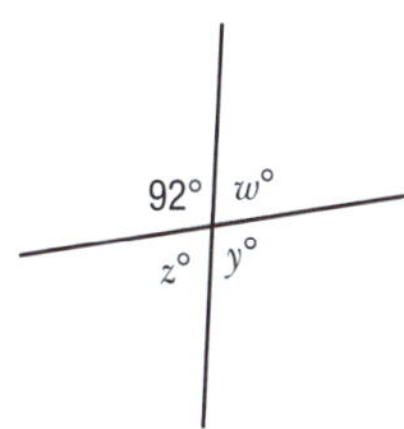

7 What is the value of each pronumeral in the following diagrams?

a

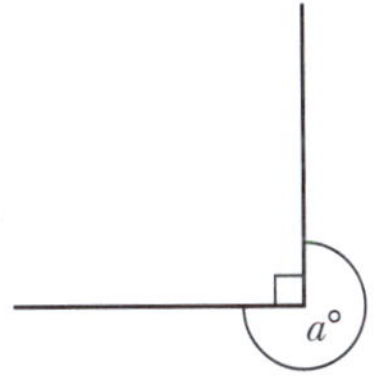

b

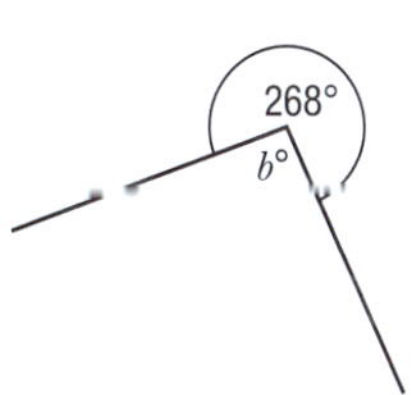

c

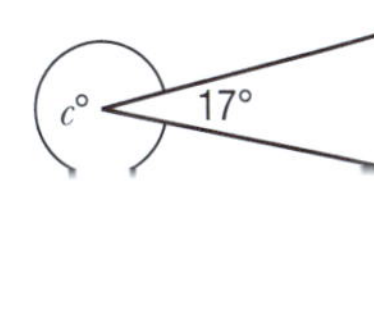

d

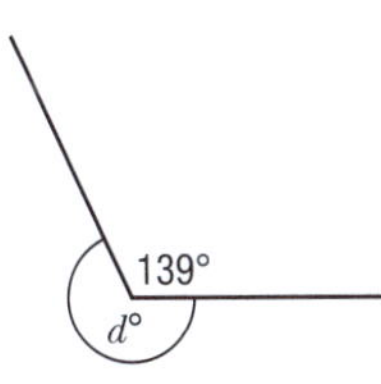

e

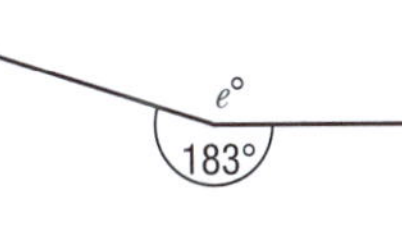

f

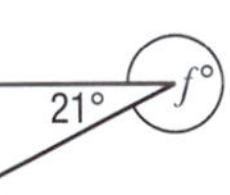

g

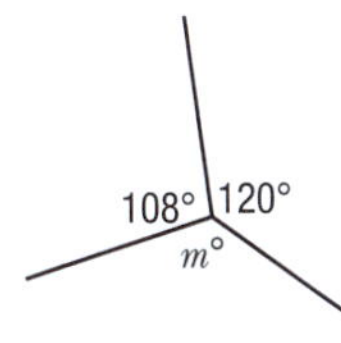

h

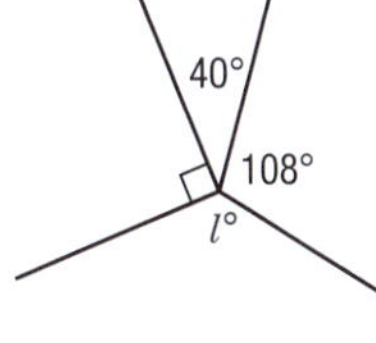

i

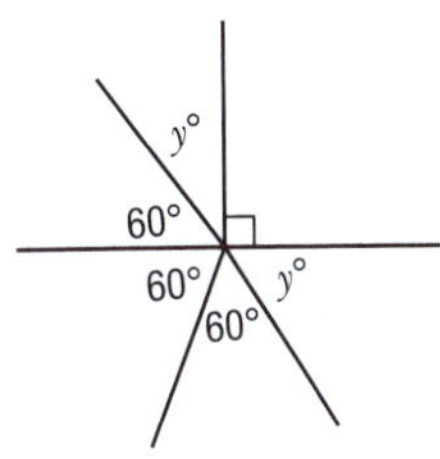

j

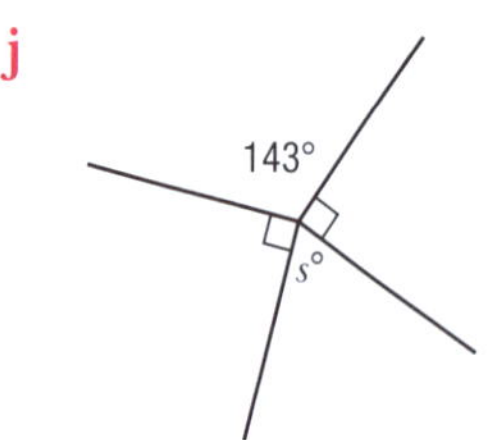

k

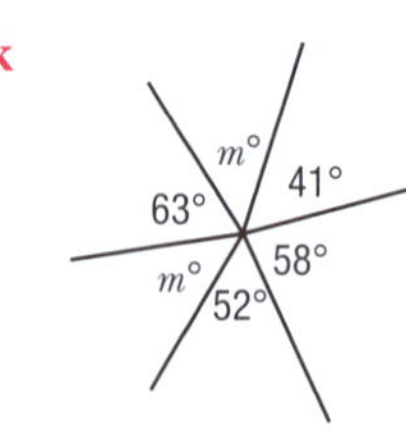

l

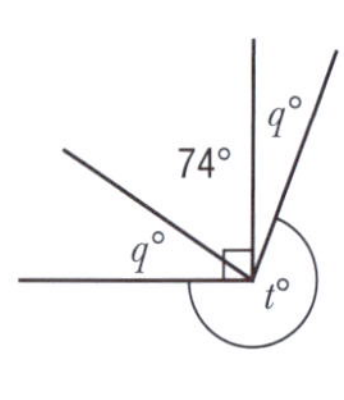

Lesson 3

Parallel and perpendicular lines

Perpendicular lines intersect at right angles. The angle between the lines is marked as shown in the diagram on the right.

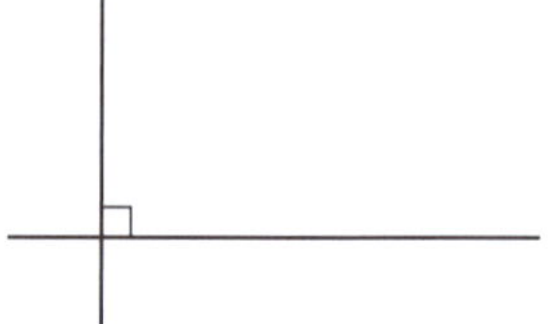

Parallel lines are straight lines in the same plane which are always the same distance apart and, so, do not intersect. Parallel lines are marked with matching arrowheads as shown.

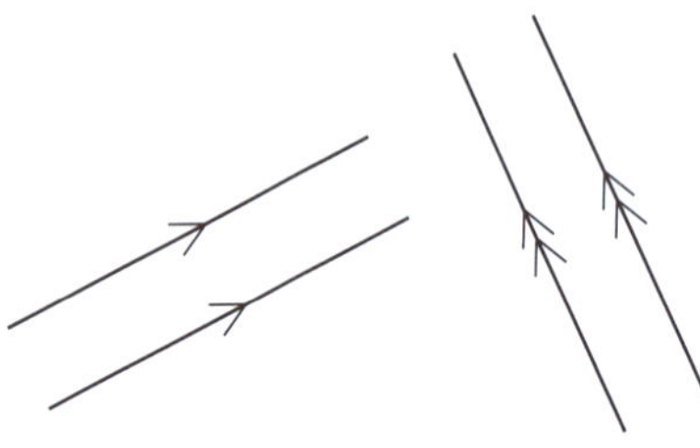

A third line which crosses two other straight lines is called a **transversal**.

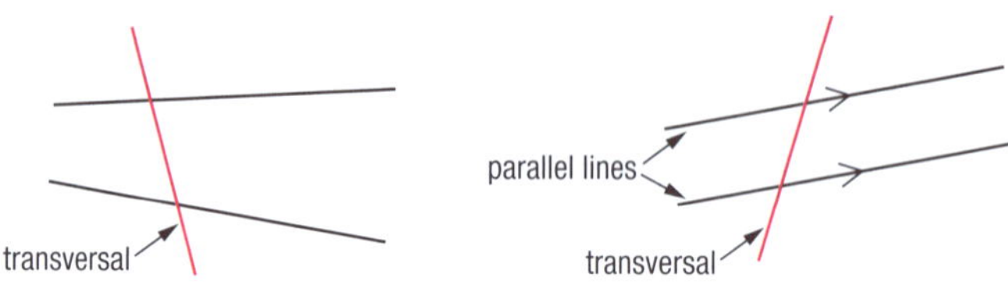

Challenge

Investigation: Transversals

a On the right are two straight lines cut by a transversal. Use your protractor to measure all eight angles formed. What do you notice?

b Draw another pair of non-parallel straight lines cut by a transversal and measure the angles formed. Is there a relationship between any of the angles? Write a statement about this.

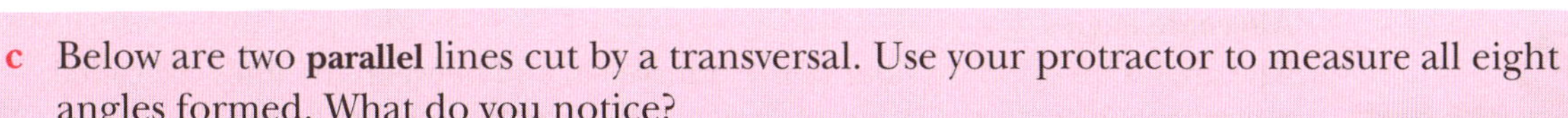

c Below are two **parallel** lines cut by a transversal. Use your protractor to measure all eight angles formed. What do you notice?

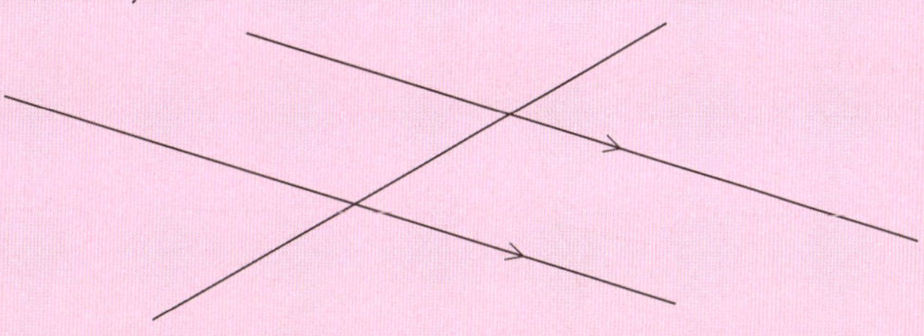

d Draw another pair of parallel straight lines cut by a transversal and measure the angles formed. Is there a relationship between any of the angles? Write a statement about this.

Angles formed by a transversal

There are three special pairs of angles formed when a transversal crosses two lines. These are called:

- **corresponding** angles
- **alternate** angles
- **co-interior** angles.

Corresponding angles

Corresponding angles have the same relative position (that is, they are both on the same side of the transversal and both on the same side of the lines). In the diagrams below, the pairs of corresponding angles are labelled:

- a and e
- b and f
- c and g
- d and h

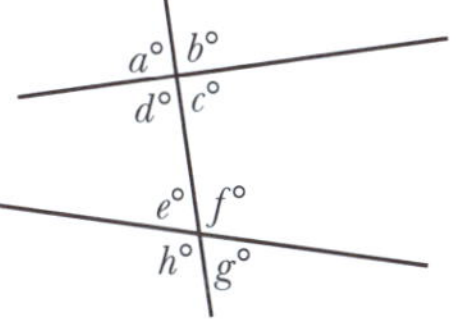

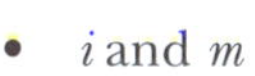

- i and m
- j and n
- k and o
- l and p

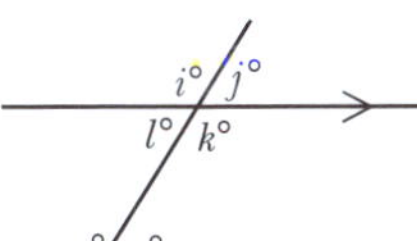

When a pair of parallel lines is cut by a transversal, the corresponding angles are equal.

Help box

A clue for locating corresponding angles is to think of the letter 'F':

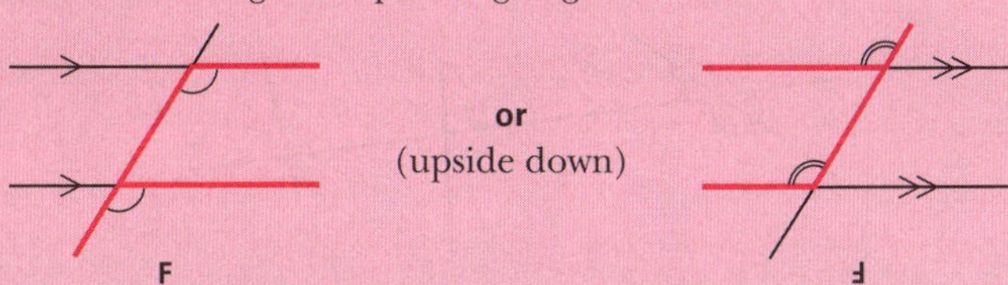

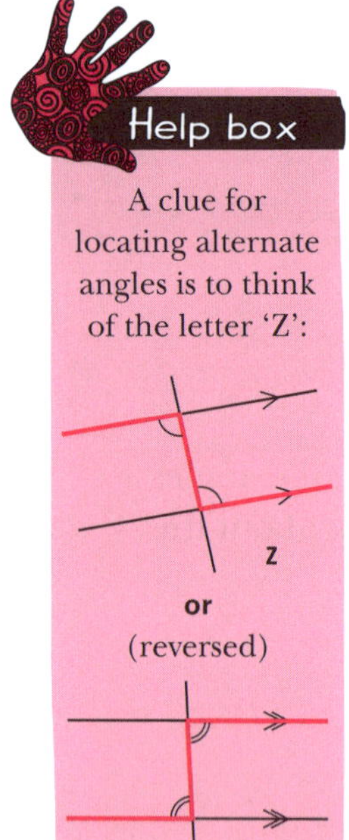

Help box

A clue for locating alternate angles is to think of the letter 'Z':

Z

or

(reversed)

Z

Alternate angles

The angles on opposite sides of the transversal and between the lines are called **alternate angles**. In the diagrams below, the pairs of alternate angles are:

- d and f
- c and e

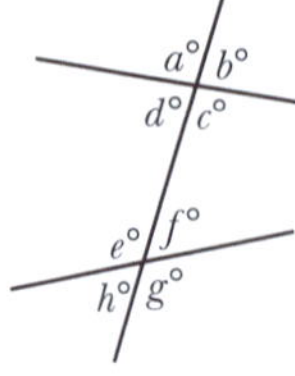

- l and n
- k and m

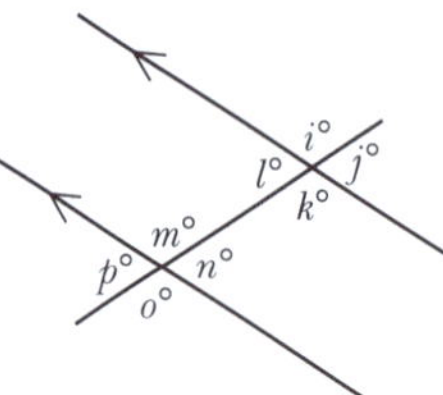

When a pair of parallel lines is cut by a transversal, the alternate angles are equal.

Co-interior angles

Co-interior angles are on the same side of the transversal and between the parallel lines. In the diagram below, the co-interior angles are:

Help box

A clue for locating co-interior angles is to think of the letter 'C':

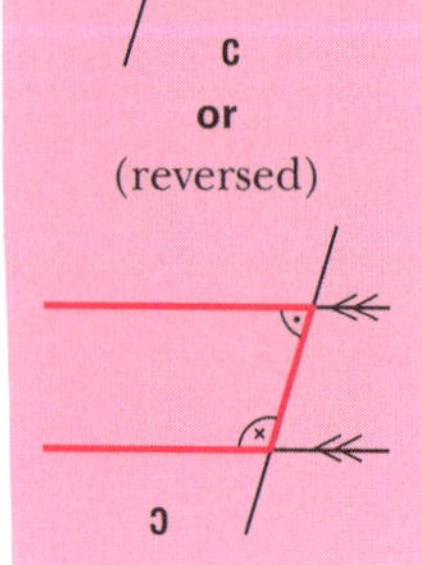

C

or

(reversed)

C

- v and w
- u and x

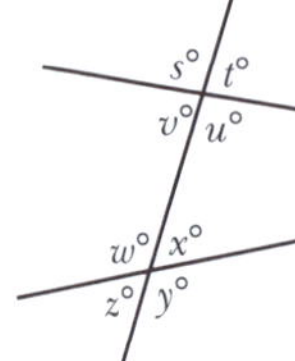

- n and o
- m and p

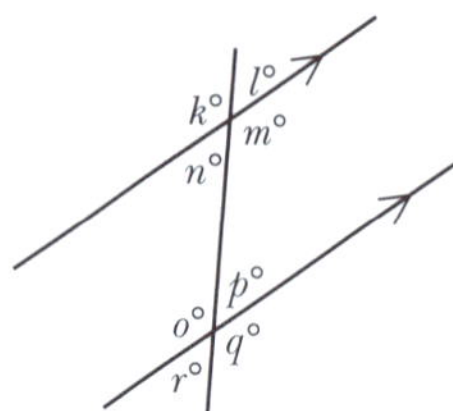

When a pair of parallel lines is cut by a transversal, the co-interior angles are supplementary (add to 180°).

EXERCISE

1 Refer to the diagram below and classify each of the following angle pairs as either corresponding, alternate, co-interior or vertically opposite.

- **a** a and p
- **b** r and w
- **c** r and x
- **d** z and s
- **e** b and q
- **f** a and c
- **g** x and z
- **h** w and s
- **i** c and p

2 Refer to the diagram on the right and classify each of the following angle pairs as either corresponding, alternate, co-interior or vertically opposite.

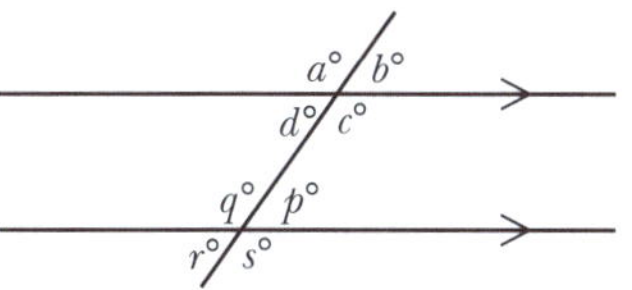

- **a** a and q
- **b** d and q
- **c** r and d
- **d** a and c
- **e** c and s
- **f** p and c
- **g** d and p
- **h** q and s
- **i** c and q

3 In Question 2, identify which pairs of angles are:

- **a** equal
- **b** supplementary

4 Find the value of the pronumerals in each of the following diagrams:

a

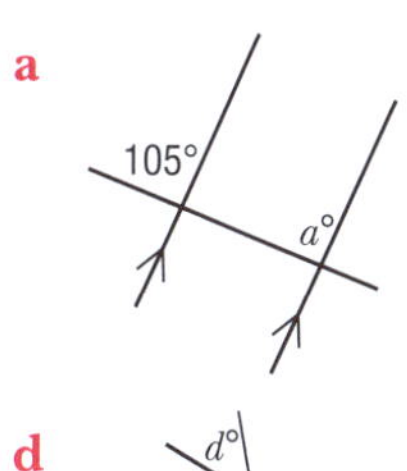

b 120° b°

c 110° c°

d

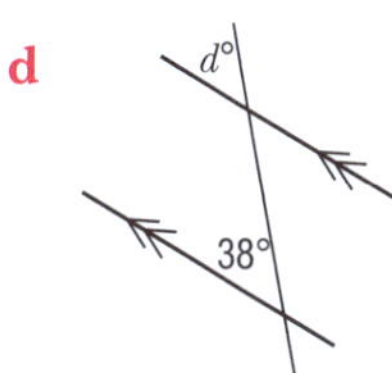

e

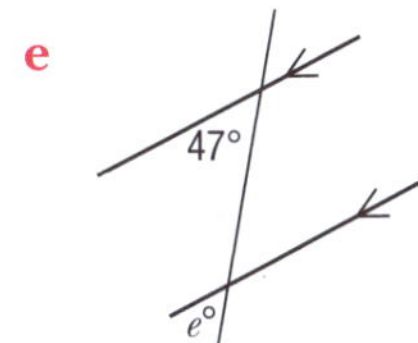

f

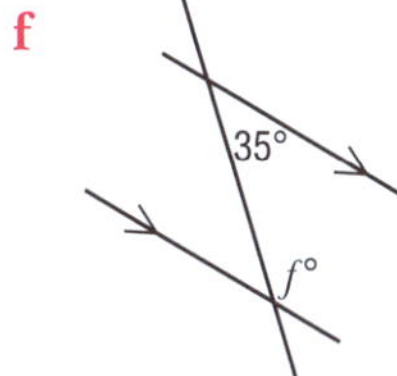

g

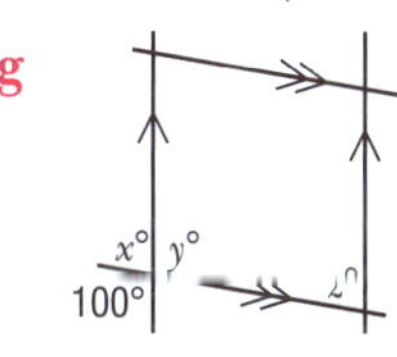

h

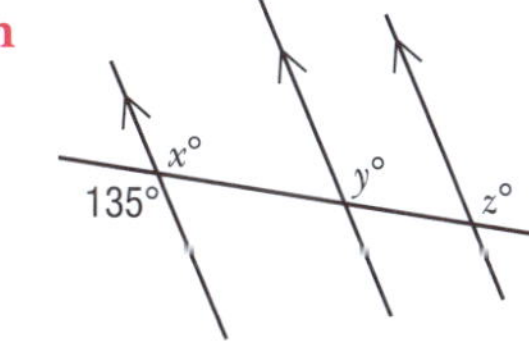

Congruent and similar figures

The tessellations you studied in Lessons 5 and 6 of 'Properties of plane figures' (see pages 267 to 270) contained shapes which were replicas of each other (that is, they were exact copies).

Polygons which have exactly the same shape and size are called **congruent figures**. The following two figures are **congruent**:

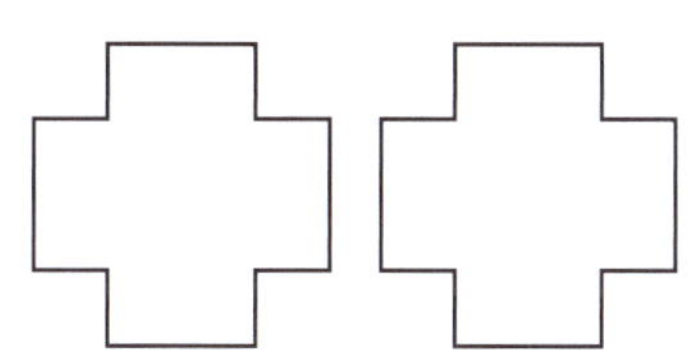

We have also completed scale drawings in the past. In the case of a scale drawing, a figure is enlarged or reduced to create another figure which has the same shape but a different size. In that case, we say that the figures are **similar**. Similar figures have measurements which are **in the same ratio**.

The following two figures are **similar**:

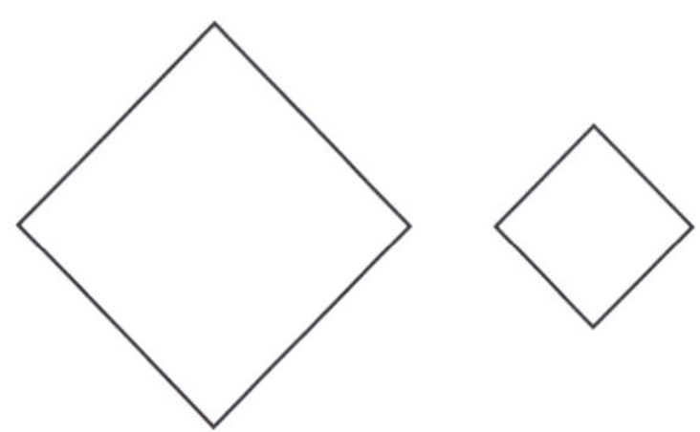

Challenge

Investigation: Congruent or similar?

a Measure the sides and angles of the following triangles. Copy and complete the tables below to record your results.

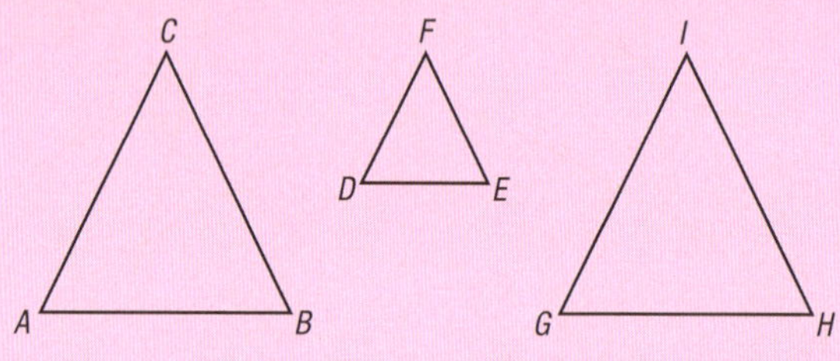

Angle	A	B	C	D	E	F	G	H	I
Size (°)									

Side	AB	BC	AC	DE	EF	DF	GH	HI	GI
Length (mm)									

b Which pair of triangles has measurements that are exactly the same?

c Copy and complete the statement: $\triangle$___ and $\triangle$___ are **congruent**.

d **i** Compare the sides of $\triangle ABC$ and $\triangle DEF$: Copy and complete the following:

$\frac{AB}{DE} = \frac{BC}{EF} = \frac{AC}{DF} =$ ___

ii What conclusion can you make about the ratio of the sides of $\triangle ABC$ and $\triangle DEF$?

e Copy and complete the statement:

$\triangle ABC$ and $\triangle DEF$ are ________ because their sides are in the same ratio.

EXERCISE

1 Which of the following pairs of figures are:

i congruent? ii similar? iii neither?

a

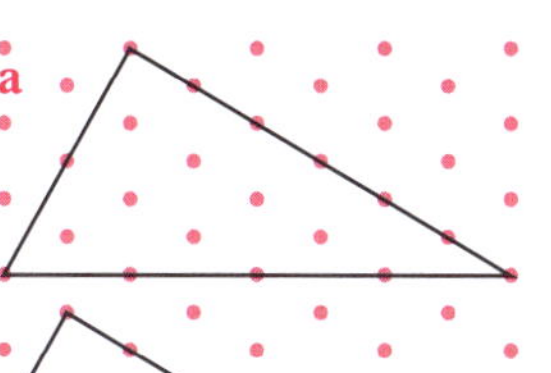

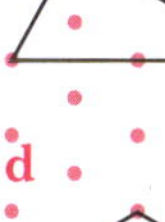

b

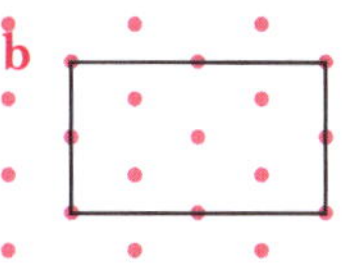

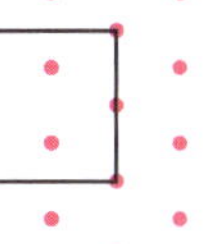

c

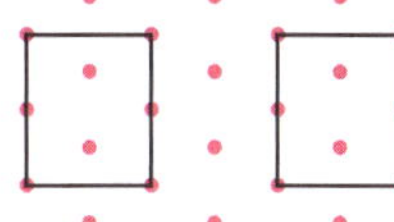

d

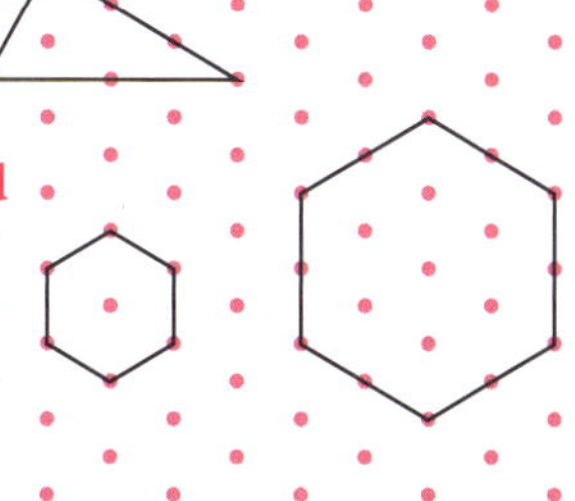

e

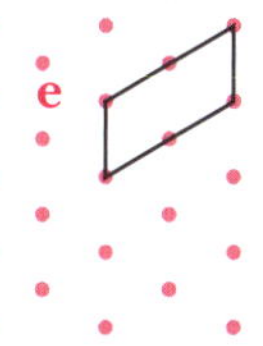

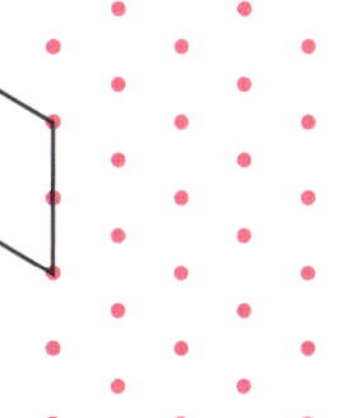

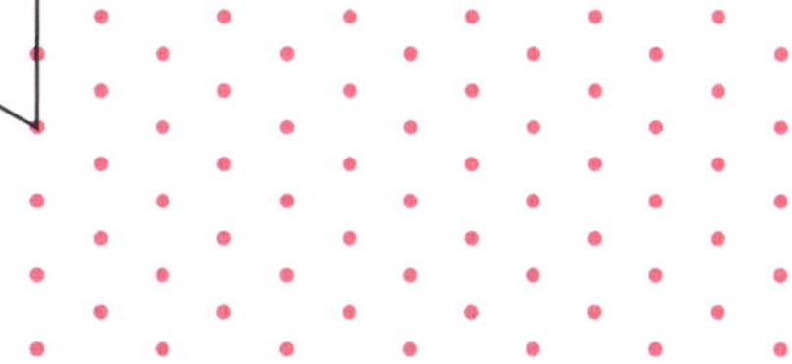

2 Draw a figure which is:

a congruent to the figure below

b similar to the figure below.

3 Draw a figure which is similar to this rectangle but has dimensions three times as long.

4 a 'All squares are similar to other squares.' True or false?

b Draw some examples to explain your conclusion.

SURFACE AREA AND VOLUME

The perimeter of plane figures reviewed

We have encountered **perimeter** in previous years and in previous sections of this textbook.

Some perimeter facts

- The perimeter of a closed, **plane figure** is the distance around the outside of the figure.
- A plane figure is a figure that has only two dimensions: length and width.
- The perimeter of a figure is a length measurement. In the metric system, lengths are usually measured in **millimetres (mm)**, **centimetres (cm)**, **metres (m)** or **kilometres (km)**.
- On a diagram, if sides are marked in the same way, those sides are the same length.
- Lengths must be expressed in the same unit before they can be added.
- The following conversion diagram helps us convert units of length in the metric system of measurement.

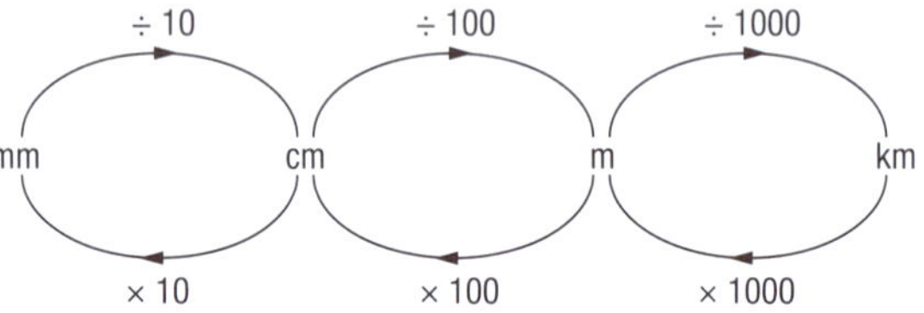

Example

Find the perimeter, in metres, of the closed figure below.

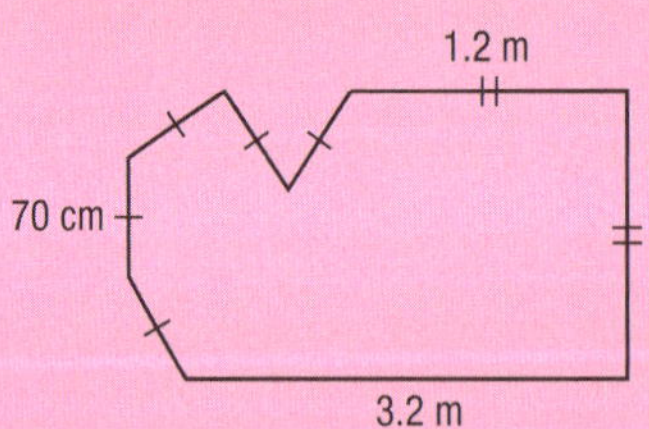

Answer

There are five sides of length 70 cm. They are all marked in the same way:

—+—

The answer is to be given in metres, so we will convert 70 cm to metres by dividing 70 by 100 to express this length as 0.7 m.

There are two sides of length 1.2 m marked:

—++—

There is one side marked 3.2 m. So:

The perimeter of the figure

$= 5 \times 0.7 + 2 \times 1.2 + 3.2$

$= 9.1$ metres

The perimeter of a regular polygon

A **regular polygon** has sides that are all equal in length (and all interior angles are equal in size). So, finding the perimeter of a regular polygon with n sides is simply a matter of multiplying the length of the side by n.

Example

Find the perimeter, in metres, of a regular octagon of side length 30 cm.

Answer

An octagon has eight sides.

Perimeter = 8 × 30

= 240 cm Divide by 100 to convert to metres.

= 2.4 metres

EXERCISE

1 Find the perimeter of each of the following figures. Give your answers in metres.

  a a square of side length 1.2 m
  b a regular pentagon of side length 80 cm
  c a regular hexagon of side length 275 mm
  d an equilateral triangle of side length 34 cm
  e a regular decagon of side length 320 mm

2 Draw a diagram of each of the following figures and calculate the perimeter of each one.

  a a rectangle of dimensions 30 cm × 25 cm
  b a rhombus of side length 40 mm
  c a quadrilateral with two sides of length 1 m, one side of 45 cm and one of 65 cm
  d a kite with sides of length 1.4 m and 80 cm
  e a heptagon with four sides of length 40 cm and the remainder of the sides of length 50 cm
  f a trapezium with parallel sides of length 30 cm and 40 cm and the other two sides of length 13 cm

3 Find the perimeter of each of the following figures, giving your answers in centimetres.

  a A patchwork mat made from regular octagons and squares of side length 20 cm.

  b A badge made from the sides of a regular hexagon, of side length 14 mm, being extended to form a six-pointed star. All triangles are equilateral triangles.

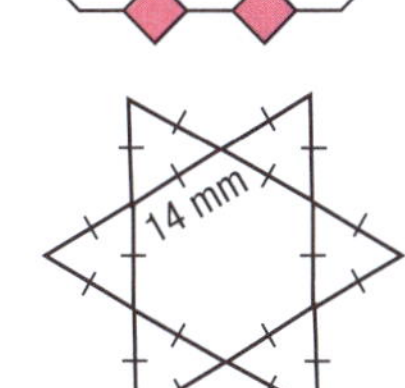

  c A table-top composed of a rectangle of dimensions 3.6 m × 2 m with half of a regular hexagon on each end.

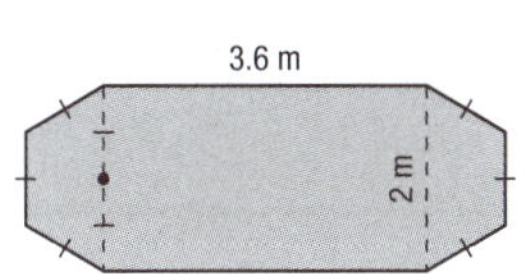

d A windmill shape cut from a regular hexagon of side length 80 cm. The centre is in the shape of a regular hexagon with side length 10 cm.

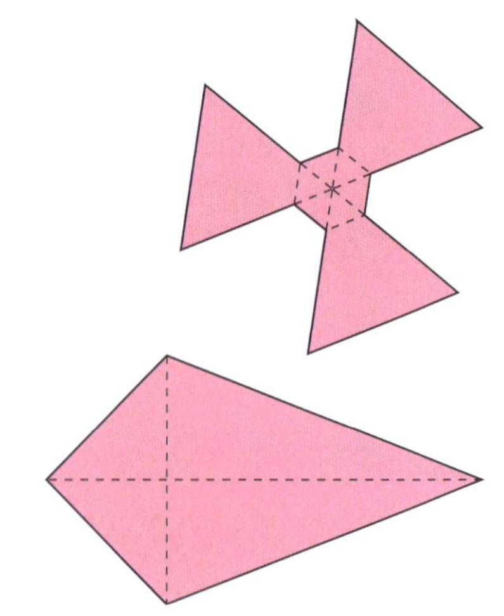

e A kite which has shorter sides that are 60 cm long and longer sides that are 2.5 times the length of the shorter sides.

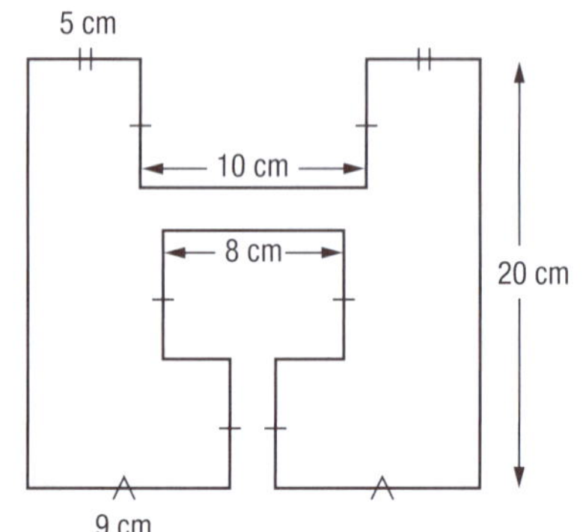

4 Find the perimeter of each of the following figures:

a

10 cm

8 cm

5 cm

b

5 cm

10 cm

8 cm

20 cm

9 cm

5 A floor plan for a modern building is shown below.

a Find the perimeter of this building, in metres:

i including the verandah

ii excluding the verandah.

b Why do you think builders require measurements to be in millimetres rather than in metres or centimetres?

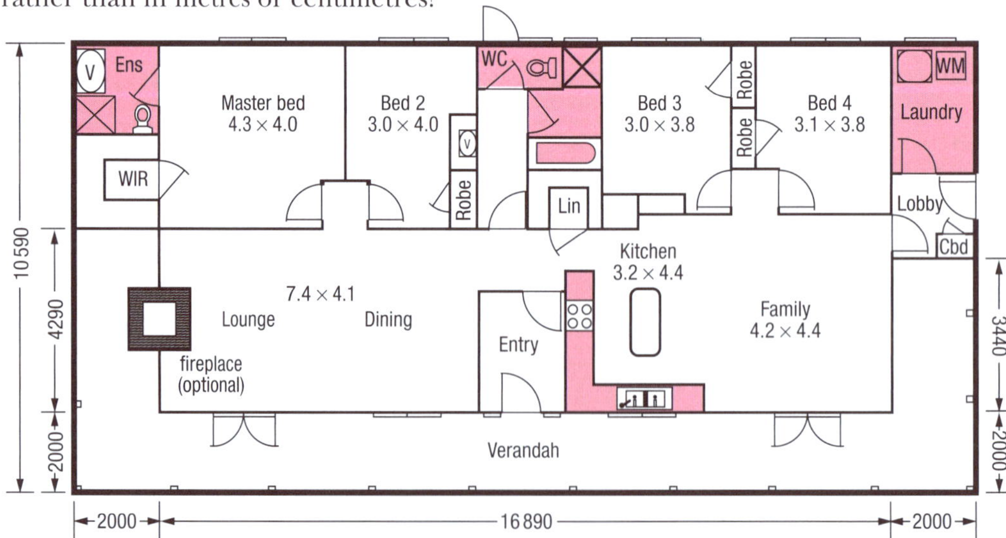

The perimeter (circumference) of a circle and of sectors of circles

Lesson 2

Some circle facts

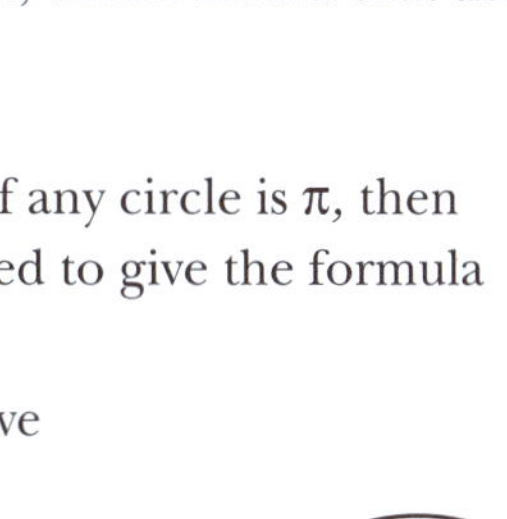

- For a circle, the points that make-up the figure are all the same distance from a single point, called the centre of the circle.
- The distance from the centre of the circle to the circle is called the **radius**, r.
- The distance from a point on the circle to another point on the circle, going through the centre, is called the **diameter**, d.
- The length of the diameter of a circle is twice the length of the radius. $d = 2r$.
- The perimeter of a circle is called the **circumference** of the circle.
- For every circle, the ratio between the circumference and the diameter is the same value, which we call **pi**.
- The symbol for pi is π.
- The value of π is 3.141 592 . . . It is a non-recurring decimal, which means that the decimal places never finish nor have a repeating pattern.
- Approximations that are often used for π are 3 or $\frac{22}{7}$.
- If the ratio between the circumference and the diameter of any circle is π, then we have the equation $\frac{C}{d} = \pi$. This equation can be transposed to give the formula for the circumference of a circle as $C = \pi d$.
- We can substitute $d = 2r$ into $C = \pi d$ to obtain the alternative formula $C = 2\pi r$.
- An **arc** of a circle is a fraction of a circle.
- Since the angle at the centre of a circle is one revolution (360°), the length of an arc that has an angle (subtends an angle) of θ° at the centre of the circle is given by:

  $$\text{Arc length} = \frac{\theta}{360} \times 2\pi r$$

- The perimeter of a sector of a circle = arc length + two radii

  $$= \frac{\theta}{360} \times 2\pi r + 2r$$

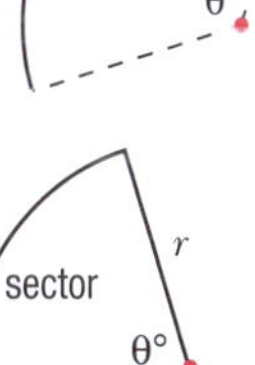

Example

Find the circumference of a circle that has a radius of 70 cm. Use $\pi = \frac{22}{7}$.

Answer

Substituting in the formula $C = 2\pi r$.

$C = 2 \times \frac{22}{7} \times 70$

$= 2 \times 22 \times 10$

$= 440$ cm

Example

Find the perimeter of the figure below. Use $\pi = \frac{22}{7}$.

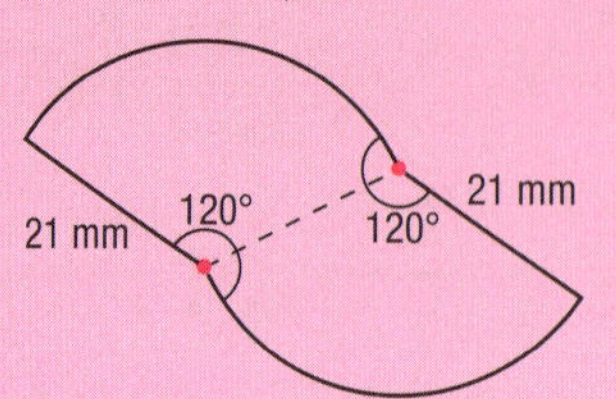

Answer

The perimeter of the figure consists of two arcs (that subtend an angle of 120° at the centre) plus two radii.

$$\text{Perimeter} = 2 \times \frac{\theta}{360} \times 2\pi r + 2r$$
$$= 2 \times \frac{120}{360} \times 2 \times \frac{22}{7} \times 21 + 2 \times 21$$
$$= 88 + 42$$
$$= 130 \text{ mm}$$

Formulas

Circumference: $C = 2\pi r$ or $C = \pi d$

Arc length: $\text{Arc length} = \frac{\theta}{360} \times 2\pi r$

The perimeter of a sector of a circle: $\text{Perimeter} = \frac{\theta}{360} \times 2\pi r + 2r$

EXERCISE

Use $\pi = \frac{22}{7}$ for these questions.

1 Find the circumference, in centimetres, of a circle with:
 a a radius of 56 cm
 b a diameter of 42 mm

2 Find the length, in millimetres, of an arc that subtends an angle of 280° at the centre of a circle of radius 45 mm.

3 Find the perimeter of each of the following sectors of circles:

a

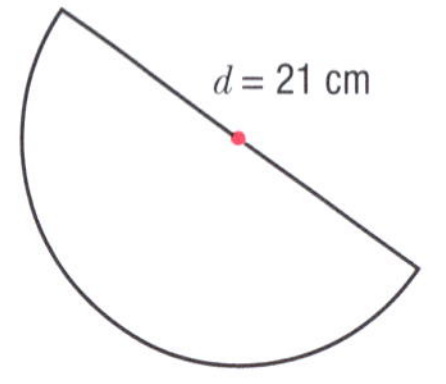

b

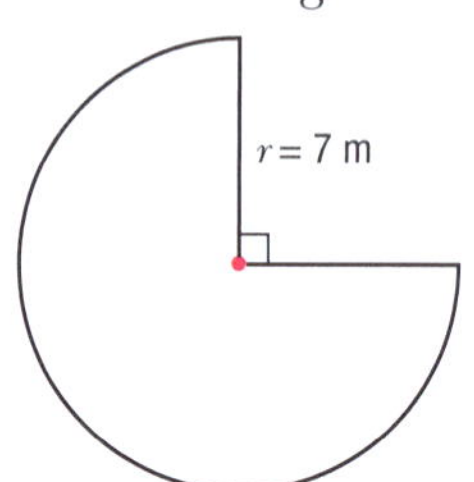

c

$r = 12$ cm

49°

4 Find the perimeter of each of the following figures:

a

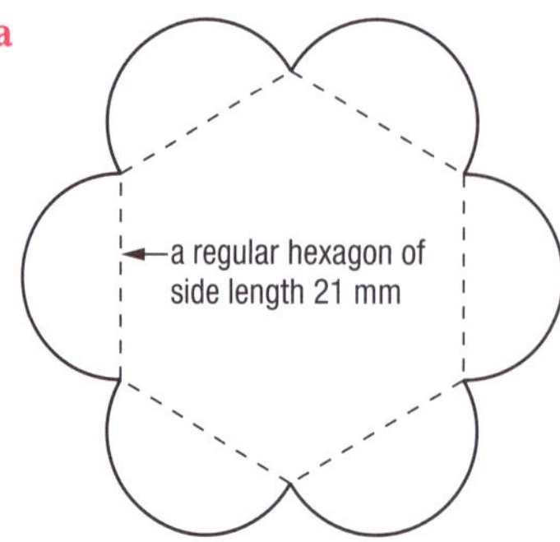

b

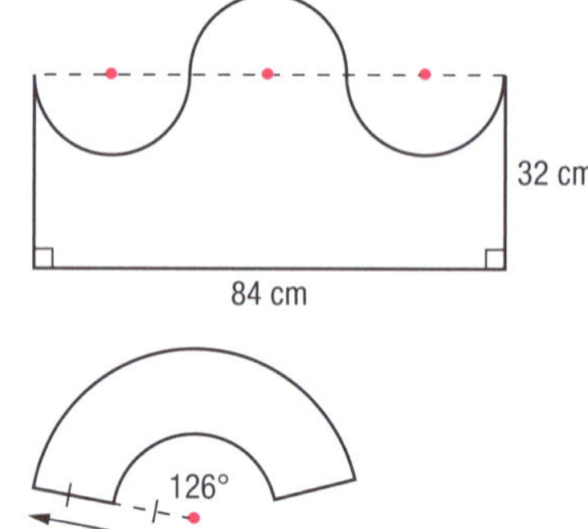

c

126°

2 m

5 Calculate the length of the pulley belt in the diagram on the right.

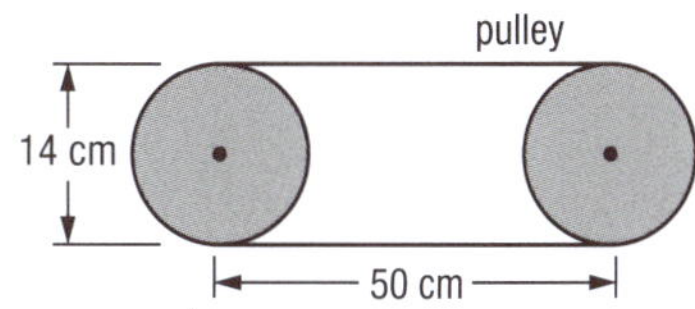

6 Stephen rode his bicycle for 1.1 km. If the wheel of his bicycle has a diameter of 70 cm, how many times did the wheel rotate on the journey?

7 A pulley belt goes around a pulley as shown in the diagram on the right. The wheels are of radius 21 cm and 14 cm and the length of the belt that is not in contact with a wheel is 67 cm on each side. Find the total length of the pulley, to the nearest centimetre.

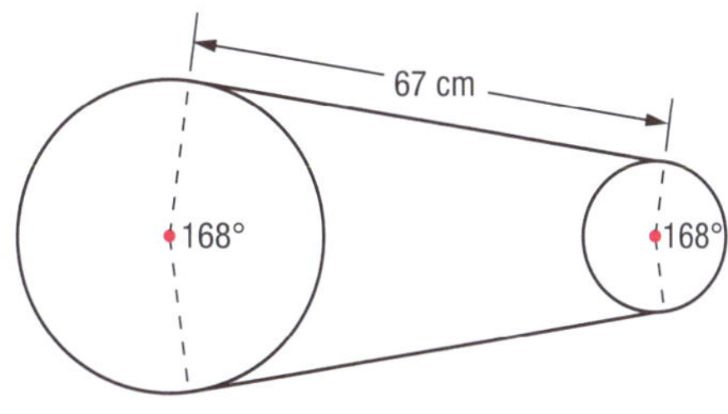

8 Bamboo poles with a diameter of 10 cm are to be transported to a building site. For the journey, the poles are strapped together in groups of three or four, as shown in the diagrams below.

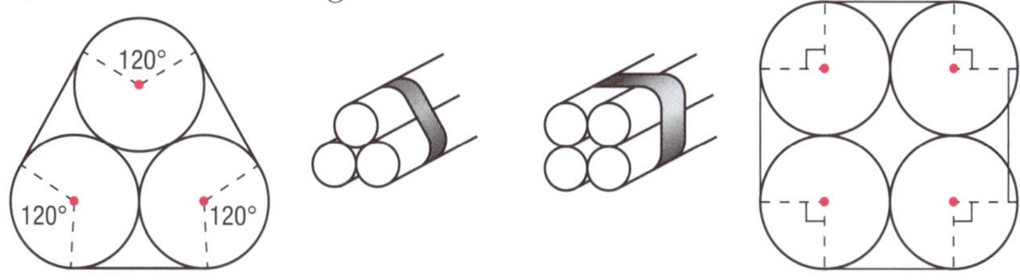

a Calculate, to the nearest centimetre, the length of a single strap that binds the poles together in groups of:

i three **ii** four.

b The bundles of poles each require three straps. If 96 poles are to be transported how would you bundle them so that you use a minimum amount of strapping?

The areas of polygons

Some area facts

- The area of a closed figure is the amount of surface covered by that closed figure.
- In the metric system, the units of area are **square metres** (**m^2**), **square centimetres** (**cm^2**), **square millimetres** (**mm^2**) and **square kilometres** (**km^2**).
- Areas must be expressed in the same unit before they can be added.
- The following conversion diagram helps us convert units of area in the metric system.

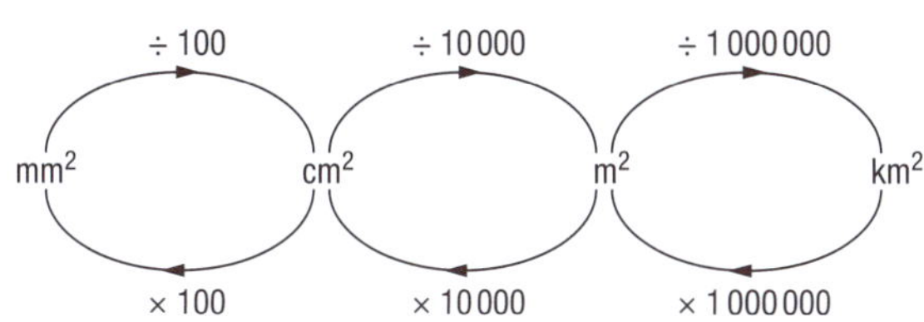

- The area of a triangle is given by the formula:

$$\text{Area} = \frac{1}{2} \times \text{base} \times \text{perpendicular height}$$

$$A = \frac{1}{2}bh$$

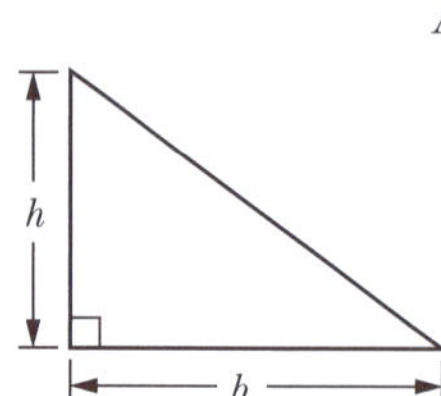

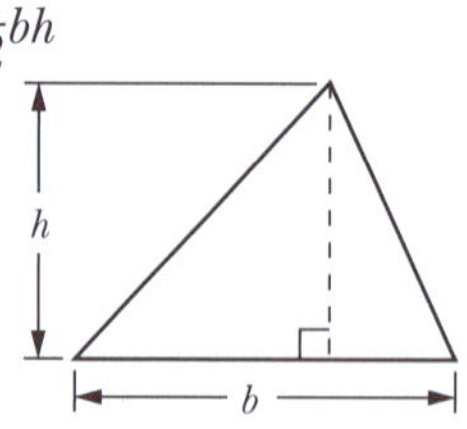

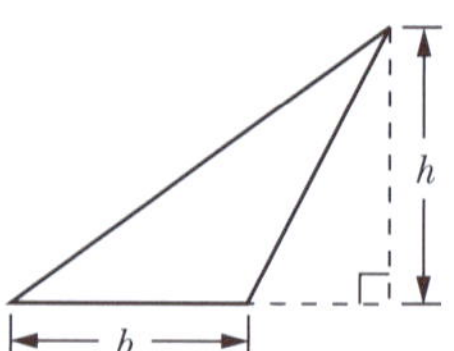

- The areas of quadrilaterals are calculated as shown:

Rectangle

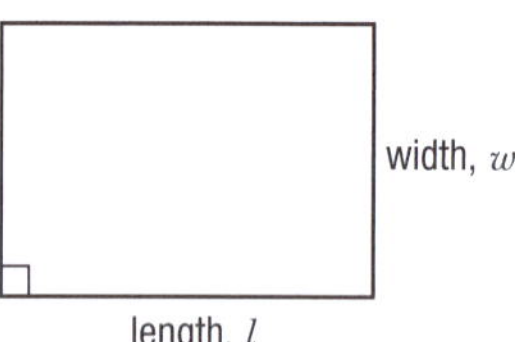

Area = length × width
$= l \times w$

Special case: Square

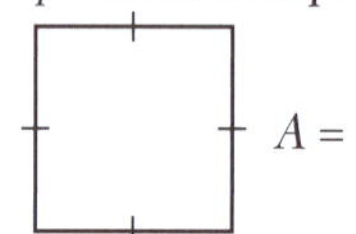

Parallelogram

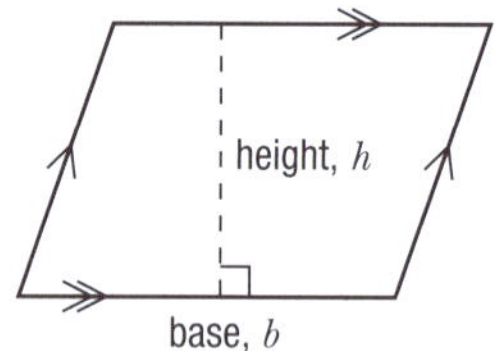

Area = base × height
$= b \times h$

Special case: Rhombus

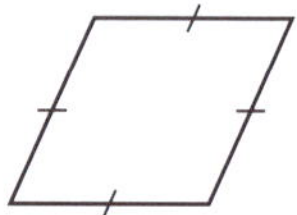

Trapezium

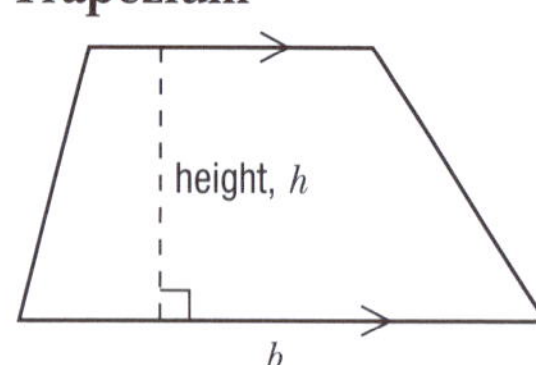

$$\text{Area} = \frac{1}{2}(\text{sum of parallel sides}) \times \text{height}$$

$$= \frac{1}{2}(a + b) \times h$$

- The area of a polygon can be found by dividing the polygon into triangles and/or quadrilaterals, finding the area of these figures and then adding their areas.

Example

Find the area of the composite figure below.

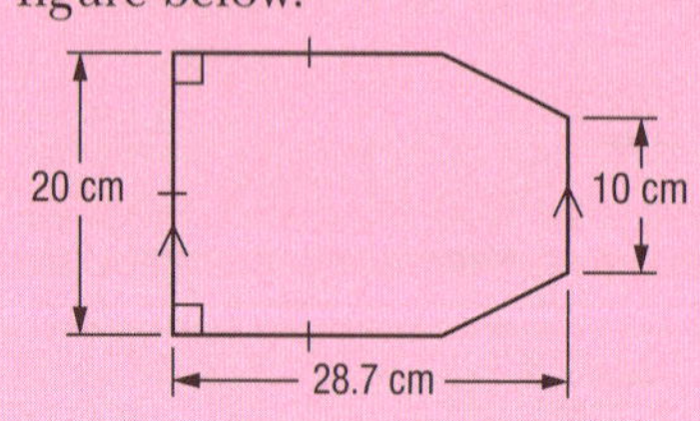

Answer

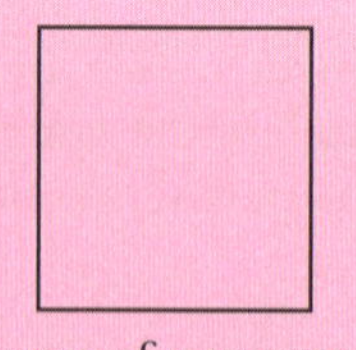
+
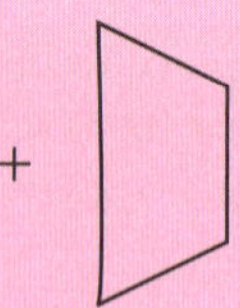

Total area = area of a square + area of a trapezium

$= 20 \times 20 + \frac{1}{2}(20 + 10) \times 8.7$

$= 400 + 130.5$

$= 530.5 \text{ cm}^2$

Example

The shaded figure below is a rhombus of side length 30 mm, with an area shaped like a right-angled triangle removed. Find the shaded area if the height of the rhombus is 26 mm and the diagonals are 30 mm and 52 mm long.

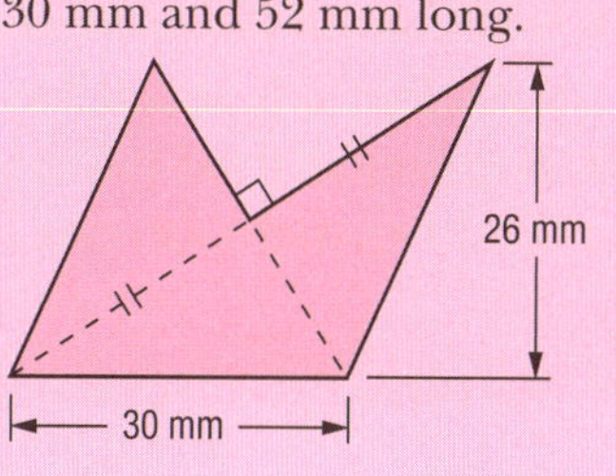

Answer

The 'base' of the triangle is: $\frac{52}{2} = 26$ mm

The 'height' of the triangle is: $\frac{30}{2} = 15$ mm

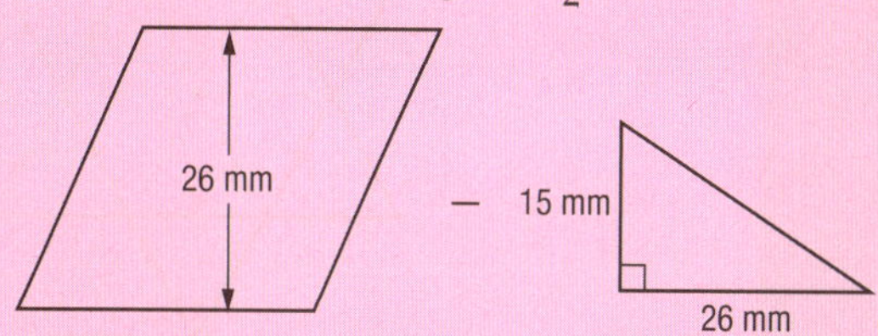

Shaded area = area of rhombus – area of triangle

$= 30 \times 26 - \frac{1}{2} \times 26 \times 15$

$= 780 - 195$

$= 585 \text{ mm}^2$

EXERCISE

1 Find the area of each of the following figures:

a

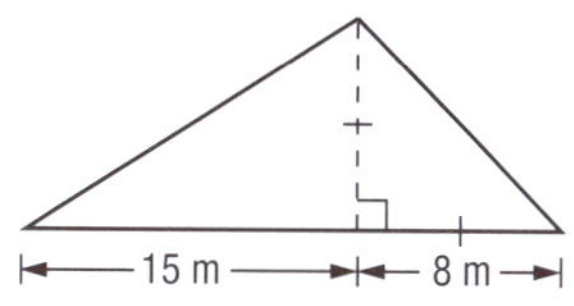

b A rhombus with diagonals of length 40 cm and 60 cm

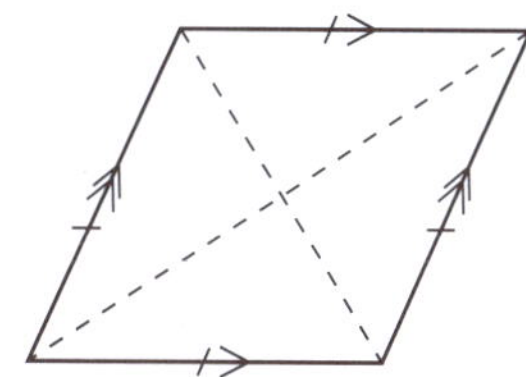

c A kite

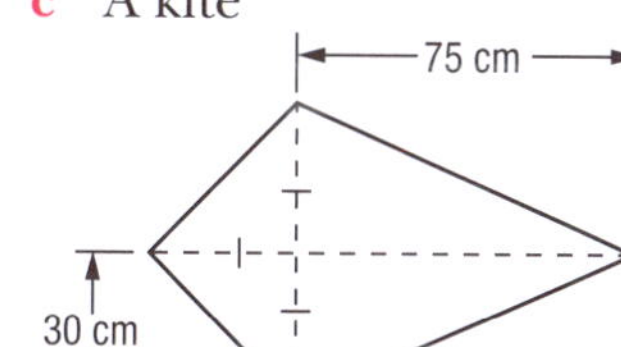

d

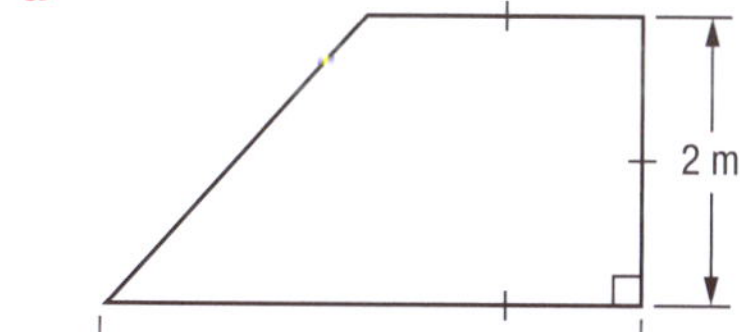

e

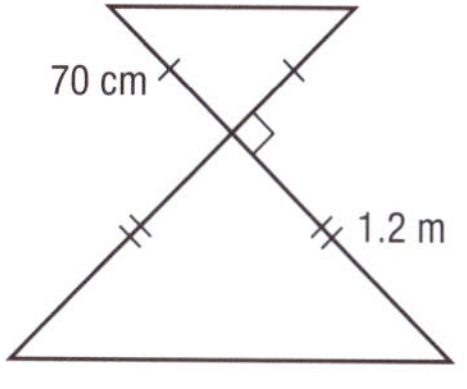

f

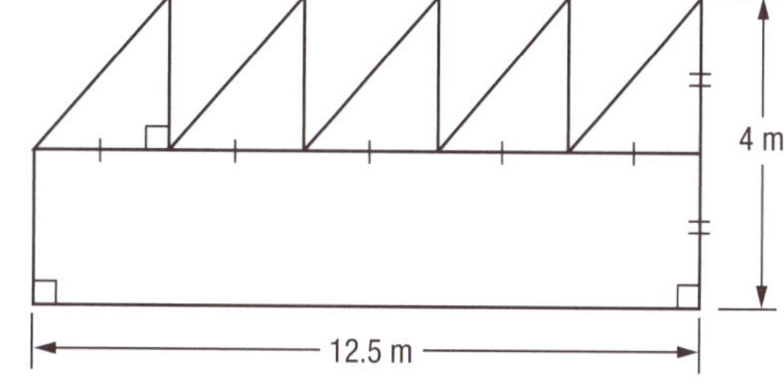

Did you know?

A **rhombus** is a parallelogram that has sides of equal length. The diagonals of a rhombus bisect each other (cut each other in half) at right angles. (90°).

Did you know?

A **kite** is a quadrilateral that has two pairs of adjacent sides (sides that are next to each other) of equal length. The diagonals intersect each other at right angles (90°).

b Carefully cut the protractor pieces into sectors of 10° and paste them on a sheet of paper, interlocking them as shown so that, finally, you have a 'parallelogram'.

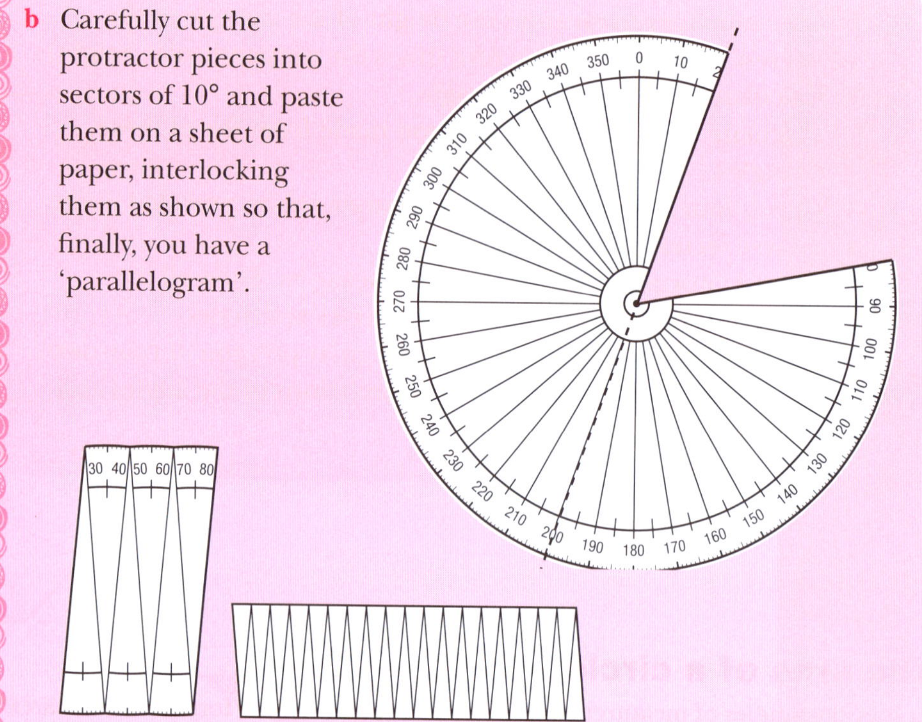

The area of the circle can now be calculated as the area of a parallelogram.
The length of this parallelogram (ignoring the fact that the edge is wavy) is half the circumference of the circle, or $\frac{1}{2}$ of $2\pi r$, which is πr.
The perpendicular height of the parallelogram is r.

Area of the circle ≈ the area of the parallelogram
= base × perpendicular height
$= \pi r \times r$
$= \pi r^2$

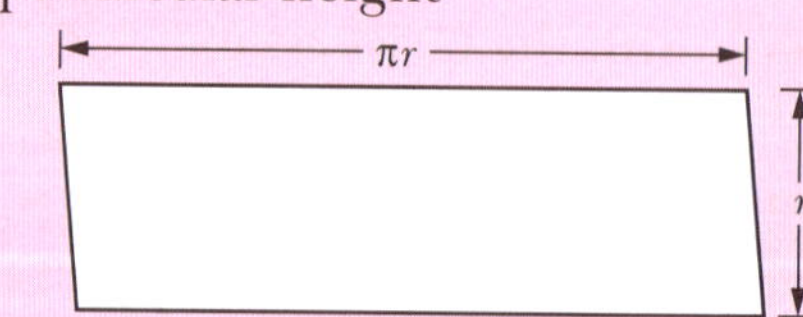

Remember
If the circle was cut into sectors of 1°, or smaller, then the parallelogram would look more like a rectangle of dimensions $2r \times r$.

Formulas:

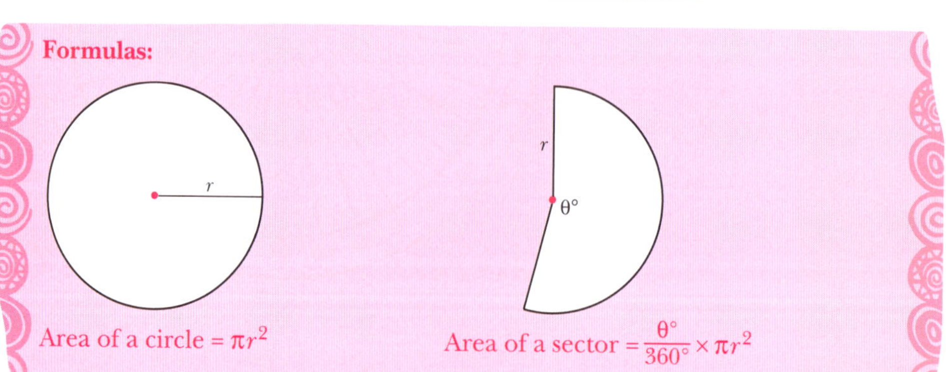

Area of a circle = πr^2

Area of a sector = $\frac{\theta°}{360°} \times \pi r^2$

Example

Find the area of the figure on the right. (Use $\pi = \frac{22}{7}$.)

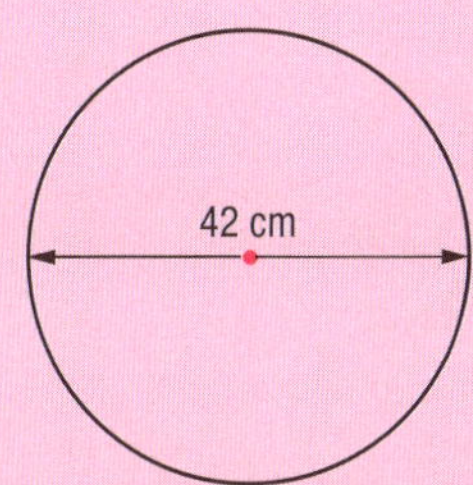

Answer

The diameter of the circle is 42 cm so the radius, r, is 21 cm.

Using the formula $A = \pi r^2$:

$$\text{Area of the circle} = \frac{22}{7} \times 21 \times 21$$
$$= 22 \times 3 \times 21$$
$$= 1386 \text{ cm}^2$$

Example

Find the area of the figure below. (Use $\pi = \frac{22}{7}$.)

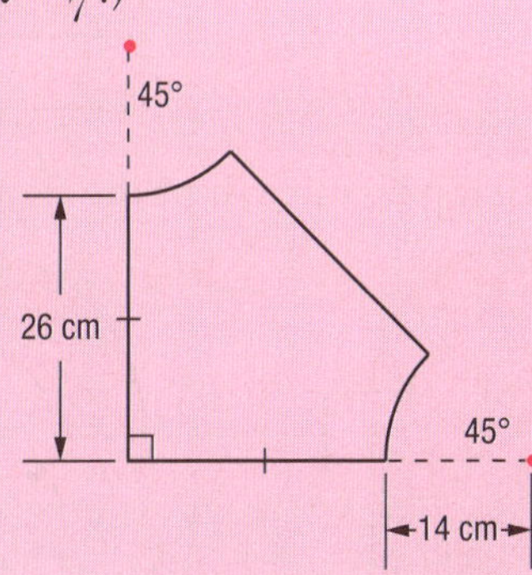

Answer

The area required = area of isosceles triangle – two arcs of a circle

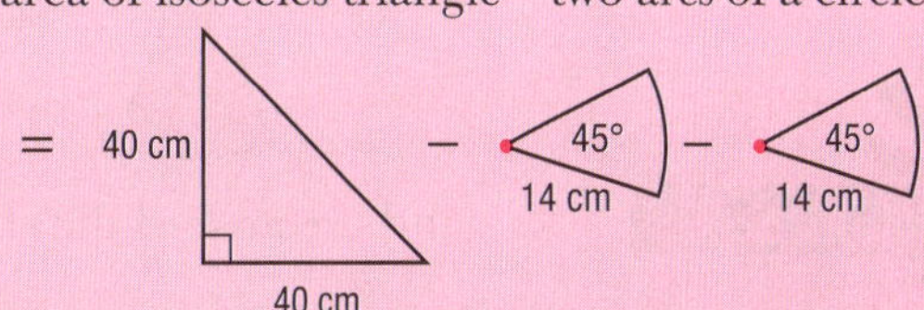

$$= \frac{1}{2} \times 40 \times 40 - 2 \times \frac{45}{360} \times \frac{22}{7} \times 14 \times 14$$
$$= 800 - 154$$
$$= 646 \text{ cm}^2$$

Help box

The calculator value of π is 3.141 592 …

A good approximation of π is $\frac{22}{7}$.

$\frac{22}{7} = 3.142\,85\ldots$ correct to two decimal places.

Using 3 as an approximation for π will sometimes give an answer that is significantly less than the exact answer.

EXERCISE

1 Calculate the areas of the circles and sectors below. Use $\frac{22}{7}$ for π.

a

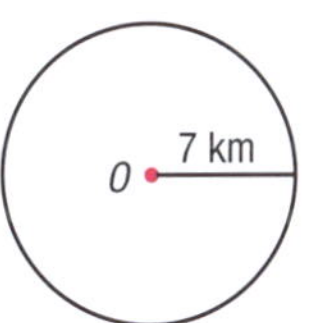

b

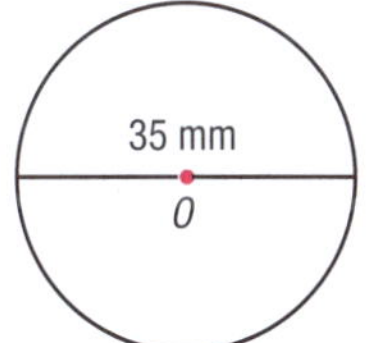

c

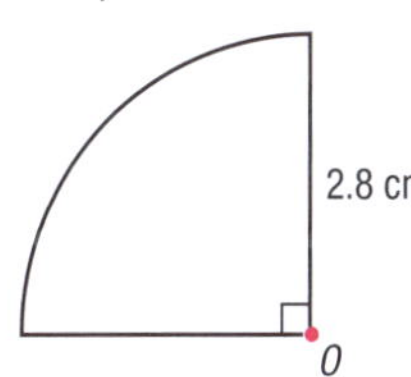

d

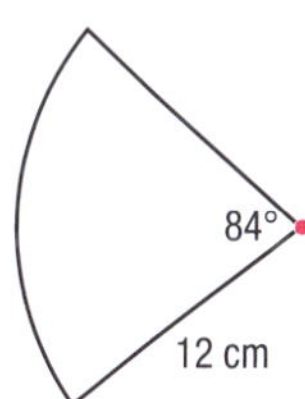

2 Find the area of this annulus. Use $\frac{22}{7}$ for π.

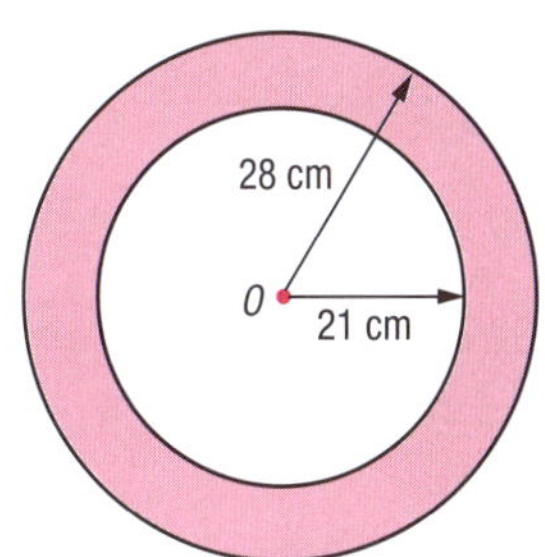

Did you know?

An **annulus** is formed by two concentric circles (that is, two circles that have a common centre).

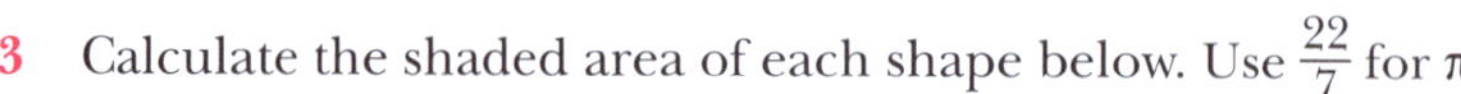

3 Calculate the shaded area of each shape below. Use $\frac{22}{7}$ for π.

a

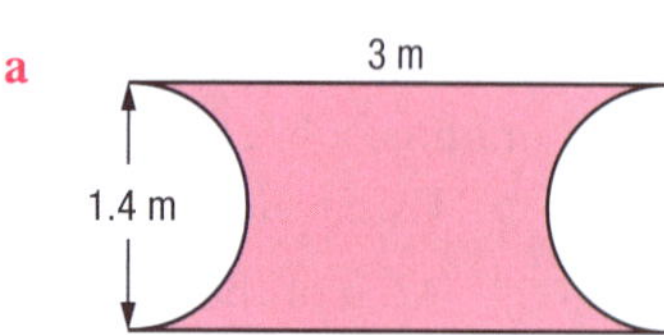

b

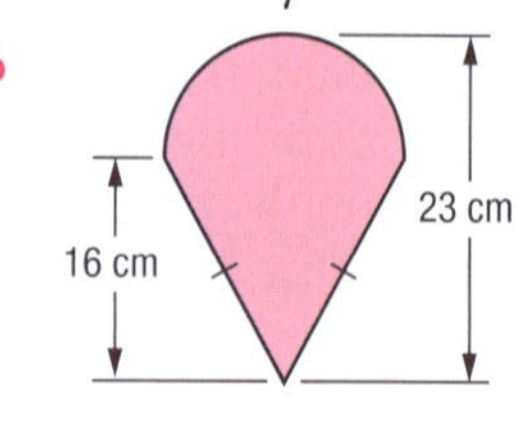

c

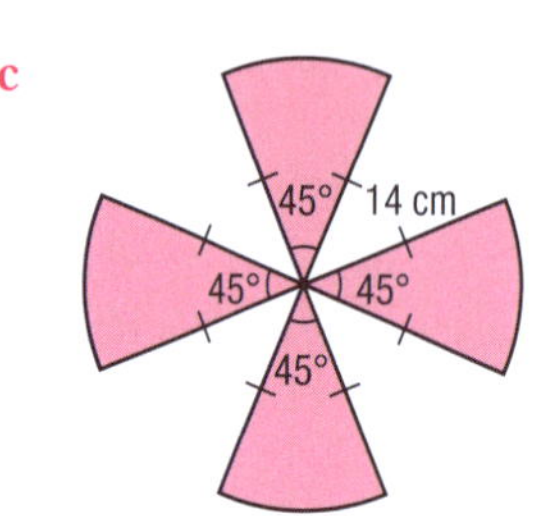

d

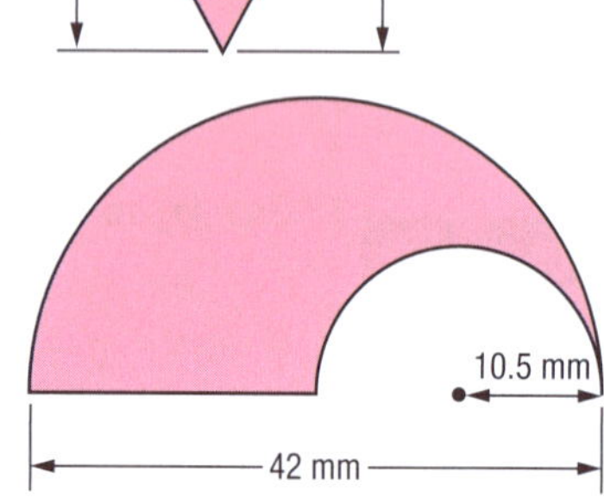

Help box

If you do not have a calculator, use 3 as an approximation for π in Questions 4, 5 and 6. (The answer in the back of the book for these questions were calculated using the calculator value for π, but answers using 3 as an approximation are also given in brackets.)

4 Find the area of the shaded part of each of the figures below.

a

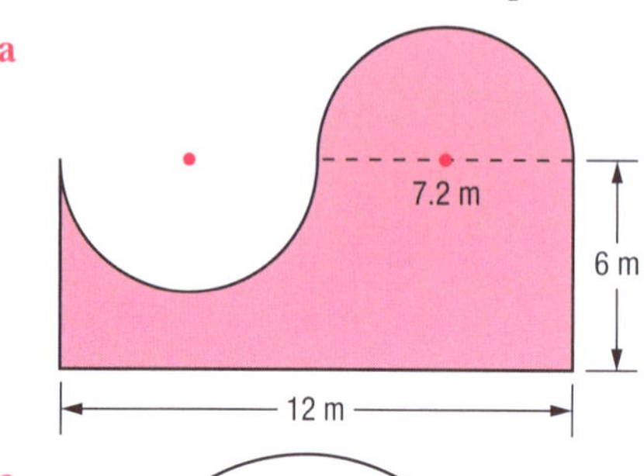

b

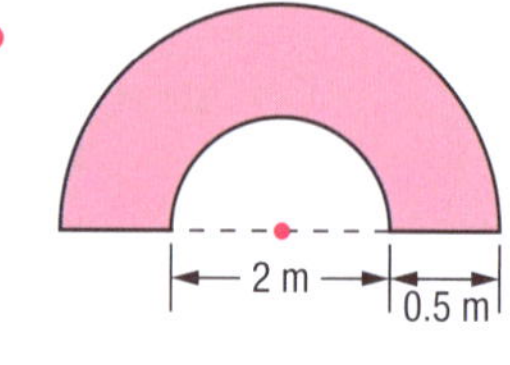

c

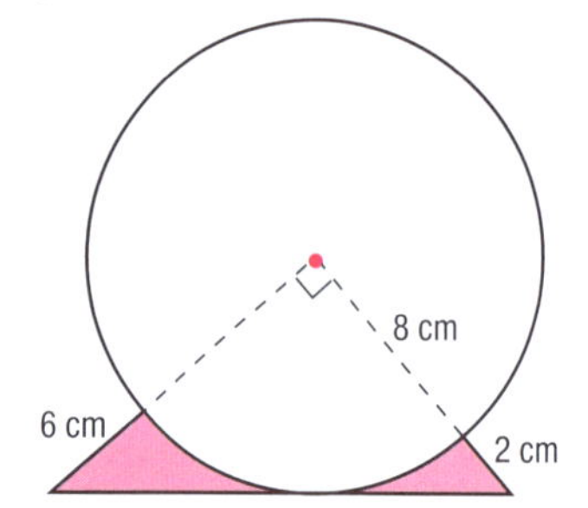

d

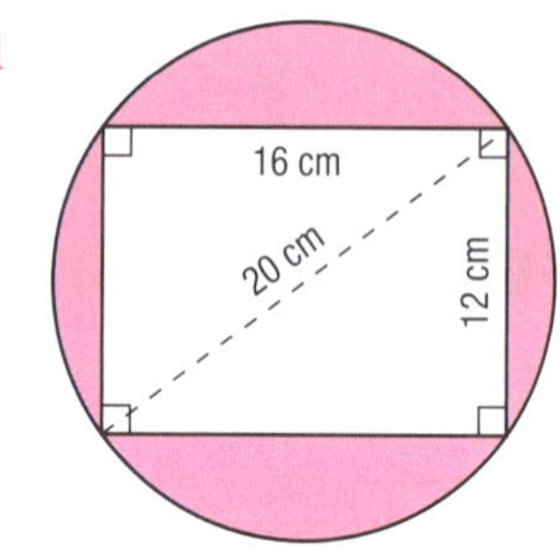

5 Athletics tracks are a standard length of 400 m. Tracks now have a special synthetic surface, but the centre of the arena is usually grassed.

a Each straight section of the track is 84.39 m long and the semi-circular ends each have a radius of 36.50 m. Calculate the area of grass which must be mowed in the centre of the arena. (Use $\pi = 3$.)

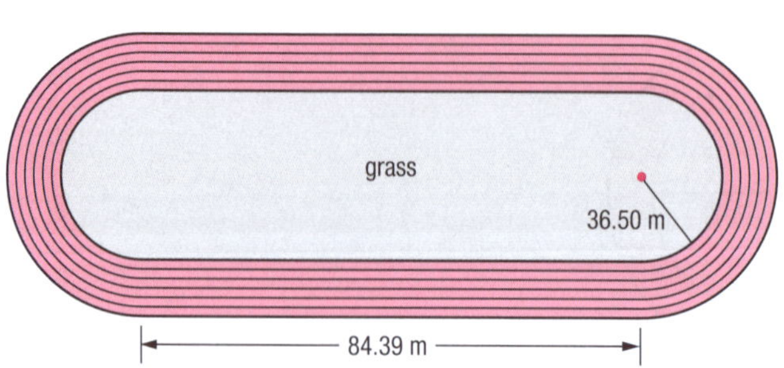

b Each lane has a width of 1.22 m, as shown on the right. Calculate the area of the inside lane.

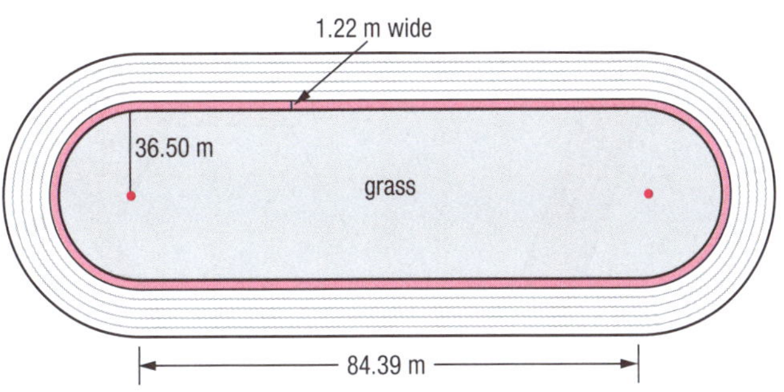

6 **a** Find the shaded area in Figure 1 below.

b Use your answer from part **a** to help you find the shaded area in Figure 2.

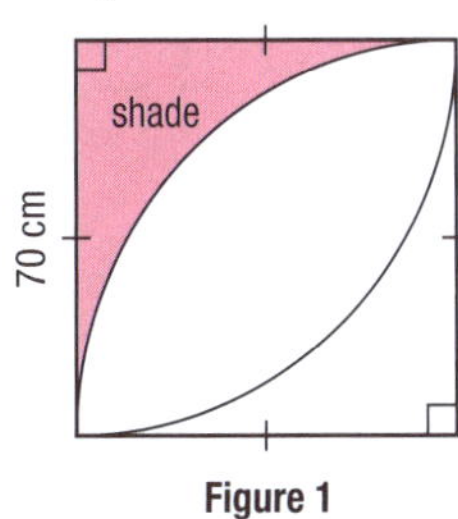

Figure 1

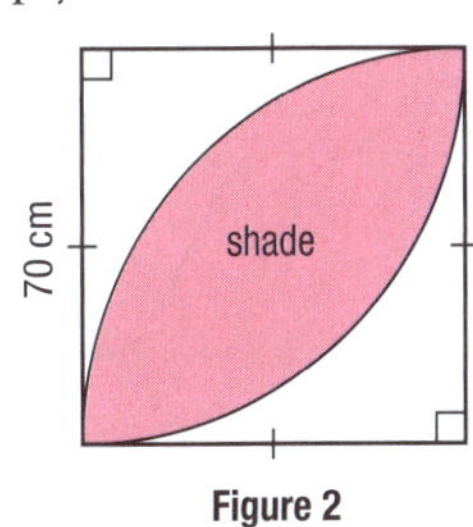

Figure 2

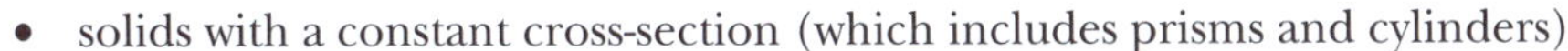

Solids and the total surface area of a prism

There are three groups of solids (three-dimensional objects) that occur frequently in our environment and with which we should be familiar:

- solids with a constant cross-section (which includes prisms and cylinders)
- tapered solids (which includes pyramids and cones)
- spheres.

Solids with a constant cross-section

Some examples of solids with a constant cross-section:

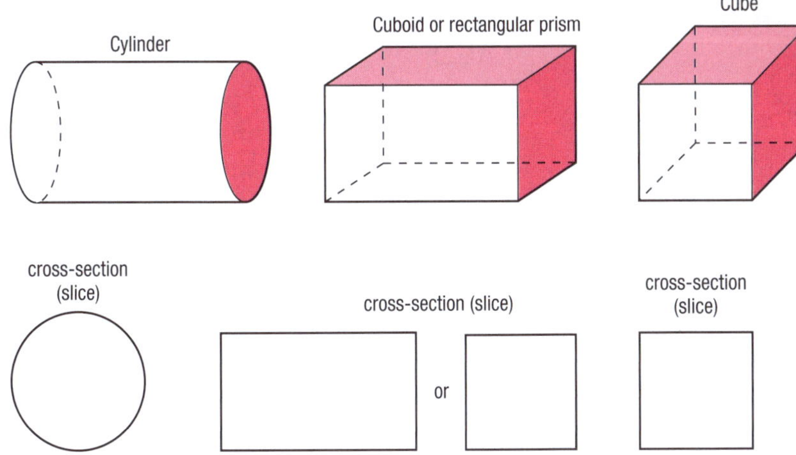

Did you know?

A **prism** is a solid with a constant cross-section that is a polygon, and sides that are rectangular. A prism is often named according to the shape of its cross-section.

Triangular prism

Hexagonal prism

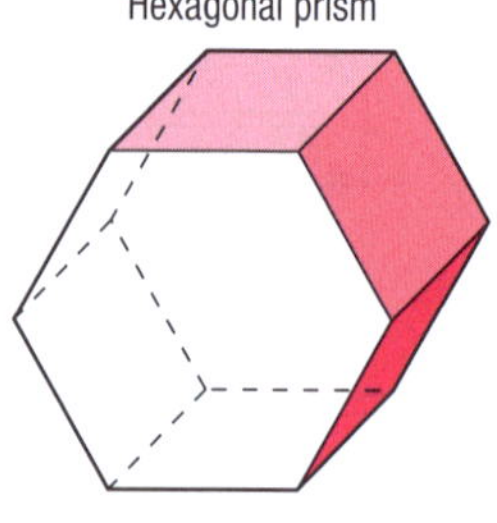

Irregular solid

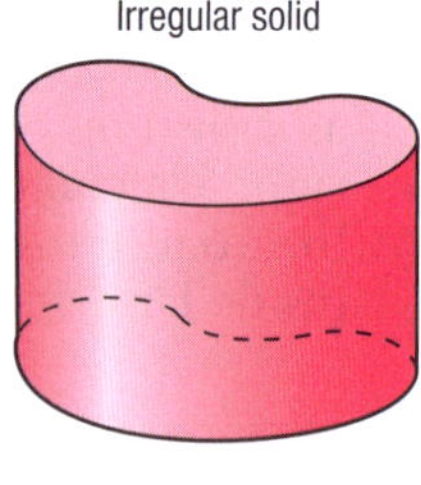

cross-section (slice)

cross-section (slice)

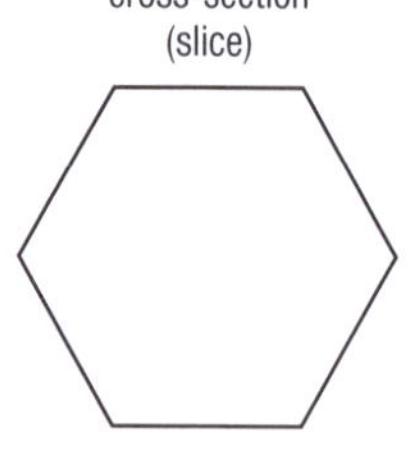

cross-section (slice)

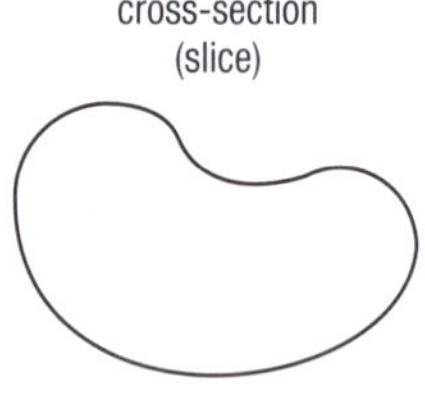

Tapered solids

Some examples of tapered solids:

Cone

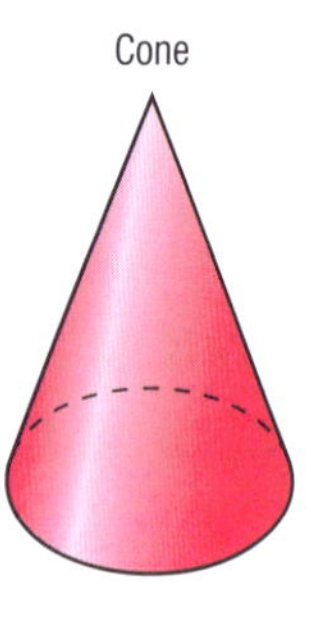

Rectangular-based pyramid

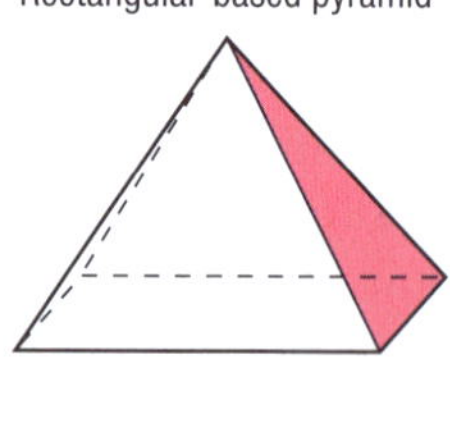

Square-based pyramid

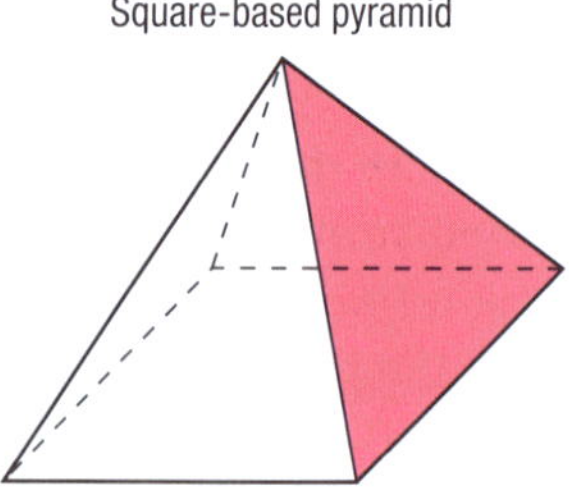

some cross-sections

some cross-sections

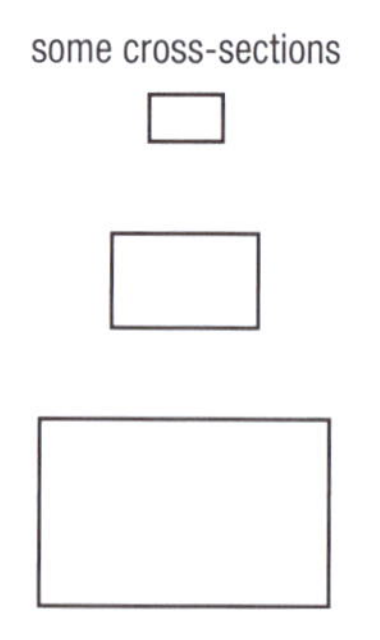

some cross-sections

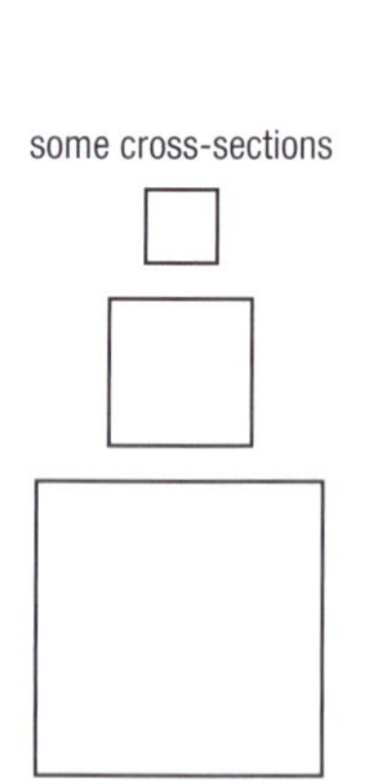

Triangular-based pyramid

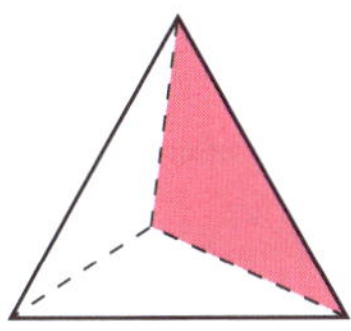

Pentagonal-based pyramid

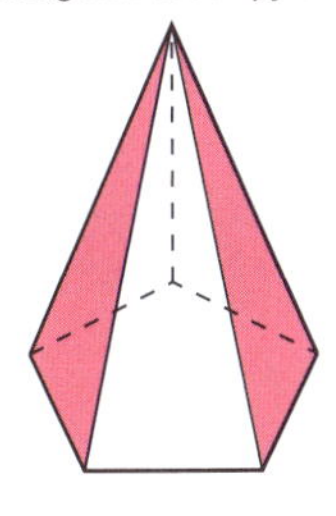

Hexagonal-based pyramid

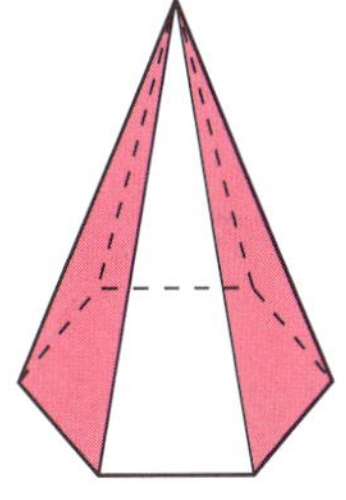

some cross-sections

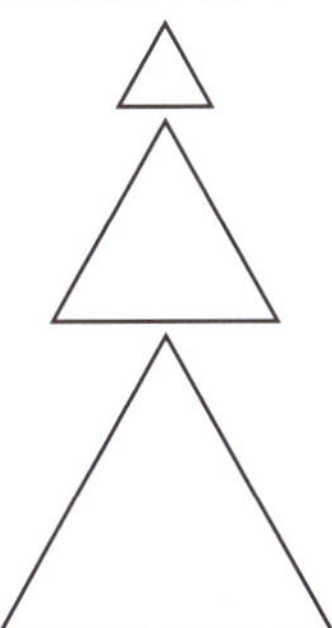

some cross-sections

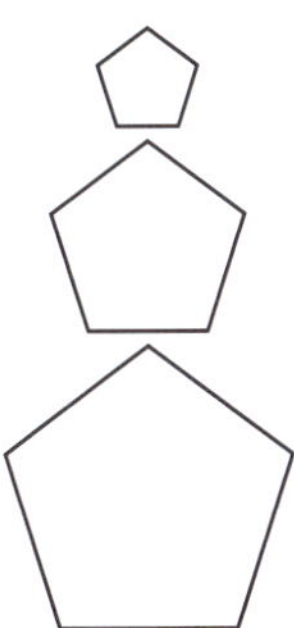

some cross-sections

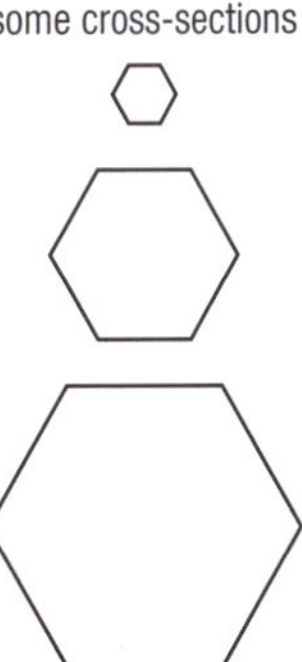

Did you know?

A **tapered solid** that has its apex directly above the centre of its base is known as a **right pyramid** or a **right cone**.

If you look carefully at the diagrams above and on the facing page, you will notice:

- a pyramid has triangular sides which come to a point (apex) and a base which is a polygon.
- a cone has a curved surface which comes to a point (apex) and a base which is a circle.

Spheres

- Every point on the surface of a sphere is the same distance from the centre of the sphere.
- The cross-sections of a sphere are all circles.
- Cross-sections of a sphere that pass through the centre of the sphere are all identical circles.
- Half of a sphere is called a **hemisphere**.

Sphere

cross-section 1

cross-section 2

Hemisphere

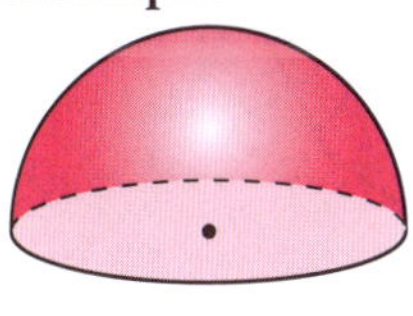

cross-section 1

cross-section 2

Faces (surfaces) of a solid

The number of faces (surfaces) of a solid varies. For example:

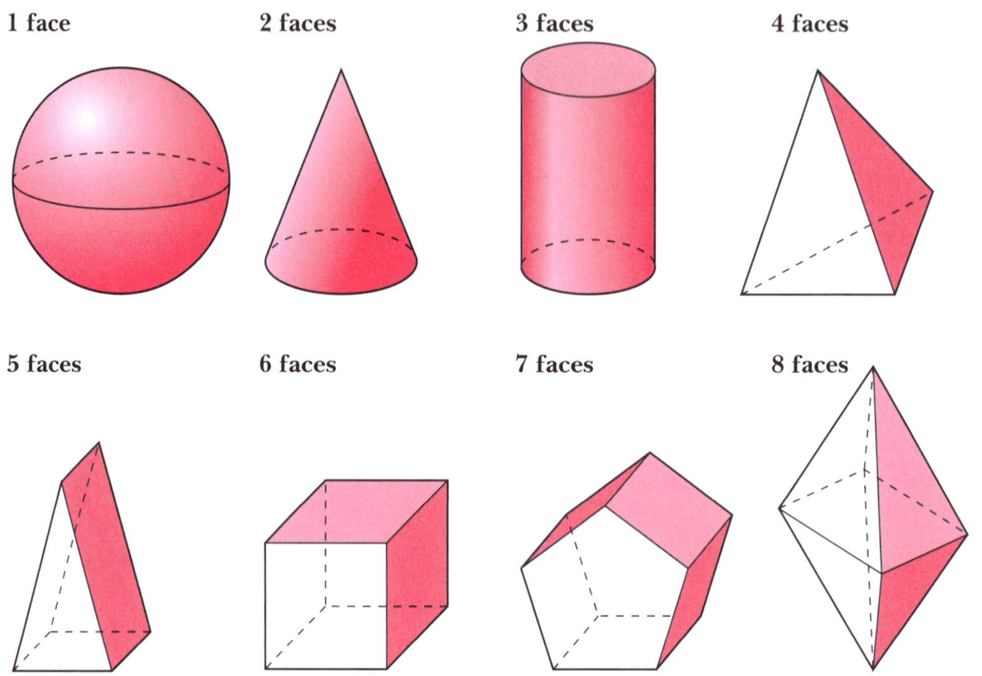

The total surface area of a prism

The **total surface area** of any solid is the sum of the areas of all the faces of the solid. It is often helpful to draw the **net** of a solid before calculating the total surface area. A net is formed by unfolding a solid to form a two-dimensional figure. A pentagonal prism and its net are shown below.

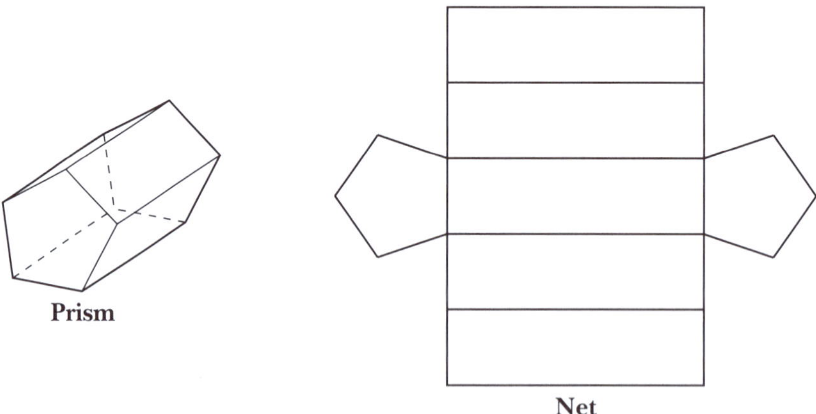

Example

Identify the number of faces of each of these solids:

a

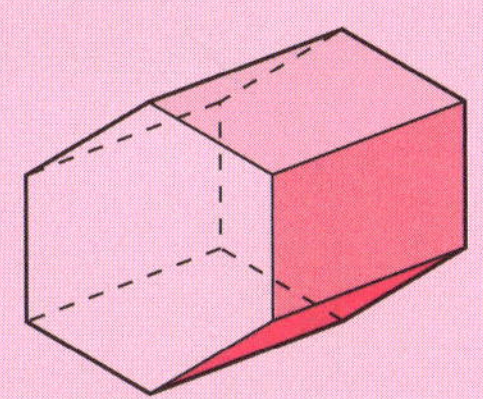

b

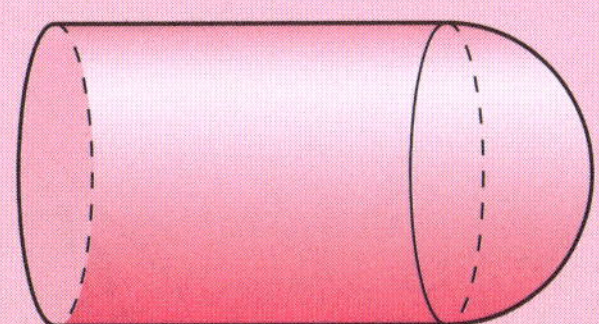

Answer

In order to identify the number of faces, we can imagine running our hands over each surface. (We **cannot** pass our hands inside the solid.) We count the surfaces as we go.

a 2 hexagonal ends
+ 6 rectangular faces
= 8 faces

b 2 curved surfaces
+ 1 flat face
= 3 faces

Example

Calculate the total surface area of the cuboid below.

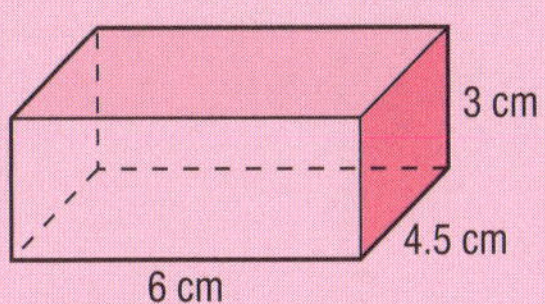

Answer

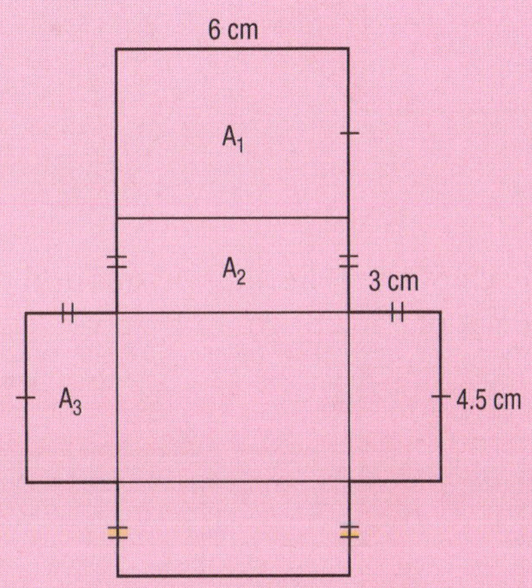

$A_1 = 6 \times 4.5 = 27$ cm^2
$A_2 = 6 \times 3 = 18$ cm^2
$A_3 = 4.5 \times 3 = 13.5$ cm^2

Total surface area $= 2 \times A_1 + 2 \times A_2 + 2 \times A_3$
$= 2 \times 27 + 2 \times 18 + 2 \times 13.5$
$= 117$ cm^2

So, the total surface area of the cuboid is 117 cm^2.

EXERCISE

1 i Name each of these solids.
ii Identify the total number of faces for each solid.

a
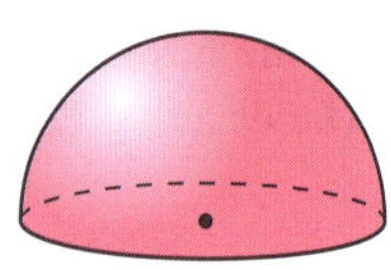
b
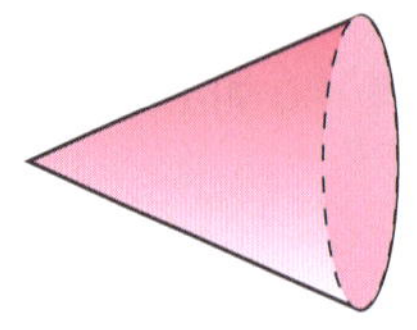
c
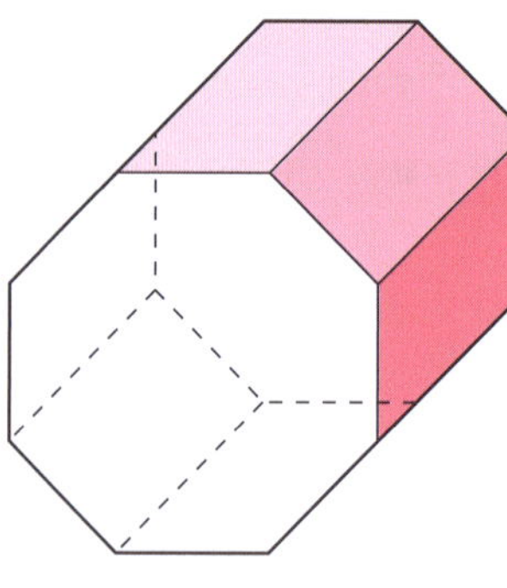
d
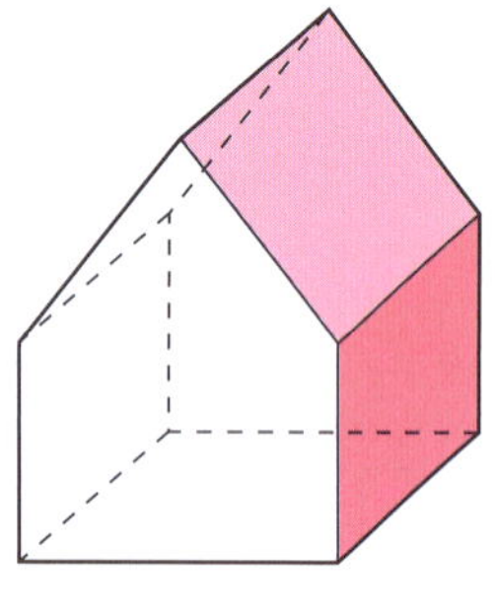
e
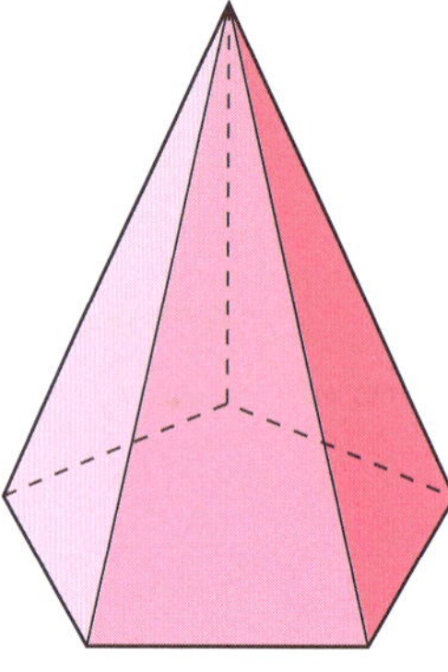
f
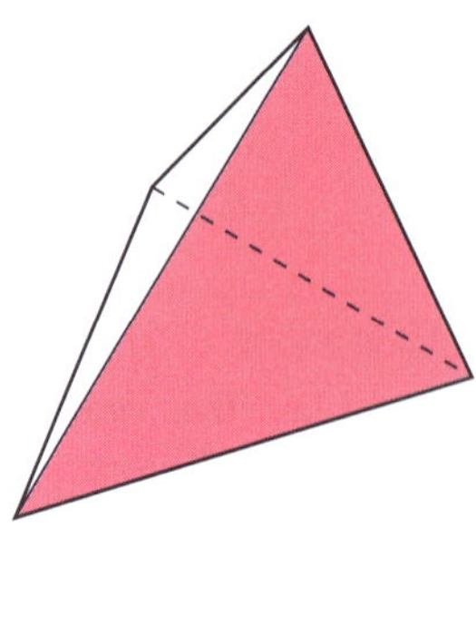

2 i Name each of the following prisms and draw its net.
ii Use the net of each prism to help you calculate its total surface area.

a
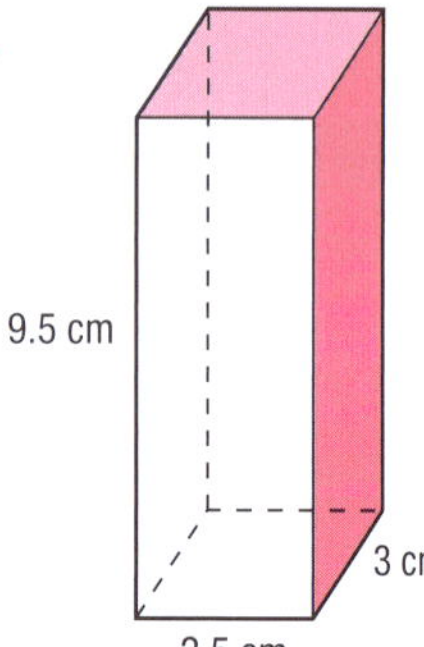

b
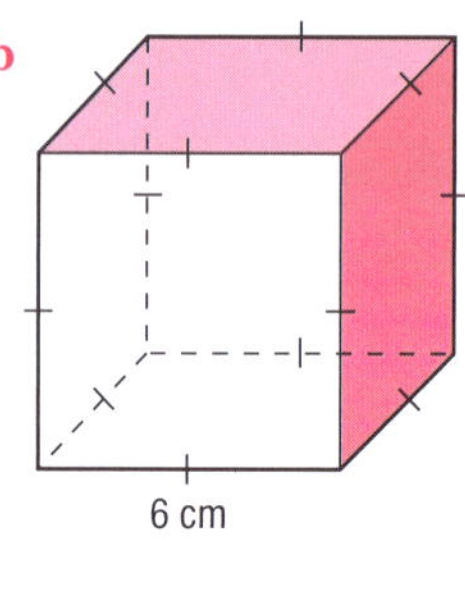

c
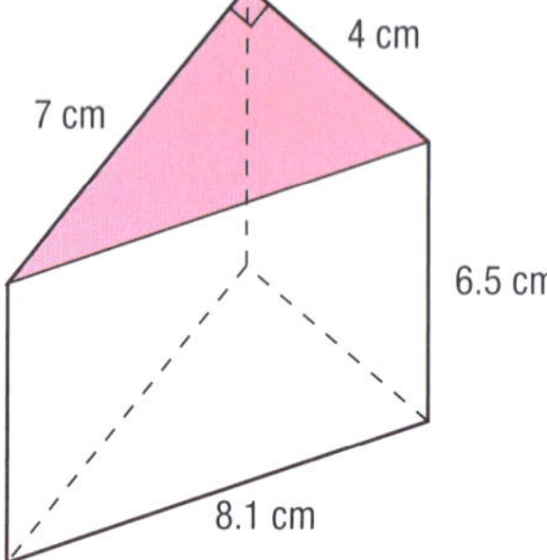

3 Calculate the total surface area of each of these prisms:

a
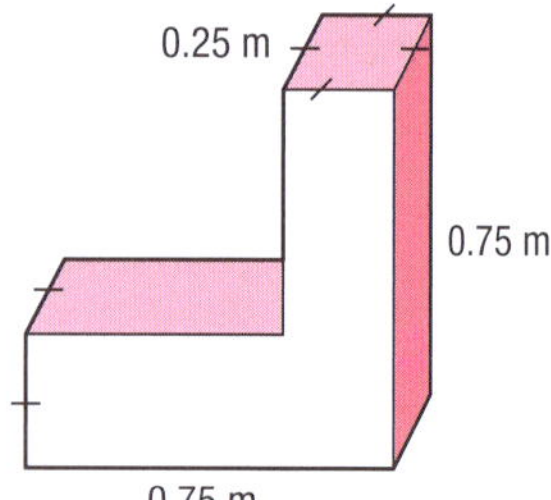

b
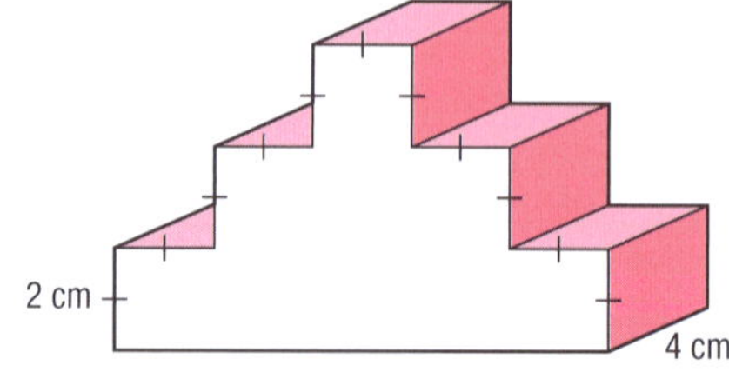

Total surface areas of cylinders

A formula for the total surface area (TSA) of a cylinder

A cylinder is a solid with a constant cross-section. We will develop and use a formula for finding its total surface area.

A cylinder has three faces, two of which are circular and one of which is curved.

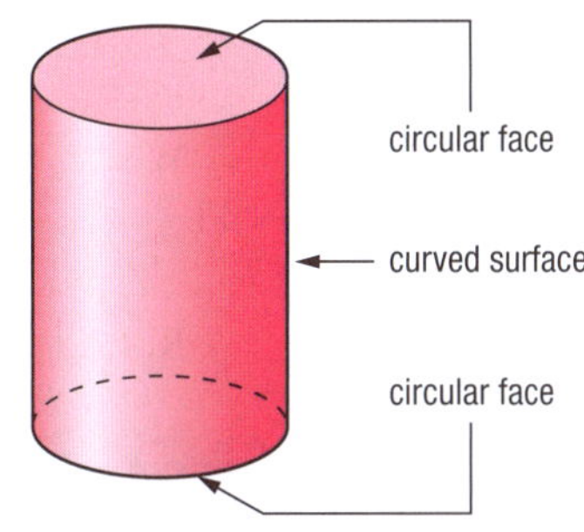

The total surface area of a cylinder is the sum of the areas of its faces.

If the cylinder's height is h units, and the radius of the circular face (base) is r units, then the distance around the circular face is $2\pi r$ units.

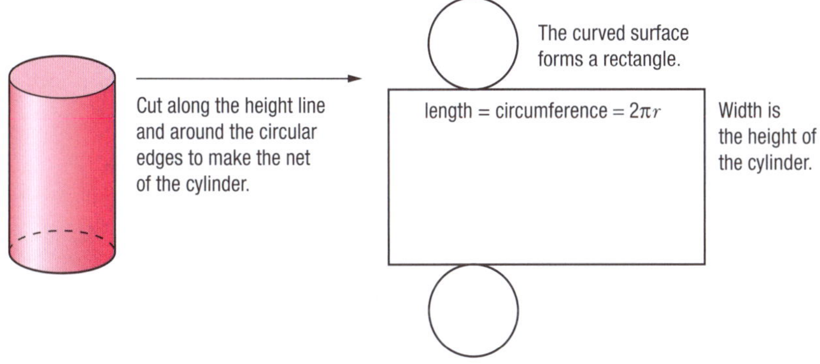

Total surface area of cylinder = 2 × area of a circle + area of a rectangle

$= 2 \times \pi r^2 + lw$ Where $l = 2\pi r$ and $w = h$.

$= 2 \times \pi r^2 + 2\pi rh$

$= 2\pi r^2 + 2\pi rh$

area of two circular faces ($2\pi r^2$); curved area surface ($2\pi rh$)

The *total* surface area of a **closed cylinder** is: $A = 2\pi r^2 + 2\pi rh$

The *outer* surface area of a **hollow open cylinder** is: $A = 2\pi rh$

Example

If $r = 5$ cm and $h = 11$ cm, calculate the total surface area of:

a the tomato soup can shown

b the paper label on the can, assuming that it does not overlap at the ends.

Answer

a Total surface area $= 2\pi r^2 + 2\pi rh$ Where $r = 5$, and $h = 11$.

$= 2 \times \pi \times 5^2 + 2 \times \pi \times 5 \times 11$

$= 157.08 + 345.58$

$= 502.66 \text{ cm}^2$ Rounded to two decimal places.

The total surface area is 502.66 cm^2.

b Curved surface area $= 2\pi rh$ Where $r = 5$, and $h = 11$.

$= 2 \times \pi \times 5 \times 11$

$= 345.58 \text{ cm}^2$ Rounded to two decimal places.

Remember

If you do not have a calculator, use 3 as an approximation for π. The answers for this Exercise, in the back of the book, have been calculated using the calculator value for π, but answers using 3 as an approximation for π are also given in brackets.

EXERCISE

1 Calculate the total surface area for each of these solids.

a

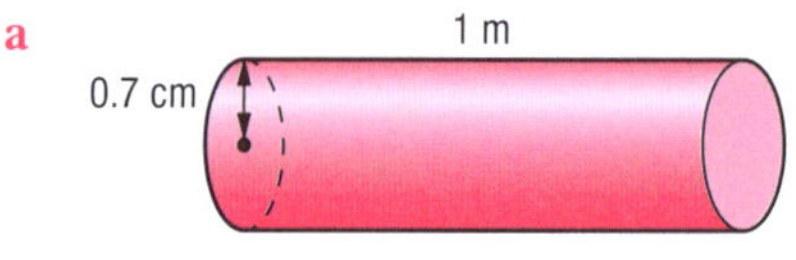

b

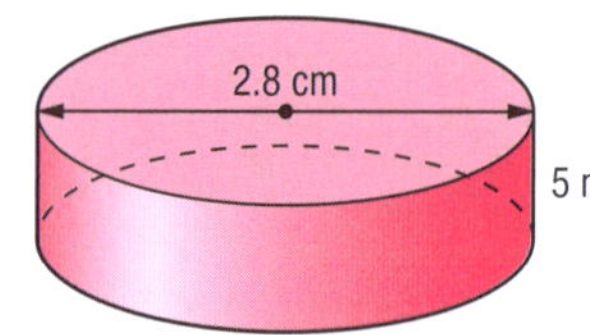

2 Calculate the total surface area and the curved surface area for each of the following:

a **b**

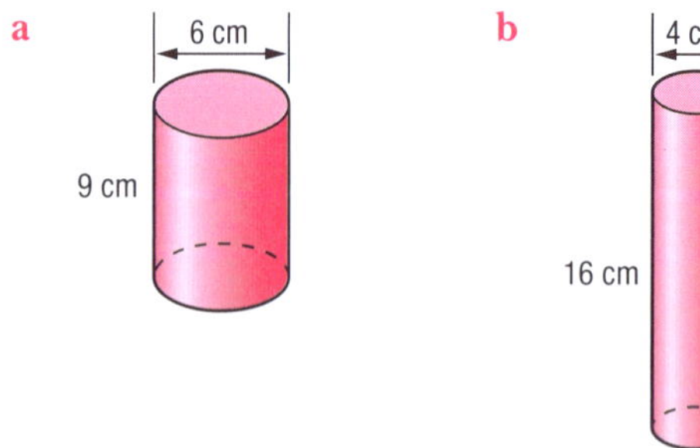

3 Which of the two cylinders in Question **2** has:

i the greatest curved surface area?

ii the greatest total surface area?

4 Calculate the total surface area of each of the following solids:

a

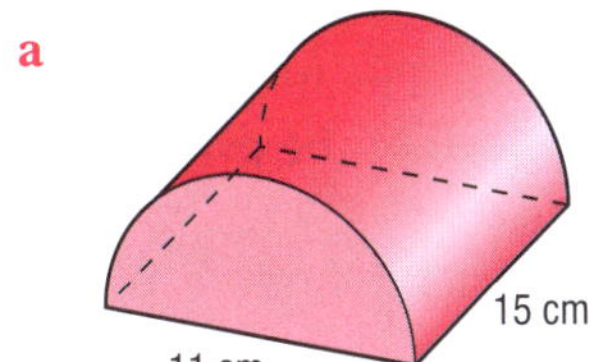

b

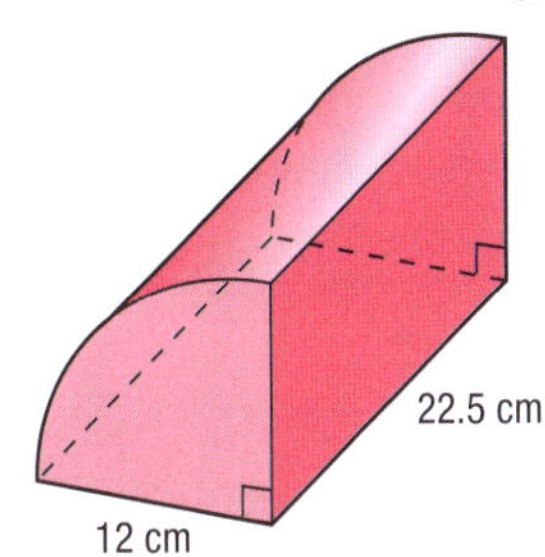

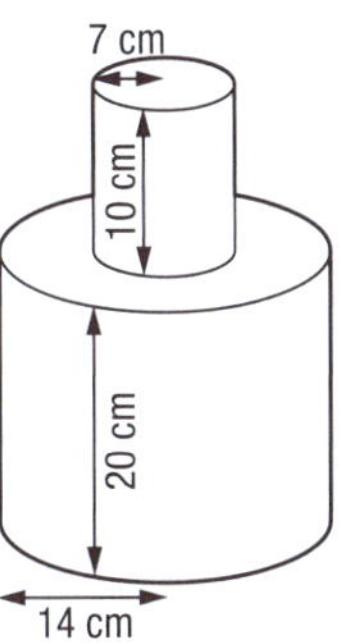

5 A solid is constructed of a closed cylinder of radius 7 cm and height 10 cm on top of another closed cylinder of radius 14 cm and height 20 cm.

a How many faces does this solid have?

b Sketch each of the faces of the solid shown, labelling them with their dimensions.

c Calculate the total surface area of the solid.

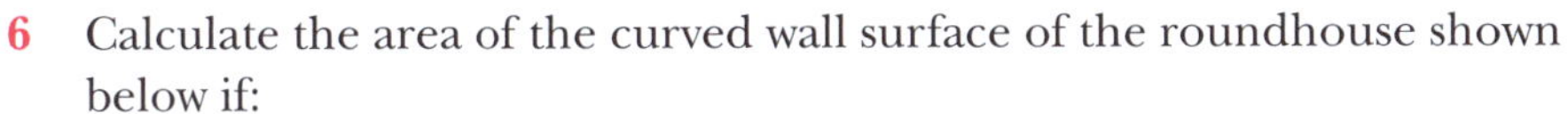

6 Calculate the area of the curved wall surface of the roundhouse shown below if:

a the inside diameter of the house is 3.5 m and the height of the walls is 1.8 m

b the inside **area** is 12 m^2 and the height of the walls is 1.8 m.

The total surface areas of pyramids, cones and spheres

Lesson 7

The total surface area of a pyramid

A square pyramid has five faces: four triangles and a square base. To find the total surface area of a pyramid, we need to first draw the net of the pyramid and then find the area of each of the faces.

Example

A right square pyramid has a square base of side length 10 cm. The height of each of the triangular faces is 12 cm. Find the total surface area of the pyramid.

Answer

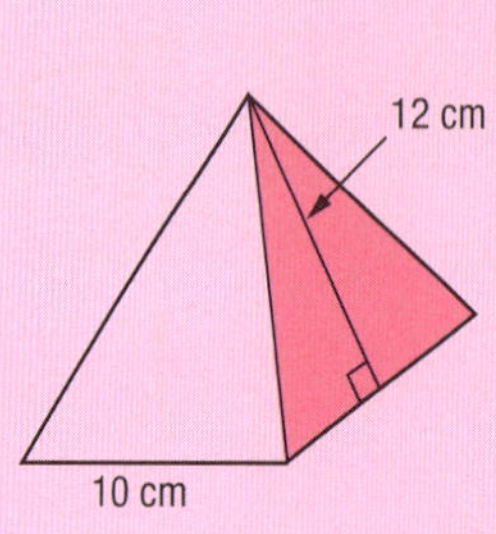

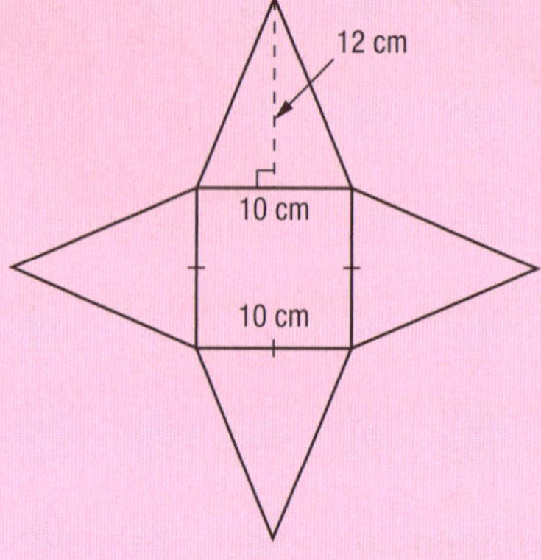

Total surface area of pyramid = area of square base + (4 × area of triangular side)

$= 10 \times 10 + 4 \times \frac{1}{2} \times 10 \times 12$

$= 100 + 240$

$= 340 \text{ cm}^2$

The total surface area of a cone

A solid cone has two faces: one circular face and a curved face which, when flattened out, is a sector of a circle.

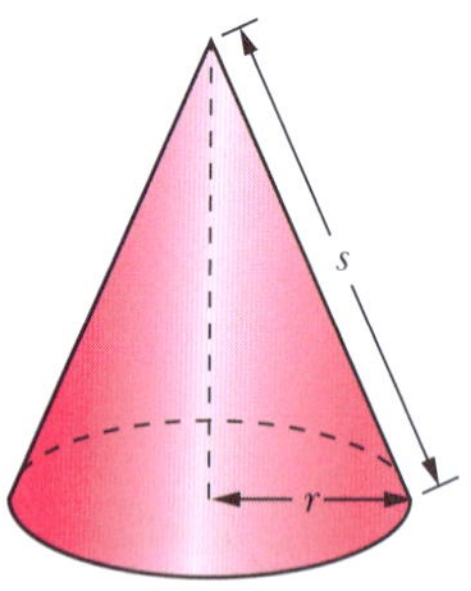

In the diagram below a cone with a **slant** height of s units and a **base radius** of r units has been opened to reveal its net.

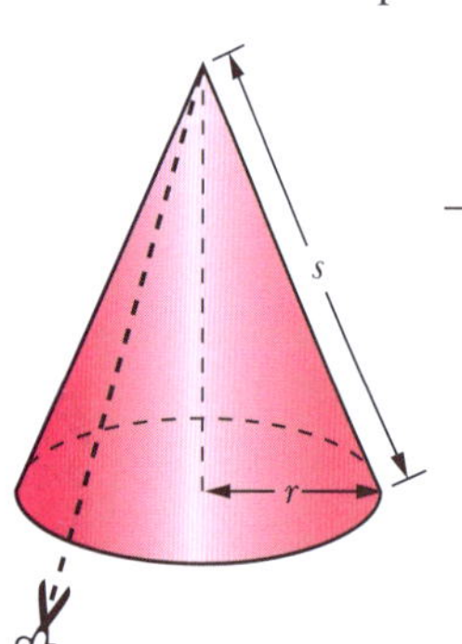

Cut along the slant edge and around the circular edge to make the **net** of the cone.

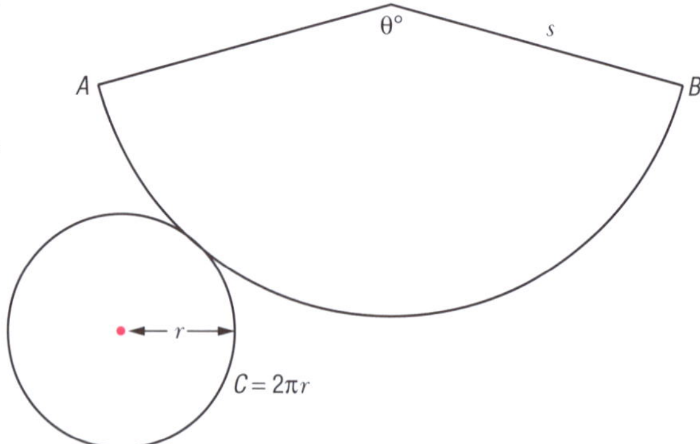

The curved surface of a cone is made from a sector of a circle with radius equal to the **slant height** of the cone.

The arc length of this sector ($\frac{\theta}{360} \times 2\pi s$) must be equal to the circumference of the base circle ($2\pi r$) for the cone to be formed.

$\frac{\theta}{360} \times 2\pi s = 2\pi r$ and so: $\frac{\theta}{360} = \frac{2\pi r}{2\pi s} = \frac{r}{s}$

Now: Surface area of closed cone = area of base circle + area of sector

$$= \pi r^2 + \frac{\theta}{360} \times \pi s^2$$

$$= \pi r^2 + \frac{r}{s} \times \pi s^2 \qquad \text{Since } \frac{\theta}{360} = \frac{r}{s}.$$

$$= \pi r^2 + \pi rs$$

Surface area of a closed cone = $\pi r^2 + \pi rs$

area of the circular base (for πr^2); curved surface area (for πrs)

Remember

- It is not necessary to draw the net of a cone when using its total surface area formula.
- Do not confuse the slant height with the perpendicular height.

Example

Calculate the total surface area of a solid cone with a radius of 60 mm and a slant height of 10 cm.

Answer

For this cone: $r = 6$ cm, and $s = 10$ cm

Total surface area = $\pi r^2 + \pi rs$

$= 301.59 \text{ cm}^2$ Rounded to two decimal places.

The total surface area is 301.59 cm^2.

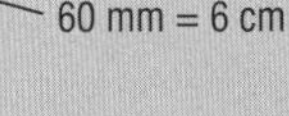

The surface area of a sphere

The level of mathematics required to develop the formula for the surface area of a sphere is beyond this level of mathematics, so the formula is given below without any explanation.

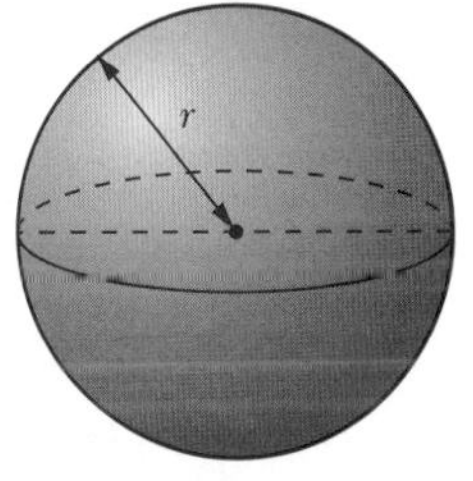

Surface area of a sphere = $4\pi r^2$

Example

Find the surface area of a sphere of diameter 1.2 m, to two decimal places. (If you do not have a calculator, use 3 as an approximation for π.)

Answer

The diameter of the sphere is 1.2 m, so the radius is: $r = \frac{1.2}{2}$ $= 0.6$ m

The area is: $A = 4\pi r^2$

$A = 4 \times \pi \times (0.6)^2$

$= 4.5238\ldots$

$\approx 4.52 \text{ m}^2$ Rounded to two decimal places.

Using $\pi = 3$: $A = 4 \times 3 \times 0.6^2 = 4.32 \text{ m}^2$

Example

Find the total surface area of a solid hemisphere of radius 35 cm.

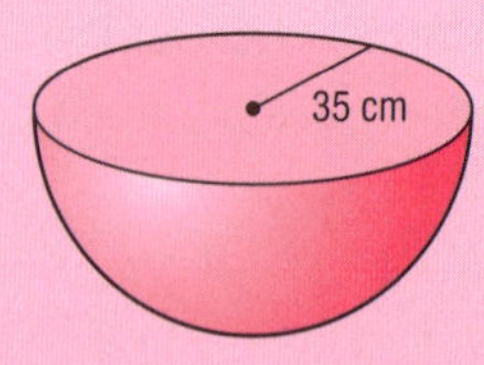

The solid hemisphere has two faces: a circle of radius 35 cm and a curved surface that is half the surface area of a sphere.

Answer

Total surface area $= \pi r^2 + \frac{1}{2} \times 4\pi r^2$ You may recognise that this simplifies to $3\pi r^2$.

$= \frac{22}{7} \times 35 \times 35 + \frac{1}{2} \times 4 \times \frac{22}{7} \times 35 \times 35$

$= 22 \times 5 \times 35 + 2 \times 22 \times 5 \times 35$

$= 11\,550 \text{ cm}^2$

Remember

If you do not have a calculator, use 3 as an approximation for π. The answers for this Exercise, in the back of the book, have been calculated using the calculator value for π, but answers using 3 as an approximation for π are also given in brackets.

EXERCISE

1 Calculate the total surface area of each of the following.

a A solid cone with base radius of 9 cm and a slant height of 20 cm.

b A solid square-based pyramid with a base length of 1.8 m and triangular faces which are each of height 3.32 m.

2 Calculate the total surface area of each of the following:

a

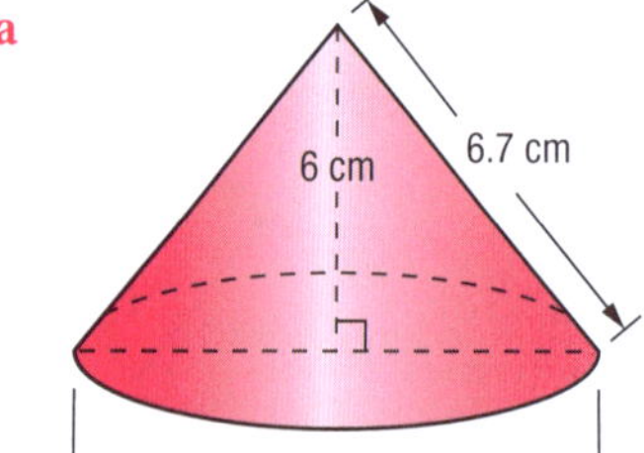

b

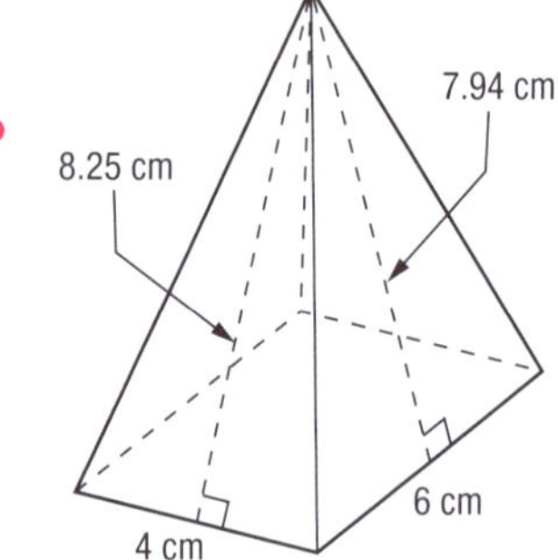

3 Find the surface area of each of the following solids, correct to two decimal places.

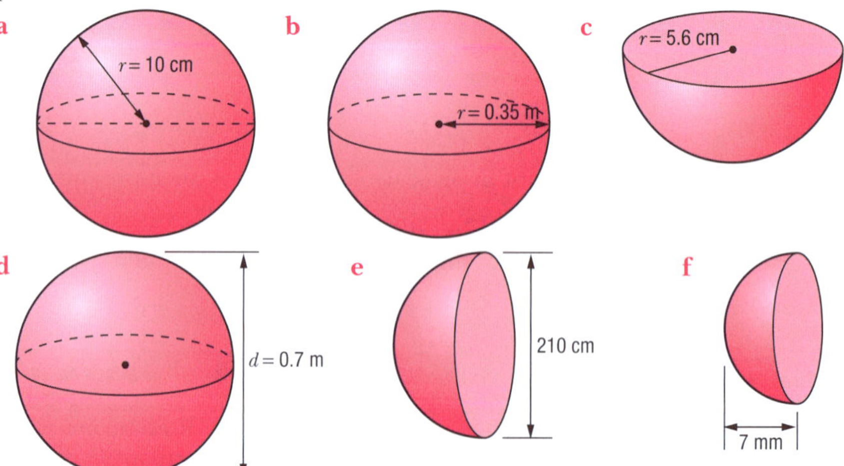

4 Find the surface area of each of the following solids, correct to two decimal places:

a

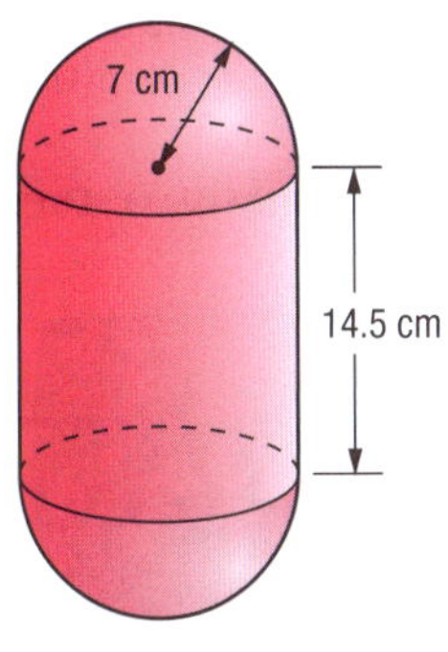

b

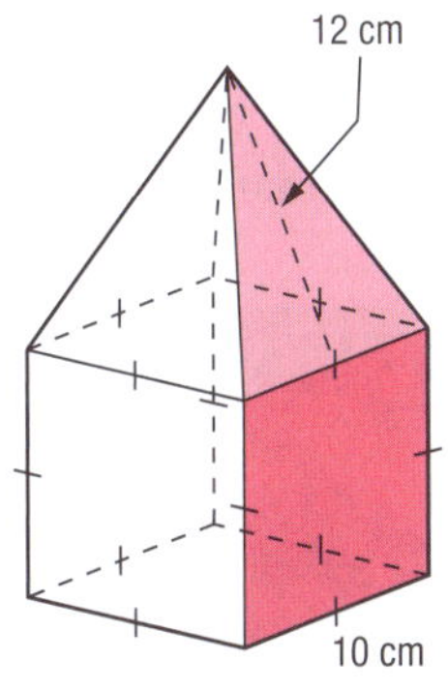

c

5 The diagram on the right is of a solid rivet constructed of a hemisphere and a cylinder.

a How many faces does this solid have?

b Sketch each of the faces, labelling each face with its dimensions.

c Calculate the total surface area of the solid.

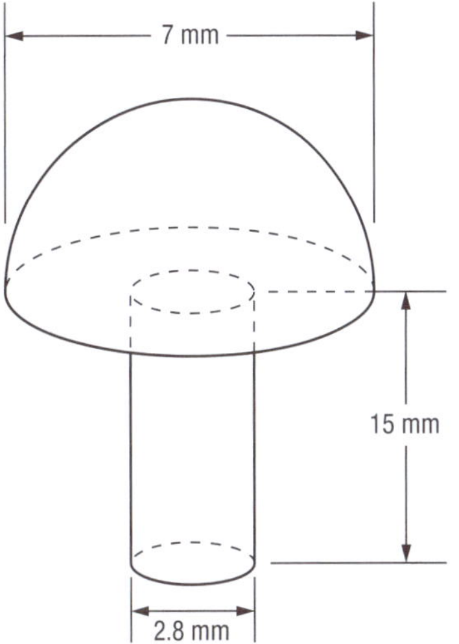

6 Calculate the surface area of the roof of this shelter, if the inside diameter of the shelter is 7 m, and the roof rafters are 4 m long.

7 a Describe the shape of the smaller basket in the photo below and draw a sketch of it.

  b Draw the shapes of all the faces of the basket (ignoring the handles). Name the shapes of the faces and estimate the relevant side lengths of the shapes.

  c Use your measurement estimates from part **c** to calculate the outer surface area of the basket.

Volume and capacity

Volume

Volume involves three dimensions: length, breadth and height. Therefore, its units are **cubic units**.

The volume of a solid is the amount of space it occupies (that is, the number of cubic units of space it takes up).

The metric units for volume are:

- cubic millimetres (mm^3)
- cubic centimetres (cm^3)
- cubic metres (m^3)

We can see from the diagrams below that one cubic centimetre occupies the same space as one thousand cubic millimetres.

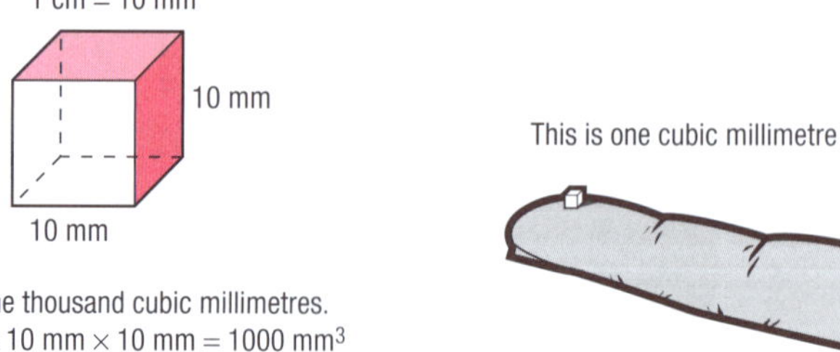

This is one thousand cubic millimetres.
$10 \text{ mm} \times 10 \text{ mm} \times 10 \text{ mm} = 1000 \text{ mm}^3$

Conversion of cubic units

This diagram helps us to convert between units of volume:

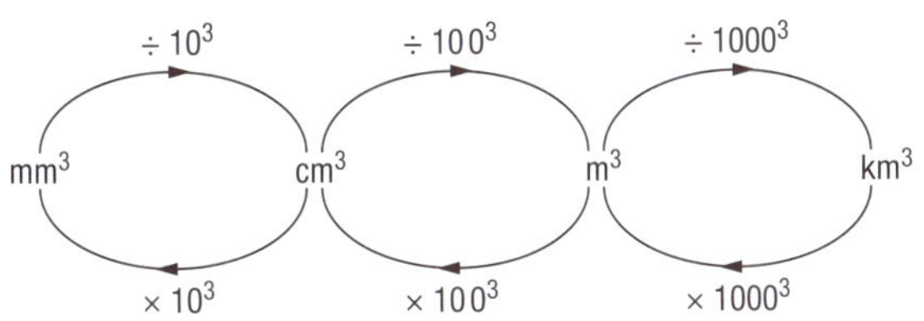

> **Remember**
>
> It is easier to convert the units of length **before** the volume is calculated.

Example

Convert:

a $12\ 345\ \text{cm}^3$ to m^3 **b** $0.15\ \text{cm}^3$ to mm^3

Answer

a *Smaller* to *larger* units, so we *divide*:

$$12\ 345\ \text{cm}^3 = 12\ 345 \div 100^3\ \text{m}^3 = 12\ 345 \div 1\ 000\ 000\ \text{m}^3 = 0.012\ 345\ \text{m}^3$$

b *Larger* to *smaller* units, so we *multiply*:

$$0.15\ \text{cm}^3 = 0.15 \times 10^3\ \text{mm}^3 = 0.15 \times 1000\ \text{mm}^3 = 150\ \text{mm}^3$$

Capacity

The **capacity** of a container is the amount of fluid or non-solid material that it can hold.

The metric units for capacity are:

- millilitres (mL)
- litres (L)
- kilolitres (kL)
- megalitres (ML)

Converting between units:

1 litre (L) = 1000 mL

1 kilolitre (kL) = 1000 L

1 megalitre (ML) = 1000 kL = 1 000 000 L

The relationships between the units of volume and units of capacity are shown below.

If we assume that the internal measurements of an object are equal to its external measurements, then the volume of the object directly relates to its capacity.

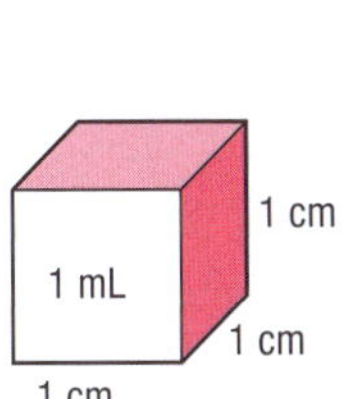

Volume = $1\ \text{cm}^3$
Capacity = 1 mL

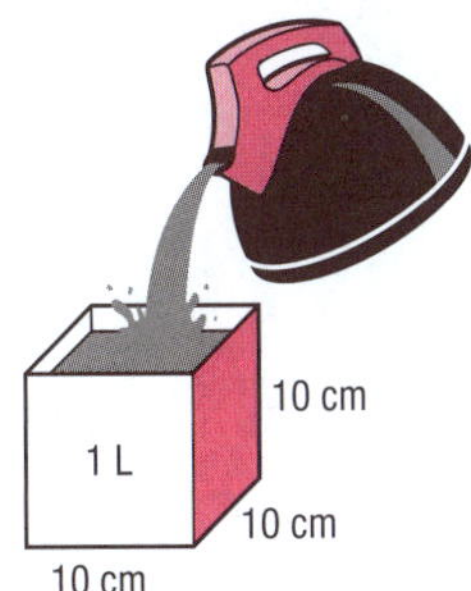

Volume = $1000\ \text{cm}^3$
Capacity = 1 L

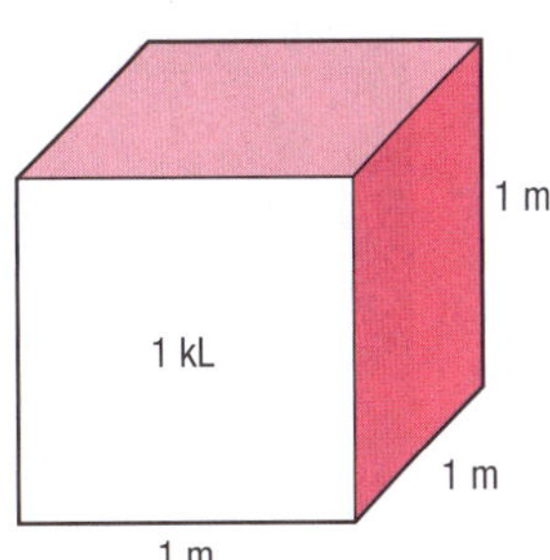

Volume = $1\ \text{m}^3$
Capacity = 1 kL

Example

Convert:

a 1.13 L to millilitres **b** 650 L to kilolitres **c** 2.575 L to cm^3

Answer

a *Larger* to *smaller* units, therefore we *multiply*:
$1.13 \text{ L} = 1.13 \times 1000 \text{ mL}$
$= 1130 \text{ mL}$

b *Smaller* to *larger* units, therefore we *divide*:
$650 \text{ L} = 650 \div 1000 \text{ kL}$
$= 0.650 \text{ kL}$

c Litres to cm^3 therefore we *multiply*:
$2.575 \text{ L} = 2.575 \times 1000 \text{ cm}^3$
$= 2575 \text{ cm}^3$

Remember

If you do not have a calculator, use 3 as an approximation for π. The answers for this Exercise, in the back of the book, have been calculated using the calculator value for π, but answers using 3 as an approximation for π are also given in brackets.

EXERCISE

1 The volume occupied by a cylindrical can is 603 186 mm^3. What is the can's volume in cm^3?

2 A truck delivers 1 cubic metre of sand to a building site. A worker moves the sand using a shovel that holds 900 cm^3. How many shovelsful of sand will the worker move?

3 Convert:
a 7.6 L to cm^3 **b** 5 000 000 cm^3 to kilolitres **c** 3.256 ML to m^3

4 Morgan is watering his garden using a bucket that has a capacity of 9 litres. He fills the bucket from a tank that has a volume of 1.8 m^3. How many bucketsful of water are in the tank?

5 A bottle contains 1.25 litres of soft drink. What is this volume in cubic centimetres?

6 Kennedy has a tank that holds 0.25 m^3 of oil. He is going to sell the oil at market in 1 litre containers. How many 1 L containers can he fill from his tank if it is full?

7 The photo below shows the information painted on the side of a shipping container.
a What is the capacity of this shipping container, in cubic metres?
b What is the capacity of this shipping container, in litres?

The volume of prisms and cylinders

Prisms and cylinders have a constant cross-section.

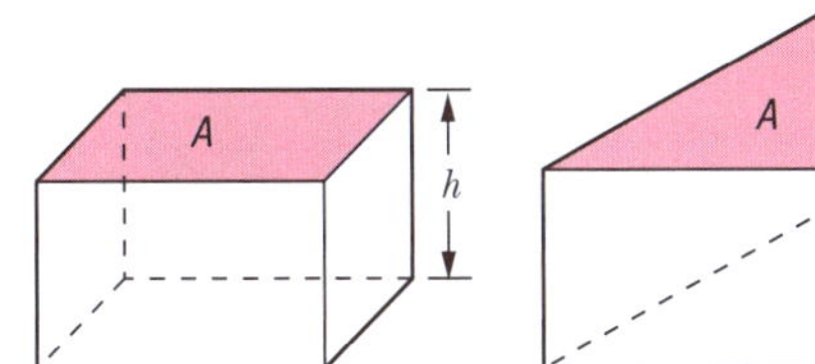

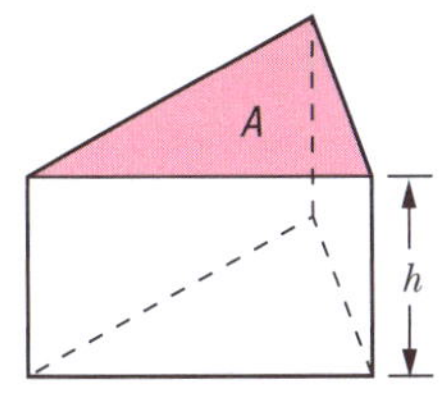

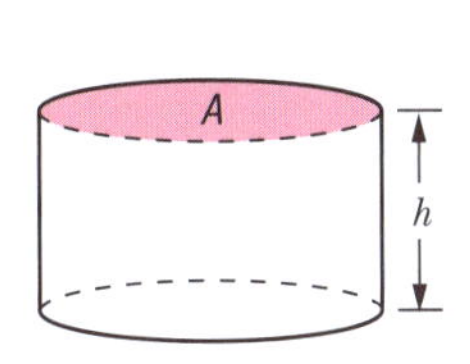

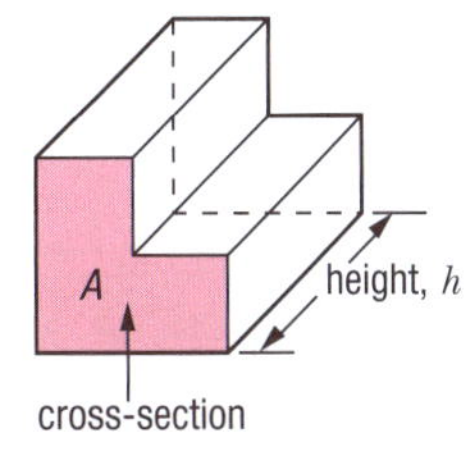

Volume of a solid with a constant cross-section = area of cross-section × height
$= A \times h$

Example

Calculate the volume of this cuboid.

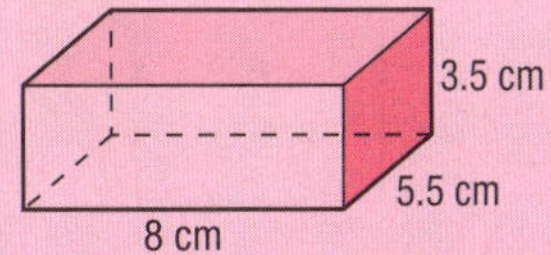

Answer

$\text{volume}_{\text{prism}}$ = area of base × vertical height
$= (8 \times 5.5) \times 3.5$
$= 154 \text{ cm}^3$

Example

Find the capacity of this rainwater tank, in litres.
The internal measurements of the tank are shown.

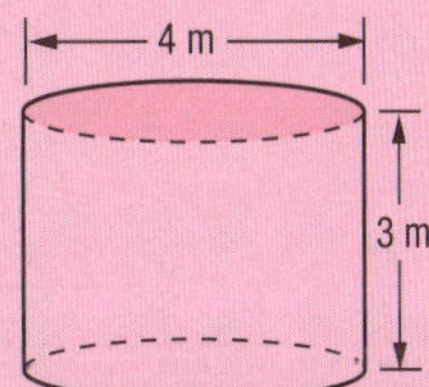

Answer

The diameter is 4 m: $r = 2$

$\text{volume}_{\text{prism}}$ = area of base × vertical height
$= \pi r^2 \times h$
$= (\pi \times 2^2) \times 3$
$= 37.6991 \text{ m}^3$ Rounded to two decimal places. Using a calculator.

Capacity = 37.6991 kL
= 37 699.1 L
≈ 37 700 L

Remember

If you do not have a calculator, use 3 as an approximation for π. The answers for this Exercise, in the back of the book, have been calculated using the calculator value for π, but answers using 3 as an approximation for π are also given in brackets.

EXERCISE

1 For each of the following:

i sketch the constant cross-section of the solid

ii find the area of the constant cross-section

iii calculate the volume of the solid.

a

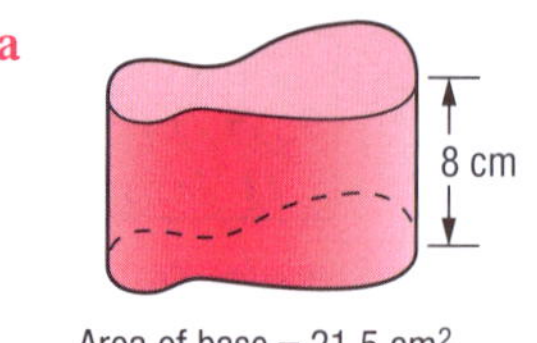

b

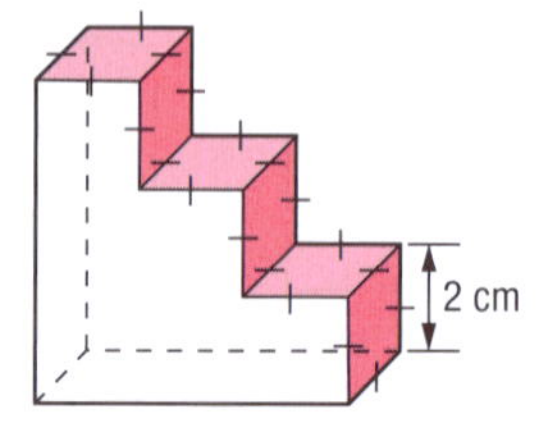

c

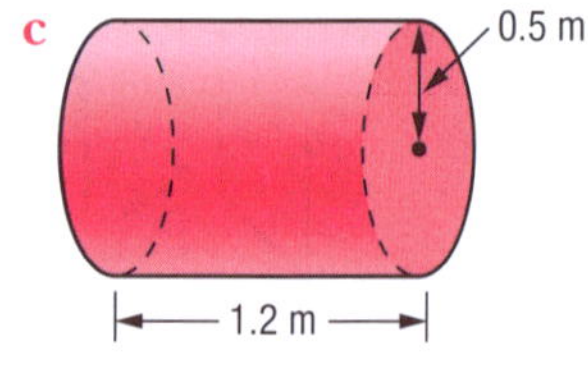

d

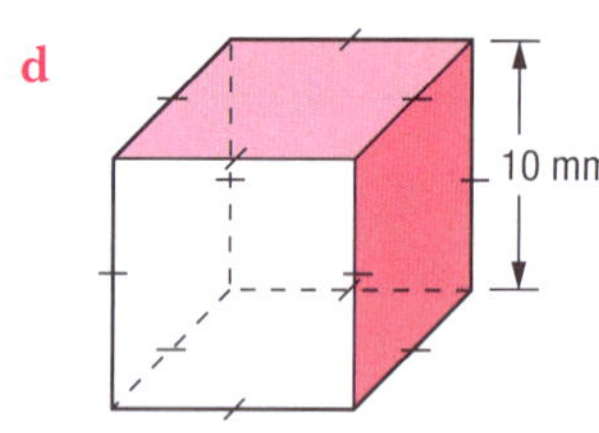

e

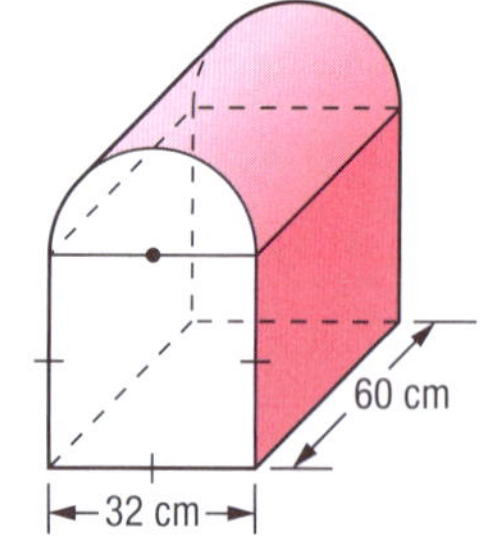

f

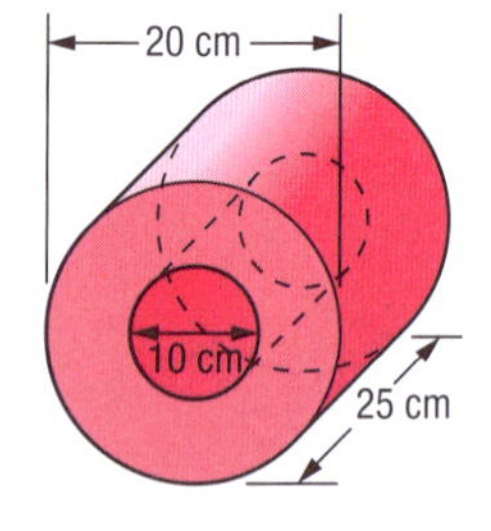

g

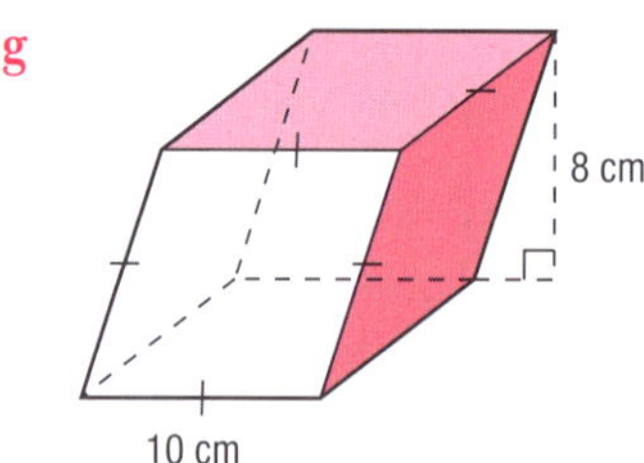

2 For each of the following:

i sketch the constant cross-section of the container

ii find the area of the constant cross-section

iii calculate the capacity of the container.

a

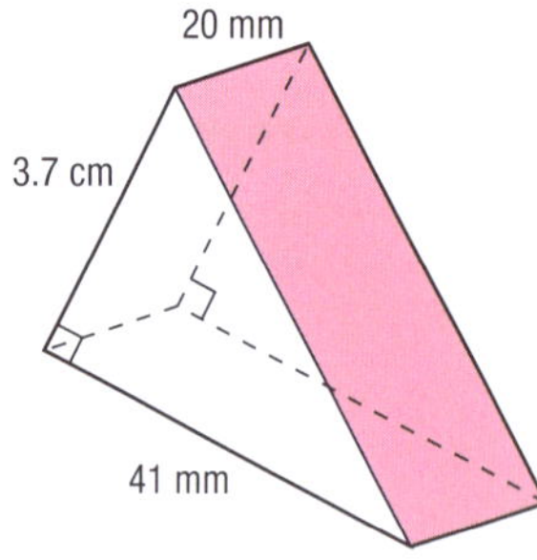

b

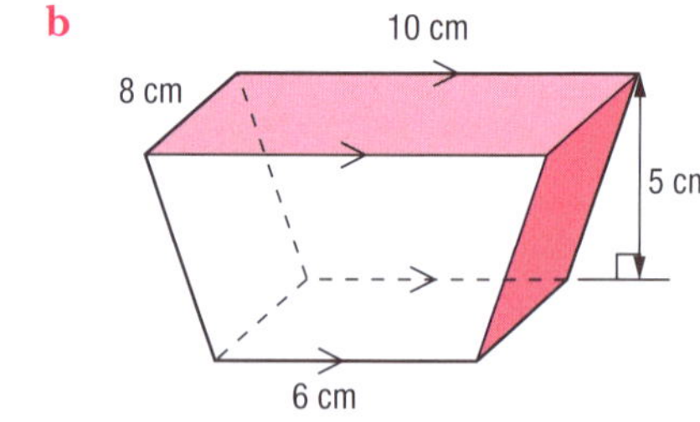

c

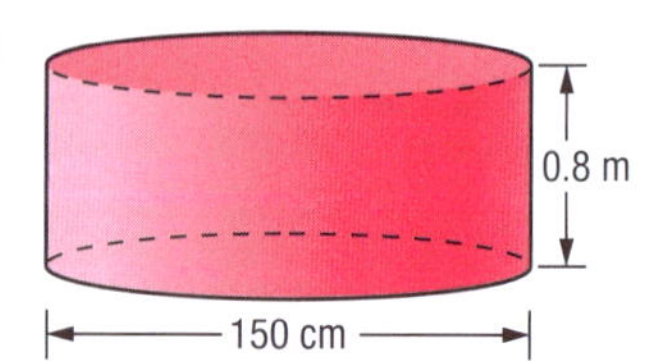

3 Wallace has made a feedbox in the shape shown in the diagram below. The slant sides of the feedbox have inside dimensions of 25 cm × 80 cm and they are at right-angles to each other. Find the capacity of the feedbox, in litres.

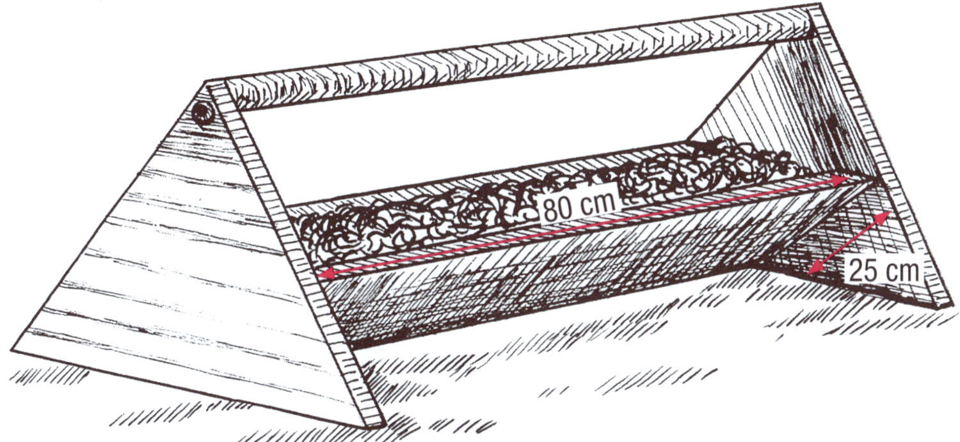

4 A swimming pool has a rectangular upper surface measuring 25 m × 10 m but is deeper at one end, as shown in the diagram.

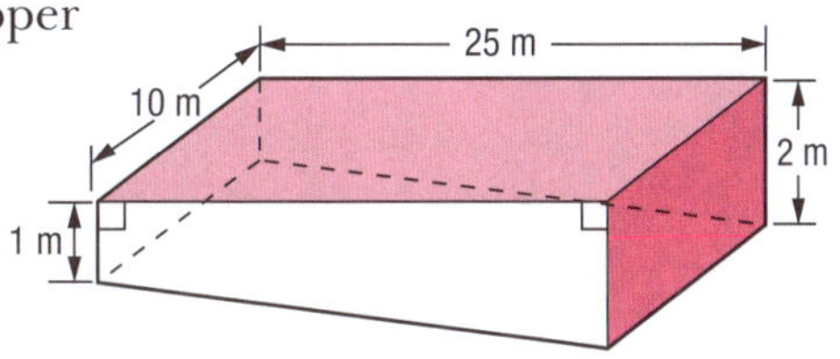

a Draw the constant cross-section of this swimming-pool.
b Calculate the area of the cross-section you drew in part a.
c Calculate the capacity of the swimming pool in kilolitres.
d If the swimming pool is filled with water to within 30 cm of the top calculate, in kilolitres, the amount of water it takes to fill the pool.

5 A food container for animals is made from half of a large cylindrical metal container, as shown in the diagram on the right. If the original metal container had circular ends of diameter 56 cm and the length of the container was 85 cm, find the capacity of the food container, in litres.

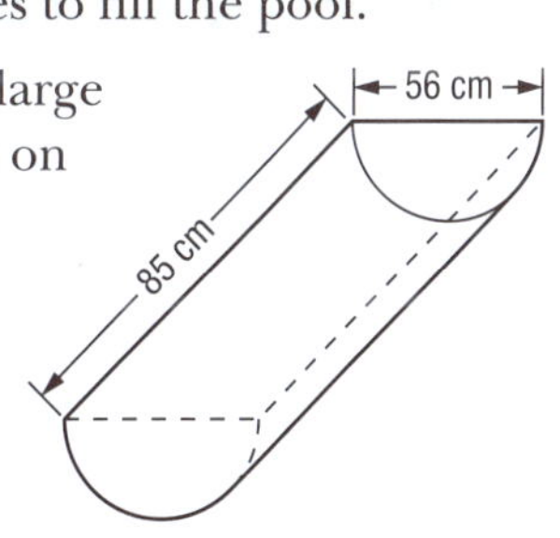

6 Moses uses a bamboo pole to carry liquid. The inside diameter of the pole is 12.6 cm and the length is 1.85 m. Find the capacity of this liquid carrier, in litres.

7 A length of cylindrical bamboo is used as a container for cooking oil. If the inside diameter of the bamboo is 84 mm, how long, in centimetres, would the length of the bamboo need to be for it to hold 1 litre of oil?

8 Concrete blocks for large buildings are partly hollow and have the shape shown in the diagram on the right. The outside dimensions of the block are: length 40 cm, width 20 cm and height 25 cm, and the 'walls' of the block are 3 cm thick.

- a Draw the constant cross-section of this block, marking all the measurements on the diagram.
- b Calculate the area of the cross-section, in square centimetres.
- c Calculate the amount of concrete needed to construct the block.
- d i In the wall of a building, what amount of space would be taken up by one of these blocks?

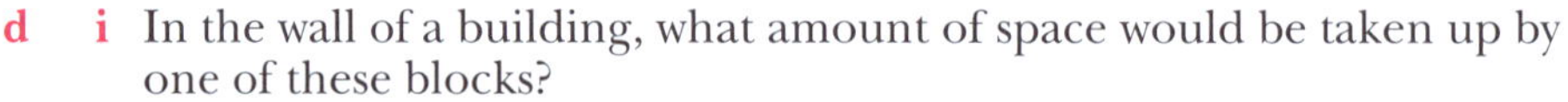

  ii What percentage of this space is concrete?
- e If 1000 of these building blocks are to be produced what volume of concrete, in cubic metres, will be required.

9 A shipping container, pictured below, has an inner capacity of 33.2 cubic metres. If its inside width is 2.35 metres and its inside height is 2.38 metres, what is its inside length, in metres?

Lesson 10

Volume of pyramids, cones and spheres

The volume of pyramids and cones

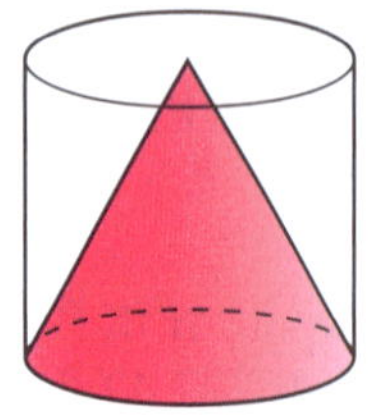

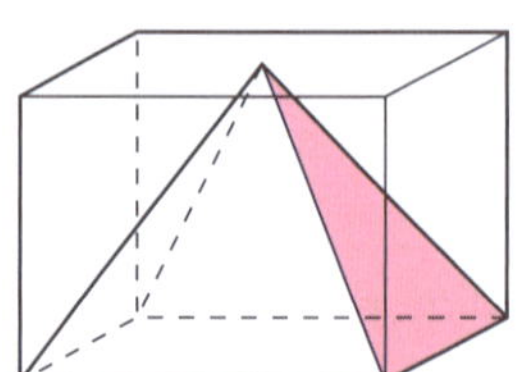

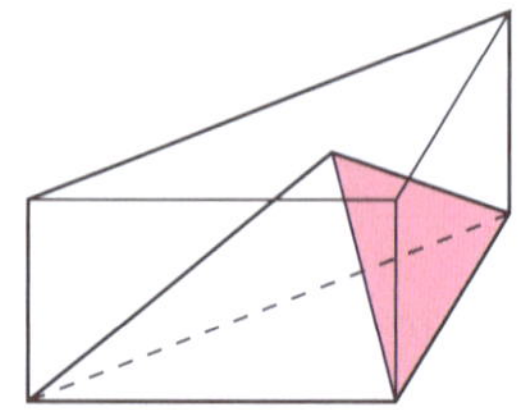

Volume of a pyramid or cone $= \frac{1}{3} \times$ (area of base × height)

Example

Calculate the volume of this pyramid.

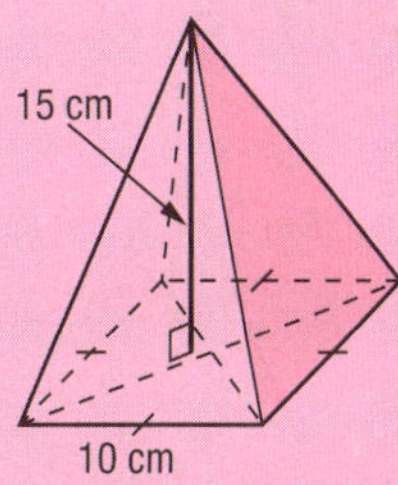

Answer

$$
\begin{aligned}
V_{\text{pyramid}} &= \frac{1}{3} \times (\text{area of base} \times \text{height}) \\
&= \frac{(10 \times 10) \times 15}{3}\ \text{cm}^3 \\
&= \frac{1500}{3}\ \text{cm}^3 \\
&= 500\ \text{cm}^3
\end{aligned}
$$

The volume of a sphere

As with the surface area of a sphere, the derivation of the formula for the **volume** of a sphere is beyond the scope of this book. But we can look at how this formula developed historically.

Archimedes (around 250 BC) discovered that the volume of a sphere is equal to two-thirds of the volume of the smallest possible cylinder that will enclose it.

Let the radius of the sphere be r units.

The smallest cylinder that will enclose this sphere has base radius r and height $2r$.

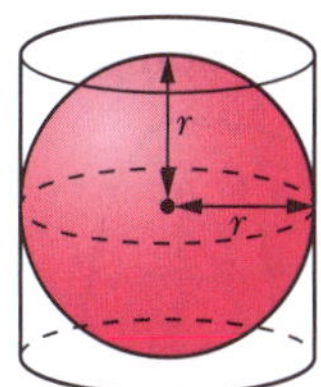

So: Volume of this cylinder = area of base × height

$$
\begin{aligned}
&= \pi r^2 \times 2r \\
&= 2\pi r^3
\end{aligned}
$$

Using Archimedes' discovery:

$$
\begin{aligned}
\text{Volume of sphere} &= \frac{2}{3} \times \text{volume of smallest cylinder that will enclose the sphere} \\
&= \frac{2}{3} \times 2\pi r^3 \\
&= \frac{4}{3}\pi r^3
\end{aligned}
$$

Volume of a sphere = $\frac{4}{3}\pi r^3$ where r is the radius

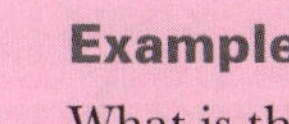

Example

What is the capacity, in litres, of a hemispherical bowl with a radius of 200 mm?

Answer

To find litres, we need to work in centimetres.

200 mm = 200 ÷ 10 = 20 cm

$V_{\text{hemisphere}} = \frac{1}{2}$ of a sphere

$$
\begin{aligned}
V &= \frac{1}{2} \times \frac{4}{3} \times \pi \times 20^3 \\
&\approx 16\,755.16\ \text{cm}^3 \\
&\approx 16\,755.16\ \text{mL} \\
&\approx 16\,755.16 \div 1000\ \text{L}
\end{aligned}
$$

Capacity ≈ 16.76 L   Rounded to two decimal places.

> **Remember**
> If you do not have a calculator, use 3 as an approximation for π. The answers for this Exercise, in the back of the book, have been calculated using the calculator value for π, but answers using 3 as an approximation for π are also given in brackets.

EXERCISE

1 Calculate the volume of each of the following solids:

a
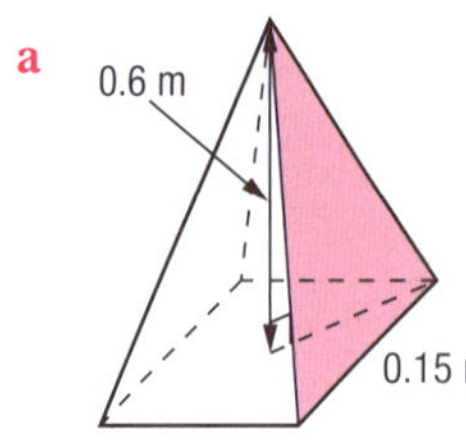

b
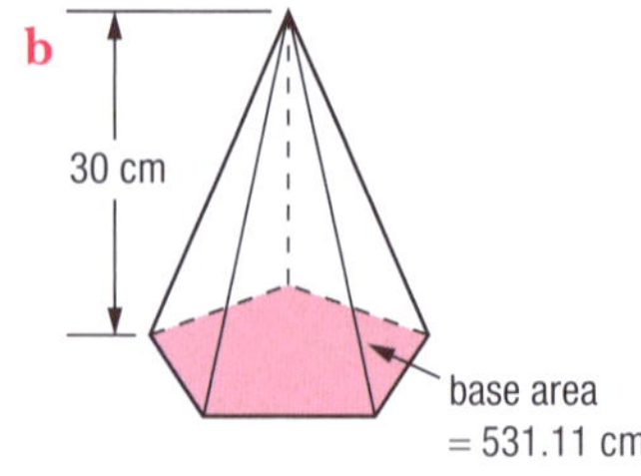

c
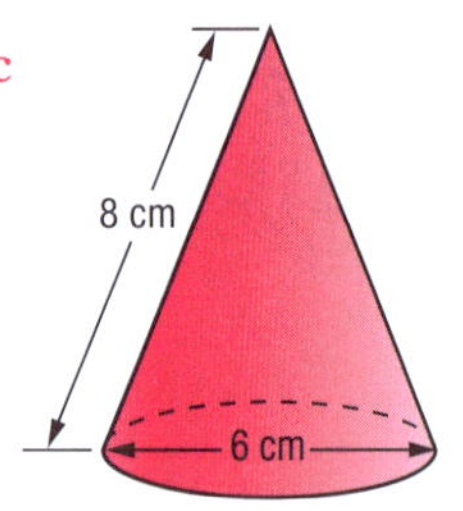

2 Calculate the volume of a triangular pyramid that has a right-angled triangle for a base and has a height of 60 mm.

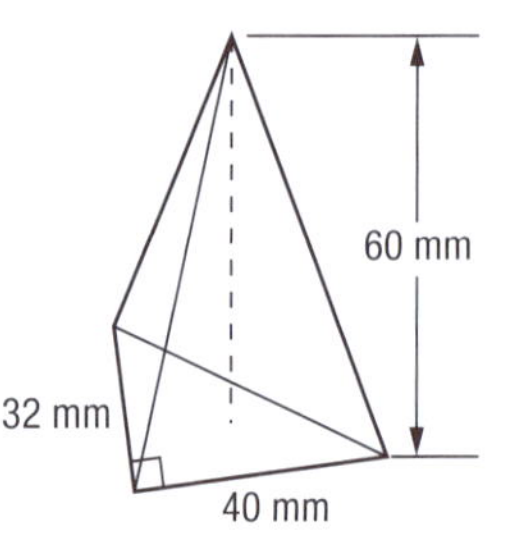

3 Calculate, to one decimal place, the volume of each of the following solids:

a
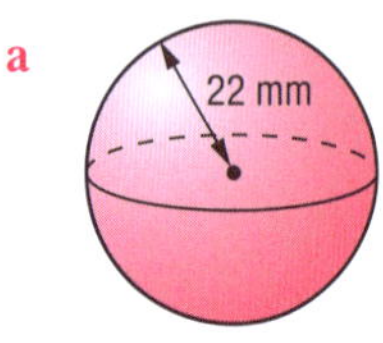

b
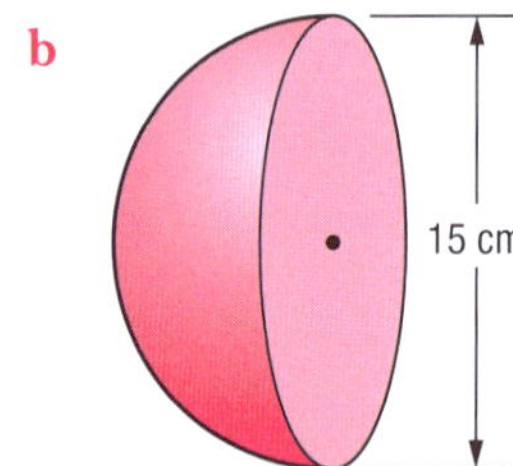

c
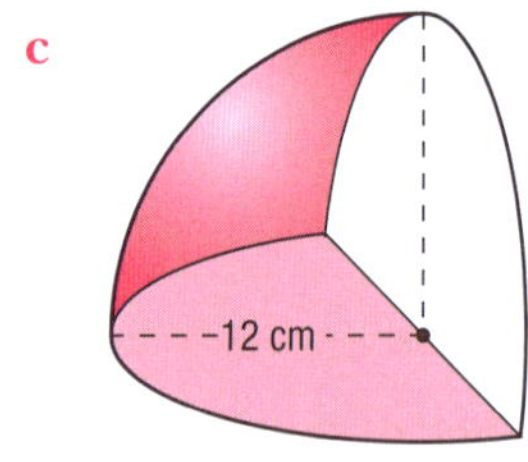

4 Find the capacity of each of the following containers, in litres:

a
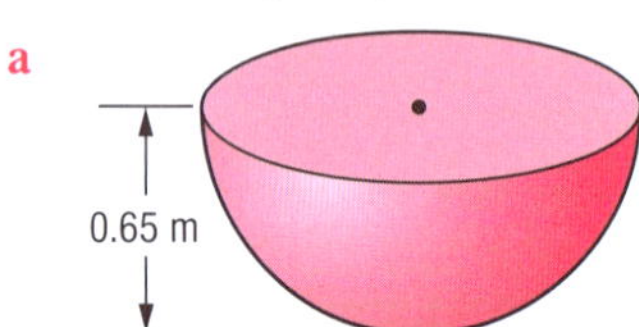

b An inverted triangular pyramid which has an equilateral triangle as its top horizontal surface.

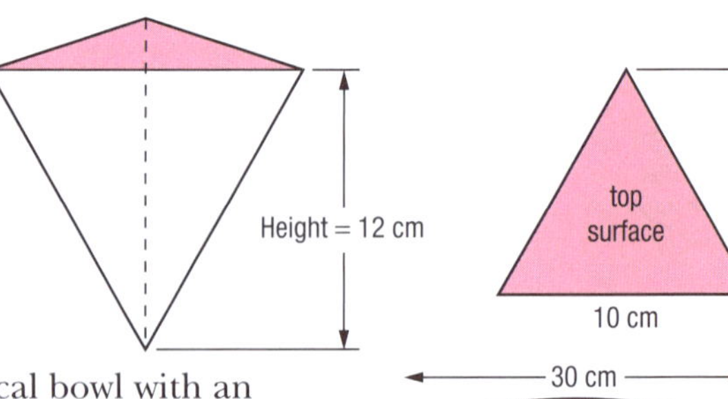

c A wooden hemispherical bowl with an outside diameter of 30 cm and walls that are 1 centimetre thick.

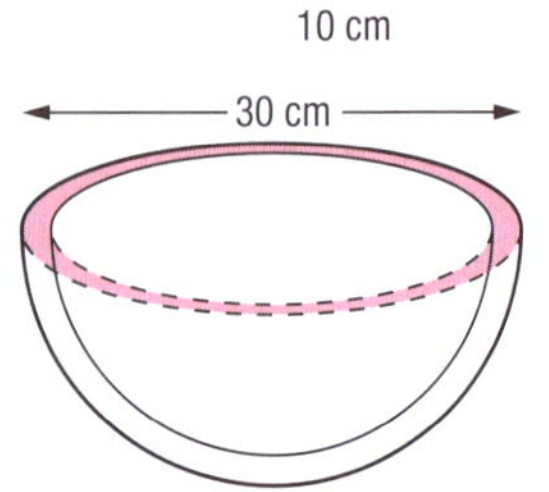

5 A conical bottle has a base with diameter 12.25 cm. If the capacity of the bottle is 1 litre, calculate the height of the bottle, in centimetres, correct to one decimal place.

6 A spherical coconut, with an outer diameter of 15 cm, has a hard shell of thickness 0.5 cm and then a layer of coconut flesh of thickness 2 cm.

   a Calculate the volume of coconut flesh inside this coconut.

   b Calculate the maximum volume of kulau (in mm^3) that could be enclosed by the coconut flesh.

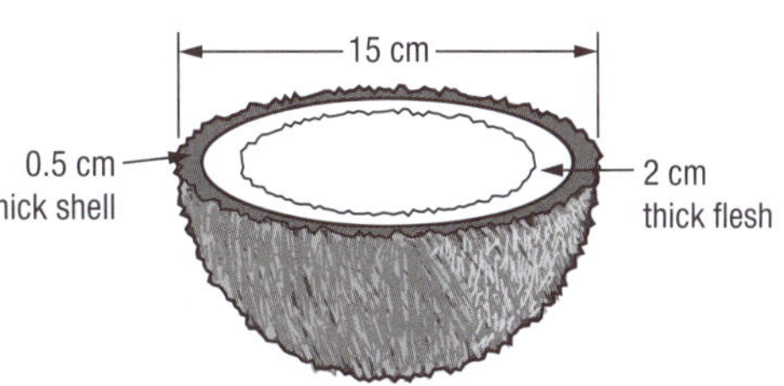

The volume of composite solids

Many solids are composites made of combinations of prisms, cones, pyramids or spheres and so their volume will be the sum of these component solids.

To calculate the volume of a composite solid:

i draw each of the component solids separately, labelling them with their dimensions

ii calculate the volume of each of the component solids

iii add (or, in some cases, subtract) the volumes.

Example

A simple building has rectangular walls of length 5 m, width 4 m and height 2.4 m. The roof is a rectangular pyramid that covers the distance between the walls and provides an overhang, as shown in the diagram below.

If the top of the roof is at a height of 3.6 metres, find the volume of the interior space of the house, assuming the roof space is hollow.

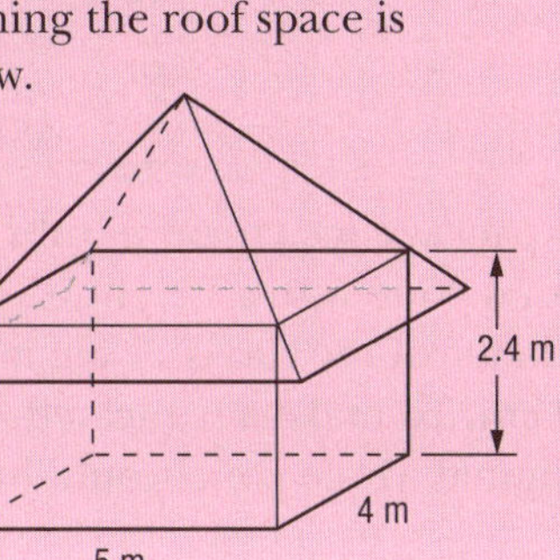

Answer

The volume of the interior space of the house is composed of a rectangular prism and a rectangular pyramid.

Volume =

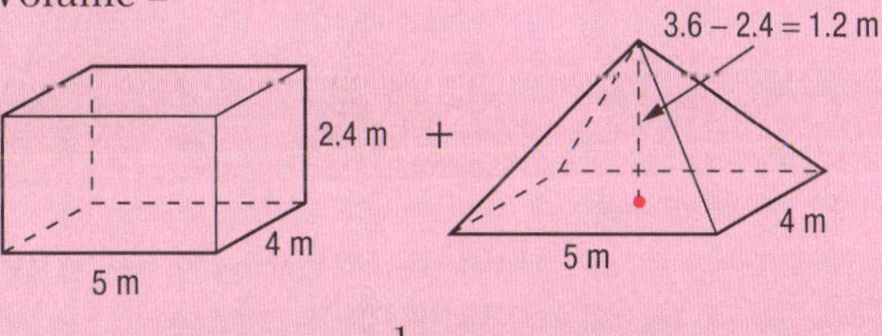

$= l \times w \times h + \frac{1}{3} \times$ (area of base × height)

$= 5 \times 4 \times 2.4 + \frac{1}{3} \times 5 \times 4 \times 1.2$

$= 48 + 8$

$= 56\ m^3$

EXERCISE

1 Find the volume of each of these solids, to one decimal place.

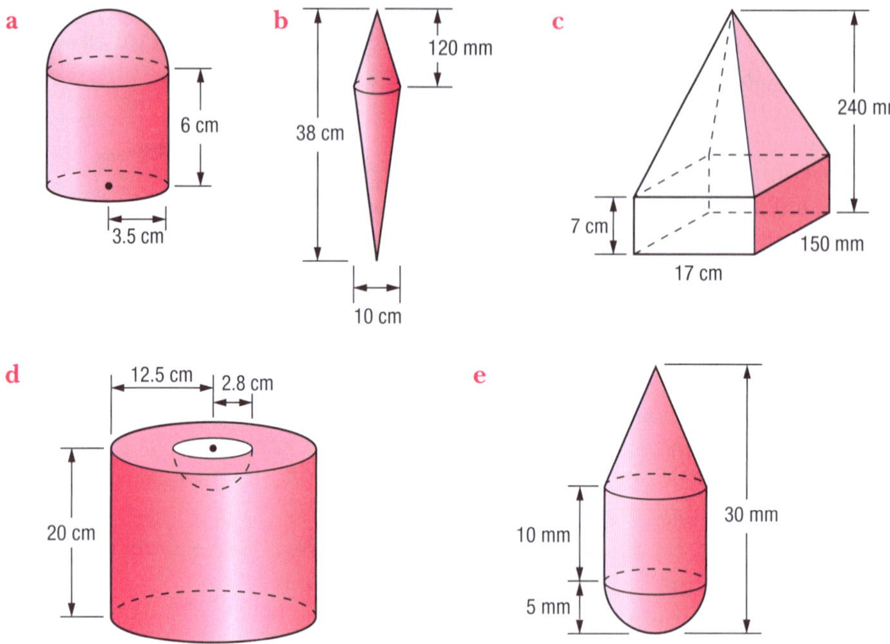

2 A truncated cone is a cone with the top cut off.

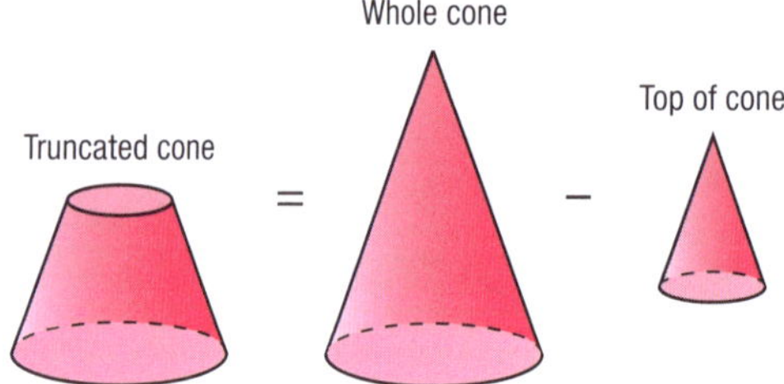

Find the volume of the truncated cone on the right.

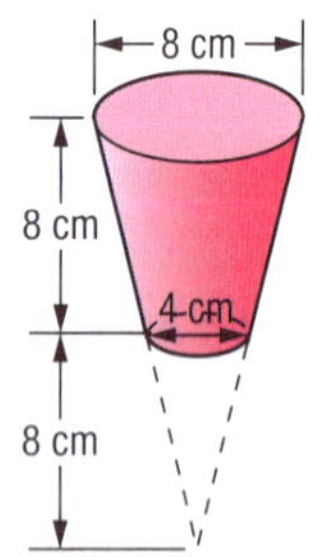

3 A manufacturer wants to package ice cream in tubs that are the shape of a truncated cone with a height of 13.5 cm. The largest diameter of the tub is 11 cm and the smaller diameter is 9 cm. To make a truncated cone of this shape the original whole cone would have been 75 cm tall.
Show that the tub will hold at least 1 litre of ice cream.

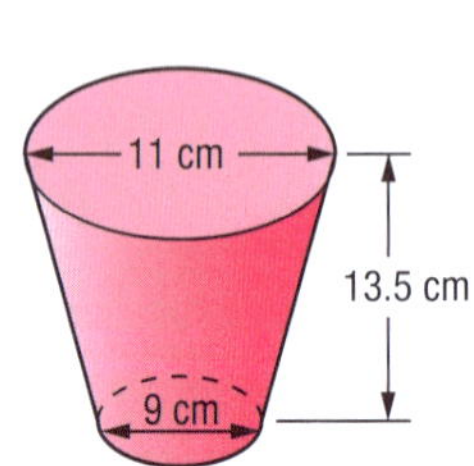

4 a 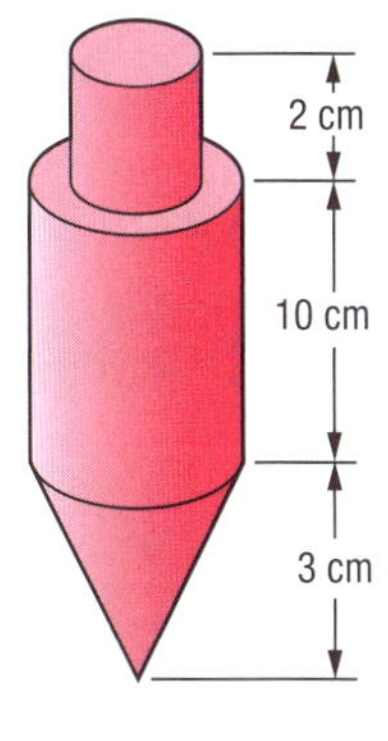

A plumb-bob is cast from steel and has the shape shown in the diagram on the left. The large cylindrical part of the plumb-bob has a diameter of 3 cm. The diameter of the small cylindrical part is 1.6 cm. The height of each part is shown on the diagram. Find the amount of metal required to cast this plumb-bob.

b If a cylindrical hole, of diameter 6 mm, is drilled in the smaller cylinder, so that the string can be tied to the plumb-bob, how much metal is removed?

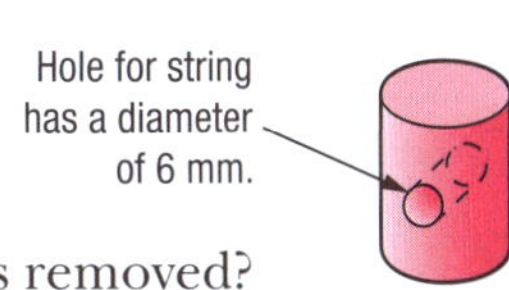

5 A skateboard ramp is constructed from concrete and has the shape shown in the diagram on the right. The curved surfaces are the shape of one-quarter of the inside of a cylinder.

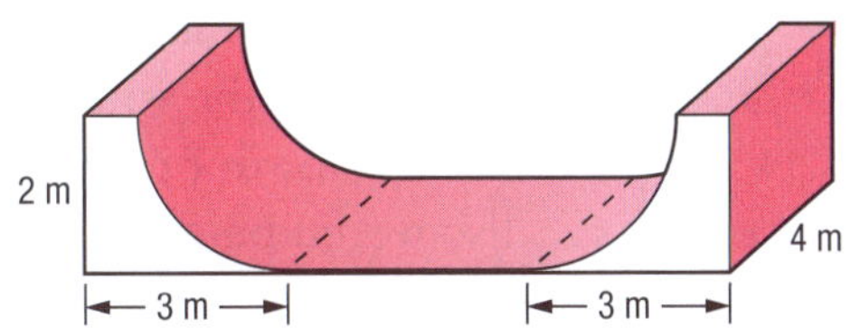

Calculate the quantity of concrete, in cubic metres, needed to construct this skateboard ramp.

Challenge

The relationship between length and volume of similar solids

Drawing each of the solids mentioned will help you with this activity.

a i Consider a (small) cube of side length 1 cm. What is its volume in cm^3?

ii Now find the volume of a cube (medium) that has a side length double the size of the cube in part i.

iii Find the volume of a cube (large) that has a side length three times the size of the cube in part i.

iv Copy and complete the following ratios for the cubes:

Length (side) Small : Medium : Large = 1 : 2 : 3

Volume Small : Medium : Large = ___ : ___ : ___

v Similar solids have corresponding angles with the same measure and their corresponding lengths are in the same ratio. Are the cubes in parts i, ii and iii similar solids?

b Repeat parts i to iii above for spheres. Give your answers in cm^3, in terms of π. For example:

A sphere of radius 1 cm will have a volume of $\frac{4}{3}\pi \times 1^3 = \frac{4}{3}\pi$ cm^3

iv Copy and complete the following ratios for the spheres:

Length (radius) Small : Medium : Large = 1 : 2 : 3

Volume Small : Medium : Large = ___ : ___ : ___

v Are the spheres in parts i, ii and iii similar solids?

c **i** Consider a cylinder of radius 1 cm and height 2 cm. What is its volume in cm^3, in terms of π?

ii Find the volume, in cm^3, of a cylinder of radius 2 cm and height 4 cm.

iii Find the volume, in cm^3, of a cylinder of radius 3 cm and height 6 cm.

iv Copy and complete the following ratios for the cylinders:

Length (radius) Small : Medium : Large = 1 : 2 : 3

Volume Small : Medium : Large = ___ : ___ : ___

v Are the cylinders in parts **i**, **ii** and **iii** similar solids?

d **i** Can you see a pattern in the results for parts **a**, **b** and **c**? How do the numbers in the length ratio relate to the numbers in the volume ratio each time?

ii Explain in your own words the relationship that you have discovered between the length and volume of similar solids.

e One way to state the relationship between the length and volume of similar solids is:

'If the scale factor for lengths in similar solids is k then the scale factor for the volume of the similar solids will be k^3.'

Use this relationship between length and volume to solve the following problem:

The two triangular pyramids illustrated here are similar solids.

i What is the scale factor of their lengths?

ii What is the scale factor of their volumes?

iii If the volume of the smaller pyramid is 700 cm^3, find the volume of the larger pyramid.

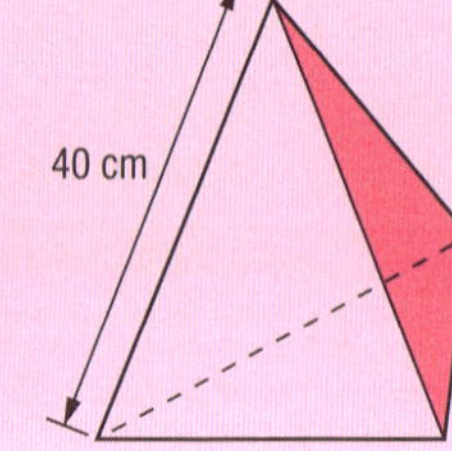

Lesson 12 Investigating the ratio of surface area to volume

Why is the ratio $\frac{\text{surface area}}{\text{volume}}$ important for three-dimensional objects?

When constructing buildings, the temperature that people inside the building will experience is a practical consideration that must be taken into account. Energy is transferred through the outside surface of the building. In cold weather, heat from inside the building can be lost through the outside surfaces. In warm weather, heat can enter the interior of the building through the outside surfaces. Minimising the outside surface area will make a building's temperature control more efficient, energy will be saved and the occupants will be more comfortable.

Does this mean that we should only build small buildings, or can we construct a building that has the least surface area for the volume that it encloses? What shape would that building be?

To explore these questions we will look at some common shapes.

Cubes

Different sized cubes are all similar to each other, but do all cubes have the same $\frac{\text{surface area}}{\text{volume}}$ ratio?

Challenge

a Copy and complete the following table:

Cube of side length (cm)	TSA (cm^2)	Volume (cm^3)	$\frac{\text{total surface area}}{\text{volume}}$
1			
2			
3			
4			
5			
10			
20			

b Repeat this exercise using spheres, which are also always all similar to each other.

c What conclusions can you make from this investigation?

Remember

TSA of a cube = 6 × (length of side)^2

Volume of a cube = (length of side)^3

Rectangular prisms

If the volume of a building in the shape of a rectangular prism, is kept constant, what would be the shape with the lowest $\frac{\text{surface area}}{\text{volume}}$ ratio.

Challenge

Consider this problem:

A builder is building a hostel with 216 rooms, each in the shape of a cube. The building can only be six 'rooms' high but can have any (whole) number of 'rooms' for its length and width.

For example, the diagram on the right shows a hostel with a length of 18 rooms, a width of 2 rooms and the fixed height of 6 rooms. (18 × 2 × 6 = 216)

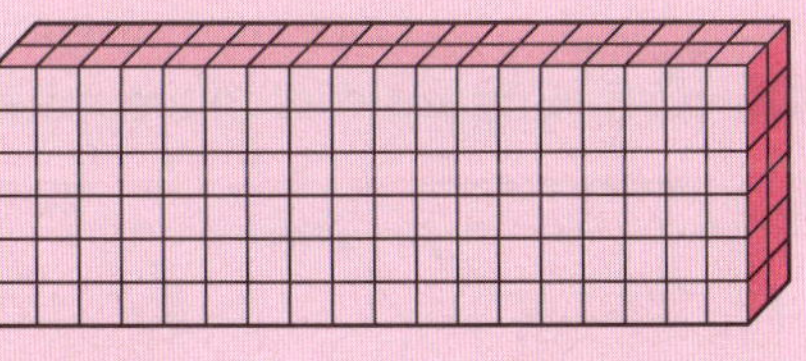

The builder wishes to build the most energy efficient building (the building that has the least surface area for the outer walls and roof).

a What other values could there be for the 'length' (in rooms) and 'width' (in rooms) for this hostel if it is to be six rooms high?

b For each of the combinations found in part **a**, calculate the outer surface area (ignore the bottom surface) and the volume, using the unit 'rooms'.

For example the 18 × 2 × 6 room hostel has:

- a volume of 18 × 2 × 6 = 216 cubic 'rooms'
- an outer surface area of:
  2 × (18 × 6) (front and back walls) + 2 × (2 × 6) (two side walls)
  + 18 × 2 (one roof)
  = 216 + 24 + 36
  = 276 square 'rooms'

c Calculate the $\frac{\text{outer surface area}}{\text{volume}}$ ratio for each of the combinations found in part **a**. For the example above, the ratio of

$$\frac{\text{outer surface area}}{\text{volume}} = \frac{276}{216}$$

= 1.28 Rounded to two decimal places.

d What are the dimensions of the hostel that results in the smallest ratio?

e Can you think of a reason why the hostel selected in part **d** had the smallest $\frac{\text{outer surface area}}{\text{volume}}$ ratio?

Another activity in which the $\frac{\text{surface area}}{\text{volume}}$ ratio could be important is the production of containers that have a given capacity. If the cost of the material that the container is made from is to be considered, then a manufacturer usually minimises the amount of material used and, hence, the cost. This means that the manufacturer would be looking for a shape that has the minimum total surface area to enclose a specified volume.

Remember

1 litre = 1000 mL
= 1000 cm^3

Challenge

Which shape, that has a capacity of 1 litre, (volume of 1000 cm^3), has the least total surface area?

a Each of the following rectangular boxes has a volume of 1000 cm^3. Find the value of x in each case and then find the value of $\frac{\text{total surface area}}{\text{volume}}$. Give your answers correct to the nearest cm^3. (Record your answers in a copy of the table below.)

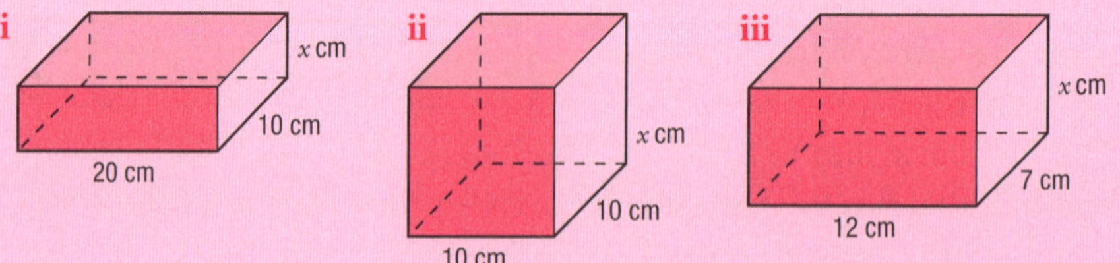

Shape	x	Volume	Surface area	$\frac{\text{surface area}}{\text{volume}}$
i		1000 cm^3		
ii		1000 cm^3		
iii		1000 cm^3		

b For which shape in part **a** is the fraction $\frac{\text{surface area}}{\text{volume}}$ smallest?

c Repeat parts **a** and **b** but, this time, use three cylinders of your choice, remembering to construct them so that their volumes are 1000 cm^3.

Choose a value for r first.

For example, if $r = 5$ cm, then substituting this in the volume formula:

$$\text{Volume} = 1000 \text{ cm}^3$$
$$1000 = \pi \times 5 \times 5 \times h$$
$$\frac{1000}{\pi \times 5 \times 5} = h$$
$$h = 12.73 \quad \text{Rounded to two decimal places.}$$

Then complete a table to record your answers for the cylinders:

Shape	r	h	Volume	TSA	$\frac{\text{total surface area}}{\text{volume}}$
i			1000 cm^3		
ii			1000 cm^3		
iii			1000 cm^3		

Remember

For a cylinder:
Total surface area $= 2\pi r^2 + 2\pi rh$
Volume $= \pi r^2 h$

d Repeat parts **a** and **b** but, this time, use the three cones given below whose volumes are 1000 cm^3. (Record your answers in a copy of the table below.)

i

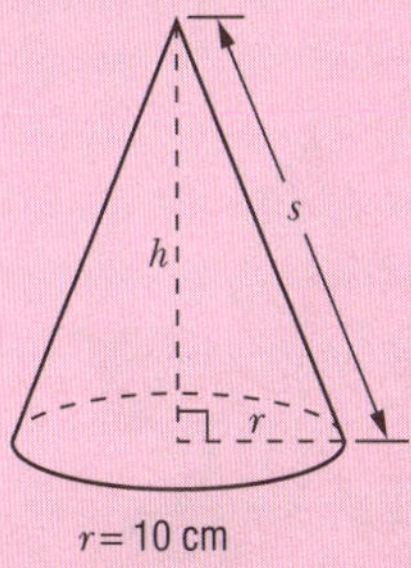

$r = 10$ cm
$h = 9.55$ cm
$s = 13.83$ cm

ii

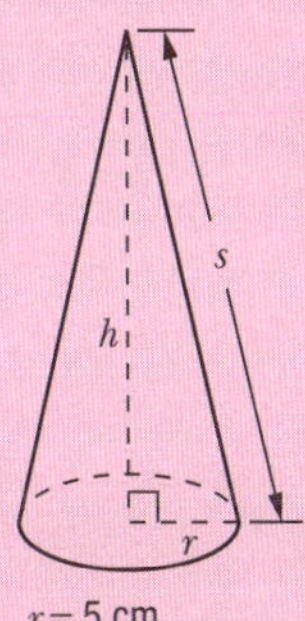

$r = 5$ cm
$h = 38.2$ cm
$s = 38.5$ cm

iii

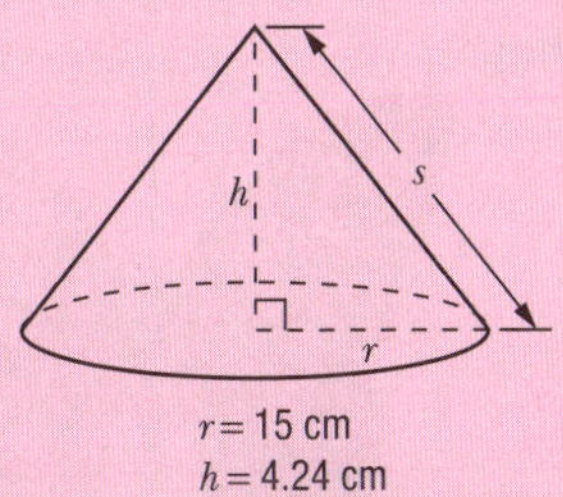

$r = 15$ cm
$h = 4.24$ cm
$s = 15.6$ cm

Shape	r	h	Volume	TSA	$\frac{\text{total surface area}}{\text{volume}}$
i			1000 cm^3		
ii			1000 cm^3		
iii			1000 cm^3		

Remember

For a cone:
Total surface area $= \pi r^2 + \pi rs$
Volume $= \frac{1}{3}\pi r^2 h$

e Only one sphere can be constructed with a volume of 1000 cm^3. Find its radius and then its surface area and $\frac{\text{total surface area}}{\text{volume}}$.

f **i** Of all the shapes that have a volume of 1000 cm^3 from parts **a** to **e** above, which shape has the smallest $\frac{\text{total surface area}}{\text{volume}}$ ratio?

ii Which shape do you suspect will always have the lowest $\frac{\text{total surface area}}{\text{volume}}$ ratio?

Remember

For a sphere:
Total surface area $= 4\pi r^2$
Volume $= \frac{4}{3}\pi r^3$

OPTION A: CONSTRUCTION

Geometrical facts and concepts are important in many aspects of our lives. They are used when a manufacturer makes storage containers for food, when a draftsperson draws plans for a building, and when an artist designs a stained glass window.

In this option, you will be making accurate constructions of 2D and 3D shapes and exploring their properties before undertaking an individual assignment which puts these skills into practice.

Lesson 1

Types of symmetry

Some shapes are pleasing to look at because they appear 'balanced' and symmetrical. There are examples of symmetry all around us. Symmetry can be seen in the natural world (for example in butterfly wings, leaves, or shells), as well as in the objects we create and construct in two dimensions and three dimensions, such as billums and buildings.

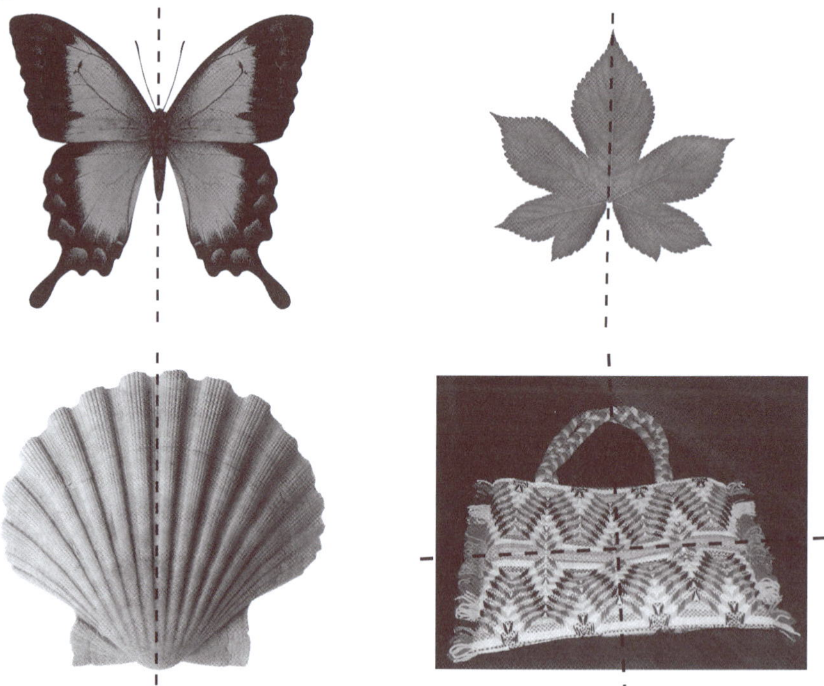

If one side of a shape or object is the mirror image of the other, the object has **bilateral symmetry**. If the shape could be folded its two halves would match. The fold line is an **axis of symmetry**. Figure A, on the next page, has four axes of symmetry but Figure B only has one axis of symmetry, and Figure C has none.

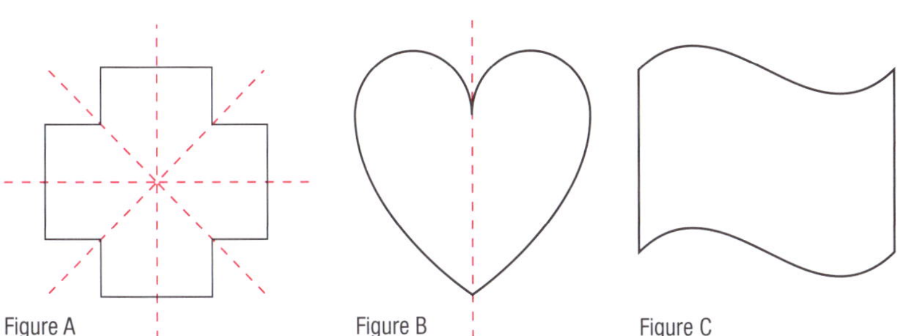

A second type of symmetry is **rotational symmetry**. For this type of symmetry, the object looks the same after it has been rotated through an angle less than 360°. A simple example is an equilateral triangle, which matches its original shape three times when rotated about its centre for one complete revolution. It is said to have rotational symmetry of **order 3**.

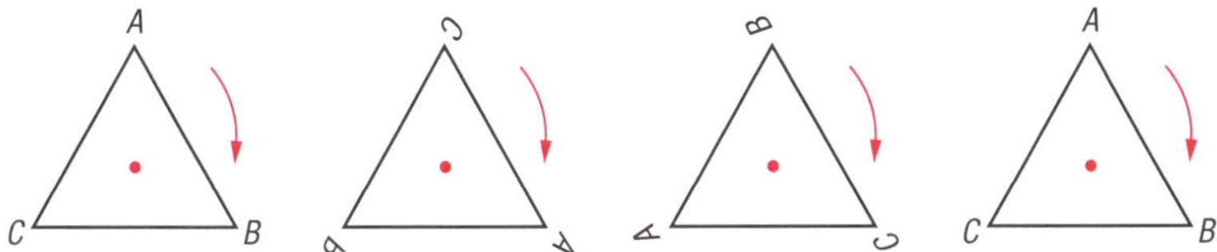

A simple way to investigate the order of rotational symmetry of a shape is to trace a copy. Then, using a pin or pressing a pencil point at the centre of the tracing, slowly rotate the tracing over the shape and observe how many times the tracing fits over the original within a full revolution.

Some shapes have both bilateral (fold) symmetry and rotational (point) symmetry.

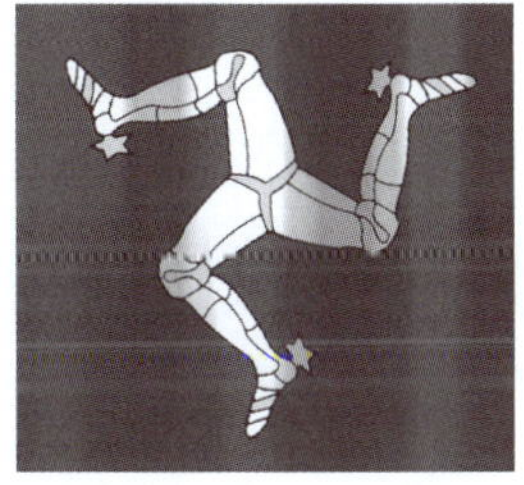

Figure D

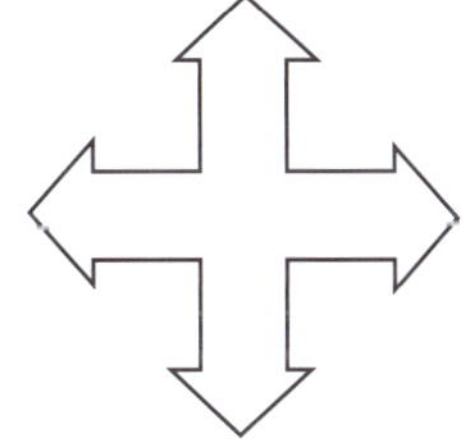

Figure E

The central design in Figure D (the flag of the Isle of Man in the United Kingdom) has no axis of symmetry but has rotational symmetry of order 3. Figure E has four axes of symmetry as well as rotational symmetry of order 4.

EXERCISE

1 How many axes of symmetry does each of the following shapes have?

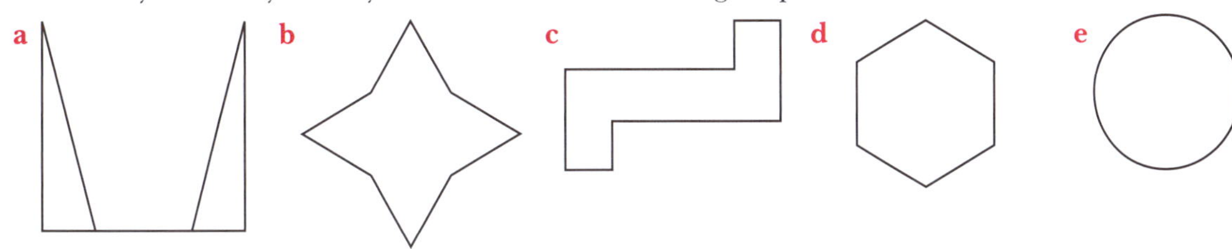

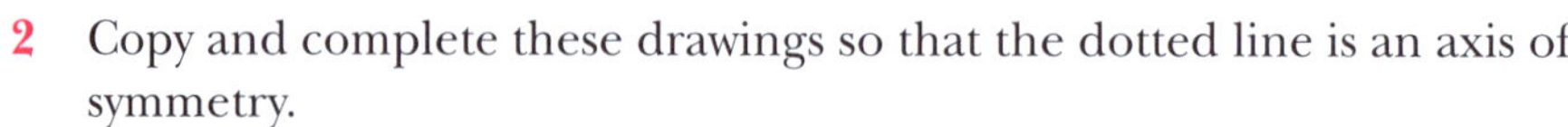

2 Copy and complete these drawings so that the dotted line is an axis of symmetry.

a

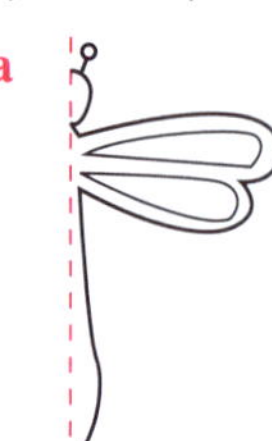

b

c

3 Sketch each of the following plane figures with the symmetry indicated:

- **a** a triangle with only one axis of symmetry
- **b** a quadrilateral with four axes of symmetry
- **c** a quadrilateral with two axes of symmetry
- **d** a pentagon with five axes of symmetry
- **e** a pentagon with only one axis of symmetry

4 Which of these shapes have rotational symmetry? For those that do, state the order of symmetry.

a **b**

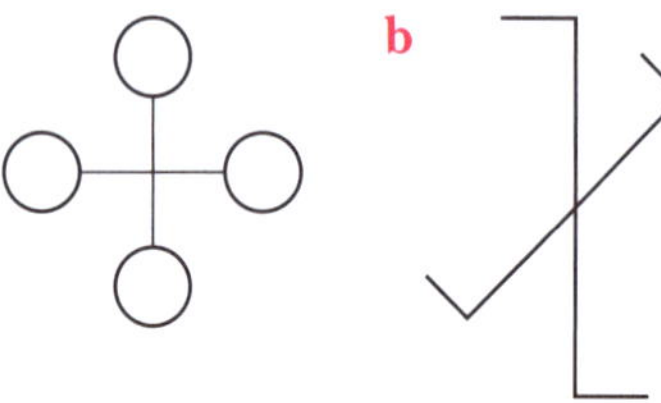

c

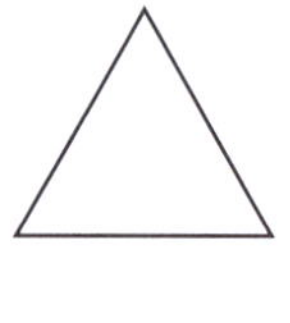

d

e

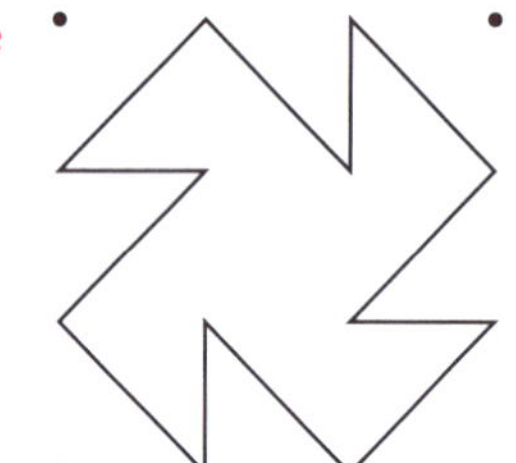

f

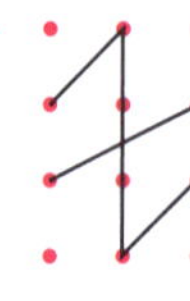

g

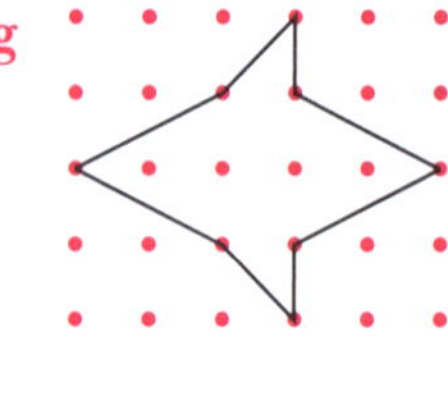

5 Draw a shape which has rotational symmetry of order:

a 2 **b** 3 **c** 4

6 Some capital letters of the alphabet, such as **C**, **E** and **M**, have an axis of symmetry, while others, such as **S**, have rotational symmetry.

- **a** Make a list of the letters of the alphabet which have an axis/axes of symmetry.
- **b** Which letters of the alphabet have rotational symmetry? State the order of symmetry for each of these letters.
- **c** '**WOW**' is a symmetrical word because an axis of symmetry can be drawn vertically through the O. Find six other words of three or four letters which are symmetrical when they are printed across or down the page.

7 Make your own symmetrical designs using one of the following methods.

a Take a square piece of paper. Fold it in half horizontally, then into quarters vertically, as shown below. Cut out shapes from the folded paper. Then open it out to show the symmetrical design. Experiment with different folds.

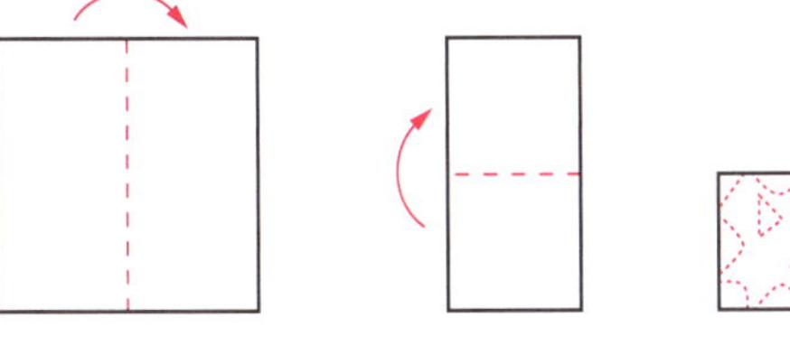

b Take a piece of grid paper or dot paper. Draw horizontal and vertical lines bisecting the page, so the page is divided into quarters. In one quarter, draw and colour a design. Reflect your design and draw its image in the other three quarters.

Constructions: Angles

Accurate drawings or diagrams are needed when constructing tools, machinery, buildings, and so on. A designer or draftsperson, builder, engineer or architect will use accurate diagrams to represent what they are designing and constructing. Special drawing tools or computer programs exist for this work; however, we can also draw many diagrams accurately using a ruler, a pencil and compasses.

Earlier, we learnt the steps for constructing a 90° angle. The following instructions are for some other useful angle constructions.

Constructing a 60° angle

Step 1: Draw a horizontal base line and label one end A.

Step 2: Place the point of your compasses at A and open the compasses to a radius smaller than the length of the horizontal base line. Then draw a large arc. Where your arc intersects the base line, label the point M.

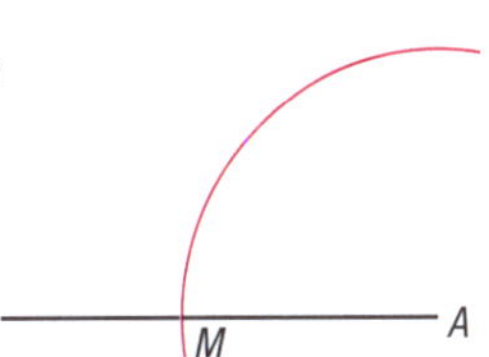

Step 3: Keeping the compasses open the same distance as for Step 2, move the point of the compasses to M and draw an arc to intersect the first arc.

Step 4: With a ruler and pencil, join A to the point of intersection of the arcs. The acute angle at A is 60°.

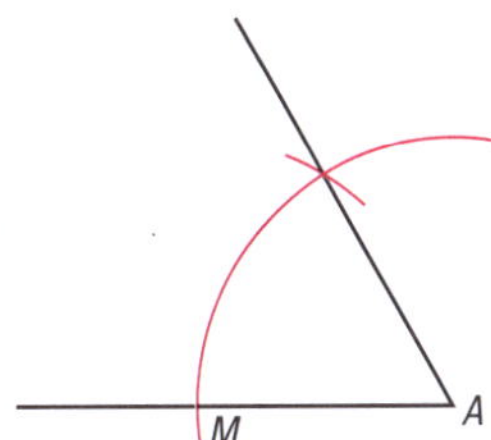

Copying an angle

We can copy an angle that has already been drawn, or can make a pair of identical angles using similar steps. Below are the steps for copying $\angle XYZ$, which involves moving between the copy and the original angle.

Step 1: Draw a horizontal line, PQ.

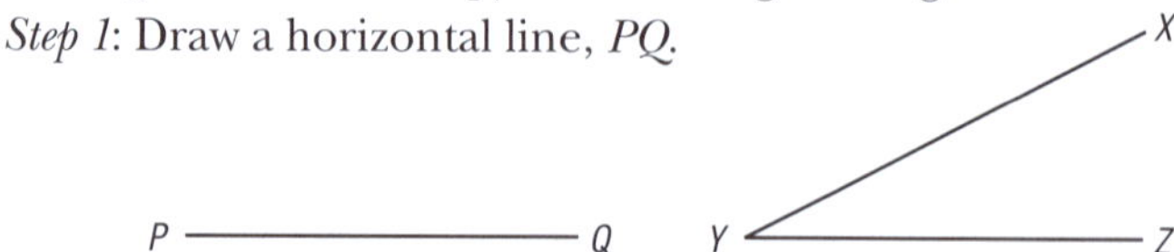

Step 2: Place the point of your compasses at Y on the original angle and open the compasses to a radius smaller than the length of YZ. Draw an arc which crosses both arms of $\angle XYZ$. Label the intersection points D and E.

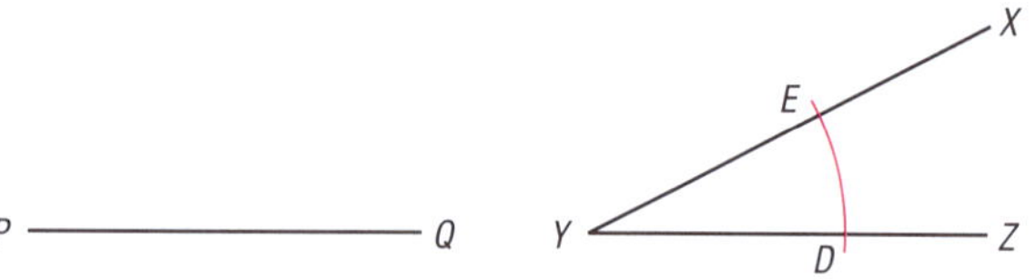

Step 3: Keeping the compasses open the same distance as for Step 2, move the point of the compasses to P and draw a large arc intersecting the line PQ at L.

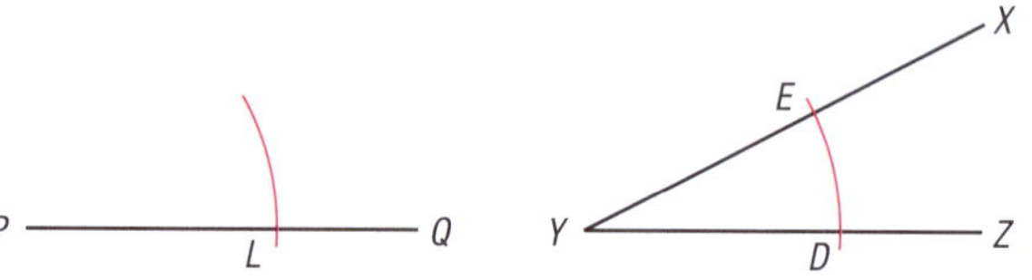

Step 4: Move the point of the compasses to D and adjust the compasses so that the pencil tip just passes through E.

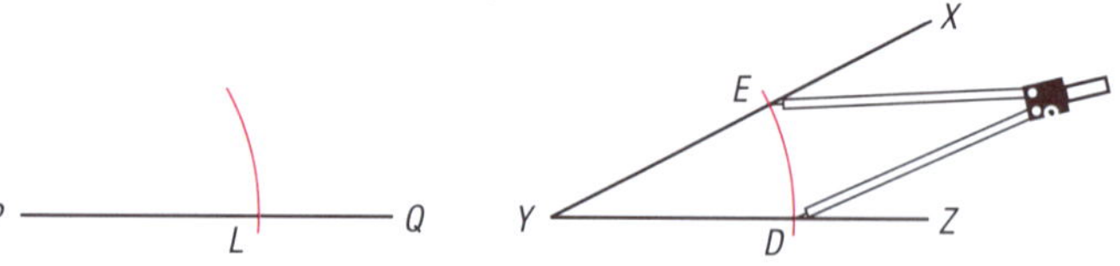

Step 5: Keeping the compasses open the same distance as for Step 4, move the point of the compasses to L and draw an arc to intersect the first arc at M.

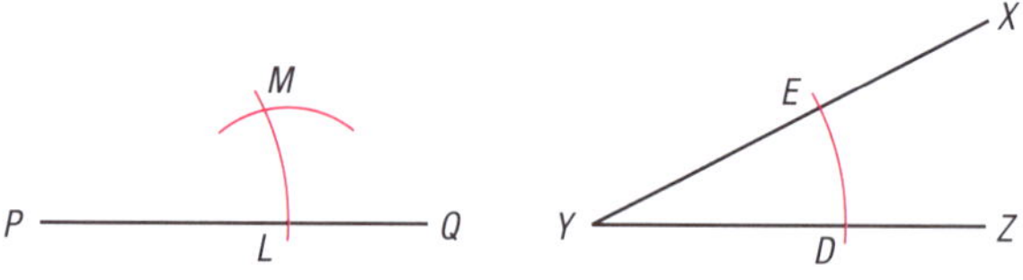

Step 6: Join M to P.

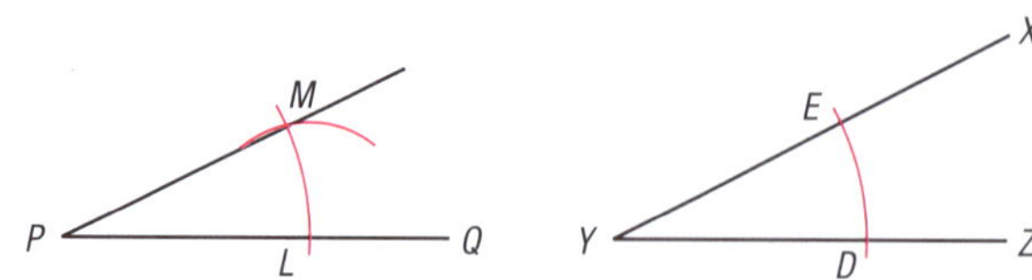

You can place one angle over the other or measure both angles to test your accuracy.

Bisecting an angle

Step 1: Use a ruler and a pencil to draw an angle. Label the angle $\angle LMN$.

Step 2: Place the point of your compasses at *M* and open the compasses to a radius that enables you to draw an arc which intersects both arms of the angle, at *F* and *G*.

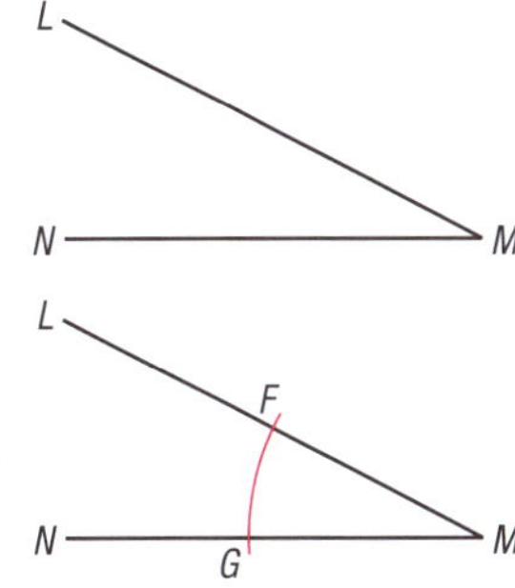

Step 3: Keeping the compasses open the same distance as for Step 2, move the point of the compasses to *F* and draw an arc. Then move the point of the compasses to *G* and draw a second arc to intersect the arc you just drew using *F* as the centre.

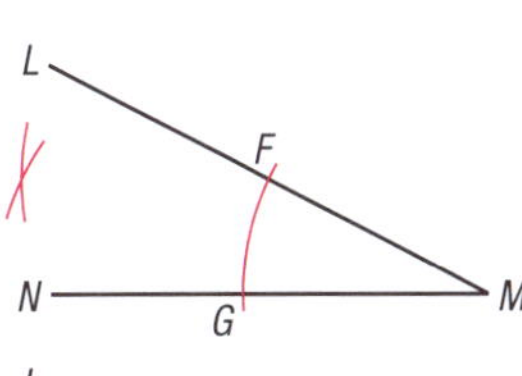

Step 4: With ruler and pencil, join the point of intersection of the arcs to *M*. Measure the two angles formed to check your accuracy.

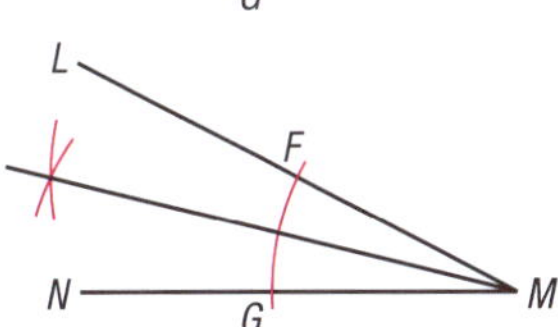

EXERCISE

1 Copy each of the following angles:

a

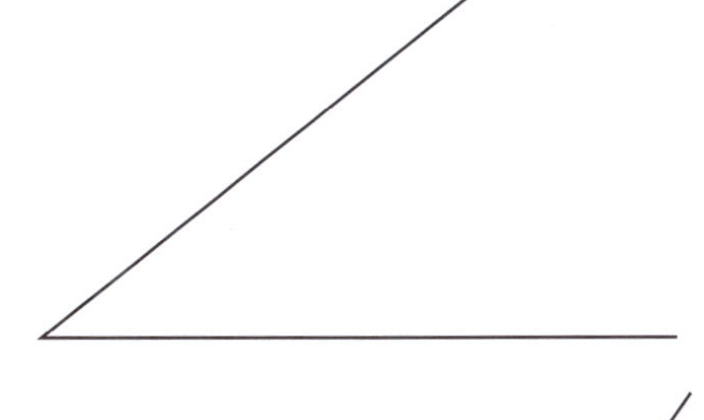

b

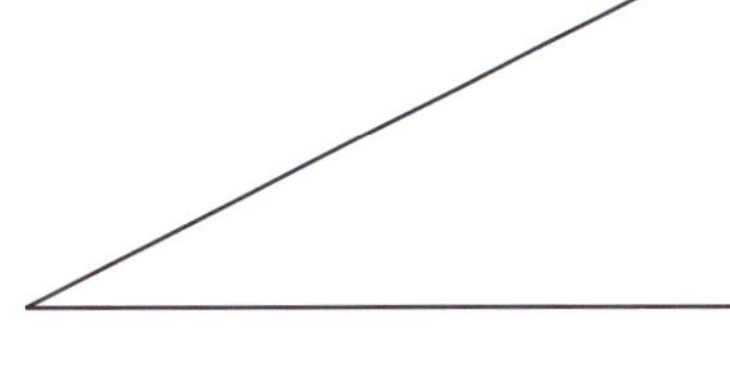

c

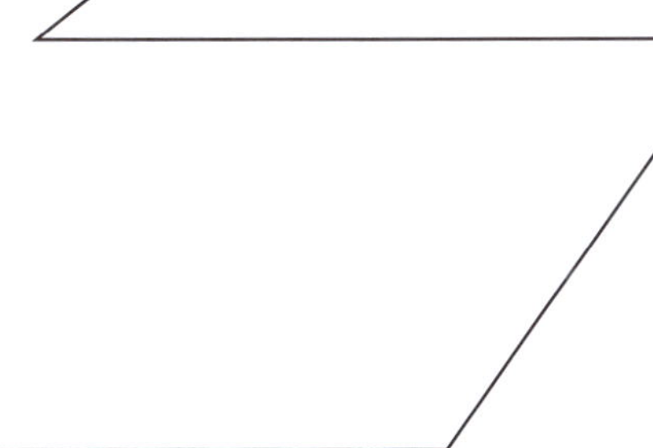

d

2 a Draw PQ that is 6 cm long. Then, using PQ as a base line, construct a 90° angle at point P and a 60° angle at point Q, as shown in the diagram on the right.

b Extend the arms of the angles, if necessary, to form a triangle and use your protractor to measure the third angle.

3 Bisect each of these angles:

a

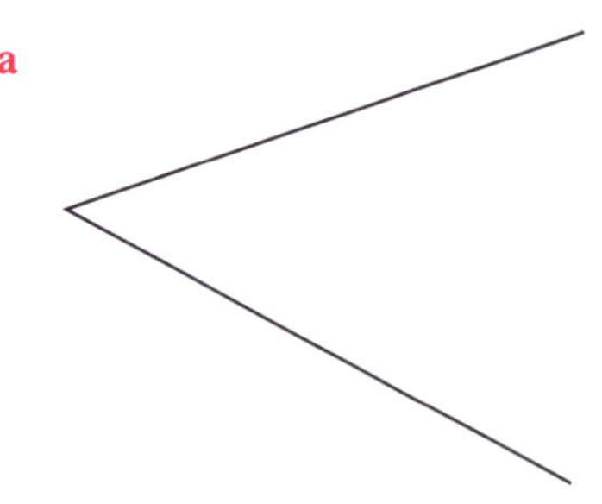

b

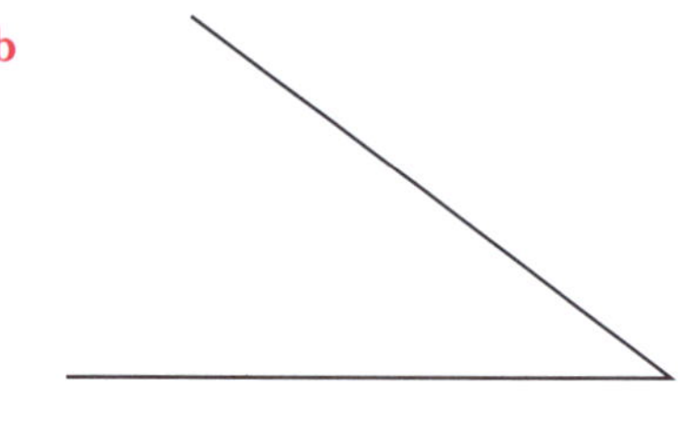

c

d

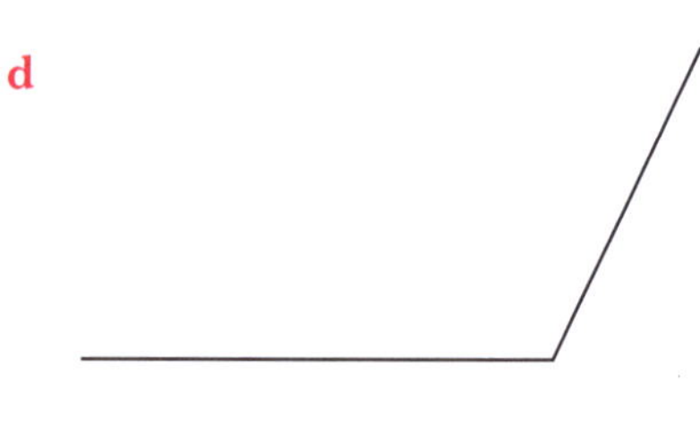

e

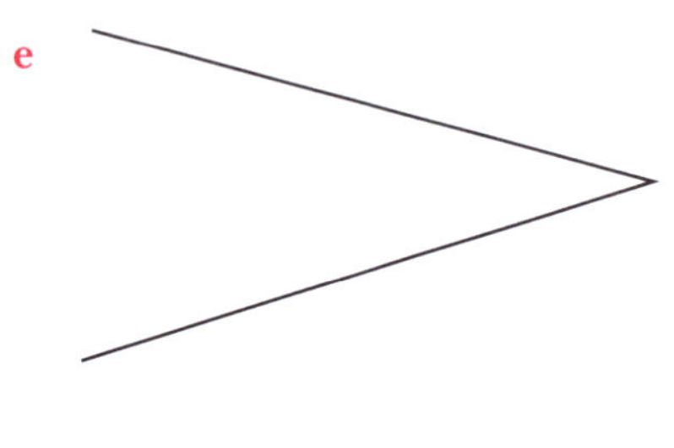

f

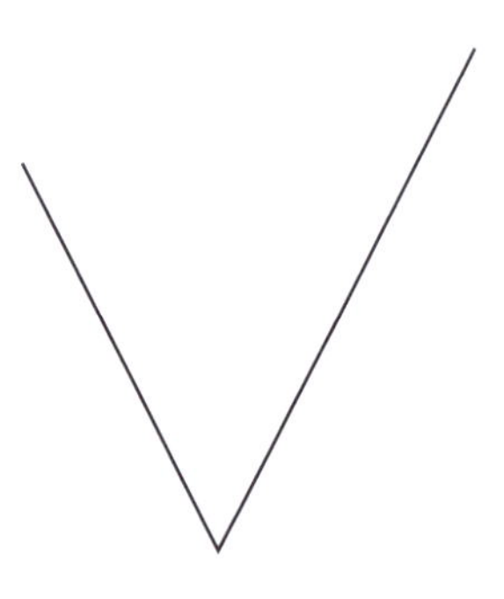

4 Construct an angle of 30° by constructing a 60° angle and bisecting it.

5 Construct a 90° angle and bisect it to construct a 45° angle.

Constructions: Lines

Constructing a perpendicular bisector

Here are the steps for constructing a line which bisects *XY* at right angles.

Step 1: Draw a horizontal line and label the end points *X* and *Y*.

Step 2: Place the point of your compasses at *X* and open the compasses to a radius that is more than halfway between *X* and *Y*. Swing the compasses above and below *XY*, marking long arcs as shown.

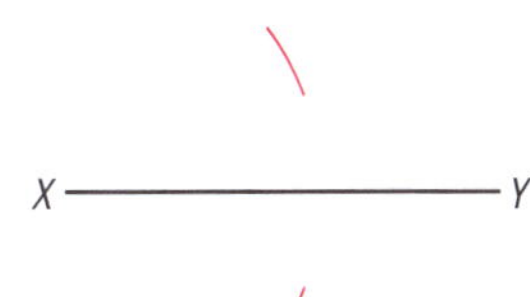

Step 3: Keeping the compasses open the same distance as for Step 2, move the point of the compasses to *Y* and mark two more arcs above and below *XY* to cross the first arcs.

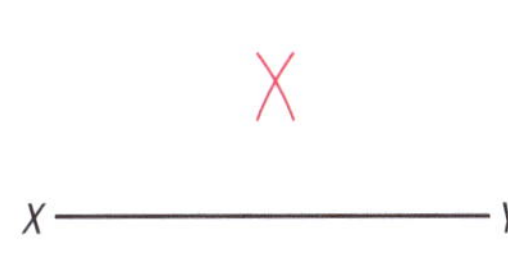

Step 4: With ruler and pencil, join the points made by the two sets of intersecting arcs. The angle made by the two lines is a right angle.

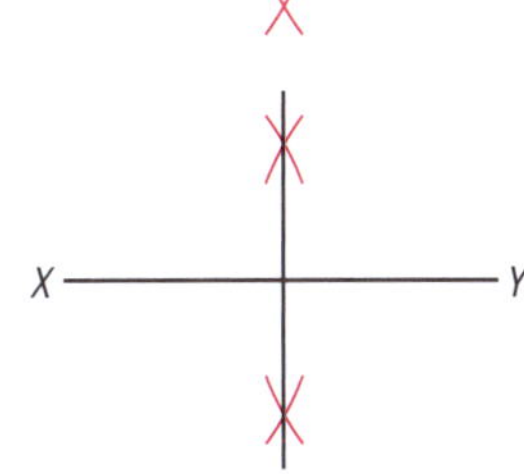

Constructing a perpendicular from a point to a line

Here are the steps for drawing a perpendicular from point *X* onto the line *AB* as shown below.

Step 1: Place the point of the compasses at *X* and draw an arc to cut *AB*, at *L* and *M*.

Step 2: Using *L* as the centre, and with the radius of the compasses at more than half *LM*, draw an arc below *AB*.

Step 3: Using *M* as the centre, and with the same radius as for Step 2, draw a second arc below *AB* to intersect the first arc, at *Y*.

Step 4: Join *XY*.

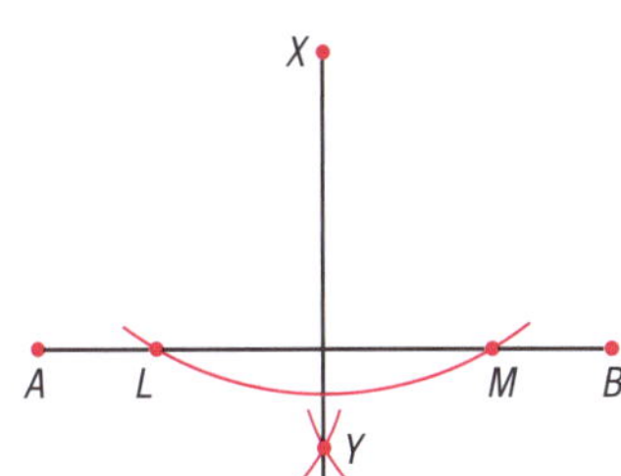

Constructing a line parallel to another line

Here are the steps for drawing a line parallel to *AB* passing through the point *X*, as shown below.

Step 1: Draw a line through *X* to cut *AB* at *Y*.

Step 2: Using *X* as the centre draw an arc that intersects at *E* the line you drew in Step 1, as shown.

Step 3: Keeping the compasses open the same distance as for Step 2, move the point of the compasses to *Y*. Draw an arc that intersects *AB* at *C*, and the line you drew in Step 1 at *D*.

Step 4: Open the compasses to the distance *CD*. Using *E* as the centre, draw another arc with this radius that intersects the arc through *E*, at *F*.

Step 5: Join *XF* and extend the line.

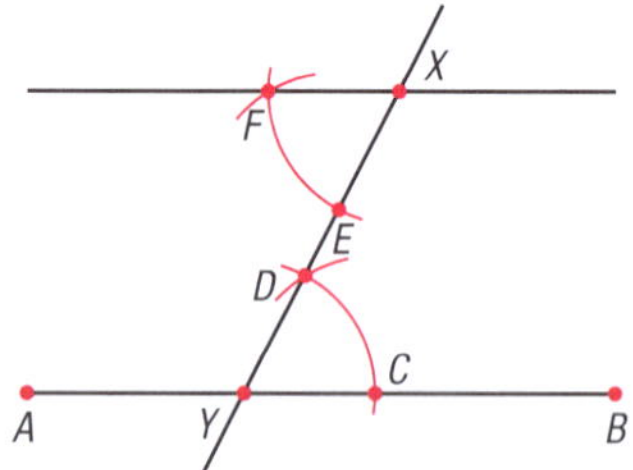

EXERCISE

1 Draw a line that is 10 cm in length. Construct its perpendicular bisector, using a ruler and compasses. Measure the line segments and the angle to check your accuracy.

2 Trace the diagram below and construct a perpendicular to the line *AB* through the point *X*.

X •

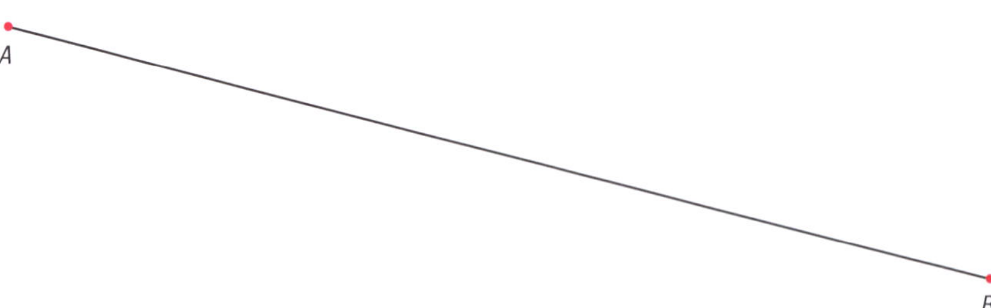

3 Trace the diagram below and construct a line parallel to the given line and passing through the point *X*.

4 Draw a scalene triangle with three acute angles. Construct the perpendicular bisectors of all three sides. (These should meet at one point, called the **circumcentre**. Using this point as the centre, you should be able to draw a circle which touches all three vertices of the triangle).

Drawing diagrams of 3D objects

Lesson 4

Two of the most common three-dimensional shapes are a cube and a rectangular prism (or cuboid, or box).

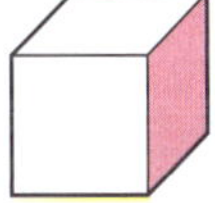

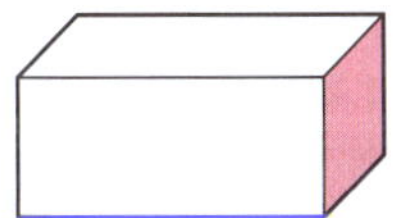

How do we draw these, and other 3D shapes, on a sheet of paper which has only two dimensions? The trick is to make it appear that the shape is standing out towards you or extending back into the page. A simple way to reproduce a rectangular prism or cube as a 2D diagram is outlined below.

Rectangular prism

Step 1: Start by drawing a rectangle, using the lines of your page as a guide.
Estimate the centre of the rectangle and mark it with a dot, as shown on the right.

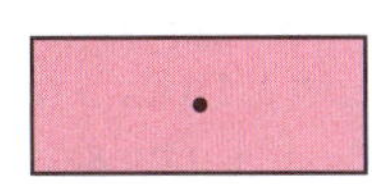

Step 2: Starting at the point marked at the centre of your rectangle, draw a second rectangle exactly the same size as the first one. Any of the four corners of this second rectangle can touch the dot at the centre of the first rectangle. So each of the four diagrams at the top of the next page is possible.

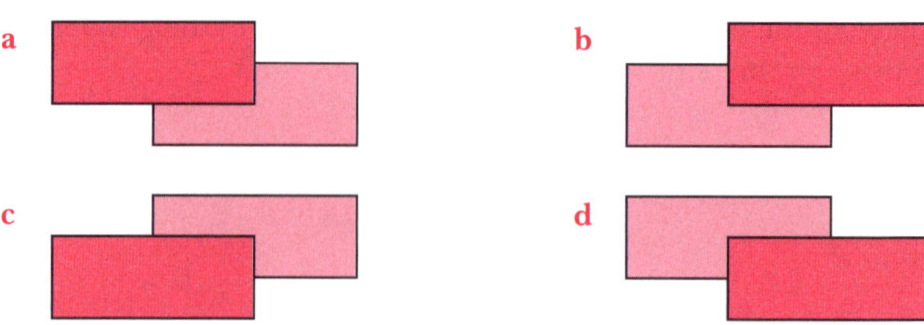

Step 3: Join the corners of the overlapping rectangles, as shown in diagram A below. You can then rub out all of the lines that would be hidden from view (see diagram B below) or show them as dotted lines (see diagram C).

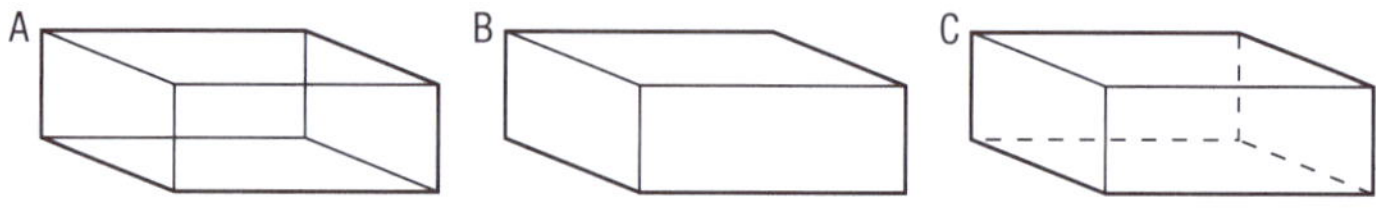

Cube

The same set of steps as given for a rectangular prism, above, can be used to draw a cube, provided we start by drawing a square and overlap it with a second square, as shown below.

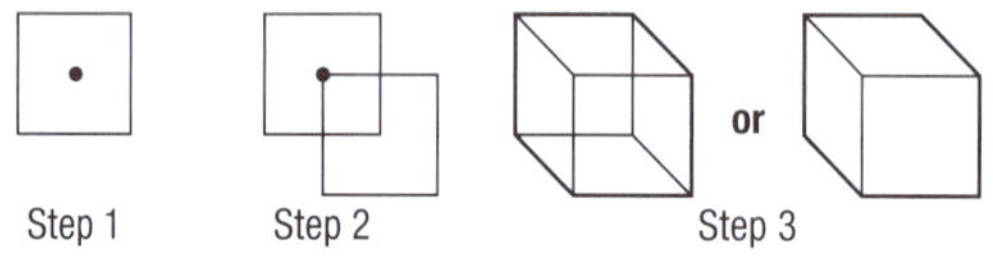

Using dot paper

Paper marked with dots in a square or triangular pattern also makes it easy to draw 3D shapes.

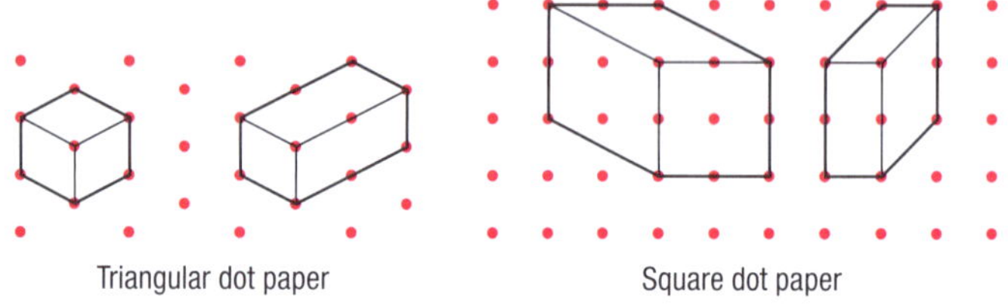

Following are some of the other solids you need to be able to sketch neatly.

Triangular-based pyramid

Step 1: Draw a triangle with its apex off-centre. (The side that will be hidden 'behind' the pyramid should be dotted.)

Step 2: Choose a point in the middle of the triangle and draw a vertical dotted line from it.

Step 3: Rule lines to join the top of the vertical line to each of the corners of the triangle.

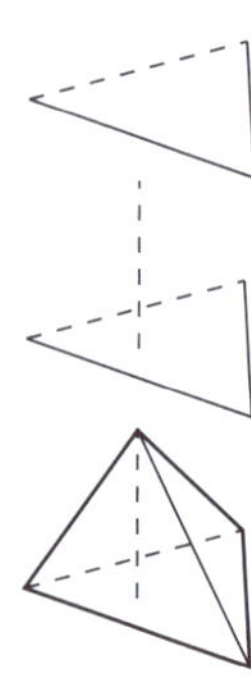

Square-based or rectangular-based pyramid

Step 1: Draw a parallelogram. (Two sides should be dotted, as shown.)

Step 2: Draw a vertical dotted line from the centre of the parallelogram.

Step 3: Rule lines to join the top of the vertical line to each of the corners of the parallelogram. (One of the lines should be dotted, as shown.)

Cone

Step 1: Draw an ellipse, with its top half dotted.

Step 2: Draw a horizontal dotted line through the ellipse.

Step 3: Draw a vertical dotted line from the centre of the ellipse.

Step 4: Rule lines to join the top of the vertical line to the edges of the ellipse, as shown.

Sphere

Step 1: Draw a circle.

Step 2: Draw a horizontal dotted diameter of the circle.

Step 3: Draw an ellipse with its top half dotted, as shown.

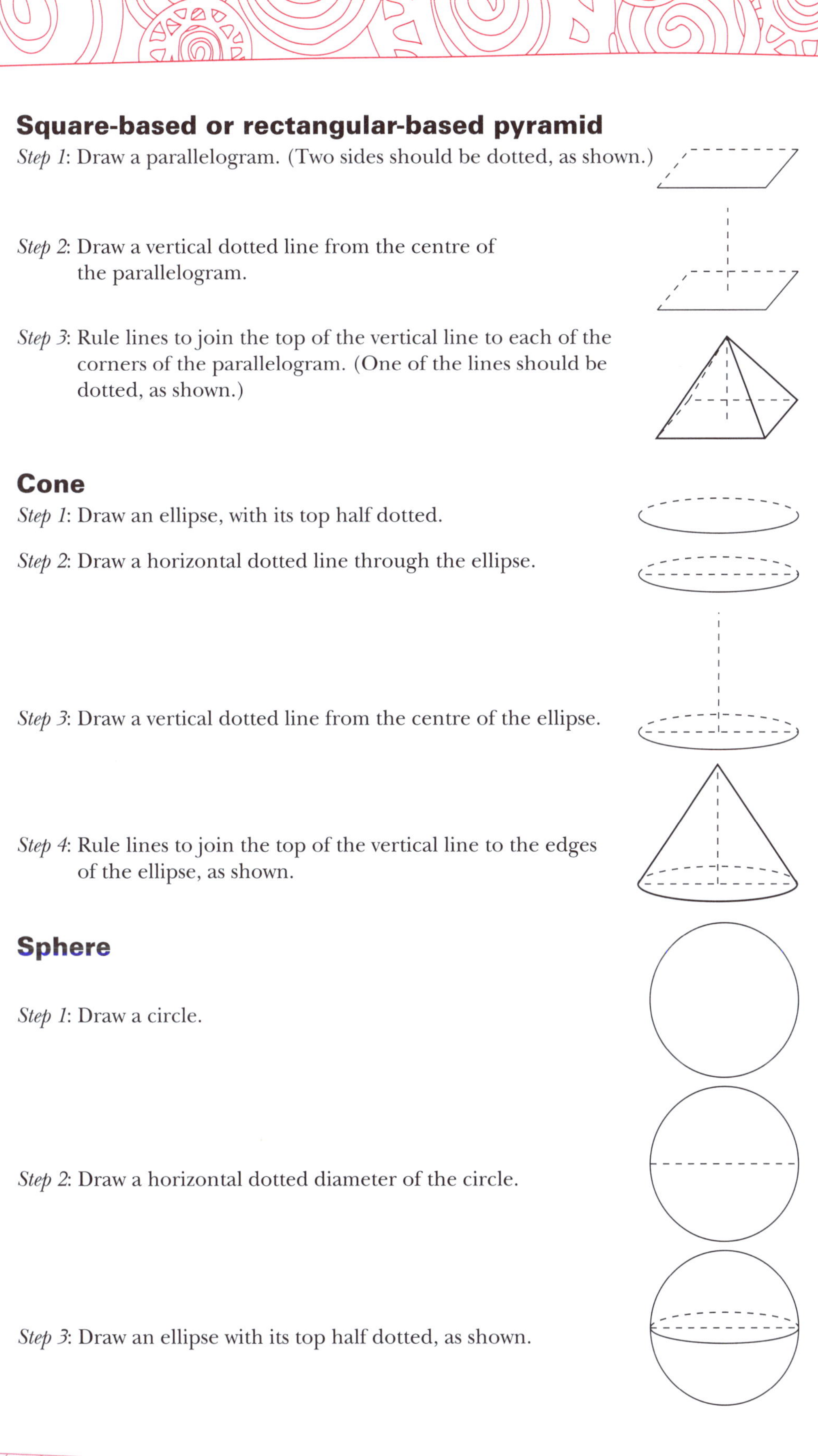

EXERCISE

1 Use a pencil and ruler to make a neat drawing of each of the following three-dimensional shapes:
 a a triangular-based pyramid
 b a square-based pyramid
 c a cone
 d a sphere
 e a hexagonal-based pyramid
 f a hemisphere (half sphere) resting on its flat surface
 g a cube with a square-based pyramid sitting on its top surface
 h a rectangular prism with a pyramid sitting on its top surface
 i a sphere just fitting inside a cube
 j a cone just fitting inside a rectangular prism

2 Each of the following diagrams is a partly completed drawing of a 3D shape. Copy each diagram and complete it.

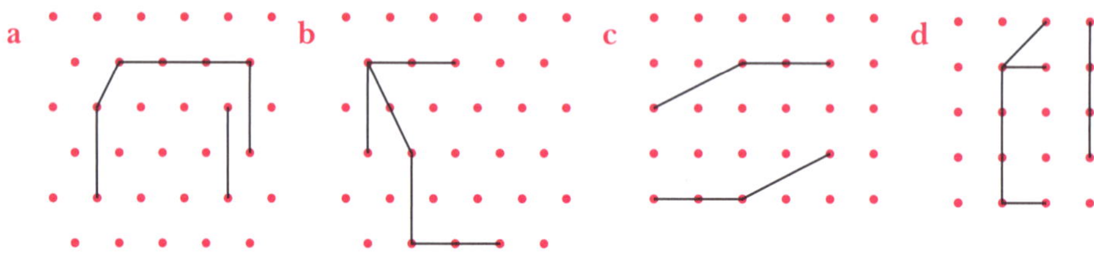

Constructing nets of solids

Calculating the surface area of solids often requires us to sketch the net of the solid. It is also sometimes necessary to construct an accurate model of a solid and this involves some of the special construction techniques covered in the last few lessons.

EXERCISE

1 To make a skewed pyramid, first construct a square with side length 6 cm, as shown in the diagram on the right. Extend the sides, as shown, to form triangles and add flaps to the edges to allow for joins (the flaps are shown as dotted lines on the diagram).

Cut out the net, crease along *AB*, *BC*, *OC* and *OA*, and crease the joining tabs. Fold to form the pyramid and glue the tabs in place.

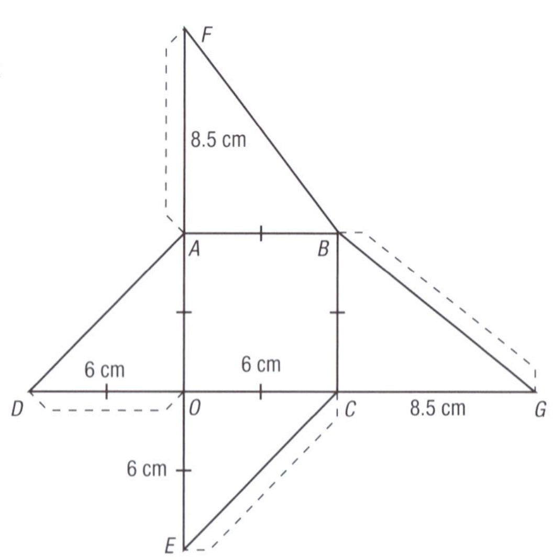

2 A tetrahedron can be accurately constructed using compasses to produce the four equilateral triangles which make up the net of a tetrahedron.

Using the same radius each time, lightly draw a set of circles, as shown below. (The centres of the circles are numbered in the order we need to construct the circles.) Join the intersection points of the circles to form four triangles. Cut around the perimeter of the set of triangles, allowing extra on some edges for joining. Fold along the lines and join the edges together.

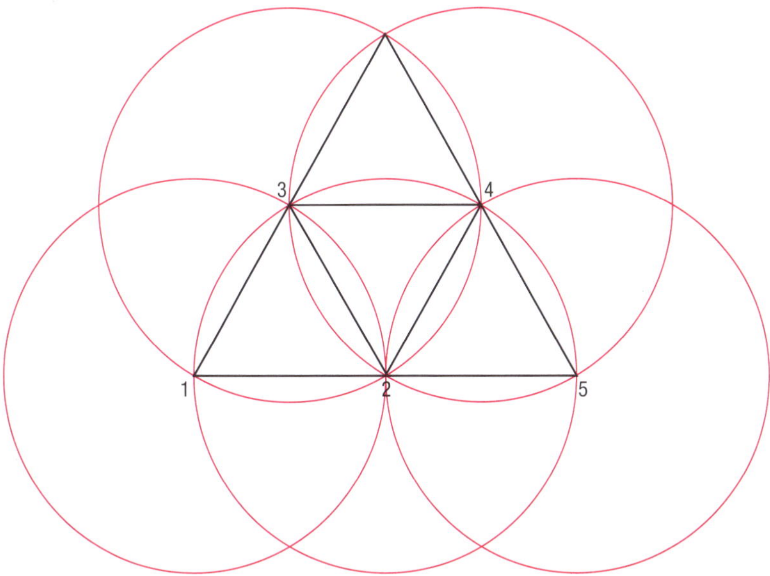

Similarity and scale drawings

We have already learned that similar figures are larger or smaller versions of each other. There are two simple methods for creating similar figures. We have seen one way of making an enlargement in our work with ratio. This involves placing a grid of small squares over the original picture, and copying the drawing onto a grid of larger squares. The **enlargement factor** or **scale factor** will determine the size of the second grid.

For example, suppose we want to enlarge the illustration on the right by a scale factor of 2. The grid over the original illustration shown is a 1 cm grid, so we will need to draw a grid with squares of 2 cm. Each part of the illustration is then transferred onto the corresponding square of the larger grid (see the top of the next page).

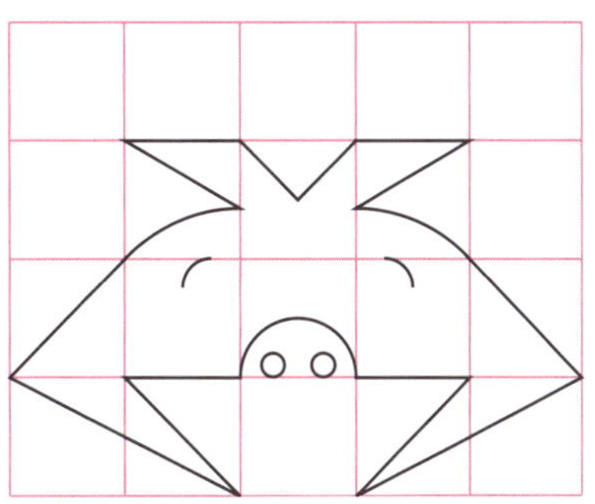

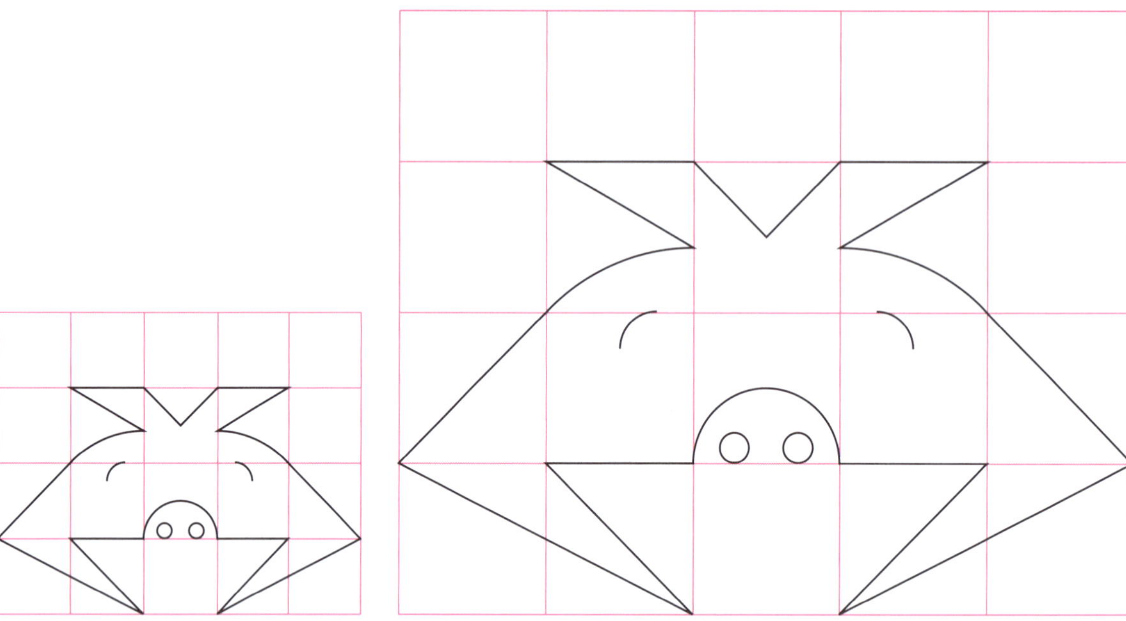

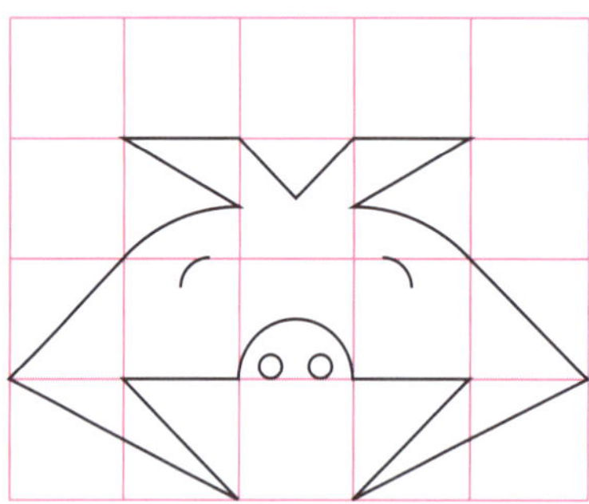

We can also **reduce** the size of a figure. This time, the second grid would consist of squares that are smaller than the squares on the original. In this case, the scale factor will be a fraction, not a whole number.

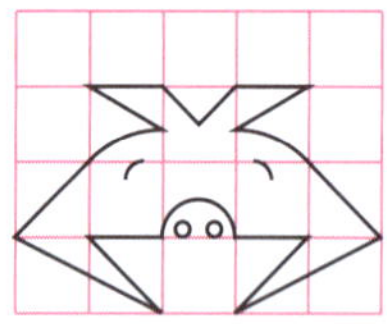

A second method for enlarging figures uses the same principle as a slide projector. In the case of a projector, the light shines through a slide (a small rectangle), and throws the image of the slide onto a projector screen. The image is a much larger rectangle than the original slide, and its size can be altered by moving the screen closer to or farther away from the projector.

To draw an image using this approach, we need to increase or decrease by the same amount the distance from a fixed point to each point on the object (see the example below). The fixed point, *O*, can be anywhere.

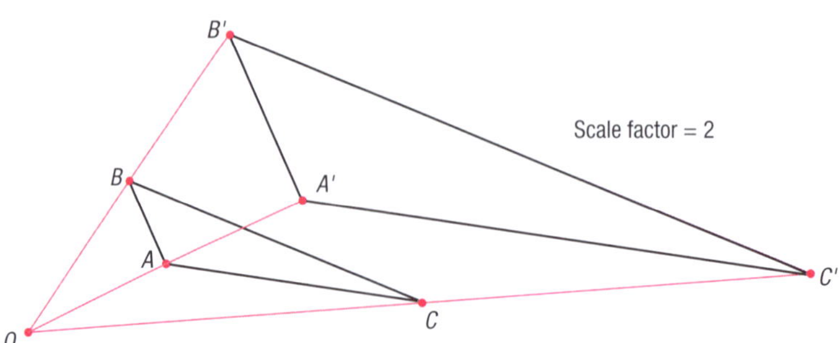

Some fractals have a property called 'self similarity'. This is present if they can be divided into parts where each part is a scaled replica of the whole. For example, in the Sierpinski triangle, each triangle is replaced by three similar triangles with side lengths half of those they are replacing.

EXERCISE

1 Choose one of the pictures below to enlarge. On an A4 or A3 sheet of paper, draw a grid with enough squares to fit the picture you have chosen, using the biggest squares possible. Draw the enlargement.

a

b

2 Copy each of the following shapes. Draw the image of each shape under the scale factor given, using O as the centre of enlargement or reduction.

a scale factor of $\frac{1}{2}$

b scale factor of 2

c scale factor of 3

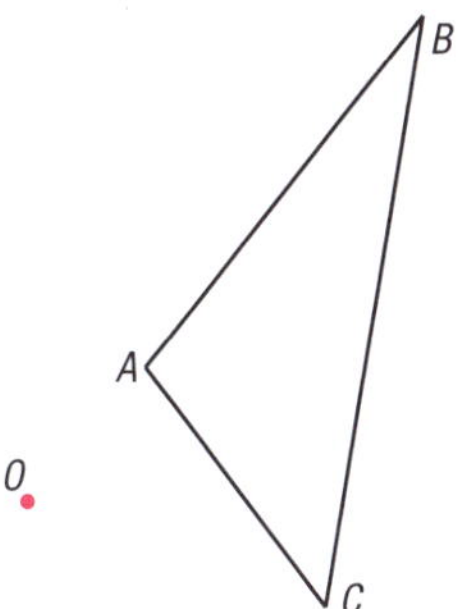

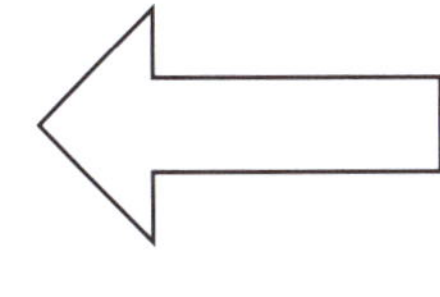

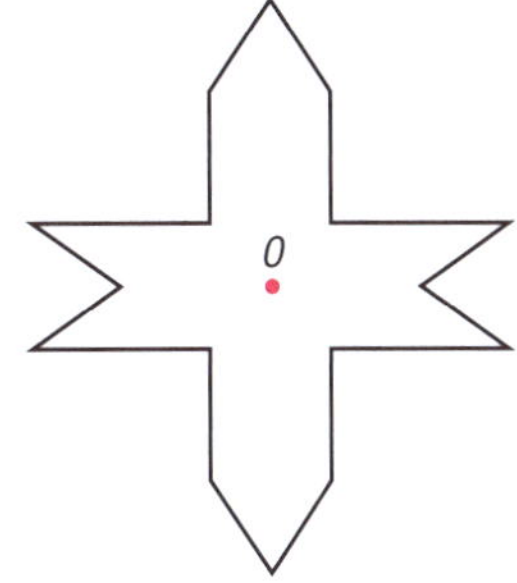

3 Choose one of the illustrations below. Trace the illustration and rule a 0.5 cm grid over your tracing. Use a 2 cm grid to enlarge the drawing.

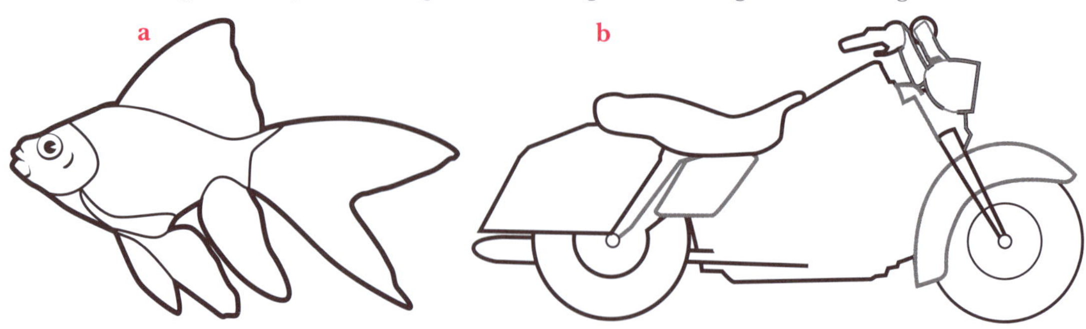

Challenge

a **Just for fun**

In all of the enlargements or reductions you have completed you have been keeping the shape of the original. This investigation considers what happens when you copy a picture onto a different form of grid which distorts the original shape.

For example:

 on 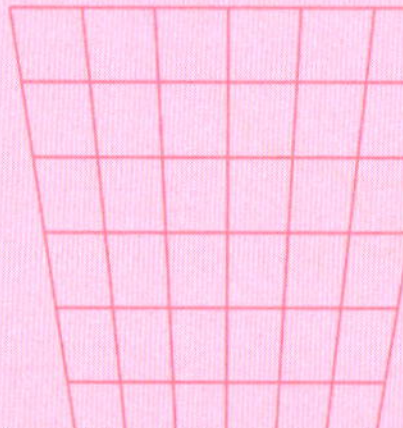 becomes 

What to do:

i On ordinary squared grid paper, draw a picture of your own choice.

ii Redraw your picture on a different type of grid, for example:

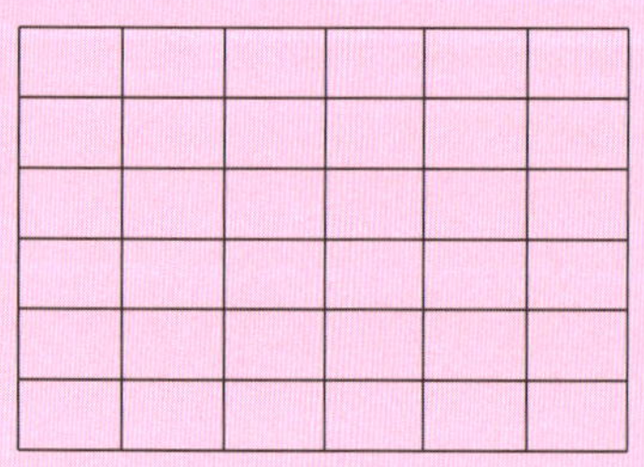

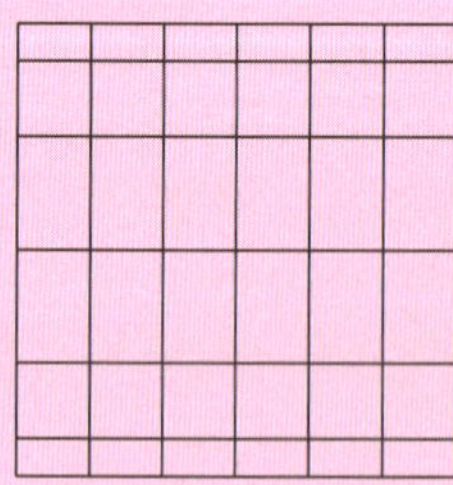

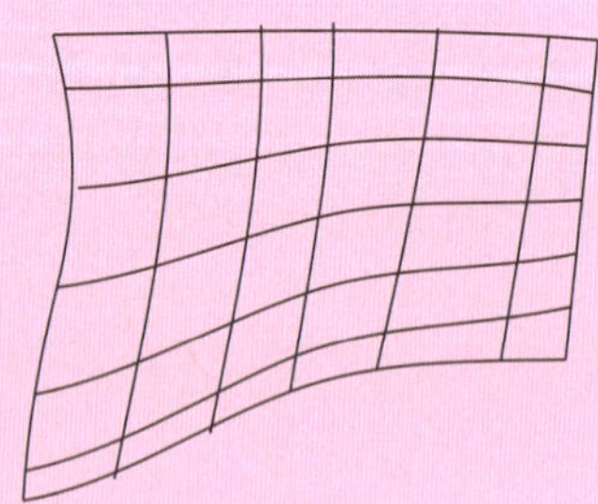

Using scales and maps

One of the most common applications of scale drawings is maps, because it is not practical to use actual measurements. Different scales are used, depending on the area being scaled down. On a map of a whole country, for example, 1 cm might represent 500 km, while on a local town map 1 cm might represent 1 km. Builders and architects also use scale drawings before constructing houses.

EXERCISE

1 Make a scale drawing of your classroom floorplan showing the main items of furniture. Start by measuring the dimensions of the room and deciding on a suitable scale to fit this onto a page of your workbook.

2 Here is a rough sketch of a block of land showing the position of the house. Using a scale of '1 cm equals 2 m', draw a scale diagram of the block of land and house.

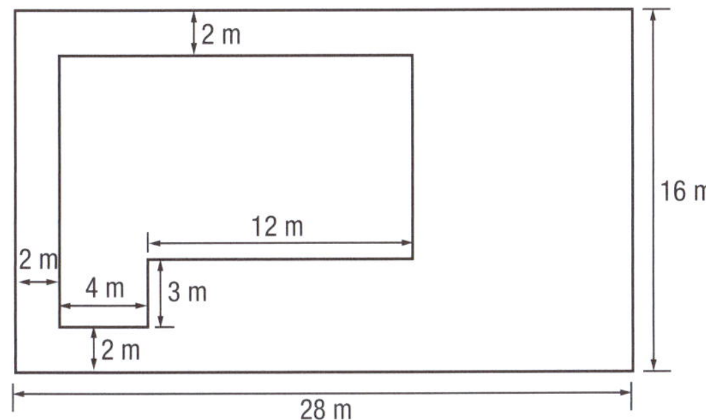

3 Draw a scale diagram of:
   a a square with sides of 300 m, using a scale of 1 cm = 50 m
   b a circle with a radius of 28 km, using a scale of '1 cm represents 1 000 000 cm'.

4 A scale diagram of a building is drawn with a scale of 1 : 1000.
   a If, on the drawing, the height is 9 cm and the width is 4 cm, what are the actual height and width of the building, in metres?
   b If the actual height of the door to the building is 2.8 m, what will its height be on the scale drawing?

5 The diagram below represents the shape of the triangular roof of a building. Draw an accurate scale drawing and use it to determine:
   a the length of the rafter *AB*
   b the perpendicular height of *A* above *BC*.

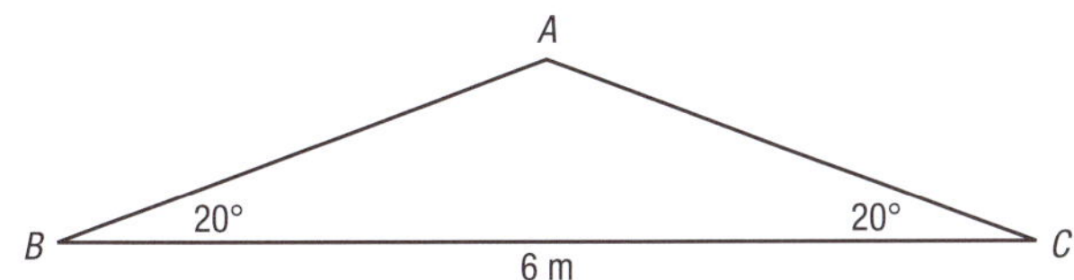

6 The scale on the map of Papua New Guinea below is 1 cm = 100 km. Use the map to calculate, to the nearest 100 km, the distance between:

a Kerema and Port Moresby
b Lorengau and Wewak
c Daru and Arawa
d Tari and Lae
e Rabaul and Madang

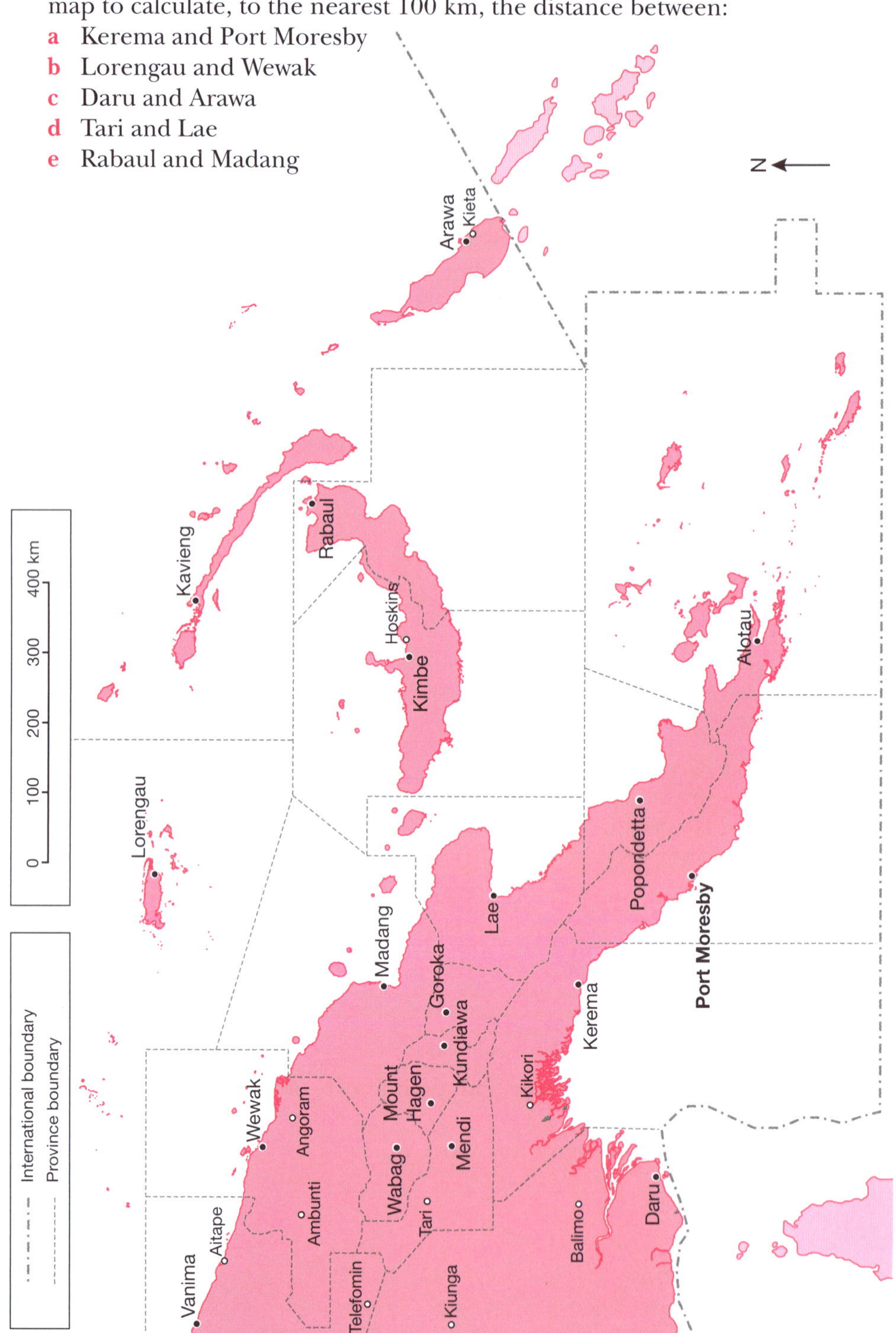

7 If two towns are 4.2 cm apart on the map from Question 6, what is the actual distance between them?

OPTION B: DEDUCTIVE REASONING

In geometry, we often rely on carefully constructed diagrams to support theories about angles, lines and the properties of shapes. For example, if we measure the interior angles of several triangles, we can come to the conclusion that: 'The sum of the interior angles of a triangle is 180°.'

Many geometric theories, such as this, can also easily be proved using deductive reasoning.

In this option, we will look at proofs, and prove some of the common geometric theories that we have used in previous work on geometry. We will also practise using deductive reasoning for solving geometry problems.

History of deductive reasoning in geometry

Euclid was a mathematician and teacher who lived in Alexandria around 300 BC. He wrote many books on mathematical theory, but the work for which he is most famous, called *Elements*, contains a complete study of geometry, including proofs using deductive reasoning. Although Euclid was probably not the first person to prove many of the geometrical theorems he managed to organise them and clearly present them in a textbook that was used for 2000 years.

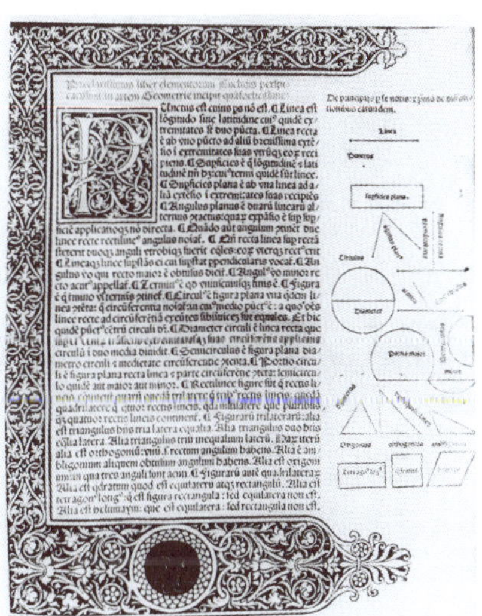

A page from Euclid's *Geometry*

Euclid

Euclid began his work on geometry by stating five simple ideas called **postulates**. One of Euclid's postulates is: 'It is possible to draw one straight line between any two points'.

To the postulates he added what he called **axioms** (facts that he thought were accepted as always being true). One of Euclid's axioms is: 'Things that are equal to the same thing are equal to each other.'

From his small collection of postulates and axioms Euclid went on to prove deductively the geometric theorems that we assume today.

Describing geometric diagrams and using conventional notation

If we are proving a geometric theorem, it is important that we are able to communicate our reasoning in a manner that other students and our teachers can easily understand.

This first lesson in this option revises the way we describe and label common shapes, and some conventional mathematical notation.

Lines

- The diagram on the right shows a line joining point *A* to point *B*.
  It is described as line segment *AB* or just *AB*.
  Sometimes this is written as $\overline{AB}$. The line ruled above *AB* indicates that a line, or line segment, joins point *A* to point *B*.
- Lines that are equal in length are marked in the same manner. On the diagram below *AB* and *BC* are the same length ($AB = BC$) because both line segments are marked with —+—.

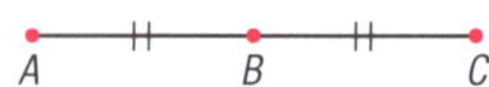

- Sets of parallel lines are marked with matching arrows. In this diagram, *AB* is parallel to *CD* and to *EF*.

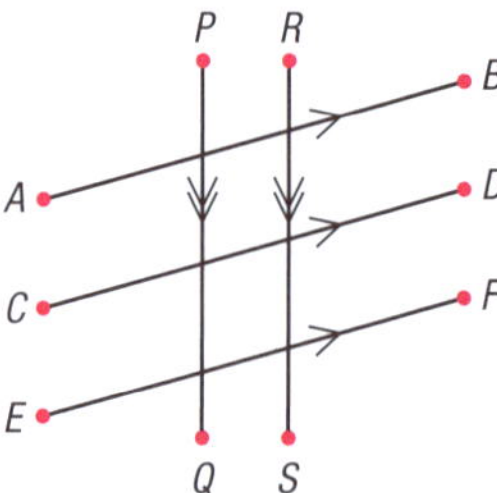

  Notation for this is $AB \parallel CD$, $AB \parallel EF$ and $CD \parallel EF$.
  If there is more than one set of parallel lines on a diagram (and their orientation is different from the other sets of parallel lines) then two arrows are used on the second set (see *PQ* and *RS* in the diagram above).
- A line that crosses a set of parallel lines is called a **transversal**.

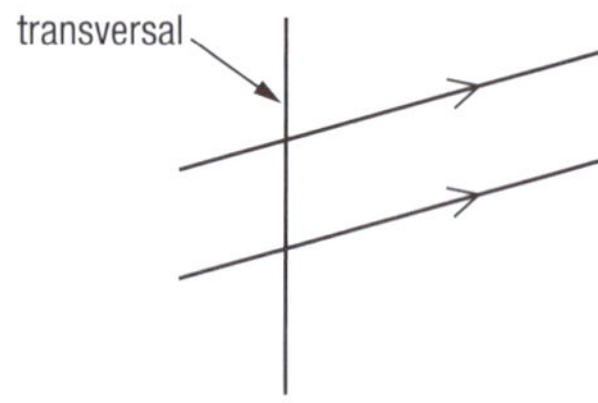

- To **bisect** a line means to cut it into two equal lengths. In the diagram below, the point *B* bisects the line *AC*.

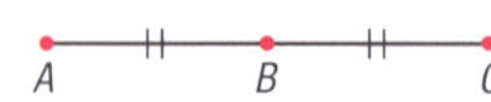

Angles between lines

Remember

In the diagram shown, B is the point of intersection of AB with BC and is always written in the middle when we use angle notation.

- An angle defines the rotation of one line from another. Angles are usually measured in degrees (°). One whole revolution is 360°.
- On diagrams, an angle is either labelled with a numeral or pronumeral representing its size (shown on the diagram on the right as $x°$), or is described by the lines that intersect at a point (the vertex) to form it.
  In the diagram on the right, the angle formed by the line segments BC and AB is written as 'the angle ABC' or, using angle notation, $\angle ABC$.

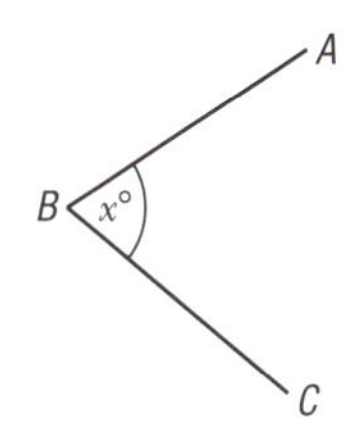

- $\angle ABC$ usually refers to the smaller of the two angles formed by AB and BC. In the diagram on the right, the other angle would be referred to as 'the reflex angle ABC'.

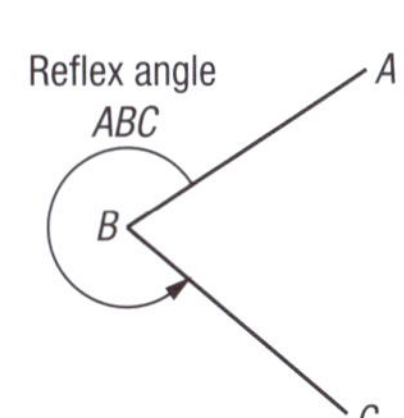

- Angles that are the same size are marked in a similar manner. In the diagram on the right $\angle BAC = \angle BCA$ because they are marked in a similar way.

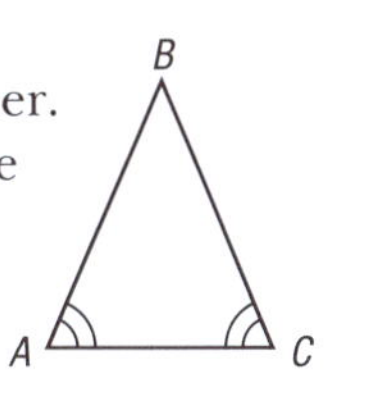

- Angles that are next to each other, and share a vertex, are referred to as **adjacent angles**. In the diagram on the right, $\angle PZQ$ is adjacent to $\angle QZR$.

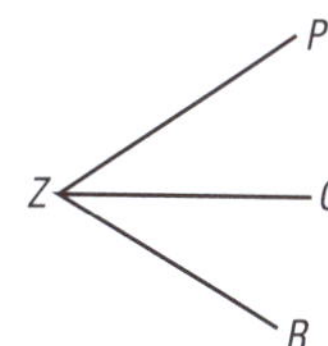

- Some ways of describing **angles**.

One revolution: 360°

180°

A straight angle: 180°

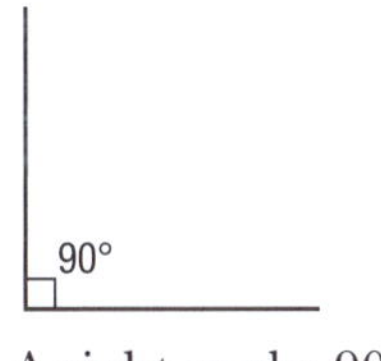

A right angle: 90°

An acute angle:
between 0° and 90°

An obtuse angle:
between 90° and 180°

A reflex angle:
between 180° and 360°

- **Perpendicular lines** intersect at right angles.
  For this diagram, we can say: 'AB is perpendicular to BC'.
  This means $\angle ABC = 90°$.
  This statement can be written as $AB \perp BC$

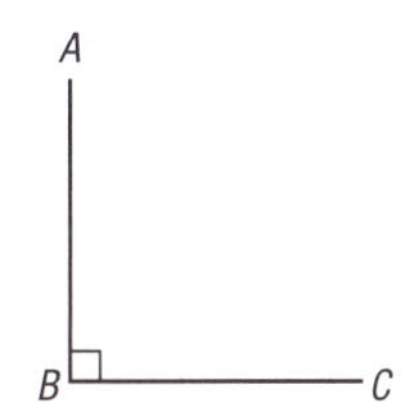

Polygons

- Polygons are closed figures that have sides that are line segments.
- Regular polygons have all their sides of equal length and their internal angles are all the same size.

Below are some common polygons.

Triangles (three sides):

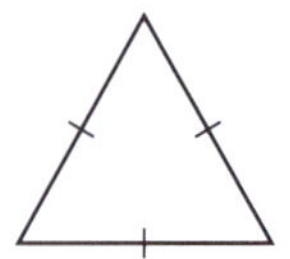
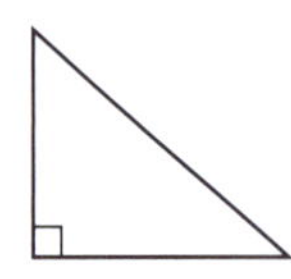
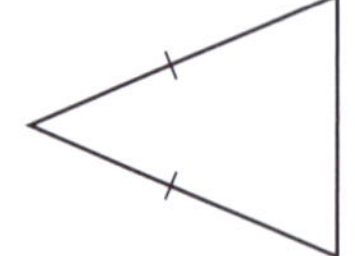
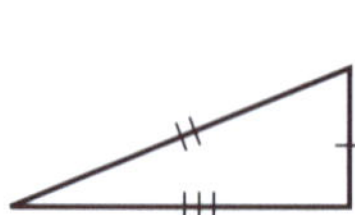
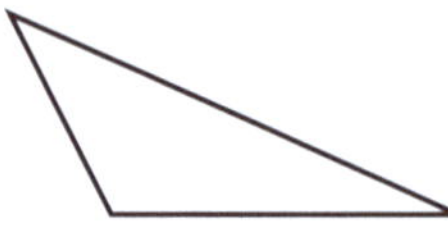

Quadrilaterals (four sides):

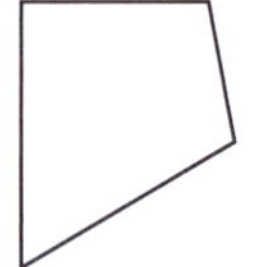
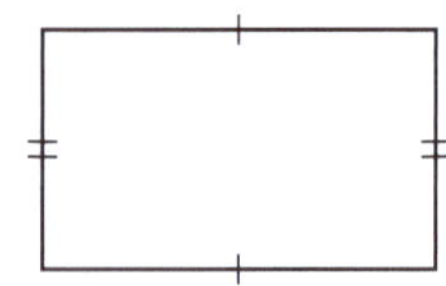
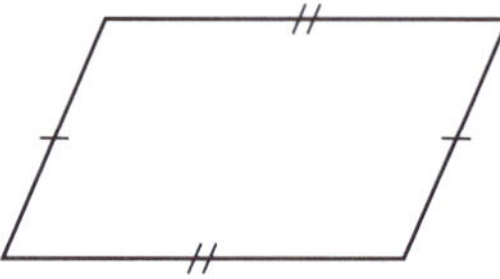
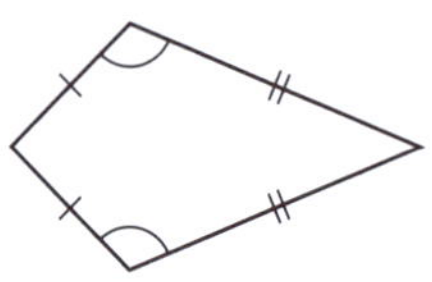
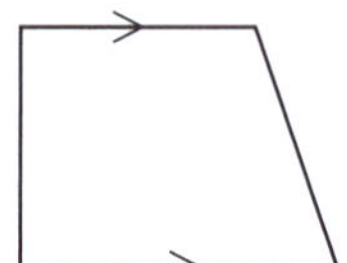

Pentagons (five sides):

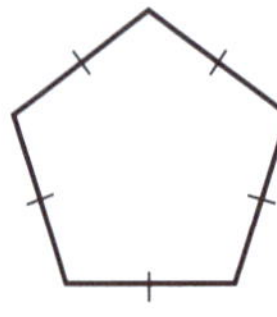
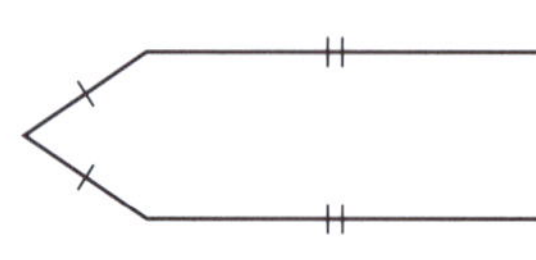
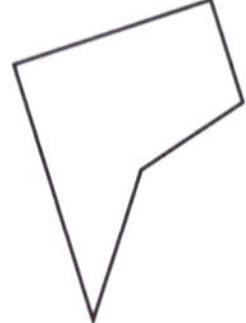

Hexagons (six sides):

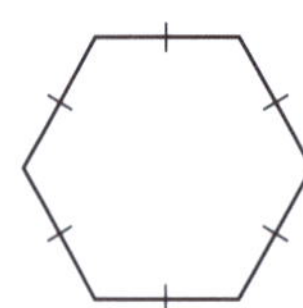
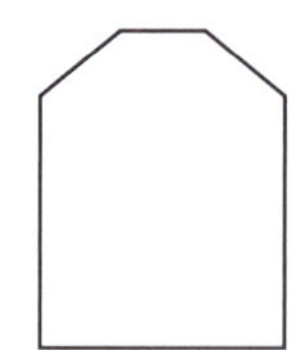
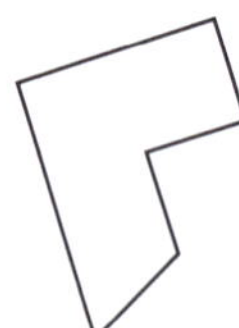

Heptagons (seven sides):

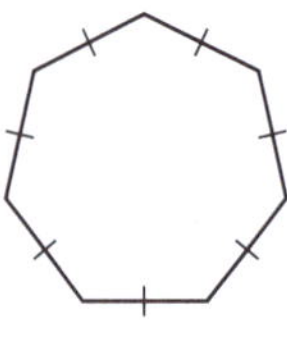
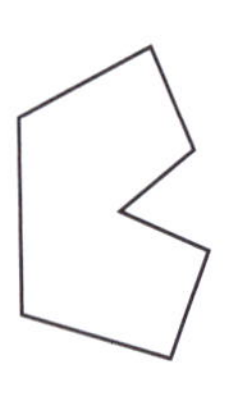

Octagons (eight sides):

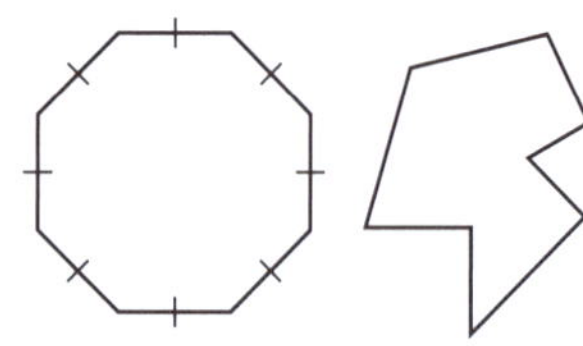

Nonagons (nine sides):

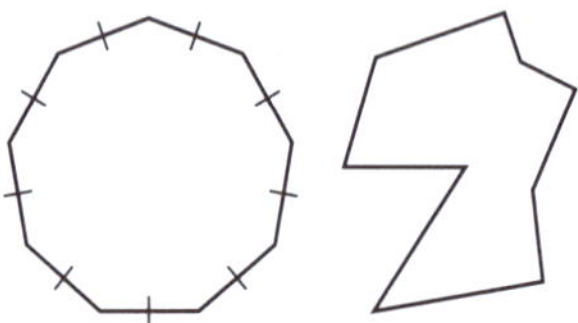

Decagon (10 sides):

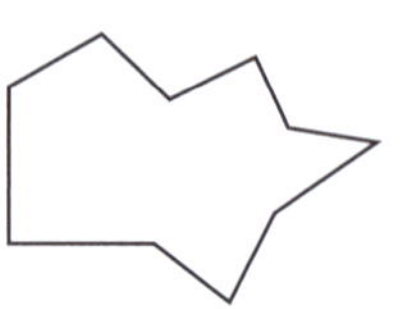

Dodecagon (12 sides):

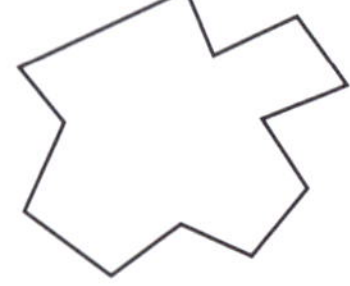

- Some ways of describing **triangles** (three-sided closed figures).

		hypotenuse
Acute-angled: All angles are between 0° and 90°.	Obtuse-angled: One angle is an obtuse angle.	Right-angled: One angle is 90°. The side opposite the right-angle is called the hypotenuse.
		60° 60° 60°
Scalene: All sides are different lengths.	Isosceles: Two sides are the same length. The angles opposite these equal sides are equal.	Equilateral: All sides are the same length. The angles are all 60°.

- Some ways of describing **quadrilaterals** (four-sided closed figures).

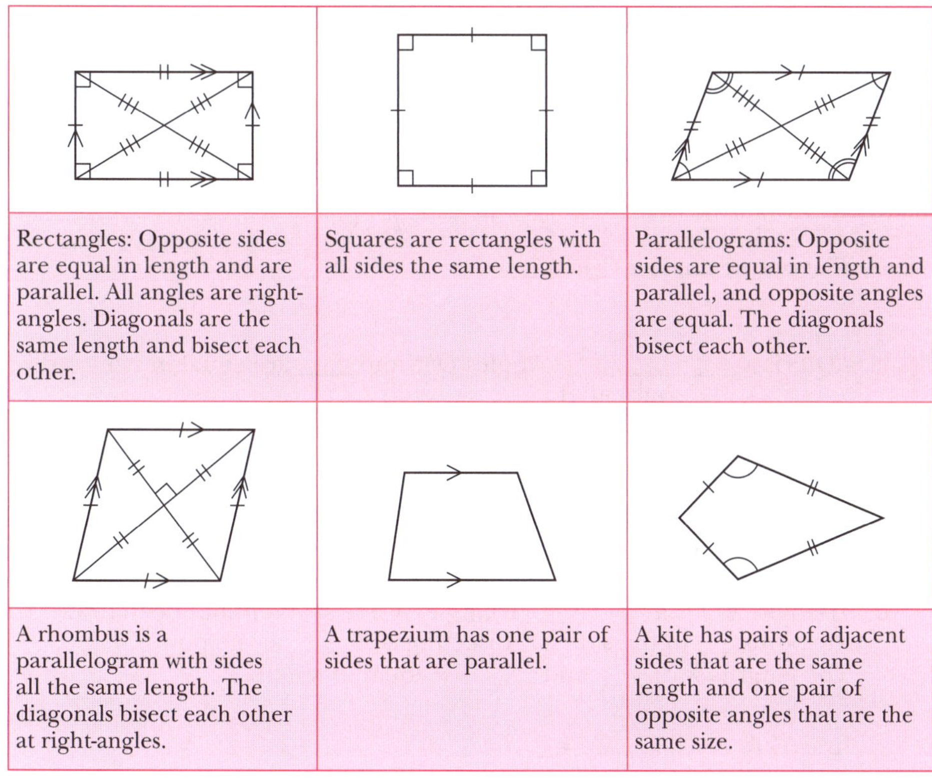

Rectangles: Opposite sides are equal in length and are parallel. All angles are right-angles. Diagonals are the same length and bisect each other.	Squares are rectangles with all sides the same length.	Parallelograms: Opposite sides are equal in length and parallel, and opposite angles are equal. The diagonals bisect each other.
A rhombus is a parallelogram with sides all the same length. The diagonals bisect each other at right-angles.	A trapezium has one pair of sides that are parallel.	A kite has pairs of adjacent sides that are the same length and one pair of opposite angles that are the same size.

Circles

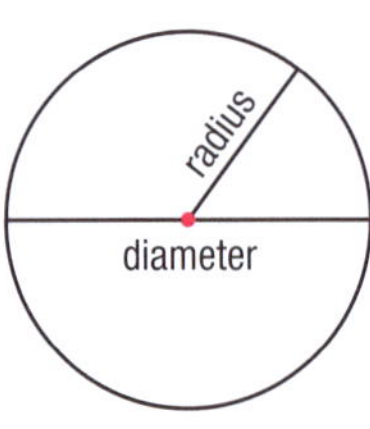

- A **circle** is a closed figure in which every point on the shape is the same distance from another point, called the **centre**.
- The **radius** is the distance from the centre of a circle to any point on the circle.
- The **diameter** is the distance from one point on a circle to another point on the circle, going through the centre. Diameter = 2 × radius.
- An **arc** is a part of a circle.
- A **sector** of a circle is a closed figure formed by an arc of a circle and two radii.

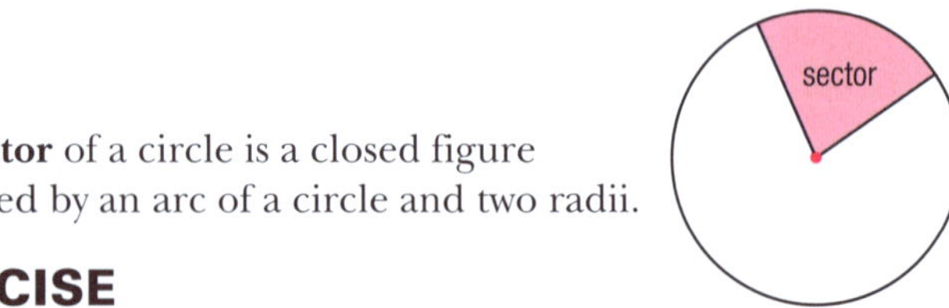

EXERCISE

1 Copy and complete the descriptions of the given diagrams.

a The line segments ___ and BC intersect at an angle of ___°.

$\angle$____ = 50°

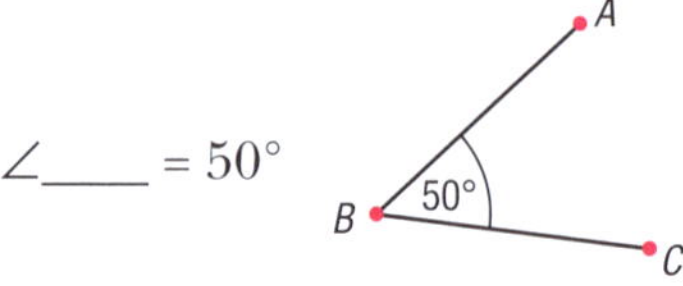

b The ______ angle formed by the line segments PQ and ___ is 220°.

The obtuse angle $\angle PQR$ = ____°

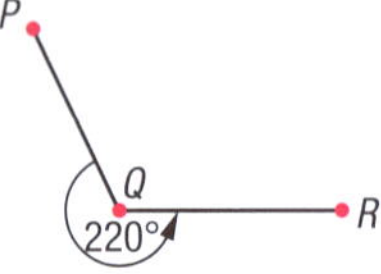

c $PQRS$ is a _____________. Angles QPS and ____ are equal in size, and angle ____ is a right angle.

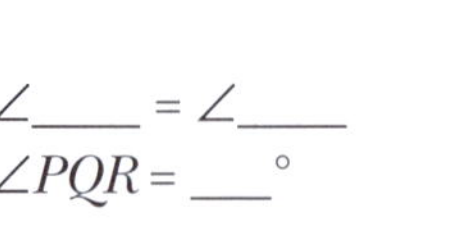

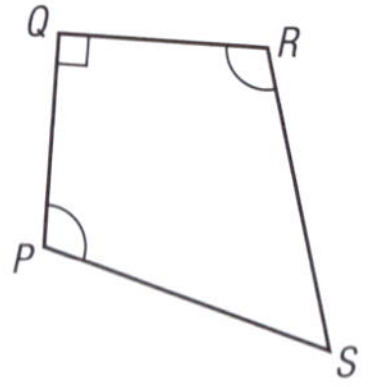

d Triangle XYZ is a _________ triangle. A line segment has been drawn from the point ___ to the side XZ, bisecting side ____ at the point M. The lengths XM, ____ and ____ are equal in length.

In $\triangle XYZ$: XM = ___ = ___

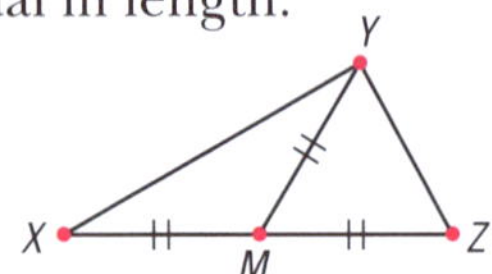

e *AB* and ___ are a pair of ________ lines.

XY is a transversal.
AB __ *CD*

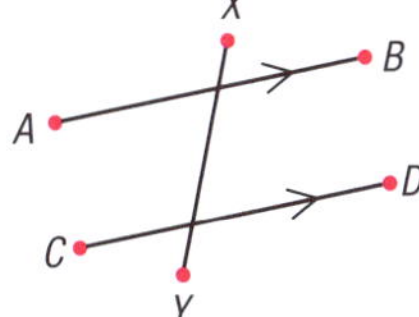

f A circle with _______ *O* has a radius of ___ cm.
The ________, *AB*, is drawn.

AO = ____ = 5 cm
AB = ___ cm.

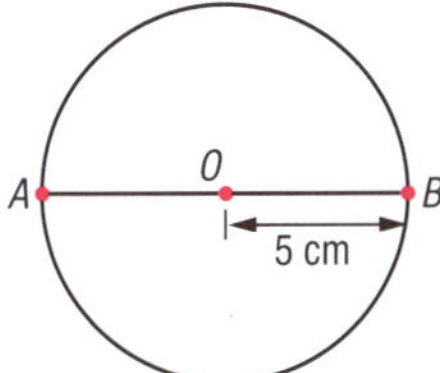

g *PQRST* is a ________ pentagon.

PQ = ___ = ___ = ___ = ___
∠*PQR* = ∠____ = ∠____ = ∠____ = ∠____

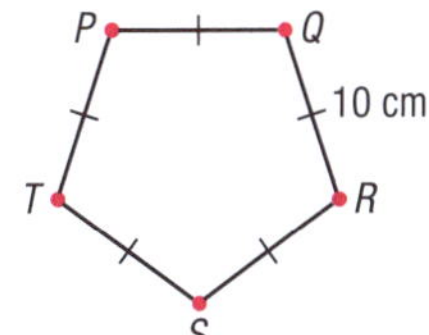

h *AB* is an ____ of a circle of radius ___ cm.
AB subtends an _____ of ___° at the centre, ___, of the circle.

∠____ = 40°
___ = ___ = 10 cm.

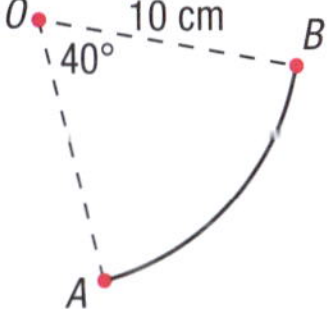

i ____ is a perpendicular bisector of ____.

AQ = ___
___ ⊥___

j *ABCD* is a ________, with side ____ parallel to side ____.
Side *AB* is ____________ to both *BC* and ___.

BC ___ *AD*
AB ___ *BC*, and *AB* ___ *AD*

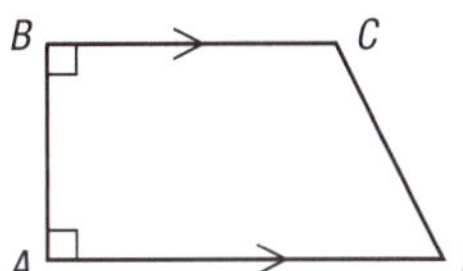

2 Describe each of the following figures, in words and symbol notation.

a

b

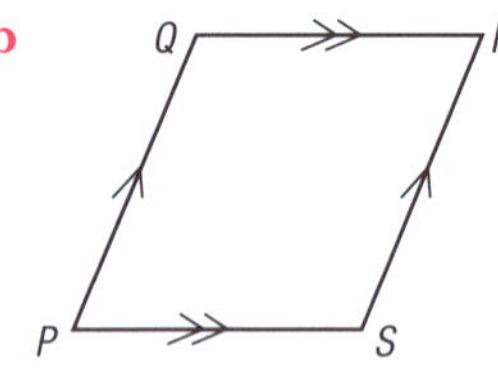

c

d

e

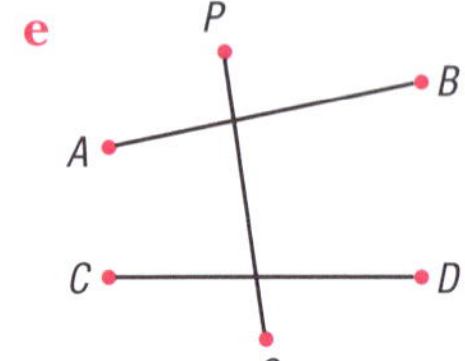

f

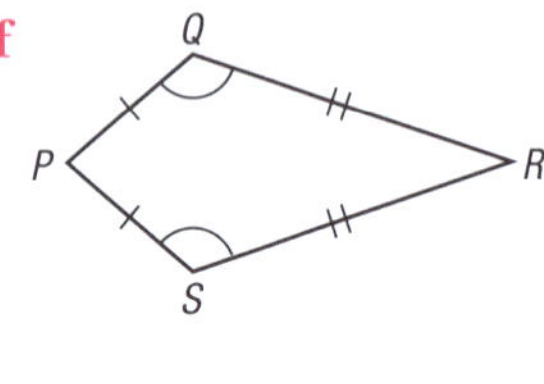

g

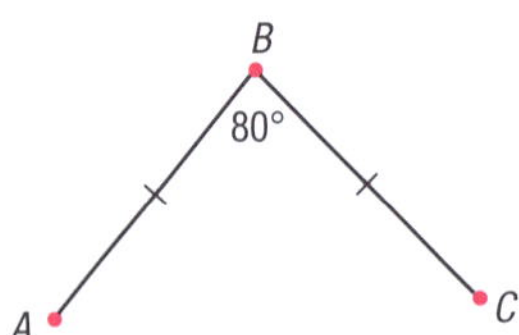

h

i

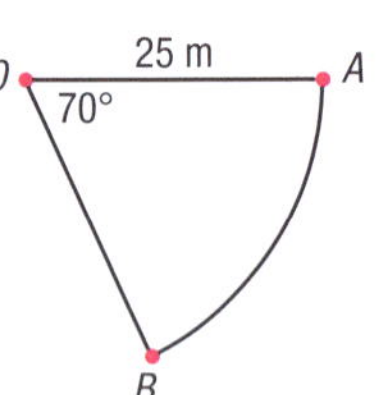

3 Draw a diagram for each of the following, including all the given and relevant information.

a a square

b two lines that intersect at an angle of 30° and one line bisects the other

c a rhombus of side length 10 cm

d a kite with pairs of sides of length 12 cm and 20 cm

e a regular hexagon of side length 2 m

f a triangle *ABC* inside a circle, so that *AC* is a diameter of the circle and vertex *B* is on the circle.

g a pentagon which has all its sides the same length

h a scalene triangle, *ABC*, in which a line from point *B* to *AC* is perpendicular to *AC*.

Angles

When Euclid wrote his book of geometric proofs he wanted to start each proof with only a few assumed facts. These facts were those that he thought were accepted to always be true by most mathematicians of the time. He called these assumed facts **axioms**.

One of Euclid's axioms is the **equality axiom**: 'Things that are equal to the same thing are equal to each other.'

In **symbols** this can be expressed as: If $a = x$ and $b = x$, then $a = b$.

We are going to use some definitions about the size of angles, with Euclid's equality axiom, and then use this information to prove some theorems involving angles.

Definitions involving the size of angles

'The sum of angles at a point is always 360°.'

Using this definition with the diagram on the right, we could say:

$$p + q + r = 360$$

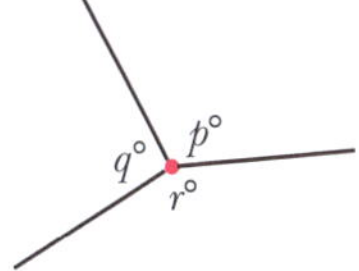

'The sum of angles on a straight line is always 180°.'

Using this definition with the diagram on the right, we could say:

$$a + b = 180$$

Angles whose sum is 180° are called **supplementary angles**. We could say, 'Angle a is the supplement of angle b.'

'The sum of angles in a right angle is always 90°.'

Using this definition with the diagram on the right, we could say:

$$p + q = 90$$

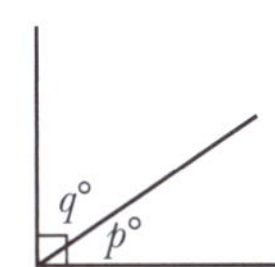

Angles whose sum is 90° are called **complementary angles**. We could say, 'Angle p is the complement of angle q.'

A proof by deductive reasoning

Proof 1

Using the axiom and definitions above to prove the theory that 'Vertically opposite angles are equal'.

In the diagram below, angles a and b are vertically opposite angles.

To prove: Vertically opposite angles are equal.

On the diagram: Prove $a = b$

Let another angle at the point of intersection of the lines be x°, as shown in the diagram.

Proof

$a + x = 180$	The sum of angles on a straight line is 180°.
and $b + x = 180$	The sum of angles on a straight line is 180°.
So: $a + x = b + x$	Using the equality axiom.
$a + x - x = b + x - x$	Subtracting x from both sides of the
$a = b$	equation.

So, it is proven that: Vertically opposite angles are equal.

Help box

Note: The reason for each step of the proof is given.

Example

Find the value of the pronumeral in the diagram below, giving a reason for your answer.

$x°$ 25°

Answer

$x + 25 = 180$	The sum of angles on a straight line is 180°.
$x + 25 - 25 = 180 - 25$	Subtracting 25 from both sides of the equation.
$x = 155$	

Example

Find the value of the pronumeral in the diagram below, by writing an equation and solving it.

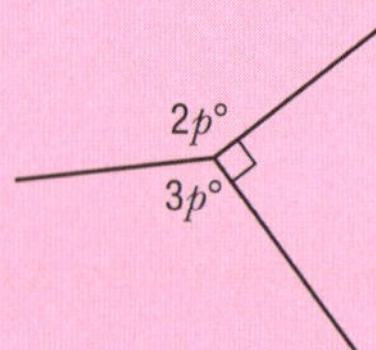

Answer

$2p + 3p + 90 = 360$	The sum of angles at a point is 360°.
$5p + 90 = 360$	Adding like terms.
$5p + 90 - 90 = 360 - 90$	Subtracting 90 from both sides of the equation.
$5p = 270$	
$\frac{5p}{5} = \frac{270}{5}$	Dividing both sides of the equation by 5.
$p = 54$	

EXERCISE

1 Write the complement of each of the following angles:
 a 21° b 83° c 11°

2 Write the supplement of each of the following angles:
 a 129° b 36° c 100°

3 From the diagram on the right, choose a pair of angles that are:
 a complementary
 b supplementary
 c equal

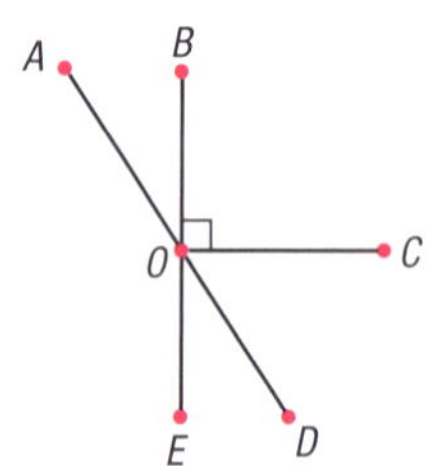

4 Find the value of the pronumeral in each of the following, giving a reason for your answer each time:

a

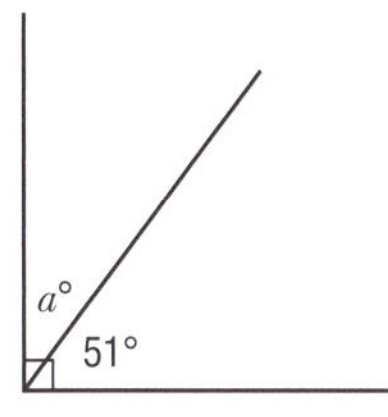

b

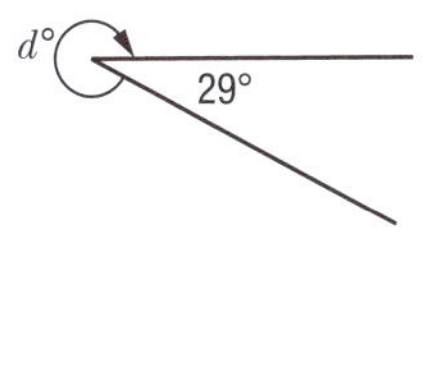

c

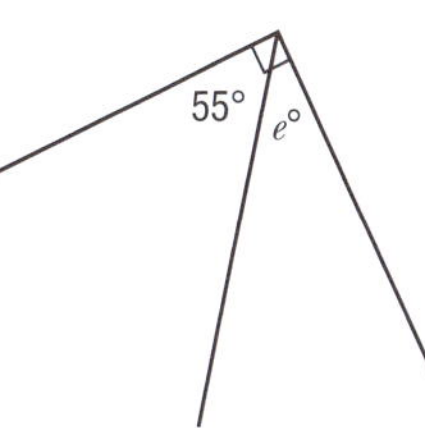

d

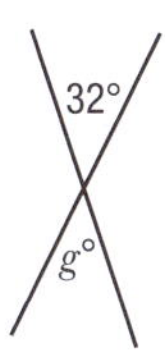

e

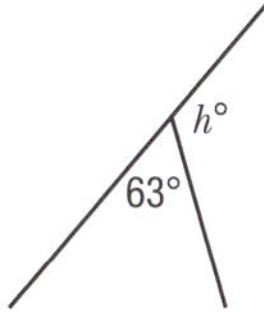

f

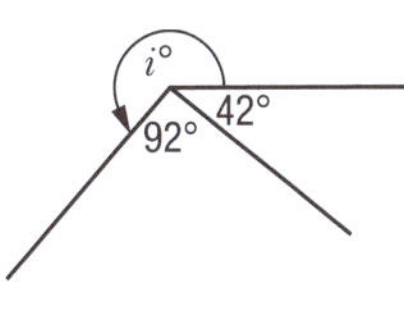

5 Form an equation for each of the following, giving a reason for your choice of each equation. Solve the equation to find the value of the pronumeral each time.

a

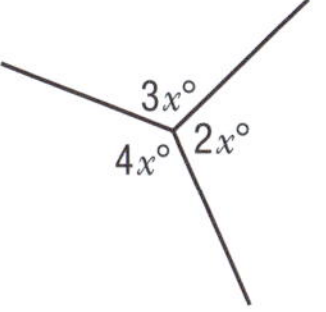

b

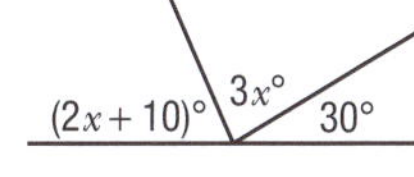

c

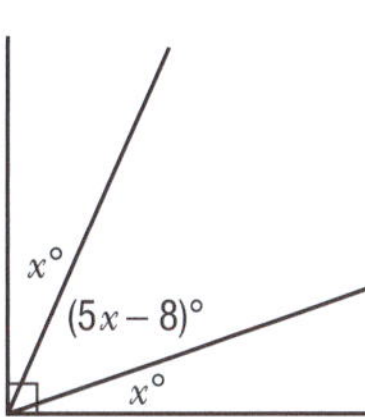

d

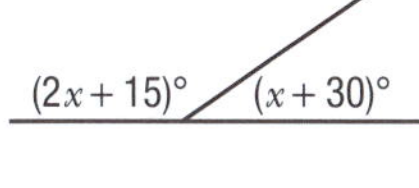

e

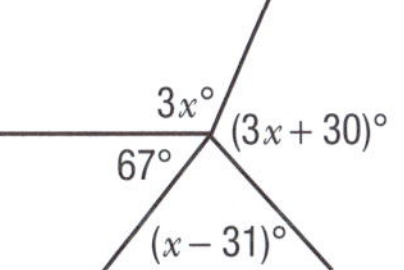

f

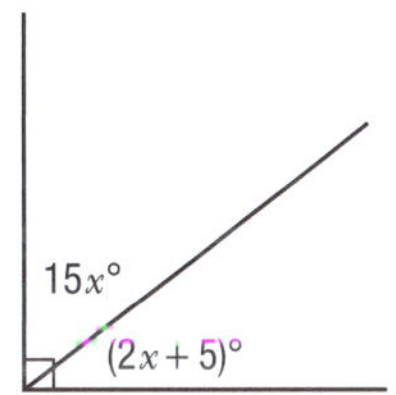

6 Two lines, AB and CD, intersect at the point O. Use deductive reasoning to prove that $\angle AOD = \angle COB$, setting your proof out in a logical order and giving reasons for your statements. (See **Proof 1** on page 353.)

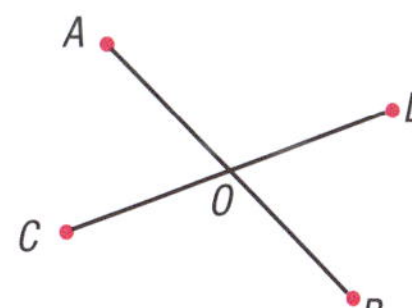

7 Use deductive reasoning to show that $\angle ZOY$ and $\angle YOX$ are complementary angles.

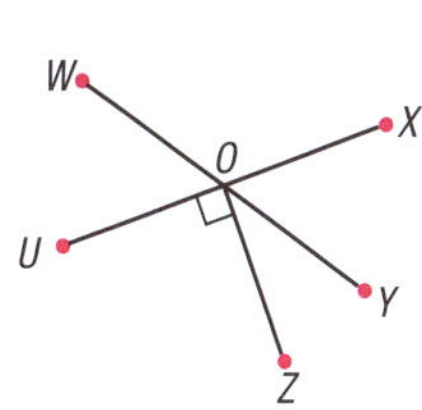

Parallel lines

To the 'equality axiom' from the previous lesson, we are going to add an axiom involving parallel lines: 'When a pair of parallel lines is crossed by a transversal, the angles in **corresponding** positions are the same size.'

For the diagram on the right, this means $a = b$.

There are two other theorems that relate to parallel lines. These can be proved from the axioms we have looked at so far, and some previously proved theorems.

Proof 2

Proving the theorem: 'When a pair of parallel lines is crossed by a transversal, the angles in **alternate** positions are equal in size'. For the diagram shown, this means $p = q$.

To prove: Alternate angles are equal in size, that is $p = q$ for the diagram given.
Let: $\angle ABM = x°$.

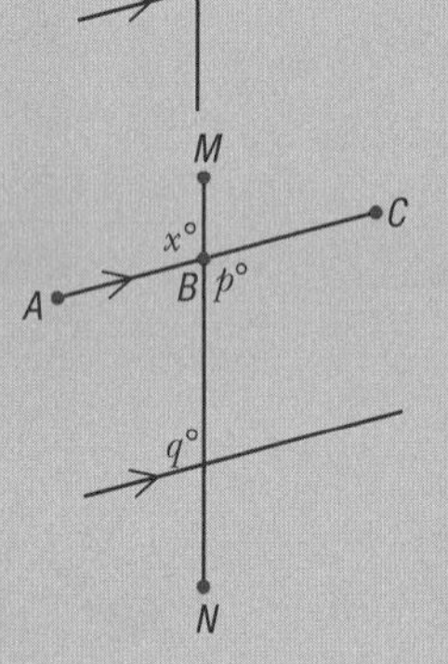

Proof

$p = x$.	Vertically opposite angles are equal.
$q = x$.	Corresponding angles are equal.
So $p = q$	Using the equality axiom.

It is proven that alternate angles are equal in size.

Proof 3

Proving the theorem: 'When a pair of parallel lines is crossed by a transversal, the sum of the **co-interior** angles is 180°.' (For the diagram shown, this means $m + n = 180$.)

To prove: Alternate angles are equal in size (that is $m + n = 180$ for the diagram given).
Let: $\angle MBC = x°$

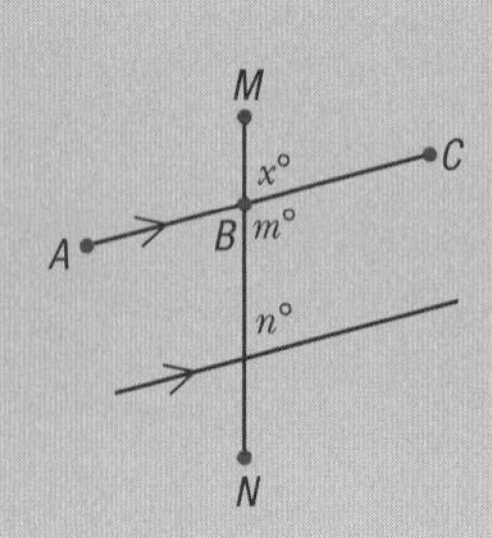

Proof

$m + x = 180$	Angles on a straight line add to 180°.
$n = x$	Corresponding angles are equal.
So $m + n = 180$	Substituting n for x.

It is proven that the sum of co-interior angles is 180°.

Example

Find the value of the pronumeral in the diagram on the right, giving a reason for your answer.

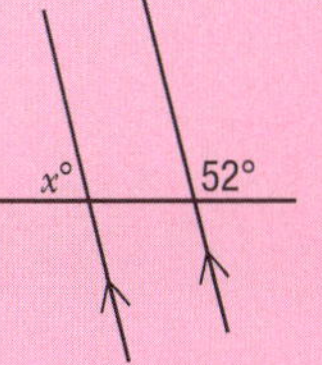

Answer

We need to label another angle, y, on the diagram, as shown.

$y + 52 = 180$	Angles on a straight line add to 180°.
$x = y$	Corresponding angles are equal.
So: $x + 52 = 180$.	
$x + 52 - 52 = 180 - 52$	Subtracting 52 from both sides of the equation.
$x = 128$	

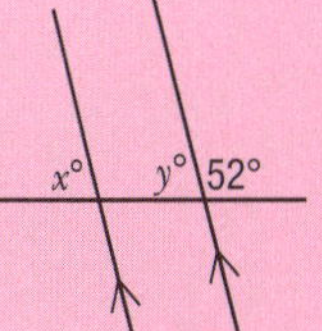

The **converse** of the parallel line axiom is also true: 'If a pair of lines is crossed by a transversal and the angles in corresponding positions are equal in size, then the pair of lines is parallel.' That is, if $a = b$ then $PQ \| RS$.

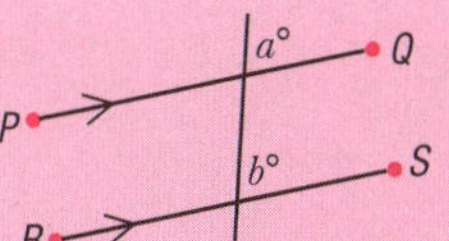

Example

Show that the lines AB and CD, in the diagram on the right, are parallel.

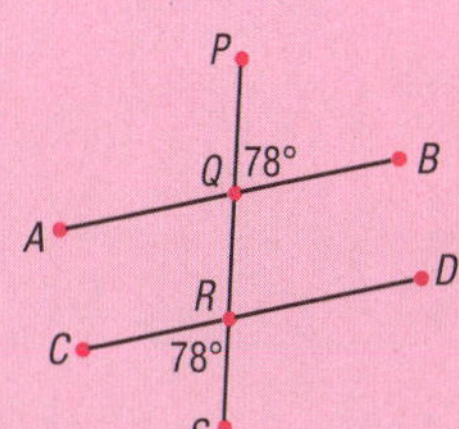

Answer

To prove: AB || CD

Given: $\angle PQB = 78°$, and $\angle CRS = 78°$

Proof

$\angle AQR = \angle PQB = 78°$	Vertically opposite angles are equal.
$\angle CRS = 78°$	Given on the diagram
So $\angle AQR = \angle CRS$	Using the equality axiom.

The angles AQR and CRS are corresponding angles on the diagram and, since these corresponding angles are equal in size, the lines AB and CD must be parallel.

EXERCISE

1 Find the value of the pronumeral in each of the following, giving reasons for your answers.

a

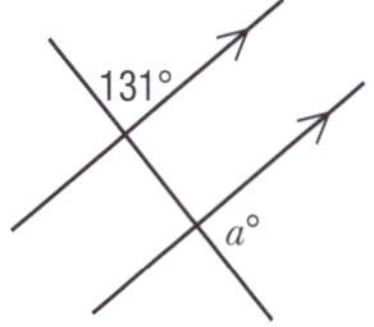

b

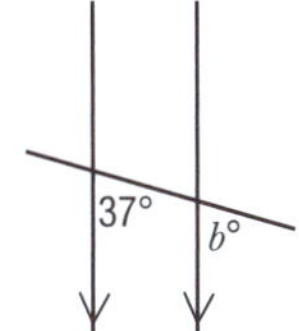

c

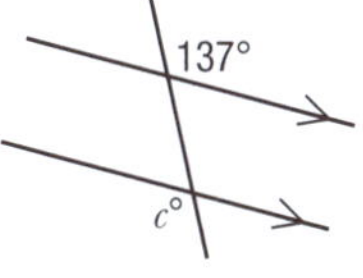

d

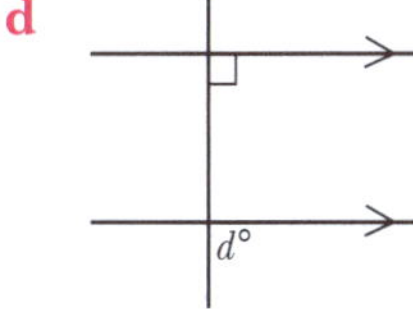

e

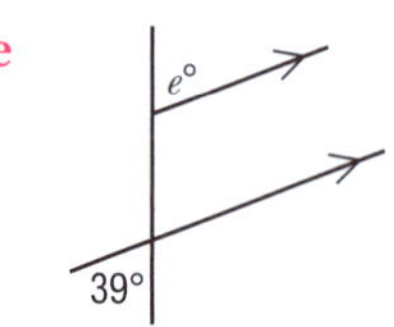

f

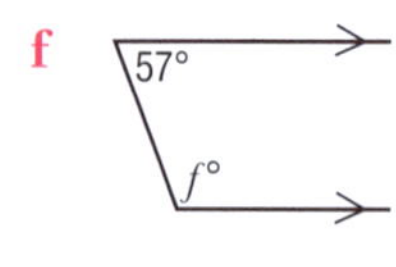

g

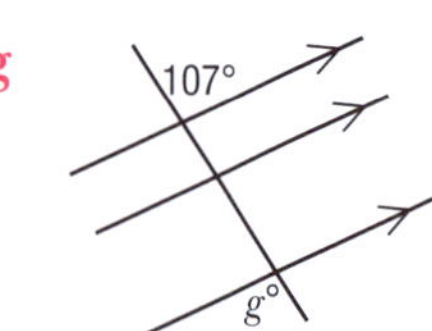

2 For each of the following:

i copy the diagram into your workbook

ii add other pronumerals, if needed

iii find the value of x, giving your reasons.

a

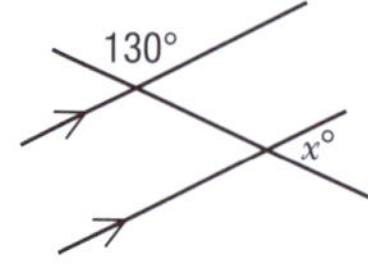

b

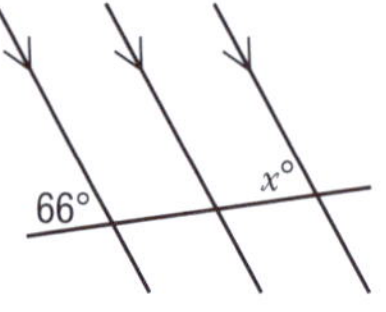

c

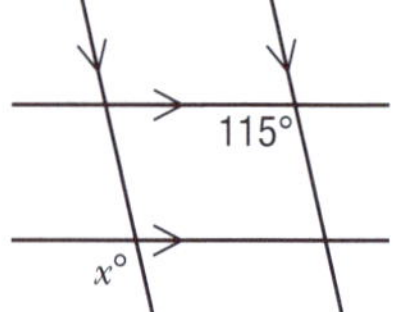

d

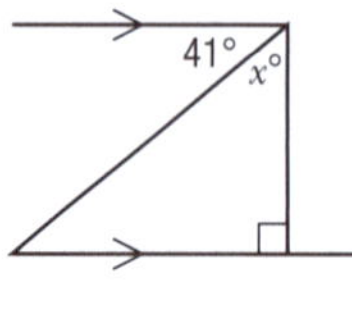

e

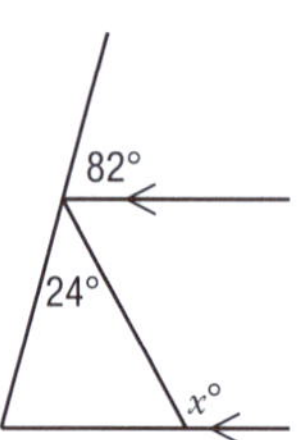

f

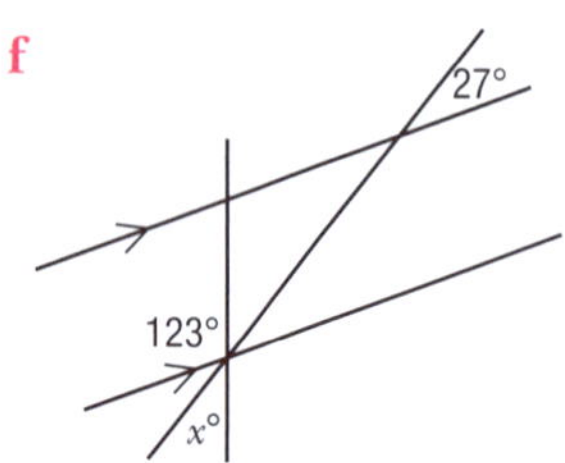

3 Form an equation for each of the following, giving a reason for your choice of each equation. Solve the equation to find the value of the pronumeral each time.

a

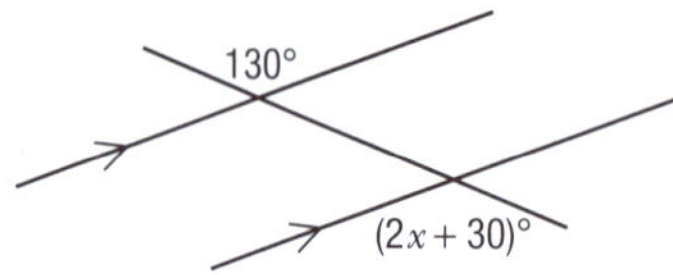

b

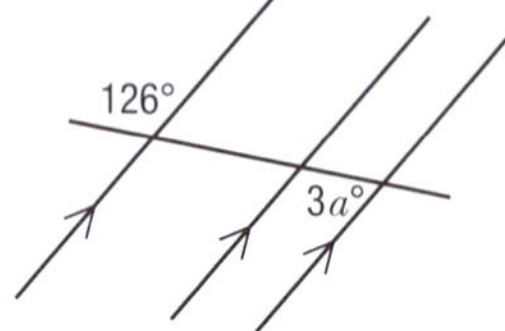

c

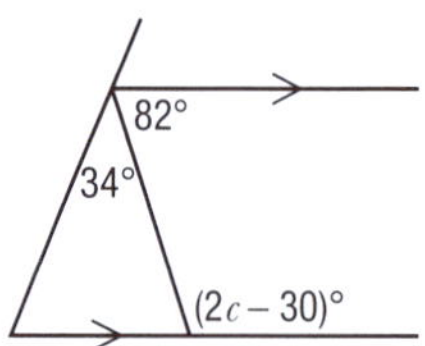

d

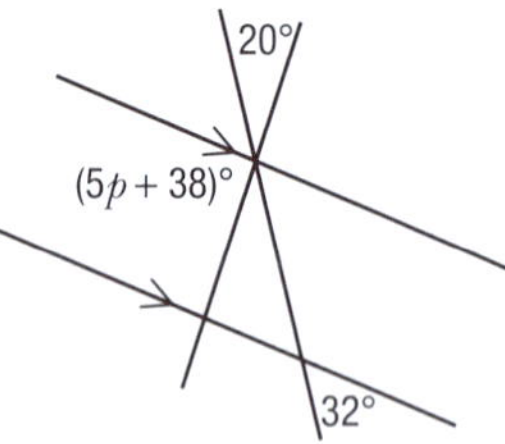

e

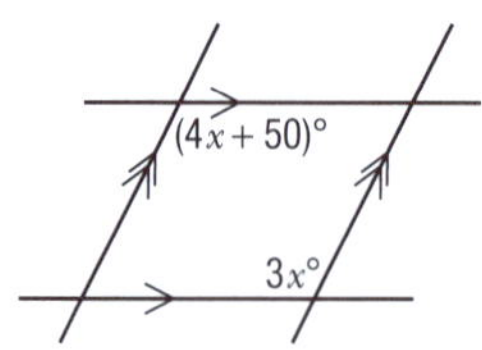

4 Determine whether the line AB is parallel to the line CD in each of the following. Give a reason for your answer.

a

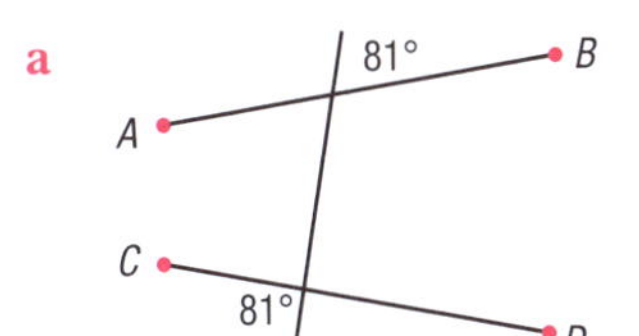

b

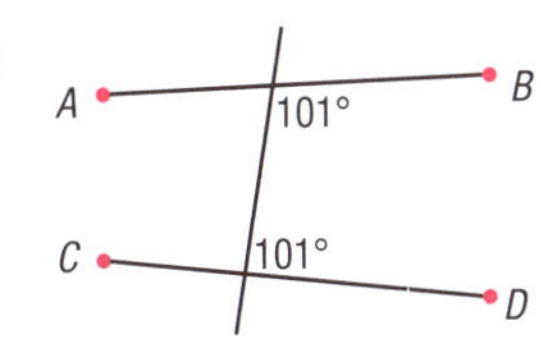

c

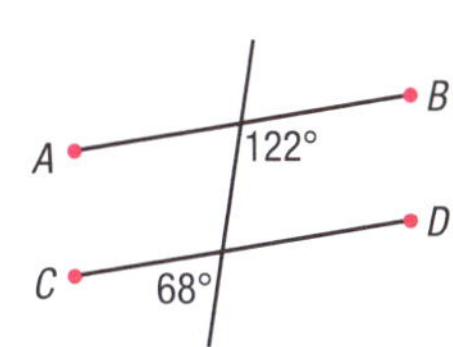

Triangles

One of the most frequently used theorems for triangles is: 'The sum of the three angles of a triangle is 180°.'

We have previously seen a demonstration of this theorem in which the three angles of a triangle are 'torn off' and are placed together to show that they form a straight angle.

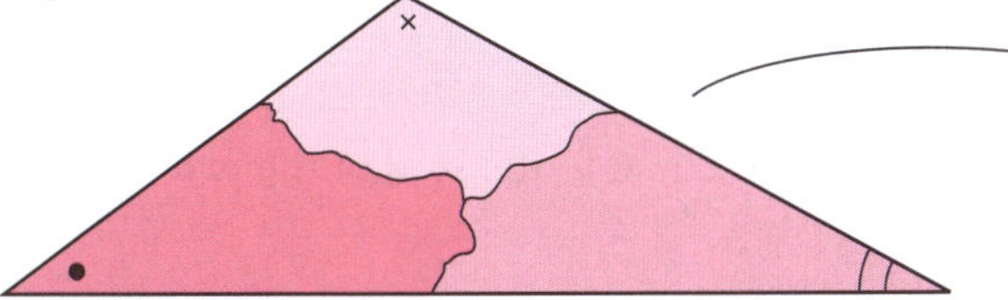

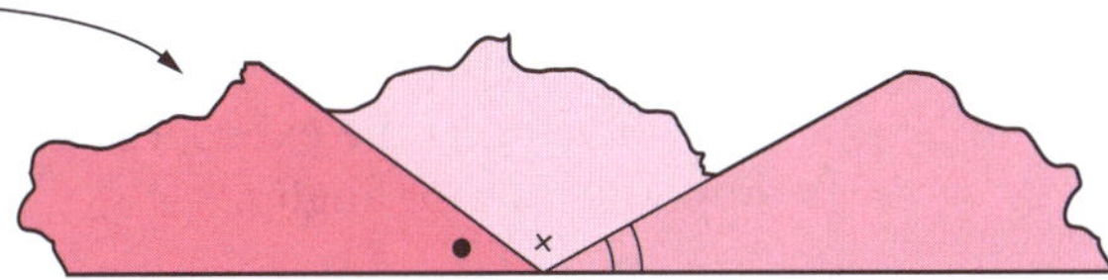

This theorem can also be simply deduced from the 'angles in parallel lines' theorems.

Proof 4

Proving the theorem: 'The sum of the three angles of a triangle is 180°.'

Construct a triangle, ABC, and a line, PQ, parallel to AC going through the point B, as shown in the diagram on the right. Label the angles in the triangle x, y and z.

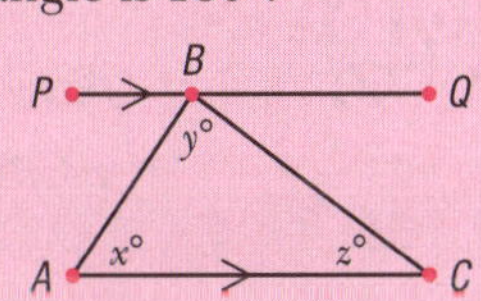

To prove: The sum of the three angles of a triangle is 180°.

In the diagram, this would be: $x + y + z = 180$

Proof

$\angle PBA + y + \angle QBC = 180$	The sum of the angles on a straight line is 180°.
$\angle PBA = x$	Alternate angles are equal.
$\angle QBC = z$	Alternate angles are equal.
So: $x + y + z = 180$	Substituting these angles.

It is proven that the sum of the three angles of a triangle is 180°.

Having deduced the theorem above we can use it to deduce another common triangle theorem: 'The exterior angle of a triangle is equal to the sum of the two opposite interior angles.'

Proof 5

Proving the theorem: 'The exterior angle of a triangle is equal to the sum of the two opposite interior angles.'

Construct a triangle, ABC, and extend side AC to the point D, so that an exterior angle BCD is produced.

Label the interior angles of the triangle x, y, and z and the exterior angle w.

To prove: The exterior angle of a triangle is equal to the sum of the two opposite interior angles.

In the diagram this would be: $w = x + y$

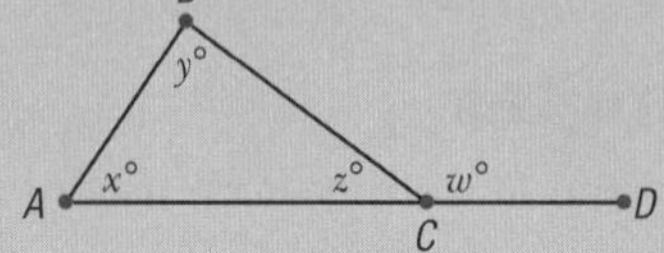

Proof

$x + y + z = 180$	The sum of the angles in a triangle is 180°.
$z + w = 180$	The sum of the angles on a straight line is 180°.
So: $x + y + z = z + w$	Using the equality axiom.
$x + y + z - z = z + w - z$	Subtract z from both sides of the equation.
$x + y = w$	

It is proven that the exterior angle of a triangle is equal to the sum of the two opposite interior angles.

Example

Find the value of a in the figure on the right, giving your reasons.

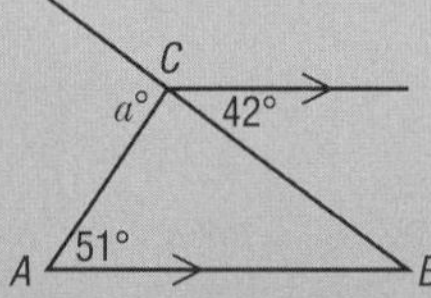

Answer

$\angle CBA = 42°$	Alternate angles are equal.
$a = \angle CAB + \angle ABC$	The exterior angle of a triangle equals the sum of the interior opposite angles.
$= 51 + 42$	
$= 93$	

So the value of a is 93.

EXERCISE

1 Determine the value of the pronumeral in each of the following diagrams, giving reasons.

a 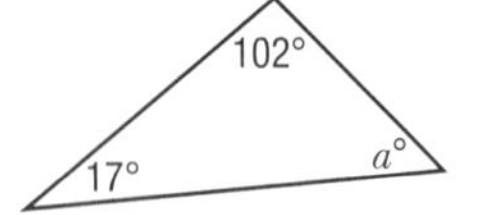

b 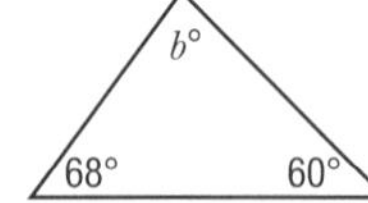

c 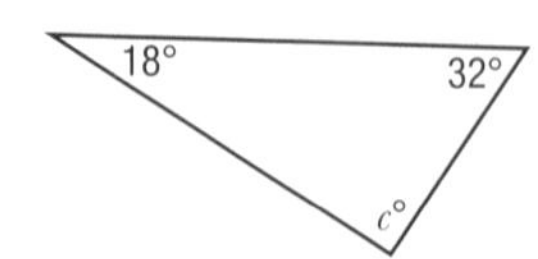

d 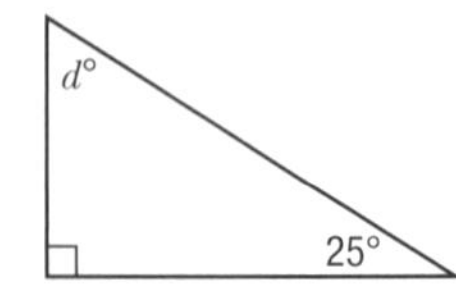

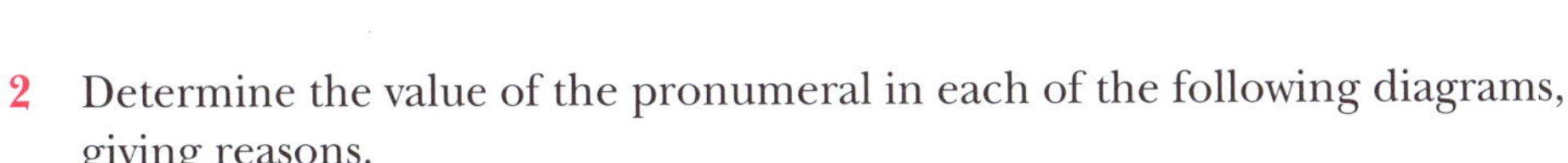

2 Determine the value of the pronumeral in each of the following diagrams, giving reasons.

a

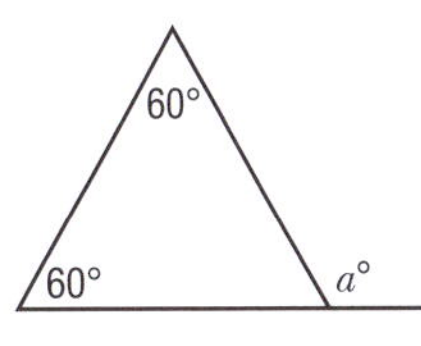

b

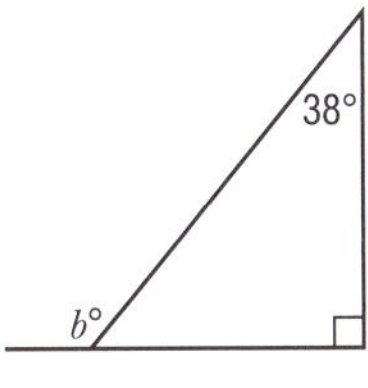

c

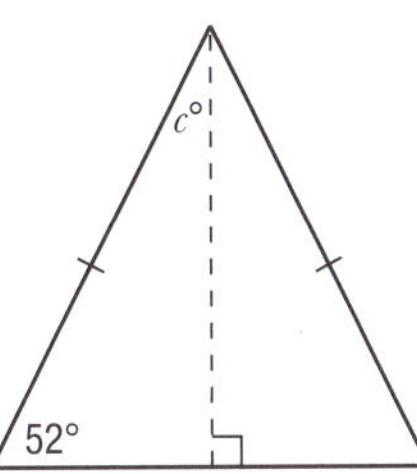

d

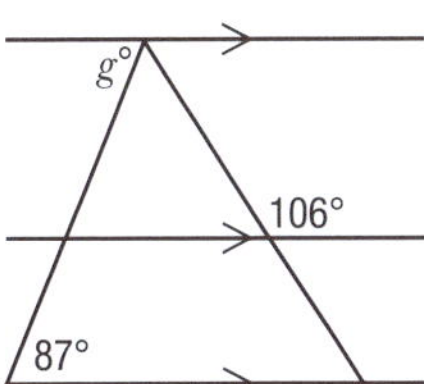

e

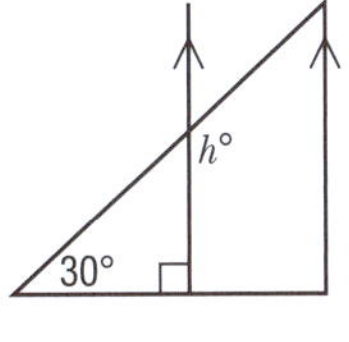

f

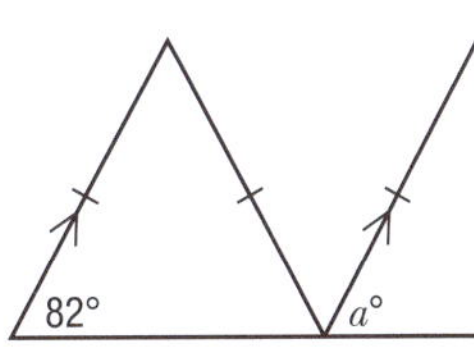

3 Form an equation for each of the following (giving a reason for your choice of each equation). Solve the equation to find the value of the x each time.

a

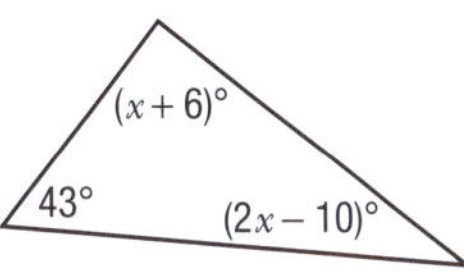

b

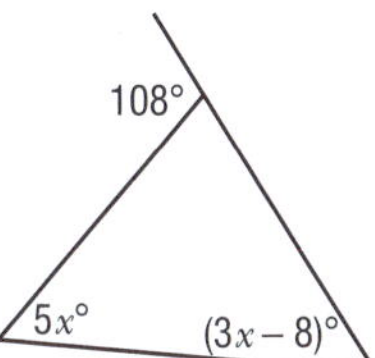

c

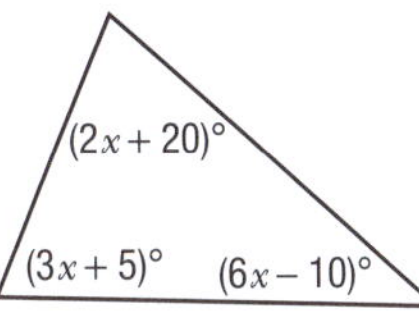

4 Show that the triangle with angles $(3x + 5)°$, $(2x + 20)°$ and $(6x - 10)°$ is an isosceles triangle.

5 Show that the triangle with angles $3x°$, $(4x - 20)°$ and $(x + 40)°$ is an equilateral triangle.

6 For the diagram on the right, show that line BC bisects $\angle ABF$, giving reasons for your answer.

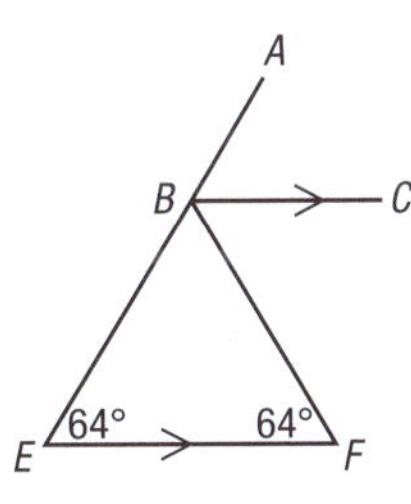

7 For the triangle on the right, set out a proof that proves the three angles of the triangle add to 180°.

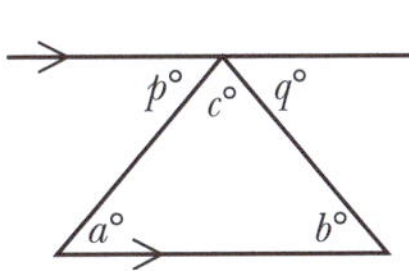

8 Set out a proof, with reasons, that shows that the angles in an equilateral triangle are all 60°.

Lesson 5

Congruent figures

Many deductive proofs of geometric theorems use properties of congruent figures.

Objects are said to be **congruent** if they are absolutely identical in size, shape and features.

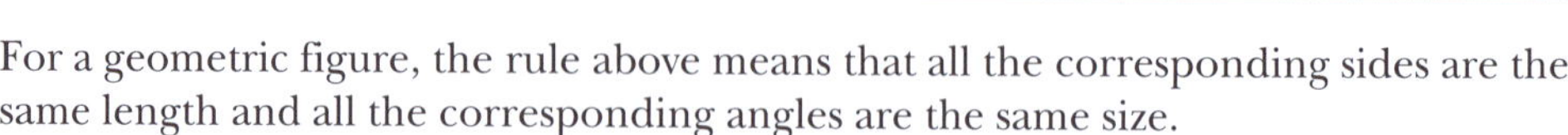

For a geometric figure, the rule above means that all the corresponding sides are the same length and all the corresponding angles are the same size.

Since most geometric figures can be divided into triangles, showing that two triangles are congruent is important to many proofs.

Congruent triangles

There are four ways of showing that two triangles are congruent. These ways are described in the rule box below.

Two triangles are congruent if and only if:

- all three corresponding sides are equal (**SSS**)
- two sides and the included angle are equal (**SAS**)
- two angles and the corresponding side are equal (**ASA**)

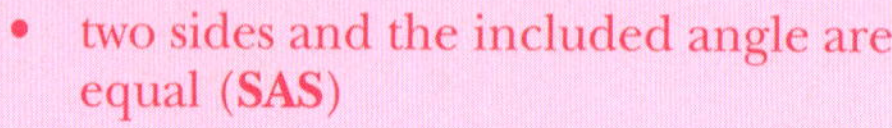

Two right-angled triangles are congruent if the hypotenuse and one pair of corresponding sides are equal (**RHS**).

Example

Show that triangles *ABC* and *PQR* are congruent triangles.

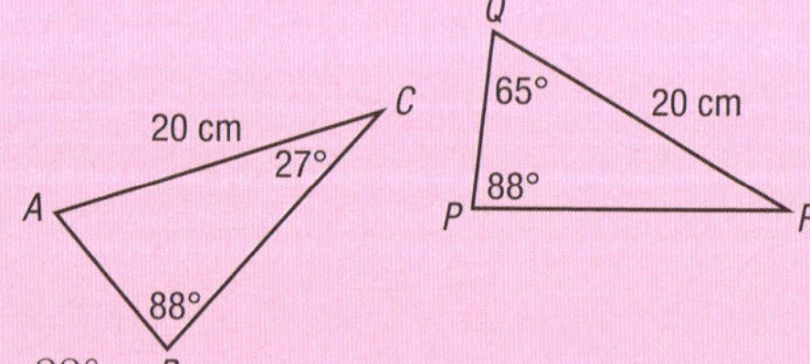

Answer

In triangle *ABC*: $\angle CAB = 180° - 27° - 88°$

$= 65°$

In triangle *PQR* $\angle QRP = 180° - 65° - 88°$

$= 27°$

So the two triangles, *ABC* and *PQR*, have angles that are the same size.

The side lengths *AC* and *QR* are equal. This means the triangles satisfy the **ASA** (two angles and the corresponding side) rule for congruence.

If triangle *ABC* is congruent to triangle *PQR*, we can write this as $\triangle ABC \equiv \triangle PQR$.

EXERCISE

1 Identify the pairs of congruent figures:

a  b  c  d  e 

f g  h  i  j 

2 Which of the following geometric figures are congruent?

a 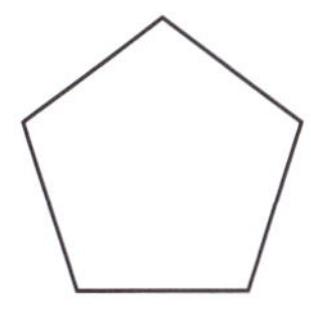 b c 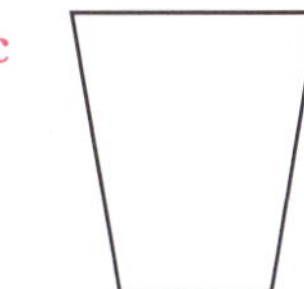 d 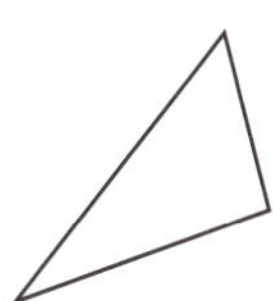 e 

f  g  h 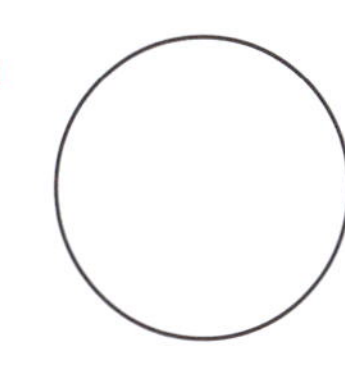 i 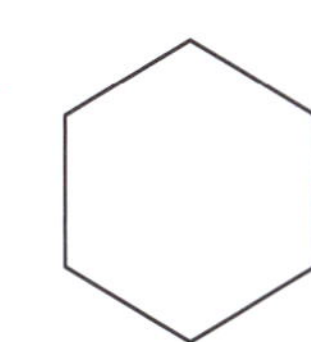 j

k 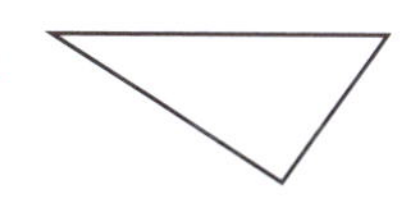 l 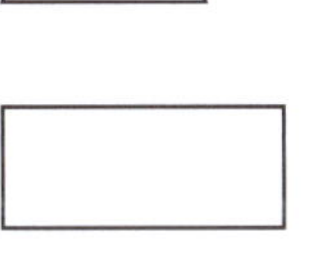 m  n 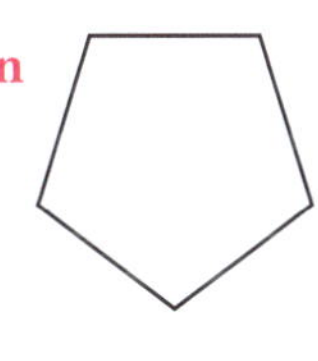 o 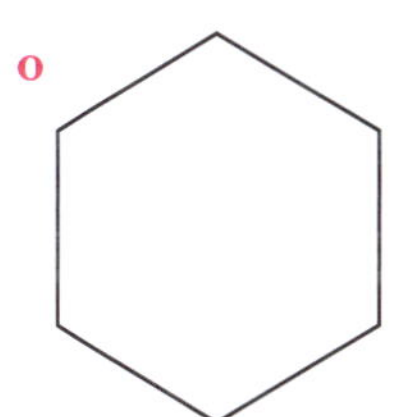

3 In each set of three triangles, two triangles are congruent. The diagrams are *not* drawn to scale. State which pair of triangles is congruent each time and give a reason (**SSS**, **SAS**, **ASA** or **RHS**).

a 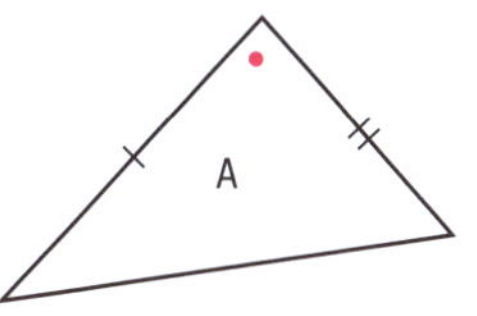

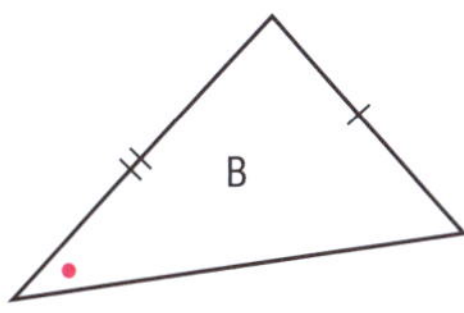

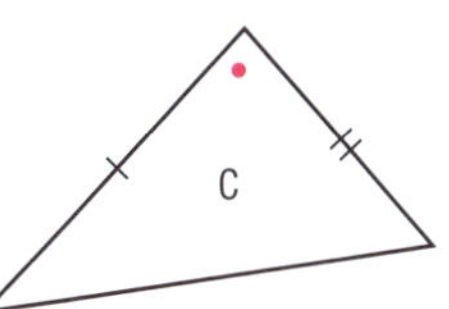

b 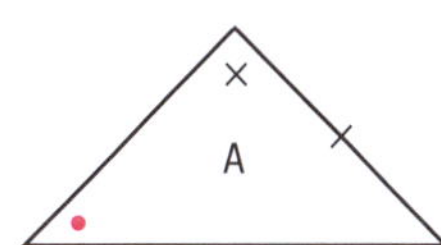

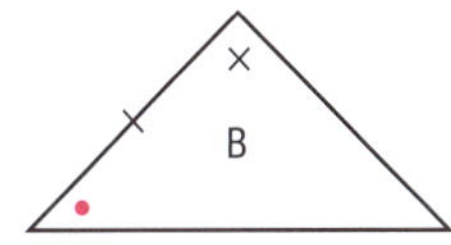

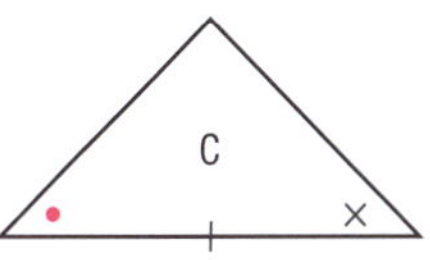

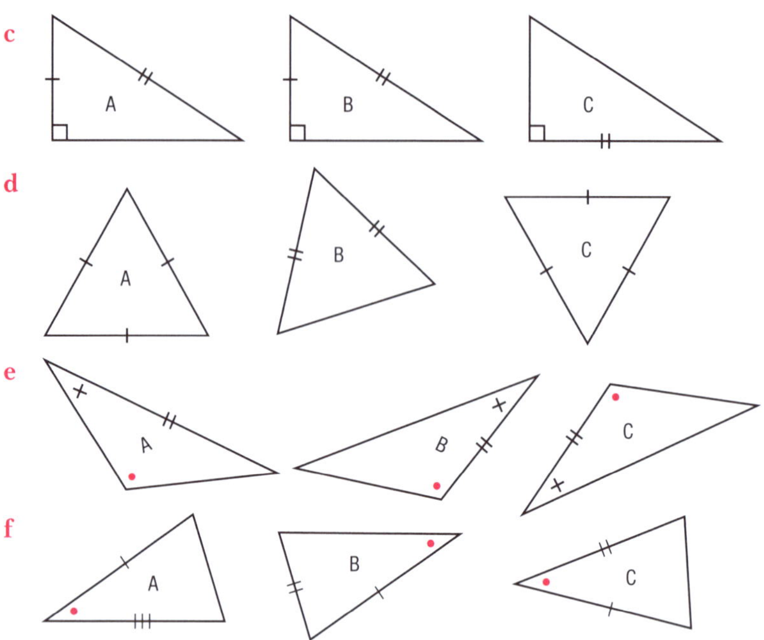

4 Find the values of the pronumerals in the following diagrams by considering the pairs of congruent triangles shown.

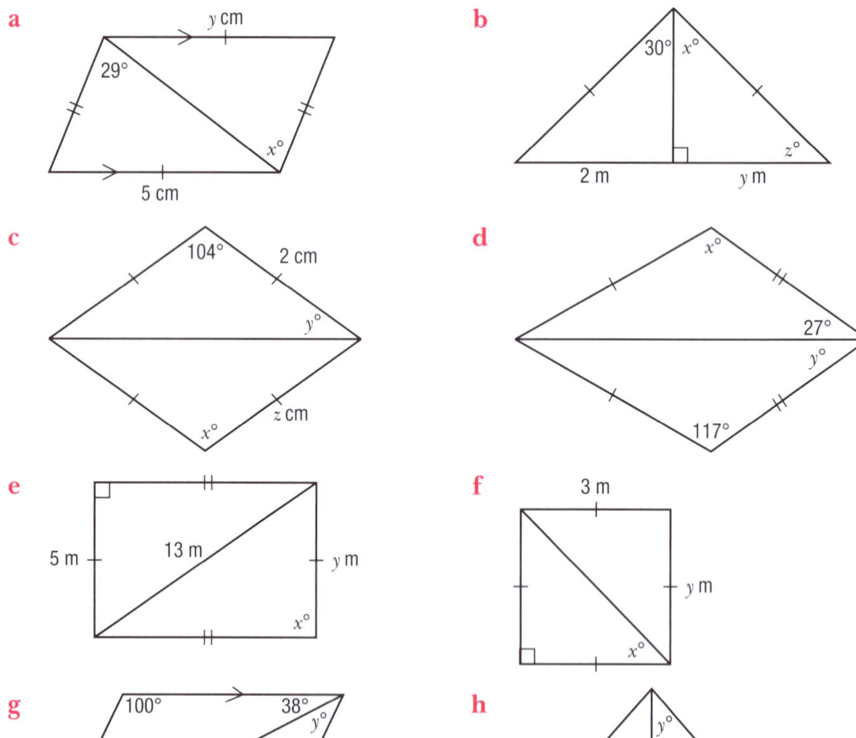

Proofs using congruent triangles

Following are some proofs of commonly used facts relating to isosceles triangles.

Proof 6

Proving the theorem: 'The angles opposite the equal sides of an isosceles triangle are equal in size.'

Construct a triangle, ABC, and a line from vertex B to the line AC, intersecting AC at the point M and bisecting angle ABC so $\angle ABM = \angle MBC$.

To prove: The angles opposite equal sides of an isosceles triangle are equal in size.

In the diagram this would be: $\angle BAM = \angle BCM$ in $\triangle ABC$

Given: $AB = BC$ Because isosceles triangles have a pair of sides of equal length.

Proof

$AB = BC$ Given.

$\angle ABM = \angle MBC$ BM bisects $\angle ABC$.

The line BM is common to both $\triangle ABM$ and $\triangle BMC$, so both these triangles have a side the length of BM.

So, it follows that $\triangle ABM \equiv \triangle BMC$ The triangles are congruent, using the **SAS** rule for congruence (two sides and the included angle).

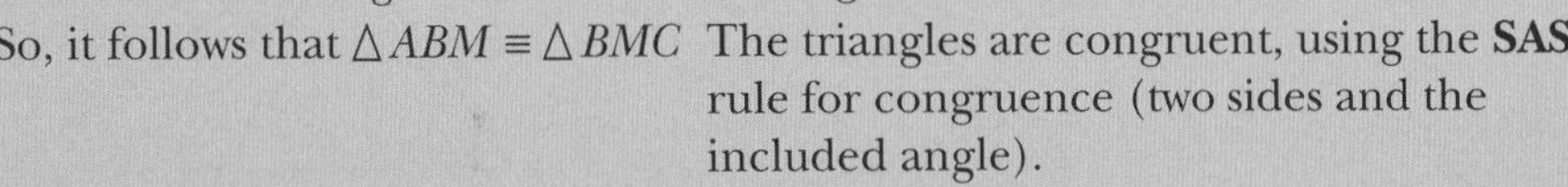

$\angle BAM = \angle BCM$ Corresponding angles in congruent triangles

It is proven that the angles opposite the sides of equal length in an isosceles triangle are equal in size.

We can also prove that the line BM is the perpendicular bisector of the line AC. See **Proof 7** below.

Proof 7

To prove: The line BM is the perpendicular bisector of the line AC in the diagram included in **Proof 6**.

We have already shown that $\triangle ABM \equiv \triangle BMC$ (**SAS**) (See **Proof 6**.)

$AM = MC$ Corresponding sides in these congruent triangles.

So point M is a bisector of AC..................**(1)**

Also: $\angle AMB = \angle BMC$ Corresponding angles in congruent triangles are equal in size.

$\angle AMB + \angle BMC = 180°$ The sum of the angles on a straight line is 180°.

$2 \times \angle BMC = 180°$ Substituting.

$\frac{2 \times \angle BMC}{2} = \frac{180°}{2}$ Dividing both sides by 2.

$\angle BMC = 90°$

and: $\angle AMB = 90°$ Angles are equal.

So $BM \perp AM$ and $BM \perp MC$....................**(2)**

(1) and **(2)** combined prove that BM is the perpendicular bisector of AC.

EXERCISE

1 Find the value of the pronumeral in each of the following, giving a reason for your answer.

a
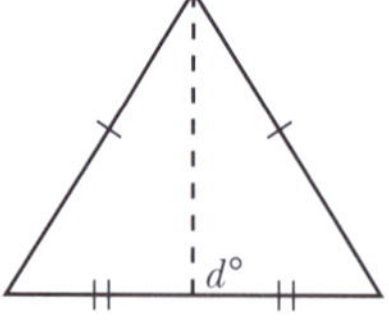

b
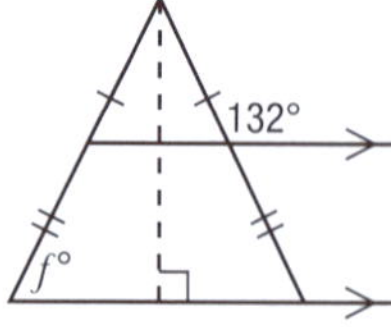

c
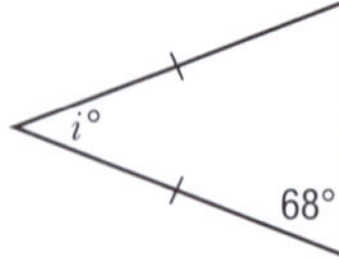

d
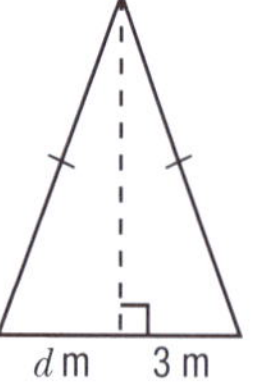

e
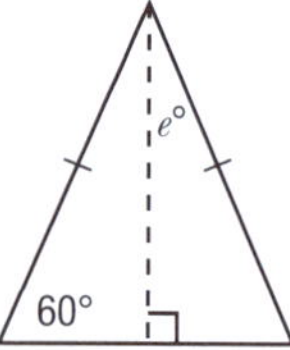

2 On the diagrams below, the length measurements are in centimetres. State which angles are equal, giving reasons.

a
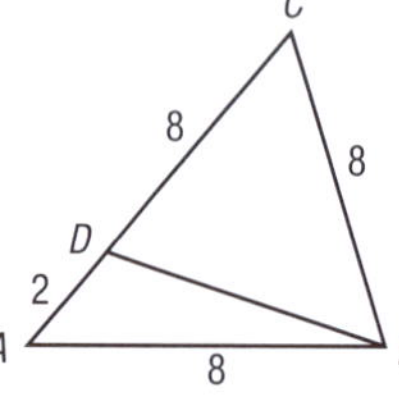

b
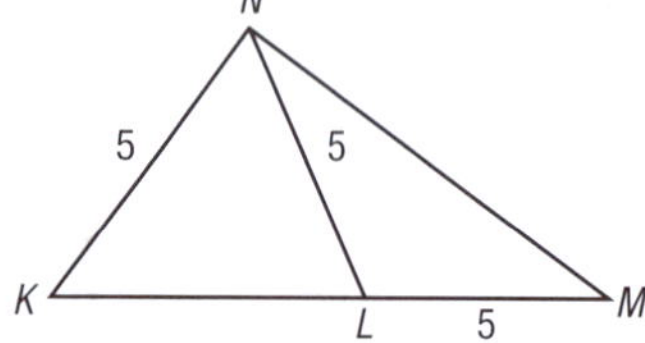

3 Find the value of x in each of the following diagrams, giving reasons for your answers.

a
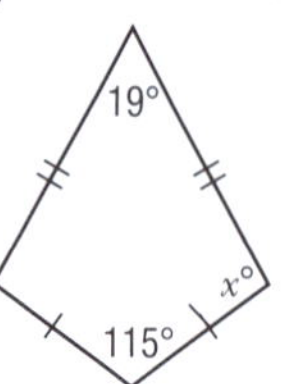

b
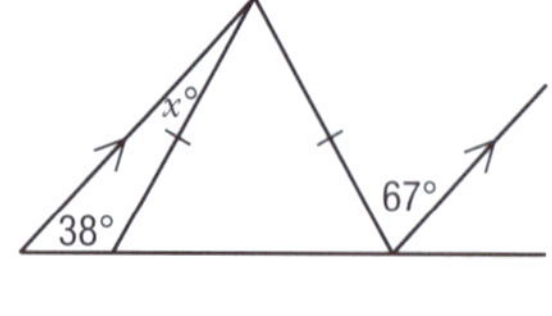

4 For the diagram on the right, prove that the line BC bisects $\angle ABF$. Set the proof out in a logical manner, giving reasons for each line.

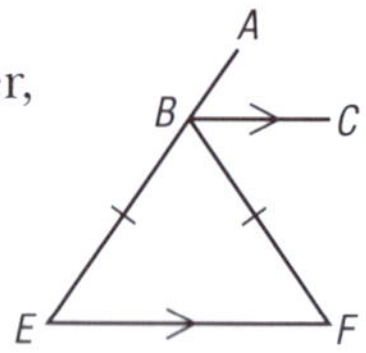

5 $\triangle ABC$ is an isosceles triangle with $AB = CB$. Prove that $\triangle BDE$ is an isosceles triangle. Set the proof out in a logical manner, giving reasons for each line.

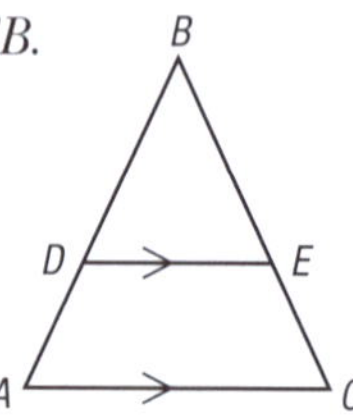

6 In $\triangle PQR$, a line is drawn from point Q to the midpoint, S, of PR.

$PS = QS = SR$

Copy and complete the following proof that $\angle PQR = 90°$.

To prove: In the diagram $\angle PQR = 90°$.

Given: $PS = QS = SR$

Let: $\angle QPS = x°$, $\angle QRS = y°$, $\angle PSQ = a°$ and $\angle QSR = b°$

Proof: $\angle PQS = x°$ Angles opposite the equal sides in an isosceles triangle.

. . . (Continue this proof.)

Quadrilaterals

In this lesson we are going to use the axioms already mentioned and the theorems that we have proved to prove some of the commonly used theorems about quadrilaterals.

Proof 8

Proving the theorem: 'The interior angles of a quadrilateral add to 360°.'

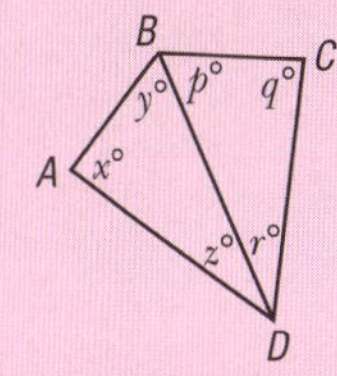

ABCD is a quadrilateral with a diagonal drawn from *B* to *D* producing two triangles, *ABD* and *DBC*. The angles in the triangles are labelled as shown.

To prove: The interior angles of the quadrilateral sum to 360°.

In the diagram this would be: $x + y + z + p + q + r = 360$

Proof

In $\triangle ABD$: $x + y + z = 180$ The sum of the angles in a triangle is 180°.

In $\triangle DBC$: $p + q + r = 180$ The sum of the angles in a triangle is 180°.

$x + y + z + p + q + r = 180 + 180$ Adding the two equations above.

So: $x + y + z + p + q + r = 360$

It is proven that the interior angles of a quadrilateral add to 360°.

Proof 9

Proving the theorem: 'The diagonals of a rectangle bisect each other'.

ABCD is a rectangle. The diagonals, *AC* and *DB*, intersect at the point *M*.

To prove: The diagonals bisect each other.

In the diagram this would be: $AM = MC$ and $DM = MB$

Given: *ABCD* is a rectangle, so opposite sides are equal in length and are parallel.

Proof

In triangles AMB and DMC:

$AB = DC$ and $AB \parallel DC$	Because $ABCD$ is a rectangle.
$\angle MAB = \angle MCD$	Alternate angles are equal.
$\angle MBA = \angle MDC$	Alternate angles are equal.
So: $\triangle AMB \equiv \triangle DMC$	(**ASA**)

The sides AM and MC are corresponding sides in these congruent triangles.
So $AM = MC$.
The sides DM and MB are corresponding sides in these congruent triangles
So $DM = MB$.
It is proven that the diagonals of a rectangle bisect each other.

Proof 10

Proving the theorem: 'The opposite angles in a parallelogram are equal in size.'

To prove: The opposite angles in a parallelogram are equal in size.

A B D C

In the diagram this would be: $\angle DAB = \angle BCD$ and $\angle ABC = \angle ADC$

Given: $ABCD$ is a parallelogram, so $AB = DC$, $AB \parallel DC$, $AD = CB$ and $AD \parallel CB$.

Proof

	$\angle DAB + \angle ABC = 180°$	The sum of co-interior angles is 180°.
So:	$\angle DAB = 180° - \angle ABC$	
	$\angle ABC + \angle BCD = 180°$	The sum of co-interior angles is 180°.
So:	$\angle BCD = 180° - \angle ABC$	
Hence:	$\angle DAB = \angle BCD$	Using the equality axiom.
Similarly:		
	$\angle ADC + \angle DCB = 180^\circ$	The sum of co-interior angles is 180°.
So:	$\angle ADC = 180^\circ - \angle DCB$	
	$\angle ABC + \angle DCB = 180^\circ$	The sum of co-interior angles is 180°.
So:	$\angle ABC = 180^\circ - \angle DCB$	
Hence:	$\angle ABC = \angle ACD$	Using the equality axiom.

It is proven that opposite angles in a parallelogram are equal in size.

Other theorems that apply to quadrilaterals, and can be proved, are:

- 'The interior angles in a rectangle are all right angles.'
- 'The diagonals of a rhombus bisect each other at right angles.'
- 'The diagonals of a kite intersect at right angles.'
- 'One pair of opposite angles in a kite are equal in size.'

EXERCISE

1 Find the values of the pronumerals in the following diagrams, giving a reason for your answer each time.

a 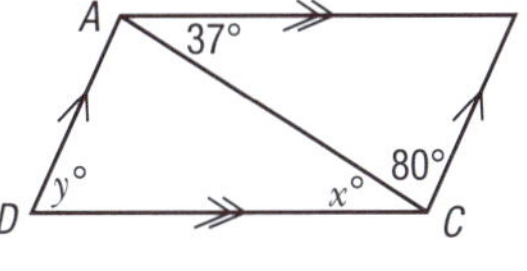

b 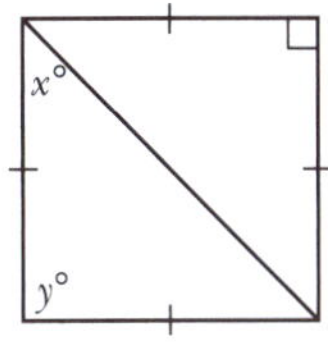

c 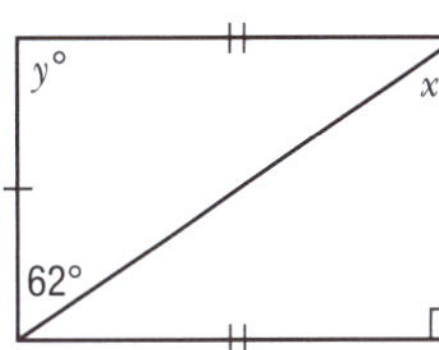

d 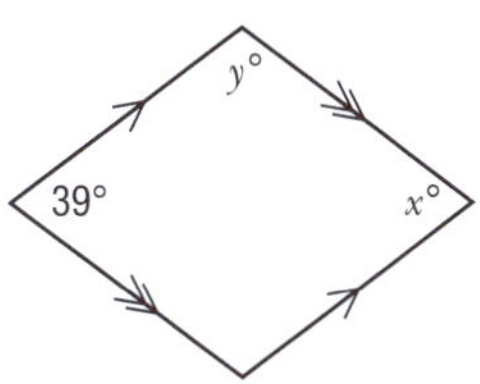

e 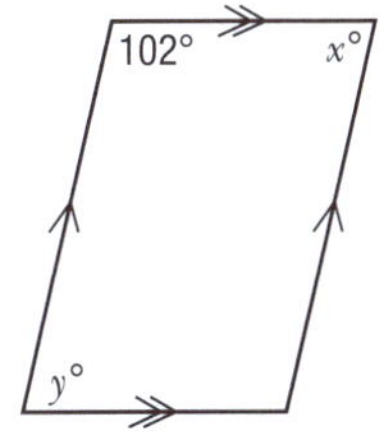

f 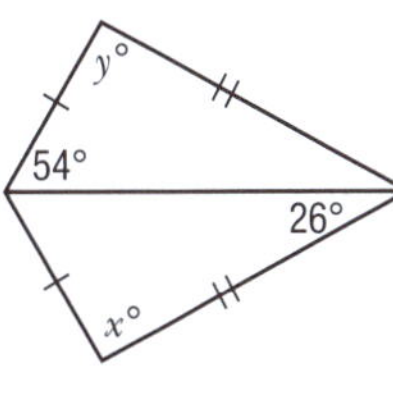

g 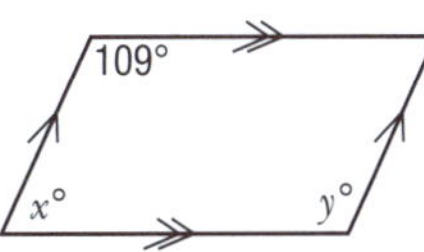

h 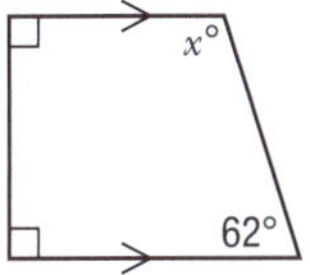

i 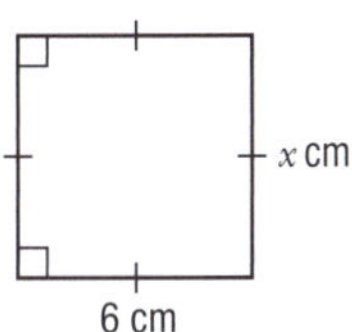

j 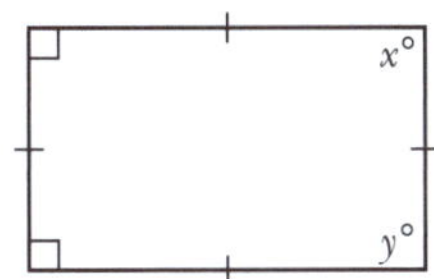

k 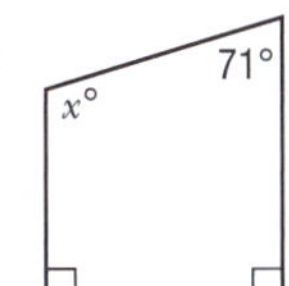

l 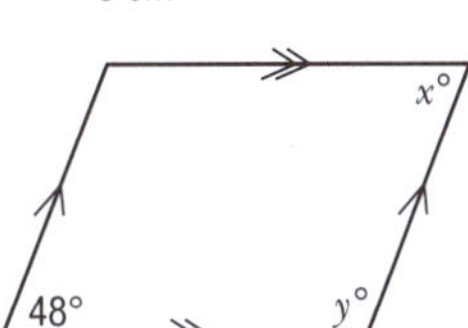

2 Form an equation for each of the following (giving a reason for your choice of each equation). Solve the equation to find the value of x each time.

a 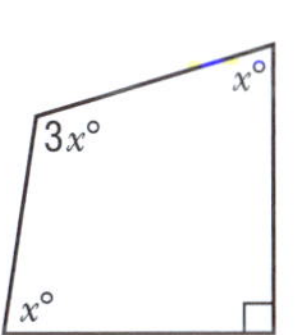

b 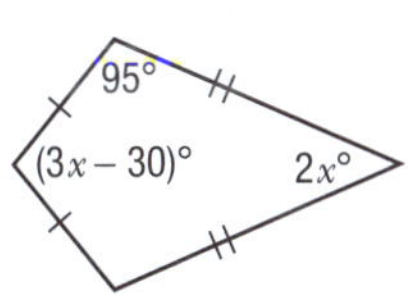

c 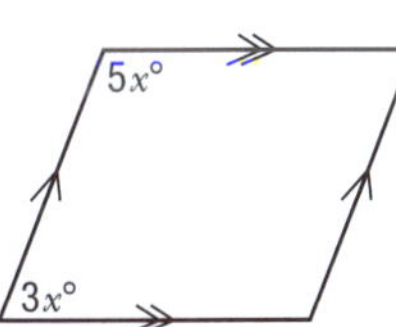

d 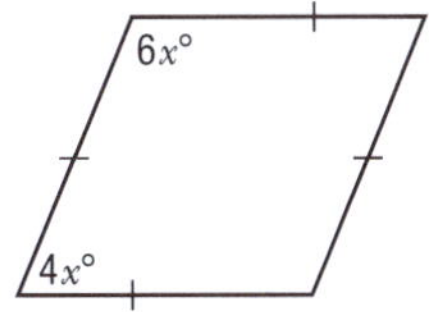

e 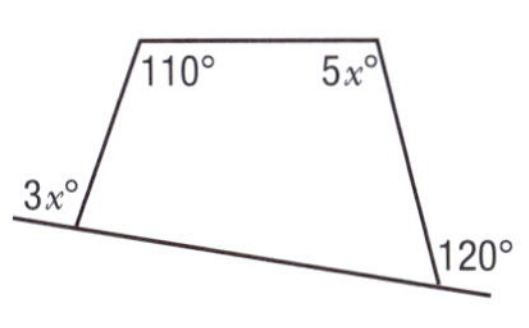

3 Comment on each of the following statements, giving reasons for your answers.

a All squares are rhombii.

b All rectangles are parallelograms.

c All squares are rectangles.

Remember

'Rhombii' is the plural of 'rhombus'.

4 Use the axioms mentioned in this unit, and the theorems that we have proved, to prove some of the commonly used theorems about quadrilaterals (listed below). Include a labelled diagram and reasons for each line of your proof. Use **Proofs 8, 9 and 10** (from Lesson 7) as a guide.

a 'The interior angles in a rectangle are all right angles.'

b 'The diagonals of a rhombus bisect each other at right angles.'

c 'The diagonals of a kite intersect at right angles.'

d 'One pair of opposite angles in a kite are equal in size.'

Lesson 8

Polygons with more than four sides

ABCDEFG is a heptagon. What is the sum of the interior angles of this heptagon?

The heptagon can be divided into five triangles, so the sum of the interior angles is $5 \times 180° = 900°$.

You will be asked to prove this theory in the exercise for this lesson.

Remember

A heptagon is a seven-sided shape.

If *ABCDEFG* is a regular heptagon, what is the size of each of its interior angles?

If *ABCDEFG* is a concave heptagon, does the theory about the interior angles still apply?

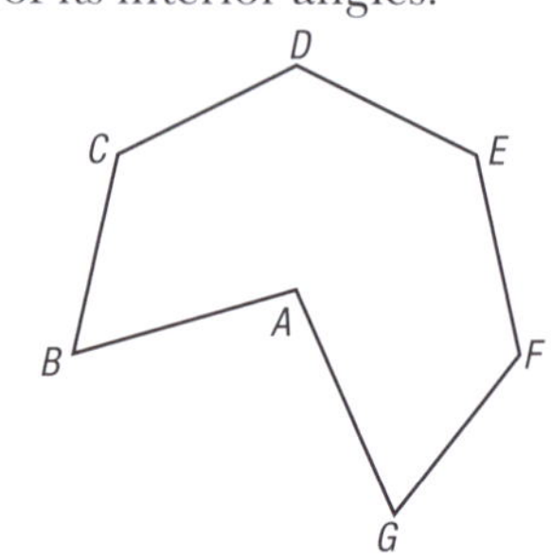

All polygons can be partitioned into triangles so theories about the sum of the **internal** angles of a polygon can be proved in a similar manner to that used in **Proof 8**.

The commonly used rule for the sum of the **exterior** angles of a polygon is: 'The exterior angles of a polygon sum to 360°.'

Proof 11

Proving, for a heptagon: 'The exterior angles of a polygon add to 360°.

ABCDEFG is a heptagon with interior angles *t*, *u*, *v*, *w*, *x*, *y* and *z*.

The sides have been extended, all in the same direction, to produce the exterior angles *a*, *b*, *c*, *d*, *e*, *f*, and *g*.

To prove: The exterior angles of a heptagon add to 360°.

In the diagram this would be: $a + b + c + d + e + f + g = 360$

Given: The sum of the interior angles of a heptagon is $5 \times 180° = 900°$.

Proof

$a + t = 180$ The sum of angles on a straight line is 180°.

For the same reason:

$b + u = 180$ $c + v = 180$

$d + w = 180$ $e + x = 180$

$f + y = 180$ $z + g = 180$

$a + b + c + d + e + f + g + t + u + v + w + x + y + z = 7 \times 180$ Adding all the equations above

But:	$t + u + v + w + x + y + z = 5 \times 180$	The sum of the interior angles of a heptagon is $5 \times 180°$.
$a + b + c + d + e + f + g + 5 \times 180 = 7 \times 180$		Substituting.
$a + b + c + d + e + f + g + 5 \times 180 - 5 \times 180 = 7 \times 180 - 5 \times 180$		Subtracting 5×180 from both sides of the equation

$a + b + c + d + e + f + g = 2 \times 180 = 360$

It is proven that the sum of the exterior angles of a heptagon is 360°.

EXERCISE

1 Prove that the sum of the interior angles of a pentagon is 540°. Set your proof out in a logical manner, including a labelled diagram and reasons for the statements in your proof.

2 **a** Can you find a rule for the sum of the interior angles of an n-sided polygon? Look carefully at the answers you already know for the sum of the interior angles of a three-sided polygon (a triangle), a four-sided polygon (a quadrilateral), a five-sided polygon (a pentagon) and a seven-sided polygon (a heptagon). Can you see a pattern?

b Write your rule in words and then in terms of n.

3 Does your rule from Question 2 apply to concave polygons? Give reasons and use diagrams to support your answer.

4 Use your conclusion from Question 1 to show that the interior angles of a regular pentagon are 108°.

5 Does the rule from **Proof 11** apply to concave polygons? Give reasons and use diagrams to support your answer.

6 *ABCDE* is a regular pentagon. A line is drawn from point *B* to point *D*, as shown in the diagram. Prove that the line *BD* is parallel to side *AE*. Set your proof out in a logical manner including a labelled diagram and give reasons for the statements in your proof.

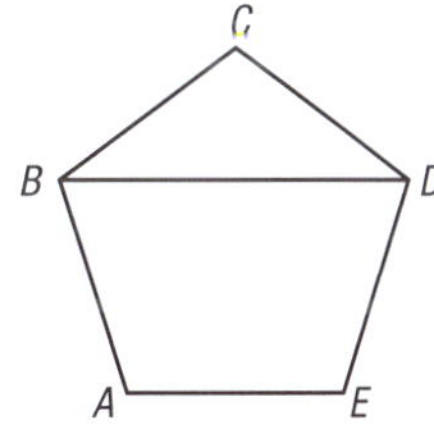

7 *ABCDE* is a regular pentagon. Lines are drawn from points *A* and *B* to point *D*, as shown in the diagram. Prove that triangle *ABD* is an isosceles triangle.

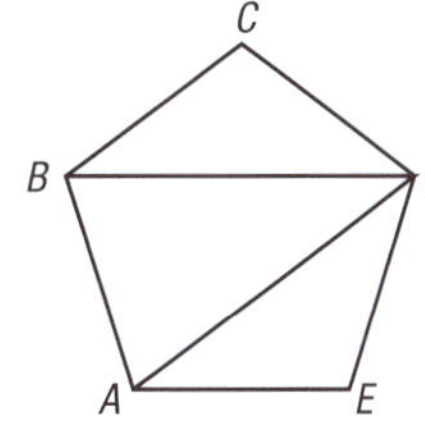

Remember

You can use any of the theorems or rules from the previous lessons.

Circles

There are many theorems that involve circles. Some of the commonly used ones are given below.

- The angle in a semi-circle is a right-angle.

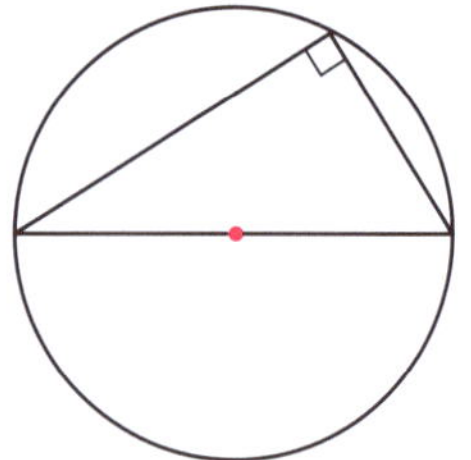

- Angles on a circle 'standing' on the same arc are equal in size.

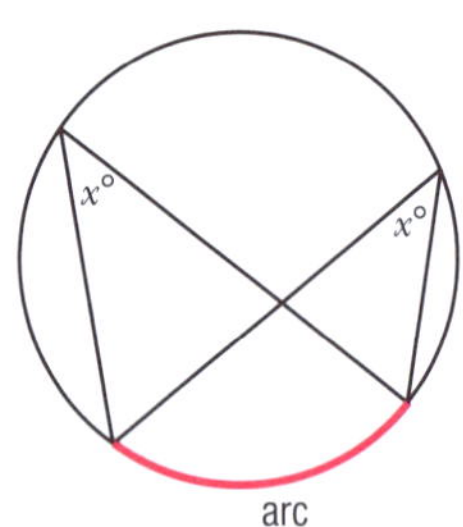

- For angles standing on the same arc, the angle at the centre is twice the size of the angle on the circle.

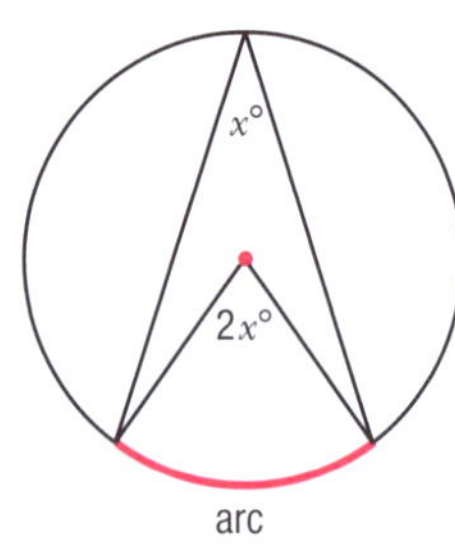

- The opposite angles of a cyclic quadrilateral sum to 180°.

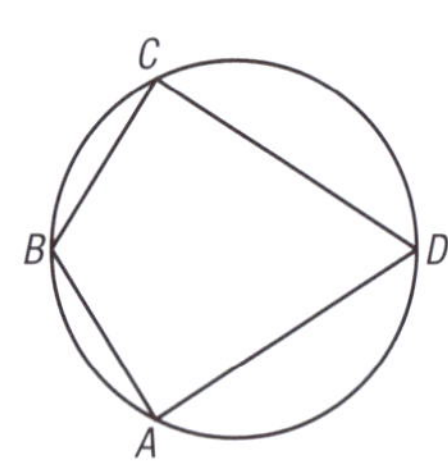

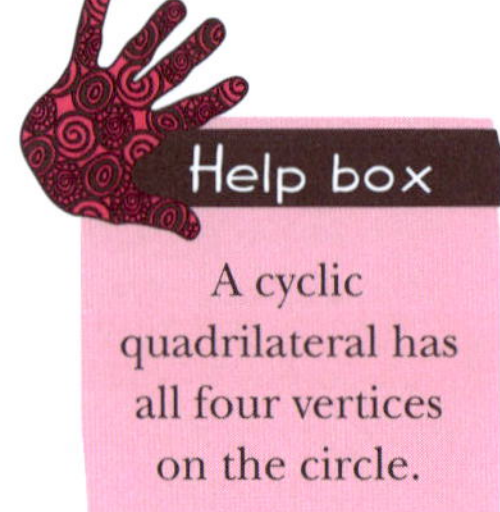

A cyclic quadrilateral has all four vertices on the circle.

Proof 12

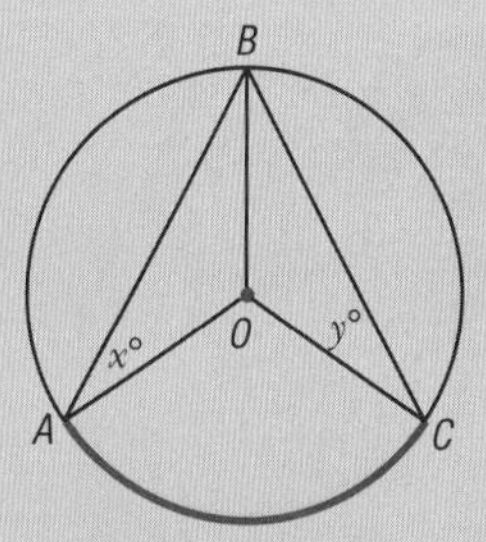

To prove: The angle at the centre is twice the size of the angle on the circle, for angles standing on the same arc.

In the diagram this would be: $\angle AOC = 2 \times \angle ABC$

Construct a line from the centre, O, to point B and label angle BAO as $x°$ and angle BCO as $y°$.

Remember

'Radii' is the plural of 'radius'.

Proof

$AO = BO = CO$ — All are radii of the circle.

$\triangle ABO$ is an isosceles triangle. — Sides AO and BO are both radii

So: $\angle ABO = x$ — Angles opposite the equal sides are equal.

$\triangle CBO$ is an isosceles triangle. — Sides CO and BO are both radii.

So: $\angle CBO = y$ — Angles opposite the equal sides are equal.

$\angle ABC = \angle ABO + \angle CBO$

$= x + y$

$\angle AOB = 180 - 2x$ — The sum of the angles in a triangle is 180°.

$\angle BOC = 180 - 2y$ — The sum of the angles in a triangle is 180°.

$\angle AOC + \angle AOB + \angle BOC = 360°$ — The sum of the angles at a point is 360°.

So: $\angle AOC = 360 - \angle AOB - \angle BOC$

$= 360 - (180 - 2x) - (180 - 2y)$ — Substituting.

$= 360 - 180 + 2x - 180 + 2y$ — Expanding.

$= 2x + 2y$ — Grouping like terms.

$= 2(x + y)$ — Factorising.

$= 2 \times \angle ABC$

It is proven that the angle at the centre is twice the size of the angle on the circle, for angles standing on the same arc.

Example

Find the value of x in the diagram giving reasons for your answer.

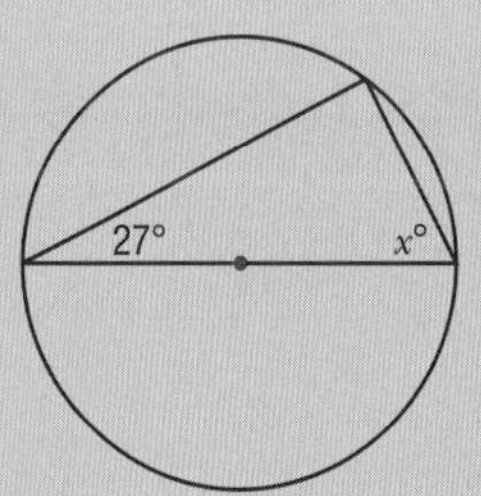

Answer

The triangle has a diameter of the circle as one of its sides so, if the angle that is not marked is half the angle at the centre, it is a right angle (90°).

$27 + x + 90 = 180$ — The sum of the angles in a triangle is 180°.

$x = 63$

Example

Find the value of the pronumerals x and y in the diagram giving reasons for your answer.

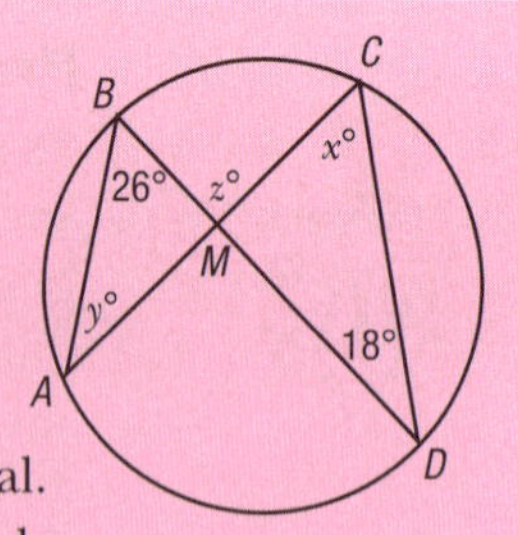

Answer

$x = 26$ Angles standing on the same arc (AD) are equal.

$y = 18$ Angles standing on the same arc (BC) are equal.

$z = 26 + y$ Exterior angle of a triangle ($\triangle ABM$) is equal to the sum of the two interior angles.

$= 26 + 18$

$= 44$

EXERCISE

1 Find the value of x in each of the following, giving a reason for your answer.

a 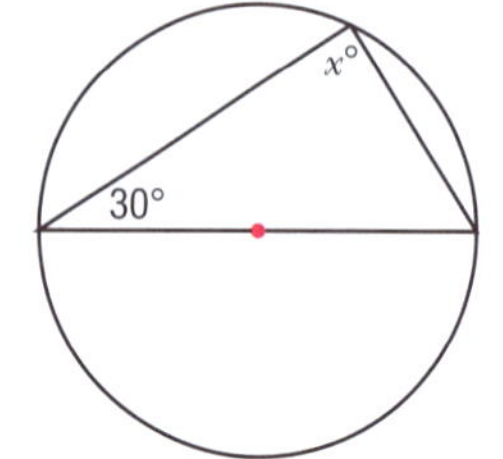

b 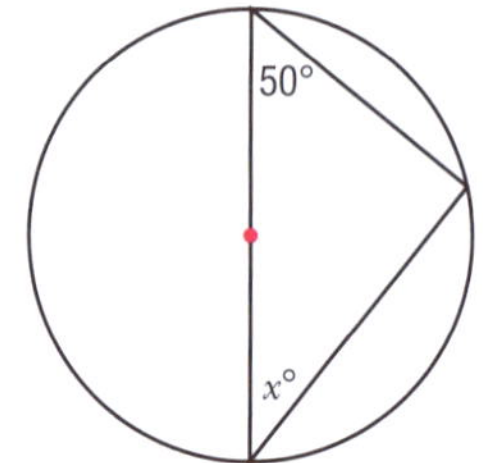

c 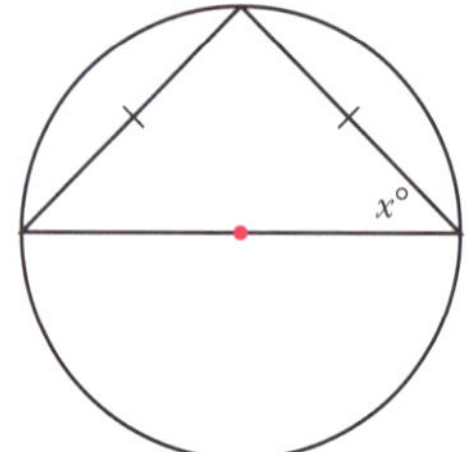

2 Find the value of the pronumeral in each of the following diagrams, giving a reason for your answer.

a 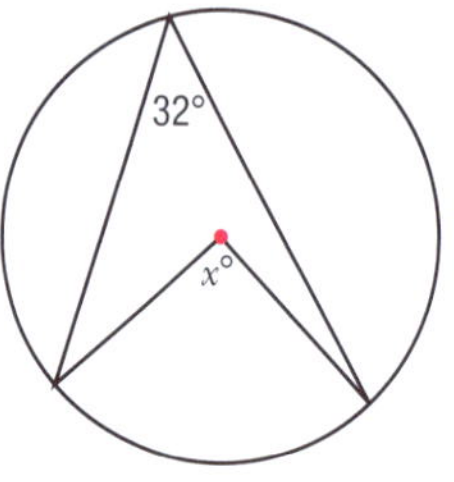

b 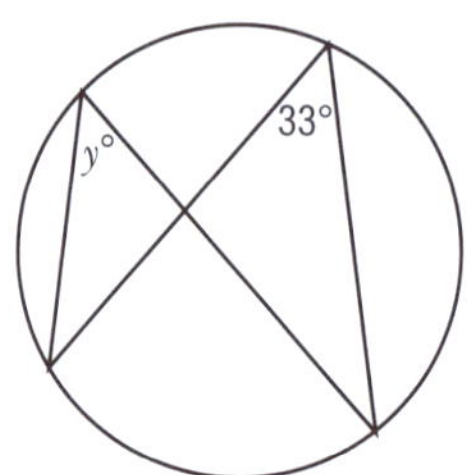

c 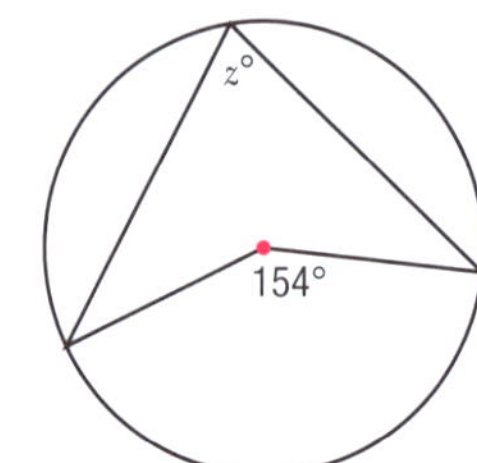

d 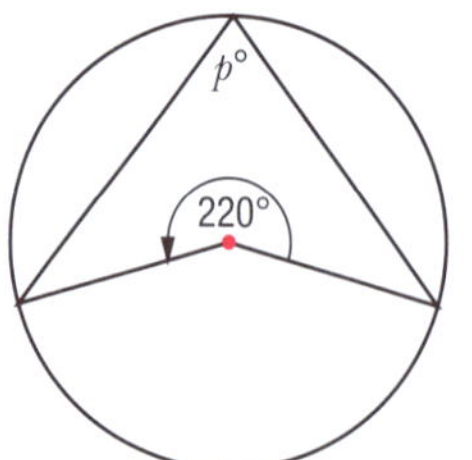

e 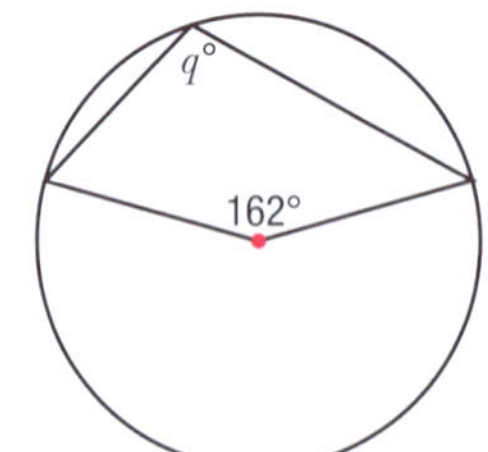

f 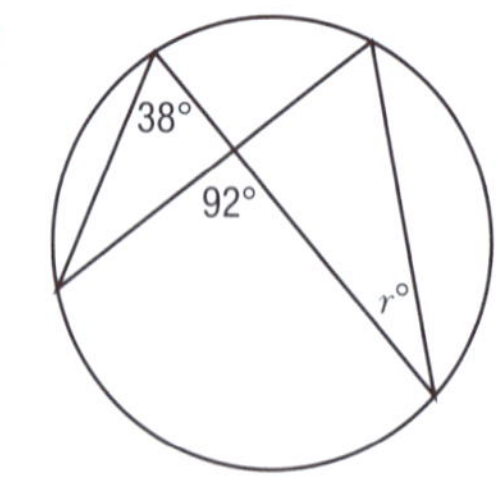

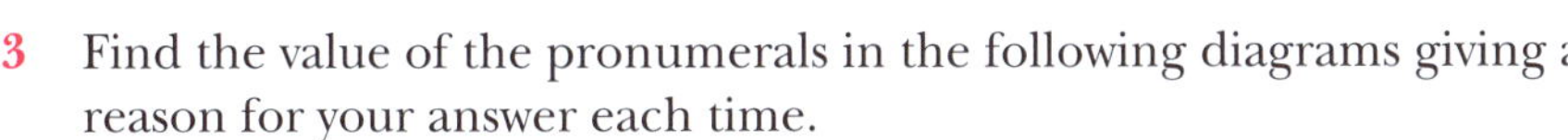

3 Find the value of the pronumerals in the following diagrams giving a reason for your answer each time.

a

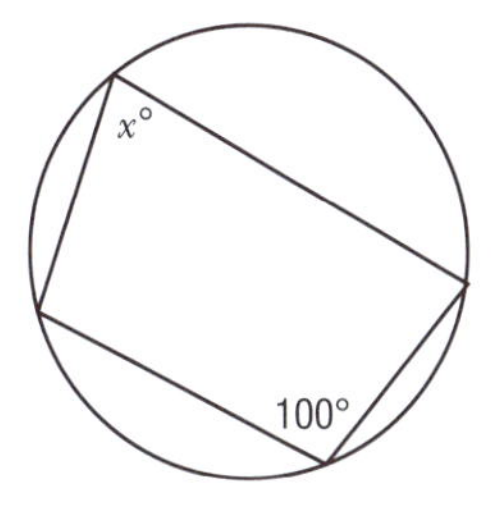

b

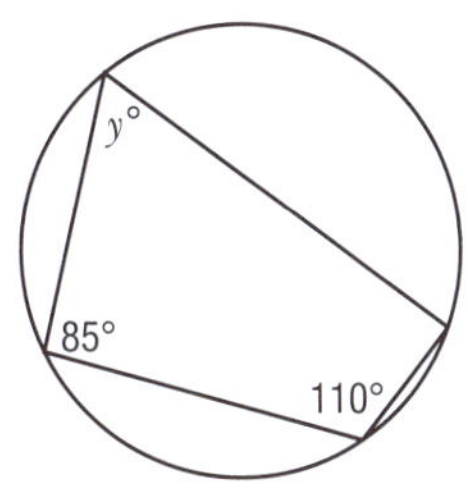

c

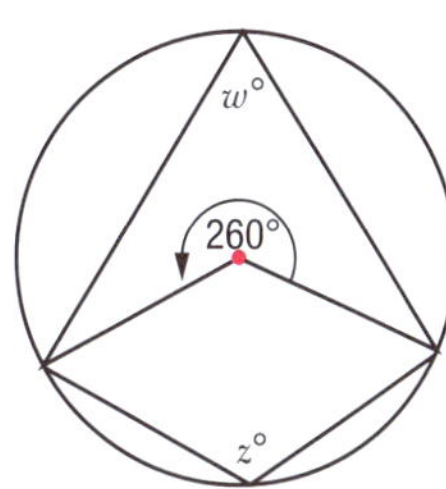

d

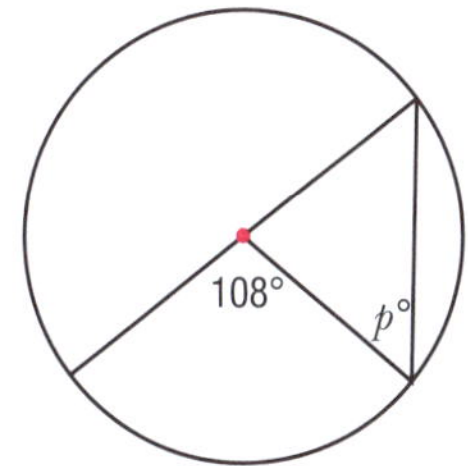

e

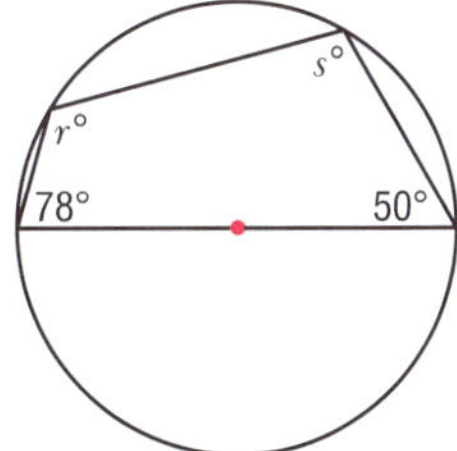

f

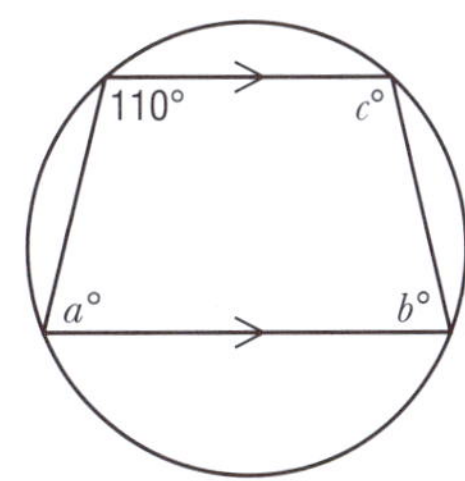

4 Prove that the angle in a semi-circle is a right-angle.
Draw a circle, as shown, in your workbook, labelling the vertices of the triangle and some angles.
Set the proof out in a logical manner, giving reasons for each line.
Use **Proof 12** on page 373 as a guide.

5 Prove that the angles on a circle 'standing' on the same arc are equal in size.
Draw a circle, as shown, but include two lines joining the ends of the arc to the centre. Label some angles on your diagram.
Set the proof out in a logical manner, giving reasons for each line and, for this proof, you can assume that the theorem proved in **Proof 12** is true.

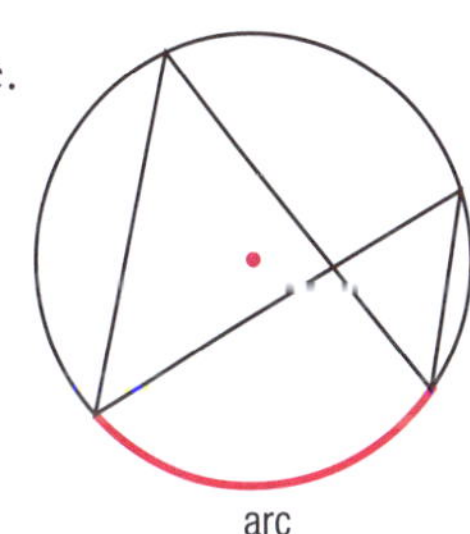

6 Prove that the opposite angles of a cyclic quadrilateral sum to 180°.
Draw a circle, as shown, but include two lines joining two opposite vertices to the centre. Label some angles on your diagram.
Set the proof out in a logical manner, giving reasons for each line and, for this proof, you can assume that the theorem proved in **Proof 12** is true.

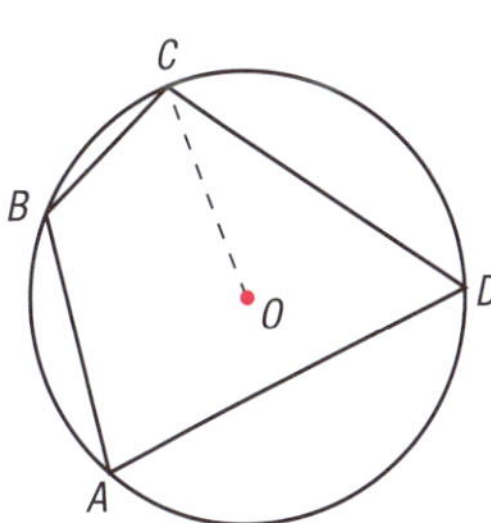

REVISION AND ASSESSMENT

2D geometry

1 Write 'True' or 'False' for each of the following statements.
   a An obtuse angle is greater than 180°.
   b The complement of 63° is 27°.
   c Vertically opposite angles add to 180°.
   d A reflex angle is larger than an obtuse angle.
   e Alternate angles add to 180°.
   f An acute angle is smaller than a right angle.
   g Angles in a revolution add to 360°.
   h The supplement of 63° is 27°.
   i A straight angle is equal to two right angles.
   j A transversal crossing parallel lines makes corresponding equal angles.
   k A polygon has five sides.
   l The sum of the angles in a rectangle is 360°.
   m The opposite sides of a kite are equal.
   n A parallelogram has two pairs of equal and parallel sides.
   o The diagonals of a rectangle are equal.
   p The exterior angles of any polygon add to 360°.

2 Draw and label a neat sketch of:
   a an equilateral triangle
   b a rhombus
   c a regular hexagon
   d a concave pentagon
   e a right-angled isosceles triangle
   f a trapezium

3 Draw and label a neat sketch of:
   a a right angle, $\angle Z$
   b vertically opposite angles, $\angle STU$ and $\angle VTW$
   c a reflex angle, $\angle WXY$
   d adjacent angles $\angle RST$ and $\angle TSU$
   e complementary angles $\angle KLM$ and $\angle MLO$
   f alternate angles, a and b
   g co-interior angles, g and h

4 What value would x need to be to make 56° and x°:
   a complementary?
   b supplementary?

5 In the diagram on the right, name one pair of:
   a corresponding angles
   b vertically opposite angles
   c supplementary angles.

6 Name this polygon and determine the sum of its interior angles.

7 Name the exterior angle in this triangle and find its size.

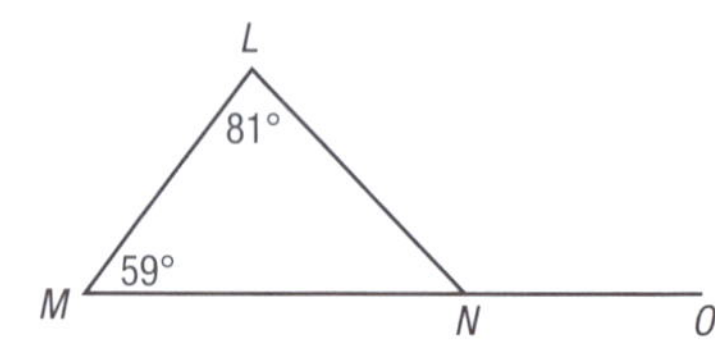

8 Find the sizes of the unknown angles.

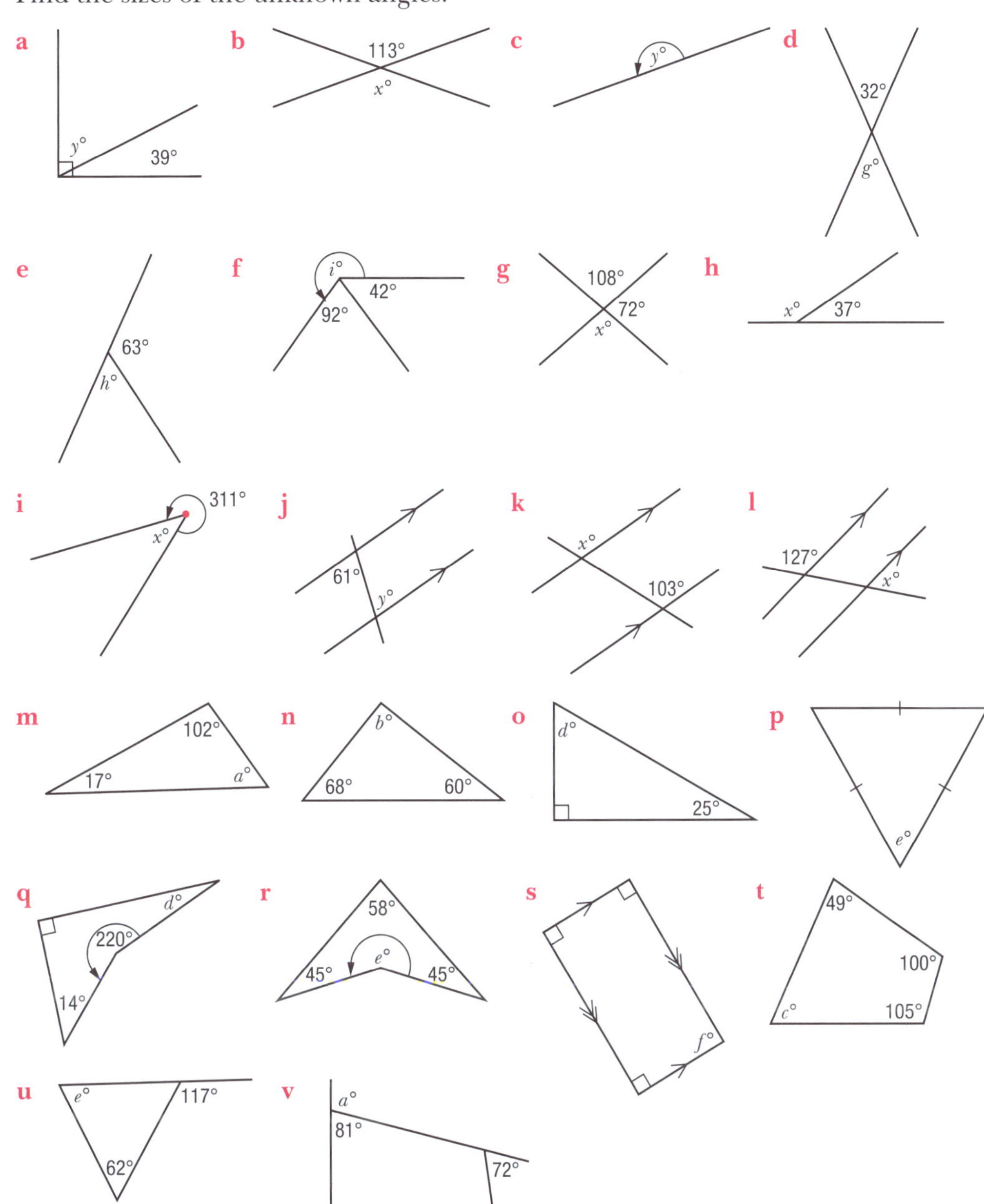

Surface area and volume

1 When converting metres to kilometres, which of the following will you need to do?

A divide by 1000

B multiply by 100

C divide by 100

D multiply by 1000

2 In a circle, which of the following is the ratio describing the number pi (π)?
- **A** the circumference to the centre
- **B** the radius to the diameter
- **C** the circumference to the diameter
- **D** the circumference to the radius.

3 An area of 2.34 square metres is equivalent to which of the following?
- **A** 234 square centimetres
- **B** 23 400 square centimetres
- **C** 234 000 square centimetres
- **D** 2 340 000 square centimetres

4 Which of the following is the area of the triangle shown?
- **A** $10 \times 12\ \text{cm}^2$
- **B** $\frac{1}{2} \times 10 \times 12\ \text{cm}^2$
- **C** $\frac{1}{2} \times 10 \times 11\ \text{cm}^2$
- **D** $\frac{1}{2} \times 11 \times 12\ \text{cm}^2$

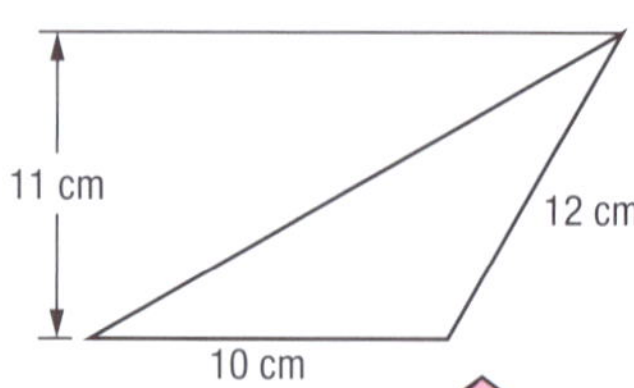

5 Which of the following is the number of surfaces on this solid?
- **A** 5
- **B** 7
- **C** 10
- **D** 15

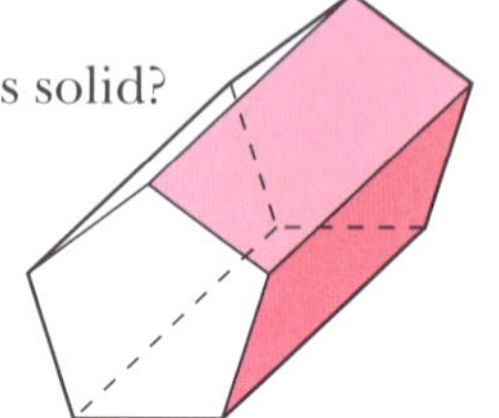

6 56 780 cubic centimetres is equivalent to which of the following?
- **A** 567.8 cubic metres
- **B** 56.78 litres
- **C** 567 800 000 cubic millimetres
- **D** 5678 millilitres.

7 Complete each of the following conversions:
- **a** 85 mm to centimetres
- **b** 50 L to millilitres
- **c** 71 cm^2 to mm^2
- **d** 45 km to centimetres

8 Find:
- **a** the perimeter of the triangle shown
- **b** the area of the triangle shown.

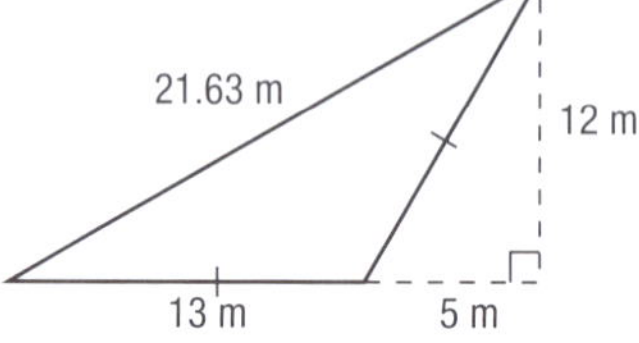

9 Find the perimeter of the figure on the right.

10 Find the area of each of the following:

a

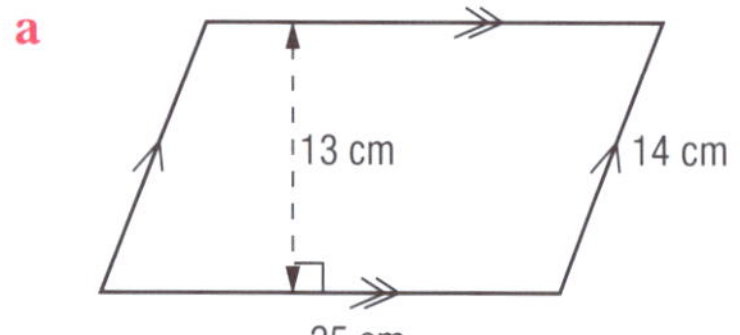

b

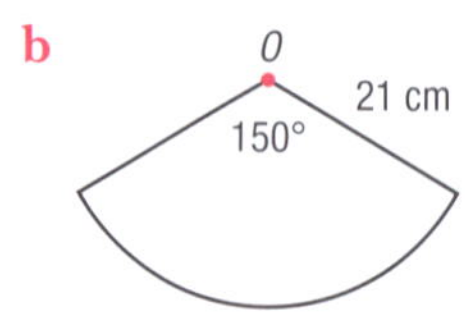

11 Find the area of eachof the following shaded region.

a

4 cm
4 cm
9 cm

b

7 cm

12 a Draw the net of the prism on the right.
b Use the net from part **a** to calculate the total surface area of the prism.

10 cm
6 cm
6 cm
8 cm

13 Find the total surface area of each of the following solids.
a a sphere of diameter 140 mm
b a cylinder of radius 35 cm and height 90 cm
c a cone with a base radius of 1 m and slant edge length of 1.8 m

14 Find the volume of each of the following solids:

a

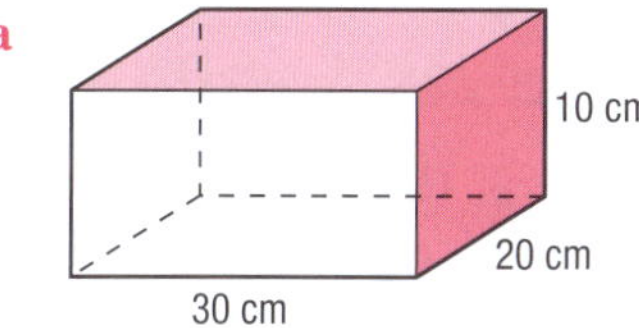

b

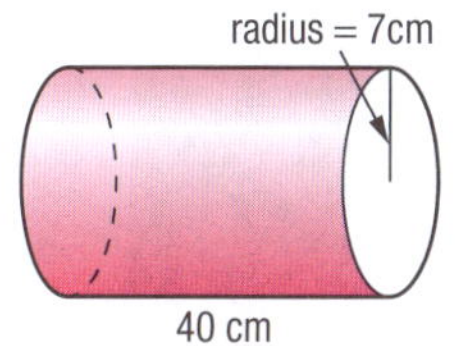

c

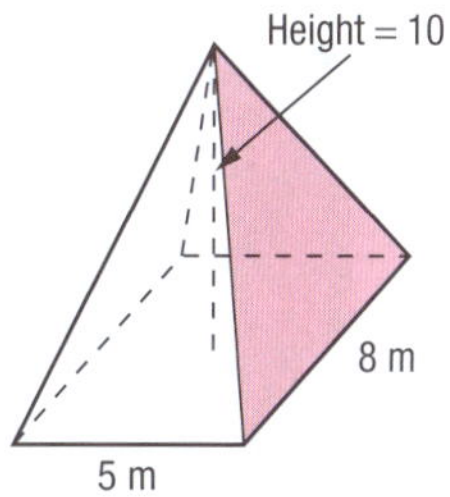

d

35 cm
14 cm

e

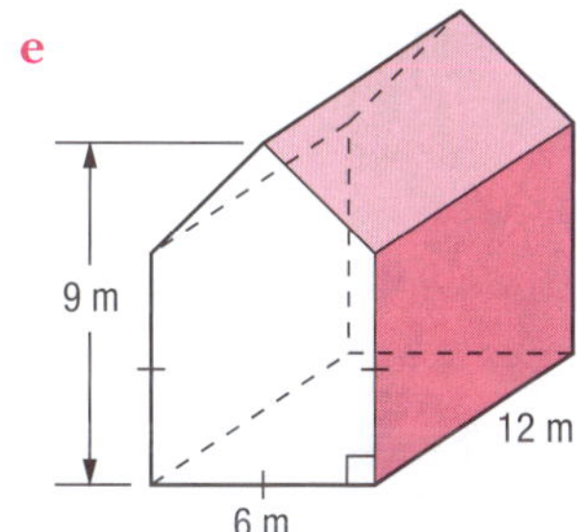

f

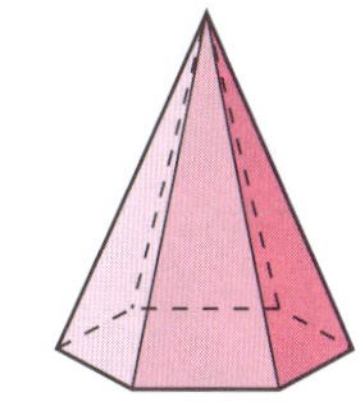

Area of hexagon base = 2.10 cm²
Height of pyramid = 3.3 cm

15 Find the capacity, in kilolitres, of a cylindrical water tank that has a height of 1.8 metres and a diameter of 2.8 metres.

16 In the shape on the right, the outer square has an area of 49 cm^2, find the area of the shaded region.

17 Find the weight, in kilograms, of the hollow cement pipe on the right, given that 1 cm^3 of the pipe has a weight of 5 grams. The outer radius of the pipe is 70 cm and the walls are 5 cm thick.

1.8 m

Remember
1 kilogram = 1000 grams

18 Find the total surface area of the figure on the right.

6.3 cm
2.8 cm
20 cm

PROJECTS ASSESSMENT TASK TWO

The assessment task for this unit is an individual investigation.

You will need to present your findings as a report, an illustrated chart, a poster, a flow-chart or an article. Your teacher will decide what the task will be and will give you the choices for presentation.

Option A: Assessment

Symmetry projects

Project 1: Tattoo designs and their use

Traditionally, many cultures have used the marking of the body by tattooing for decorative or religious purposes, as a sign of status or as a rite of passage. Examples have been found dating back hundreds of thousands of years, and the practice is used even more widely today around the world.

1 Investigate the past or present use of tattooing, or another form of traditional marking, in your area or in another part of the Pacific Islands. Illustrate examples of tattoos or markings and explain how they have been used. Describe any symmetry apparent in the designs.

2 Design a symmetrical tattoo suitable to be used on a person's face or upper back.

Project 2: Symmetry in art and design

Celtic mandala

A stained glass ceiling

Indian swastika

1 Investigate the use of symmetry in one of the following:
  a religious symbols (for example the Christian cross, the Jewish Star of David, the Shinto torii, or the Bahai star)
  b company logos

  c national flags
  d traditional weaving

2 Find at least six to 10 examples which demonstrate some form of symmetry. Write a report, including illustrations of each example, explaining its origin. For each example, describe any:
  a bilateral (fold) symmetry, showing the axes of symmetry
  b rotational symmetry, stating the order.

3 Create one example of your own which has either bilateral symmetry or rotational symmetry and explain its purpose and meaning.

Construction projects in 2D

Project 1: Line and circle designs

Many designs are based on lines drawn in angles or circles. To reproduce these, we only need a ruler, a sharp pencil and compasses.

The designs below are all based on the same technique. Follow the steps below to see how.

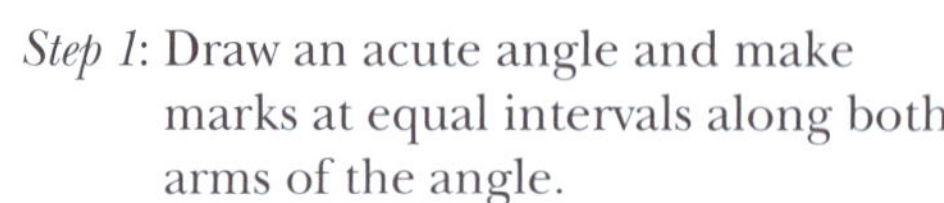

Step 1: Draw an acute angle and make marks at equal intervals along both arms of the angle.

Step 2: Number the marks in ascending order along one arm and in descending order on the other arm.

Step 3: Join matching numbers. (An example has been begun below.)

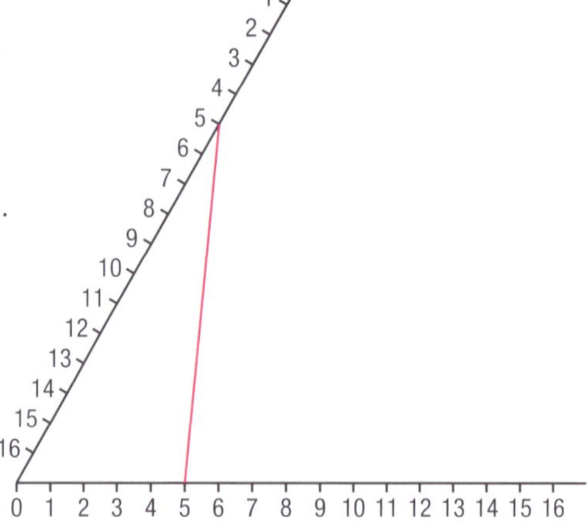

With practice, this can be done without writing the numbers along the arms of the angle.

Instead of using ruled lines, we can also create these designs using a needle and thread and sewing through the backing card, or by winding a length of string around fine nails hammered into marks along the arms of an angle drawn on a sheet of timber.

Investigation

1 Investigate the effect of changing the size of the angle (larger and smaller). Write a statement reporting your conclusions.

2 Investigate the effect of changing the spacing between the divisions on both arms (larger and smaller spacing). Write a statement reporting your conclusions.

3 Choose one of the given designs (on the right or following) and reproduce it, or invent one of your own.

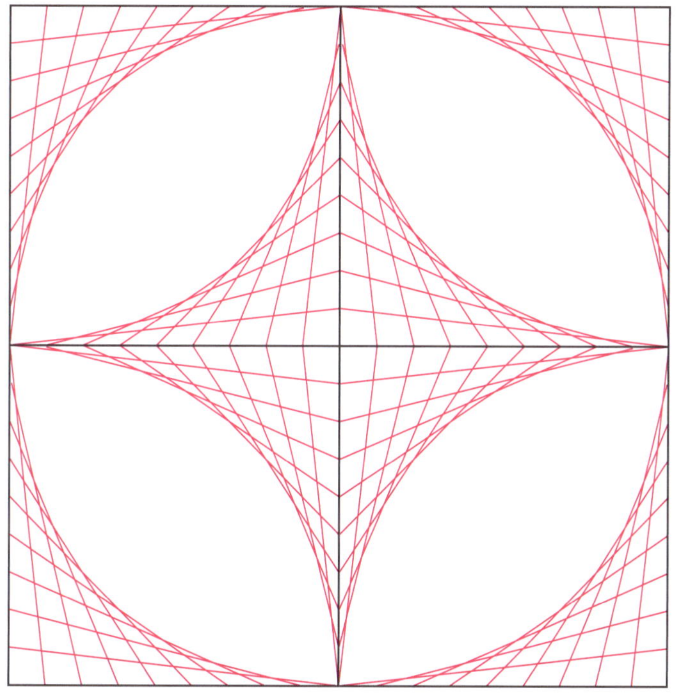

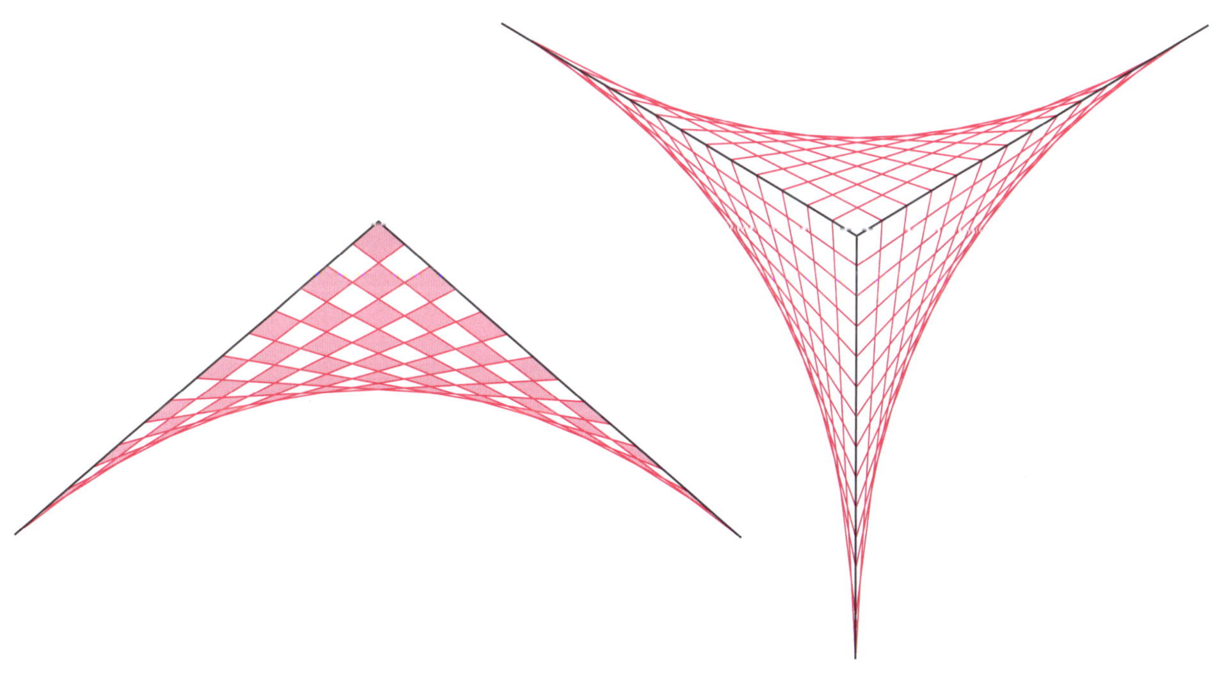

Project 2: Curve of pursuit

Imagine three dogs are sitting at the corners of an equilateral triangle. The dog at *A* sets off to chase the dog at *B* who sets off to chase the dog at *C*. After a few moments, all three dogs have moved to a new position so each chaser has to run in a slightly different direction.

The steps for drawing this 'curve of pursuit' are given below.

Step 1: Construct an equilateral triangle with sides that are 12 cm in length (or a number of centimetres that is divisible by 6).

Step 2: Mark a point one-sixth of the way along each side and join these points to make a new triangle, as shown at right.

Step 3: Continue in this way, making smaller and smaller triangles until you cannot go any further.

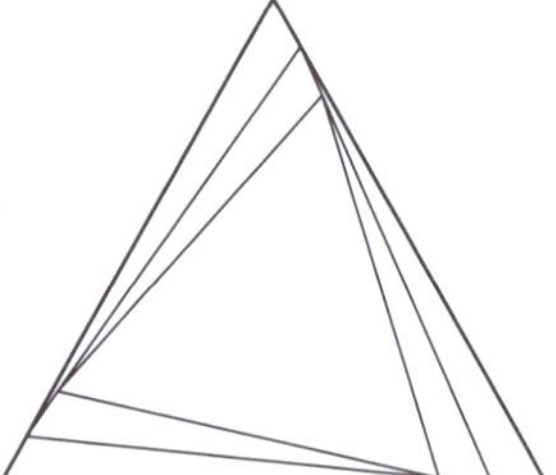

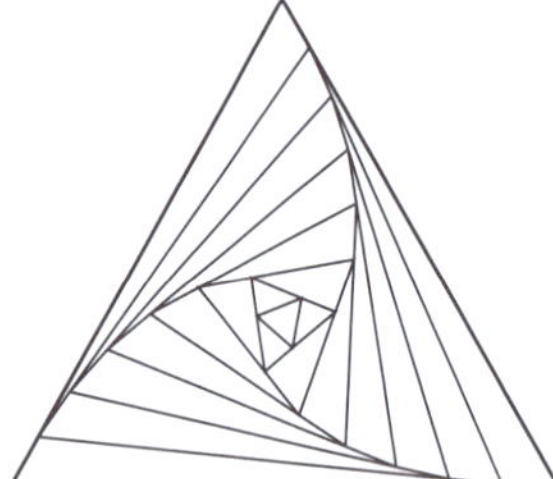

Investigation

1 What happens if we start with a scalene triangle? Describe how the result differs from the example above.

2 Draw a 'curve of pursuit' starting with a square.

3 Using either a triangle or a square as the basis, construct a 'curve of pursuit' design from coloured paper or fabric.

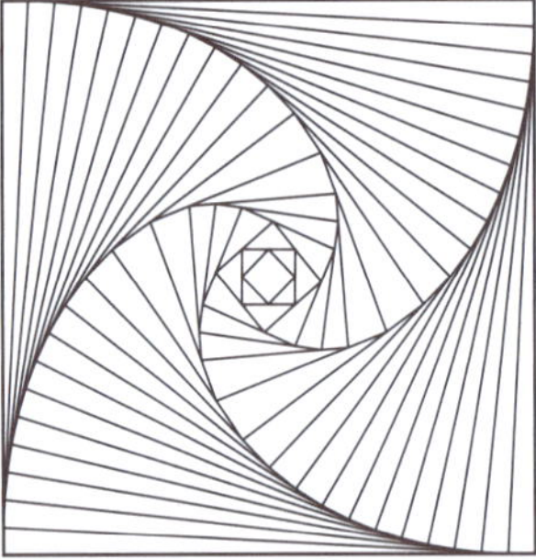

Construction projects in 3D

Project 1: Polyhedra

Here are the instructions for making a truncated tetrahedron or cuboctahedron:

Step 1: Trace the net then transfer it onto light cardboard or stiff paper. You might be able to do this more easily by making pinholes through the vertices of the tracing onto the cardboard.

Step 2: Cut around the solid lines on the outside of the net.

Step 3: Crease along all the lines, folding *away* from you.

Step 4: Lightly shape the net into a ball shape in your hands so you can see which flaps to join.

Step 5: Glue the flaps under the matching faces, leaving each section to dry before gluing the next.

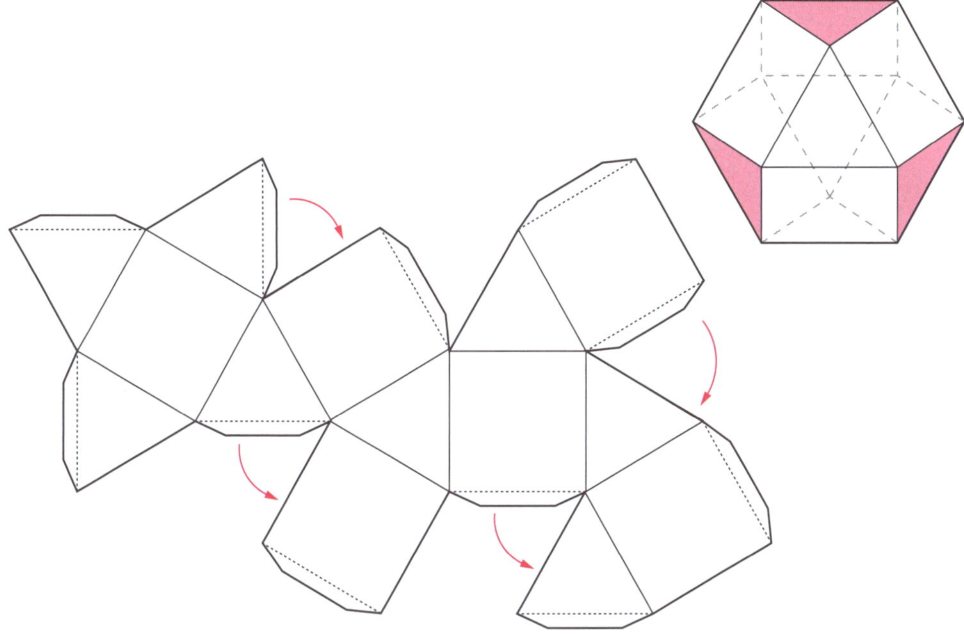

cuboctahedron

truncated tetrahedron

1 A manufacturer wants to make small boxes for individual Christmas cakes, similar to the one shown below.

The finished box is to be a hexagonal prism 6 cm high, with the sides of the hexagonal ends measuring 4 cm.

a Sketch one of the hexagonal ends. By dividing it into four triangles, work out the angle sum of the hexagon and, hence, the size of each angle.

b Sketch the net of the prism.

c On cardboard, construct the net accurately. Allow for tabs, on the sides, to fasten the bottom of the prism and, on the top, so you can fold the lid in.

3D Tessellations

Project 1: Quilting

Quilting is a popular activity in which small pieces of material are joined together to form a blanket or other item. Most quilting designs are tessellations based on one or more repeated shapes.

1 The two designs below are based on simple two-dimensional shapes but, because of the way they have been coloured, the design appears three-dimensional.

a Using coloured paper or fabric, copy one of the designs below to give a finished piece with sides of at least 20 cm.

b Create your own design for a quilt based on a tessellation.

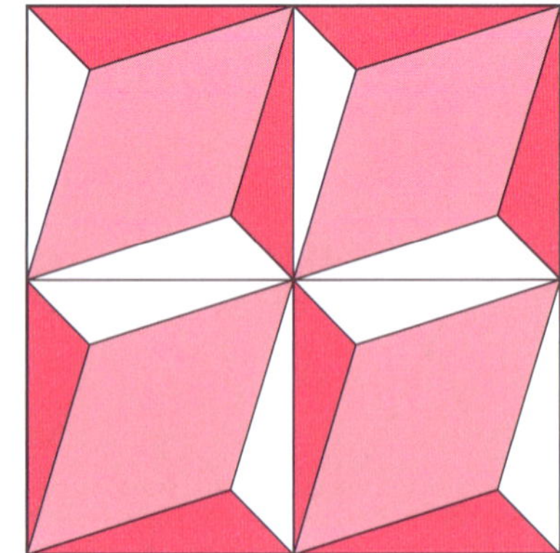

Option B: Assessment

Project 1: Investigating a hexagon

Some, or all, of the parts of this activity can be attempted. Reasons, or properly set-out proofs, should be given for any parts where you are asked to 'show'. Properly constructed and labelled diagrams should be drawn as part of your explanations.

1 **For any hexagon.**
  - a Define a hexagon, explaining each of the descriptive words that you use in your definition.
  - b Show that the sum of the interior angles of a hexagon add to 720°.
  - c Show that the sum of the exterior angles of a hexagon add to 360°.
  - d Do the rules from parts **a** and **b** apply to a concave hexagon? Give reasons for your answer.

2 **For a regular hexagon.**
  - a Define a regular hexagon.
  - b Find the size of the exterior angle of a regular hexagon, giving reasons for your answer.
  - c Find the size of the interior angles of a regular hexagon, giving reasons for your answer.
  - d *ABCDEF* is a regular hexagon, as shown.
    - i Copy the diagram and construct a line from point *A* to point *D*.
    - ii Show that the line *AD* is parallel to side *BC*.
    - iii What shape is figure *ABCD*? Give reasons for your answer.
  - e *ABCDEF* is a regular hexagon, as shown.
    - i Copy the diagram and construct lines from point *A* to point *C* and from point *F* to point *D*.
    - ii Show that triangles *ABC* and *FDE* are congruent, giving reasons for each step.
    - iii Show that line *AC* is parallel to line *FD*, giving reasons.
    - iv What shape is figure *ACDF*? Give reasons for your answer.
  - f *ABCDEF* is a regular hexagon, as shown.
    - i Copy the diagram and construct lines from point *A* to point *D*, from point *B* to point *E* and from point *C* to point *F*.
    - ii What shape(s) have been produced by constructing these lines.
    - iii Label the point where the three lines intersect as *O*. Describe the triangle *ABO*, giving reasons for your answers.

3 On the diagram on the right the sides of a hexagon have been extended to produce a six-point star. Show that the sum of the angles at the 'points' of the star is 360°.

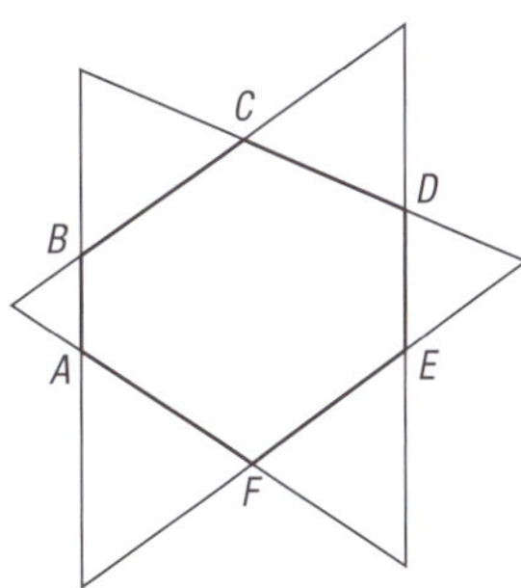

4 Design a patchwork square that includes a hexagon and at least two other polygons. Label all the angles in the polygons you use with their sizes.
Note: This activity can easily be adapted for other polygons.

Project 2: Polygons in circles

Some, or all, of the parts of this activity can be attempted. Reasons, or properly set-out proofs, should be given for any parts where you are asked to 'show'. Properly constructed and labelled diagrams should be drawn as part of your explanations.

1 Side AB of triangle ABC is a diameter of the circle.
Prove that angle ACB in the triangle is a right angle.

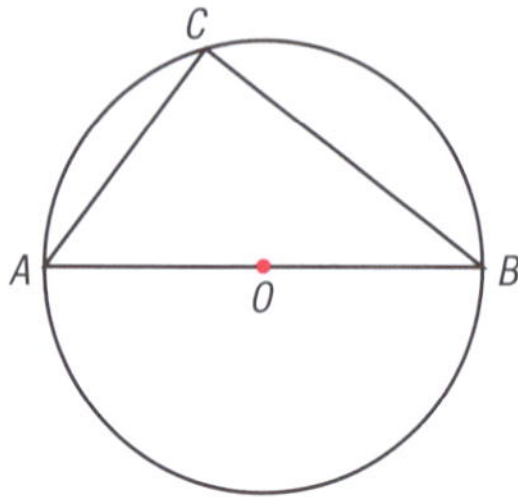

2 a i Draw and define an equilateral triangle.
ii Show that the interior angles of an equilateral triangle are each equal to 60°.

b An equilateral triangle, ABC, has its three vertices resting on a circle as shown in the diagram. The centre of the circle is the point O and a line is drawn from B, through O, to side AC of the triangle.
i Show that the line BM is perpendicular to AC.
ii Show that the line BM bisects the line AC.

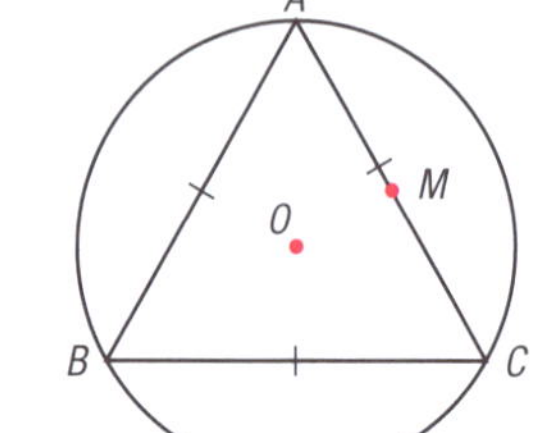

3 a i Describe a cyclic quadrilateral.
ii Define supplementary angles.

b A cyclic quadrilateral $ABCD$ is constructed in a circle, as shown in the diagram.
One side, AD, of the quadrilateral is a diameter of the circle.
Use the angles, as labelled, to prove that the opposite angles in a cyclic quadrilateral are supplementary.

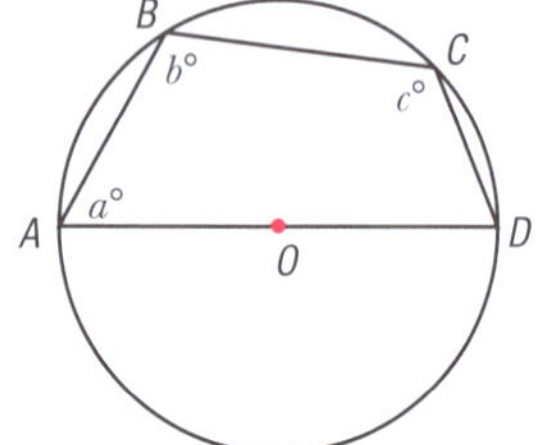

c Find the value of the pronumeral in each of the following, giving a reason for your answer.

i

85°
98°
x°

ii

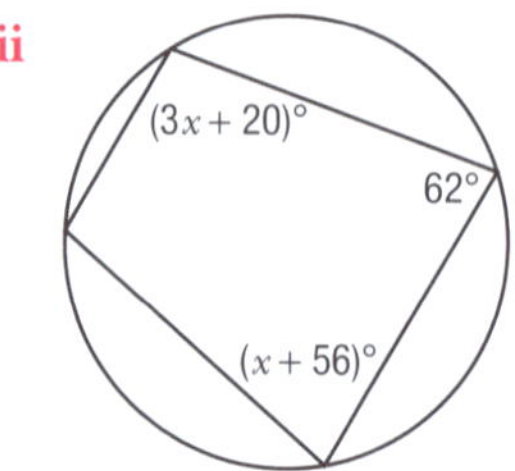

4 a Describe quadrilateral $ABCO$ in the diagram on the right, mentioning any particular features it has. (Point O is the centre of the circle.)

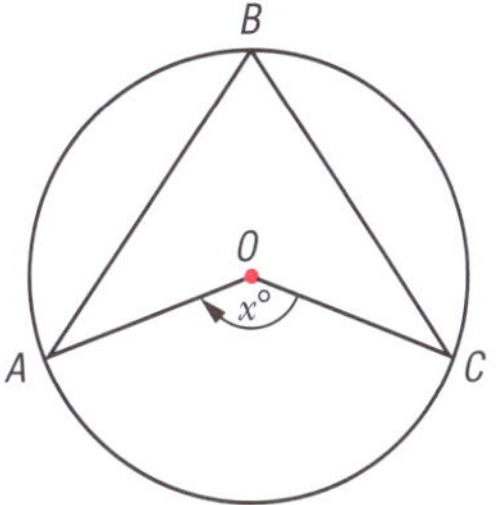

b For quadrilateral $ABCO$, prove that the angle marked $x°$ is twice the size of $\angle ABC$.

c Find the value of the pronumeral in each of the following, giving reasons for your answer.

i
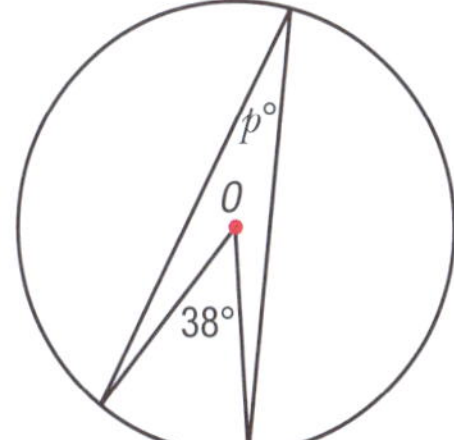

ii
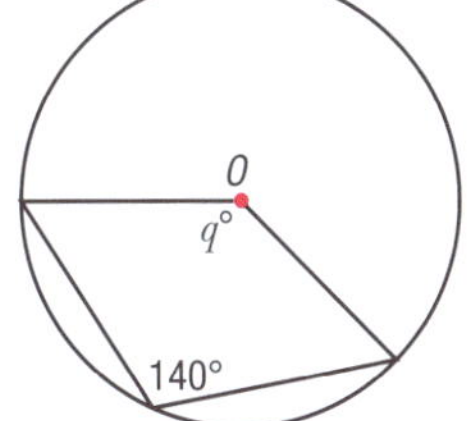

iii
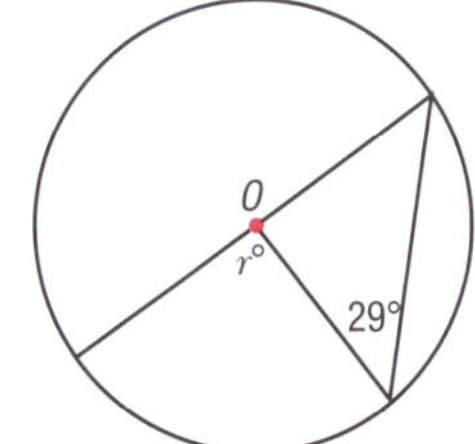

5 Prove the theorem: 'If a cyclic quadrilateral has a pair of adjacent angles that are right angles then the quadrilateral is a rectangle.'

Project 3: Parallel lines

1 Lines AB and CD are crossed by a transversal, MN, as shown in the diagram. Use the converse of the 'corresponding angles' axiom to prove that the lines AB and CD are not parallel.

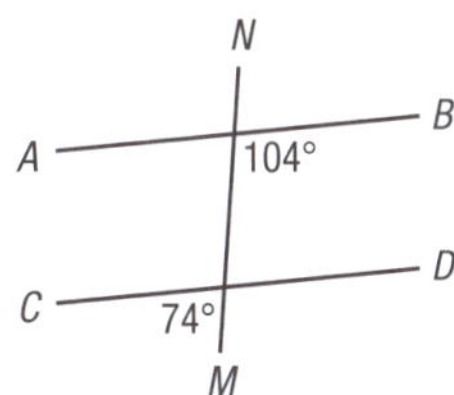

2 Copy the following diagrams, labelling where necessary. Then find the size of the angles, giving a reason for your answer each time.

a
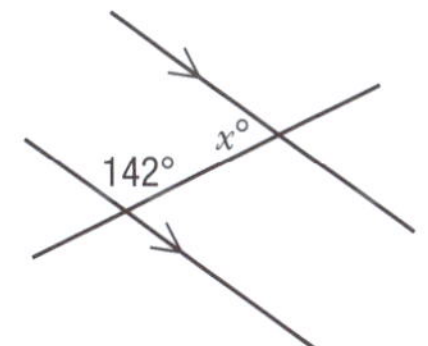

b
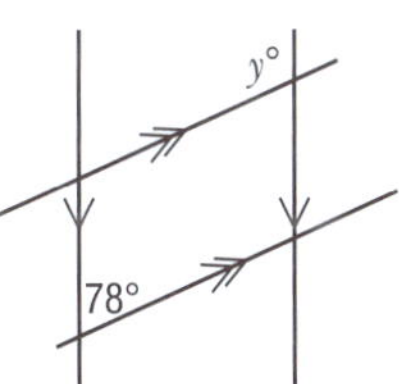

c
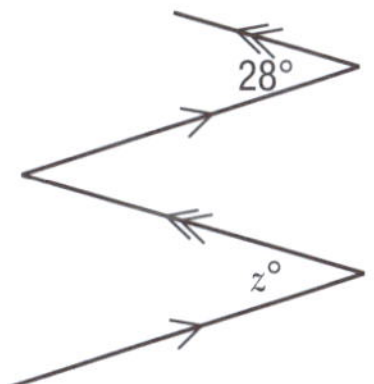

3 Construct a pair of parallel lines crossed by a transversal. Use the 'corresponding angles' axiom to prove that a pair of co-interior angles add to 180°.

4 For the diagram on the right, find the size of $\angle ABC$, in terms of x and y, giving reasons for your answer.

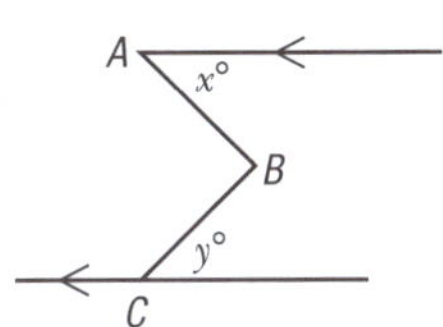

5 *ABCD* is a rhombus.

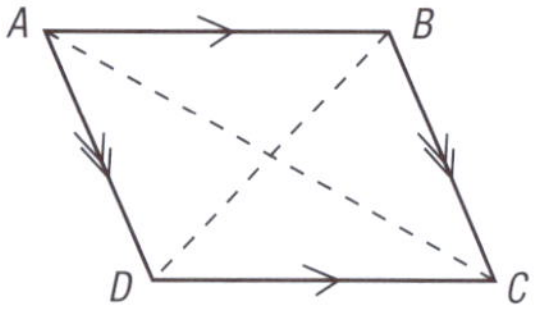

a List the properties of a rhombus.

b Prove that the diagonals of a rhombus bisect each other at right angles.

6 A quadrilateral is constructed inside a circle, so that all four vertices rest on the circle, as shown in the diagram on the right.

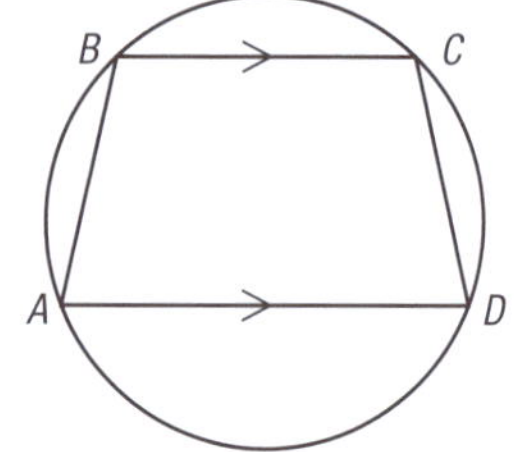

a Describe the quadrilateral.

b Write all the facts that you know about quadrilateral *ABCD*.

c Prove that side *AB* is the same length as side *DC*.

7 Prove that the diagonals of a parallelogram bisect each other.

Appendix 1

Square root table

X	0	1	2	3	4	5	6	7	8	9
0	0.0000	1.0000	1.4142	1.7321	2.0000	2.2361	2.4495	2.6458	2.8284	3.0000
1	3.1623	3.3166	3.4641	3.6056	3.7417	3.8730	4.0000	4.1231	4.2426	4.3589
2	4.4721	4.5826	4.6904	4.7958	4.8990	5.0000	5.0990	5.1962	5.2915	5.3852
3	5.4772	5.5678	5.6569	5.7446	5.8310	5.9161	6.0000	6.0828	6.1644	6.2450
4	6.3246	6.4031	6.4807	6.5574	6.6332	6.7082	6.7823	6.8557	6.9282	7.0000
5	7.0711	7.1414	7.2111	7.2801	7.3485	7.4162	7.4833	7.5498	7.6158	7.6811
6	7.7460	7.8102	7.8740	7.9373	8.0000	8.0623	8.1240	8.1854	8.2462	8.3066
7	8.3666	8.4261	8.4853	8.5440	8.6023	8.6603	8.7178	8.7750	8.8318	8.8882
8	8.9443	9.0000	9.0554	9.1104	9.1652	9.2195	9.2736	9.3274	9.3808	9.4340
9	9.4868	9.5394	9.5917	9.6437	9.6954	9.7468	9.7980	9.8489	9.8995	9.9499
10	10.0000	10.0499	10.0995	10.1489	10.1980	10.2470	10.2956	10.3441	10.3923	10.4403
11	10.4881	10.5357	10.5830	10.6301	10.6771	10.7238	10.7703	10.8167	10.8628	10.9087
12	10.9545	11.0000	11.0454	11.0905	11.1355	11.1803	11.2250	11.2694	11.3137	11.3578
13	11.4018	11.4455	11.4891	11.5326	11.5758	11.6190	11.6619	11.7047	11.7473	11.7898
14	11.8322	11.8743	11.9164	11.9583	12.0000	12.0416	12.0830	12.1244	12.1655	12.2066
15	12.2474	12.2882	12.3288	12.3693	12.4097	12.4499	12.4900	12.5300	12.5698	12.6095
16	12.6491	12.6886	12.7279	12.7671	12.8062	12.8452	12.8841	12.9228	12.9615	13.0000
17	13.0384	13.0767	13.1149	13.1529	13.1909	13.2288	13.2665	13.3041	13.3417	13.3791
18	13.4164	13.4536	13.4907	13.5277	13.5647	13.6015	13.6382	13.6748	13.7113	13.7477
19	13.7840	13.8203	13.8564	13.8924	13.9284	13.9642	14.0000	14.0357	14.0712	14.1067
20	14.1421	14.1774	14.2127	14.2478	14.2829	14.3178	14.3527	14.3875	14.4222	14.4568
21	14.4914	14.5258	14.5602	14.5945	14.6287	14.6629	14.6969	14.7309	14.7648	14.7986
22	14.8324	14.8661	14.8997	14.9332	14.9666	15.0000	15.0333	15.0665	15.0997	15.1327
23	15.1658	15.1987	15.2315	15.2643	15.2971	15.3297	15.3623	15.3948	15.4272	15.4596
24	15.4919	15.5242	15.5563	15.5885	15.6205	15.6525	15.6844	15.7162	15.7480	15.7797
25	15.8114	15.8430	15.8745	15.9060	15.9374	15.9687	16.0000	16.0312	16.0624	16.0935
26	16.1245	16.1555	16.1864	16.2173	16.2481	16.2788	16.3095	16.3401	16.3707	16.4012
27	16.4317	16.4621	16.4924	16.5227	16.5529	16.5831	16.6132	16.6433	16.6733	16.7033
28	16.7332	16.7631	16.7929	16.8226	16.8523	16.8819	16.9115	16.9411	16.9706	17.0000
29	17.0294	17.0587	17.0880	17.1172	17.1464	17.1756	17.2047	17.2337	17.2627	17.2916
30	17.3205	17.3494	17.3781	17.4069	17.4356	17.4642	17.4929	17.5214	17.5499	17.5784
31	17.6068	17.6352	17.6635	17.6918	17.7200	17.7482	17.7764	17.8045	17.8326	17.8606
32	17.8885	17.9165	17.9444	17.9722	18.0000	18.0278	18.0555	18.0831	18.1108	18.1384
33	18.1659	18.1934	18.2209	18.2483	18.2757	18.3030	18.3303	18.3576	18.3848	18.4120
34	18.4391	18.4662	18.4932	18.5203	18.5472	18.5742	18.6011	18.6279	18.6548	18.6815
35	18.7083	18.7350	18.7617	18.7883	18.8149	18.8414	18.8680	18.8944	18.9209	18.9473
36	18.9737	19.0000	19.0263	19.0526	19.0788	19.1050	19.1311	19.1572	19.1833	19.2094
37	19.2354	19.2614	19.2873	19.3132	19.3391	19.3649	19.3907	19.4165	19.4422	19.4679
38	19.4936	19.5192	19.5448	19.5704	19.5959	19.6214	19.6469	19.6723	19.6977	19.7231
39	19.7484	19.7737	19.7990	19.8242	19.8494	19.8746	19.8997	19.9249	19.9499	19.9750
40	20.0000	20.0250	20.0499	20.0749	20.0998	20.1246	20.1494	20.1742	20.1990	20.2237
41	20.2485	20.2731	20.2978	20.3224	20.3470	20.3715	20.3961	20.4206	20.4450	20.4695
42	20.4939	20.5183	20.5426	20.5670	20.5913	20.6155	20.6398	20.6640	20.6882	20.7123
43	20.7364	20.7605	20.7846	20.8087	20.8327	20.8567	20.8806	20.9045	20.9284	20.9523
44	20.9762	21.0000	21.0238	21.0476	21.0713	21.0950	21.1187	21.1424	21.1660	21.1896
45	21.2132	21.2368	21.2603	21.2838	21.3073	21.3307	21.3542	21.3776	21.4009	21.4243
46	21.4476	21.4709	21.4942	21.5174	21.5407	21.5639	21.5870	21.6102	21.6333	21.6564
47	21.6795	21.7025	21.7256	21.7486	21.7715	21.7945	21.8174	21.8403	21.8632	21.8861
48	21.9089	21.9317	21.9545	21.9773	22.0000	22.0227	22.0454	22.0681	22.0907	22.1133
49	22.1359	22.1585	22.1811	22.2036	22.2261	22.2486	22.2711	22.2935	22.3159	22.3383
X	0	1	2	3	4	5	6	7	8	9

Square Root Table (1 to 499)

X	0	1	2	3	4	5	6	7	8	9
50	22.3607	22.3830	22.4054	22.4277	22.4499	22.4722	22.4944	22.5167	22.5389	22.5610
51	22.5832	22.6053	22.6274	22.6495	22.6716	22.6936	22.7156	22.7376	22.7596	22.7816
52	22.8035	22.8254	22.8473	22.8692	22.8910	22.9129	22.9347	22.9565	22.9783	23.0000
53	23.0217	23.0434	23.0651	23.0868	23.1084	23.1301	23.1517	23.1733	23.1948	23.2164
54	23.2379	23.2594	23.2809	23.3024	23.3238	23.3452	23.3666	23.3880	23.4094	23.4307
55	23.4521	23.4734	23.4947	23.5160	23.5372	23.5584	23.5797	23.6008	23.6220	23.6432
56	23.6643	23.6854	23.7065	23.7276	23.7487	23.7697	23.7908	23.8118	23.8328	23.8537
57	23.8747	23.8956	23.9165	23.9374	23.9583	23.9792	24.0000	24.0208	24.0416	24.0624
58	24.0832	24.1039	24.1247	24.1454	24.1661	24.1868	24.2074	24.2281	24.2487	24.2693
59	24.2899	24.3105	24.3311	24.3516	24.3721	24.3926	24.4131	24.4336	24.4540	24.4745
60	24.4949	24.5153	24.5357	24.5561	24.5764	24.5967	24.6171	24.6374	24.6577	24.6779
61	24.6982	24.7184	24.7386	24.7588	24.7790	24.7992	24.8193	24.8395	24.8596	24.8797
62	24.8998	24.9199	24.9399	24.9600	24.9800	25.0000	25.0200	25.0400	25.0599	25.0799
63	25.0998	25.1197	25.1396	25.1595	25.1794	25.1992	25.2190	25.2389	25.2587	25.2784
64	25.2982	25.3180	25.3377	25.3574	25.3772	25.3969	25.4165	25.4362	25.4558	25.4755
65	25.4951	25.5147	25.5343	25.5539	25.5734	25.5930	25.6125	25.6320	25.6515	25.6710
66	25.6905	25.7099	25.7294	25.7488	25.7682	25.7876	25.8070	25.8263	25.8457	25.8650
67	25.8844	25.9037	25.9230	25.9422	25.9615	25.9808	26.0000	26.0192	26.0384	26.0576
68	26.0768	26.0960	26.1151	26.1343	26.1534	26.1725	26.1916	26.2107	26.2298	26.2488
69	26.2679	26.2869	26.3059	26.3249	26.3439	26.3629	26.3818	26.4008	26.4197	26.4386
70	26.4575	26.4764	26.4953	26.5141	26.5330	26.5518	26.5707	26.5895	26.6083	26.6271
71	26.6458	26.6646	26.6833	26.7021	26.7208	26.7395	26.7582	26.7769	26.7955	26.8142
72	26.8328	26.8514	26.8701	26.8887	26.9072	26.9258	26.9444	26.9629	26.9815	27.0000
73	27.0185	27.0370	27.0555	27.0740	27.0924	27.1109	27.1293	27.1477	27.1662	27.1846
74	27.2029	27.2213	27.2397	27.2580	27.2764	27.2947	27.3130	27.3313	27.3496	27.3679
75	27.3861	27.4044	27.4226	27.4408	27.4591	27.4773	27.4955	27.5136	27.5318	27.5500
76	27.5681	27.5862	27.6043	27.6225	27.6405	27.6586	27.6767	27.6948	27.7128	27.7308
77	27.7489	27.7669	27.7849	27.8029	27.8209	27.8388	27.8568	27.8747	27.8927	27.9106
78	27.9285	27.9464	27.9643	27.9821	28.0000	28.0179	28.0357	28.0535	28.0713	28.0891
79	28.1069	28.1247	28.1425	28.1603	28.1780	28.1957	28.2135	28.2312	28.2489	28.2666
80	28.2843	28.3019	28.3196	28.3373	28.3549	28.3725	28.3901	28.4077	28.4253	28.4429
81	28.4605	28.4781	28.4956	28.5132	28.5307	28.5482	28.5657	28.5832	28.6007	28.6182
82	28.6356	28.6531	28.6705	28.6880	28.7054	28.7228	28.7402	28.7576	28.7750	28.7924
83	28.8097	28.8271	28.8444	28.8617	28.8791	28.8964	28.9137	28.9310	28.9482	28.9655
84	28.9828	29.0000	29.0172	29.0345	29.0517	29.0689	29.0861	29.1033	29.1204	29.1376
85	29.1548	29.1719	29.1890	29.2062	29.2233	29.2404	29.2575	29.2746	29.2916	29.3087
86	29.3258	29.3428	29.3598	29.3769	29.3939	29.4109	29.4279	29.4449	29.4618	29.4788
87	29.4958	29.5127	29.5296	29.5466	29.5635	29.5804	29.5973	29.6142	29.6311	29.6479
88	29.6648	29.6816	29.6985	29.7153	29.7321	29.7489	29.7658	29.7825	29.7993	29.8161
89	29.8329	29.8496	29.8664	29.8831	29.8998	29.9166	29.9333	29.9500	29.9666	29.9833
90	30.0000	30.0167	30.0333	30.0500	30.0666	30.0832	30.0998	30.1164	30.1330	30.1496
91	30.1662	30.1828	30.1993	30.2159	30.2324	30.2490	30.2655	30.2820	30.2985	30.3150
92	30.3315	30.3480	30.3645	30.3809	30.3974	30.4138	30.4302	30.4467	30.4631	30.4795
93	30.4959	30.5123	30.5287	30.5450	30.5614	30.5778	30.5941	30.6105	30.6268	30.6431
94	30.6594	30.6757	30.6920	30.7083	30.7246	30.7409	30.7571	30.7734	30.7896	30.8058
95	30.8221	30.8383	30.8545	30.8707	30.8869	30.9031	30.9192	30.9354	30.9516	30.9677
96	30.9839	31.0000	31.0161	31.0322	31.0483	31.0644	31.0805	31.0966	31.1127	31.1288
97	31.1448	31.1609	31.1769	31.1929	31.2090	31.2250	31.2410	31.2570	31.2730	31.2890
98	31.3050	31.3209	31.3369	31.3528	31.3688	31.3847	31.4006	31.4166	31.4325	31.4484
99	31.4643	31.4802	31.4960	31.5119	31.5278	31.5436	31.5595	31.5753	31.5911	31.6070
X	0	1	2	3	4	5	6	7	8	9

Square Root Table (500 to 999)

Appendix 2

Grid paper

Appendix 3

Random number table

39634	62349	74088	65564	16379	19713	39153	69459	17986	24537
14595	35050	40469	27478	44526	67331	93365	54526	22356	93208
30734	71571	83722	79712	25775	65178	07763	82928	31131	30196
64628	89126	91254	24090	25752	03091	39411	73146	06089	15630
42831	95113	43511	42082	15140	34733	68076	18292	69486	80468
80583	70361	41047	26792	78466	03395	17635	09697	82447	31405
00209	90404	99457	72570	42194	49043	24330	14939	09865	45906
05409	20830	01911	60767	55248	79253	12317	84120	77772	50103
95836	22530	91785	80210	34361	52228	33869	94332	83868	61672
65358	70469	87149	89509	72176	18103	55169	79954	72002	20582
72249	04037	36192	40221	14918	53437	60571	40995	55006	10694
41692	40581	93050	48734	34652	41577	04631	49184	39295	81776
61885	50796	96822	82002	07973	52925	75467	86013	98072	91942
48917	48129	48624	48248	91465	54898	61220	18721	67387	66575

Answers

Unit 1: Mathematics in our community

Numbers and operations exercises

Lesson 1 (pages 4–5)

1 **a** True **b** False **c** True
d False **e** False

2 **a** > **b** < **c** > **d** > **e** <

3 **a** 3, 4, 8, 12, 15, 17
b 28, 29, 31, 35, 37, 42
c 93, 110, 134, 171, 249, 256
d 9801, 9819, 9824, 9832, 9846, 9857
e 106, 689, 732, 2314, 3478, 7506

4 **a** 18, 17, 14, 13, 11, 9
b 42, 39, 35, 29, 28, 27
c 395, 384, 381, 352, 316, 315
d 652, 601, 530, 499, 452, 316
e 12 822, 12 547, 12 471, 12 189, 12 063, 12 025

5 **a** True **b** False **c** True
d False **e** False

6 **a** $\frac{1}{9}, \frac{1}{8}, \frac{1}{5}, \frac{1}{4}, \frac{1}{3}, \frac{1}{2}$ **b** $\frac{2}{17}, \frac{2}{11}, \frac{2}{9}, \frac{2}{7}, \frac{2}{5}, \frac{2}{3}$
c $\frac{1}{10}, \frac{1}{4}, \frac{7}{20}, \frac{11}{20}, \frac{3}{5}, \frac{9}{10}$ **d** $1\frac{1}{8}, 2\frac{1}{3}, 3\frac{2}{7}, 3\frac{1}{3}, 4\frac{1}{2}, 4\frac{2}{3}$
e $1\frac{1}{7}, 1\frac{3}{8}, 1\frac{1}{2}, 1\frac{4}{5}, 1\frac{5}{6}, 1\frac{7}{8}$

7 8848, 8611, 8586, 8516, 8485, 8188, 8167, 8163, 8125, 8091

8 **a** 65, 75, 90, 105, 108 **b** 108 kg **c** 89 kg

9 Damoi, George, Fred, Noah, James and Darius

10 **a** Gai's group **b** $\frac{1}{30}$

Lesson 2 (pages 6–7)

1 **a** $1\frac{5}{12}$ **b** $\frac{1}{6}$ **c** $\frac{11}{18}$ **d** $\frac{13}{20}$ **e** 0

2 **a** $3\frac{1}{12}$ **b** $8\frac{5}{36}$ **c** $6\frac{1}{6}$ **d** $9\frac{1}{10}$ **e** $2\frac{1}{9}$

3 $4\frac{7}{12}$ 4 $8\frac{3}{5}$ litres

5 No, he only has $9\frac{29}{30}$ m 6 $6\frac{7}{24}$ litres

7 **a** $\frac{5}{8}$ **b** $\frac{1}{4}$ **c** $3\frac{3}{8}$ **d** $2\frac{16}{27}$ **e** $\frac{8}{105}$ **f** $\frac{5}{8}$

8 $2\frac{2}{3}$ kg 9 $16\frac{5}{6}$ litres 10 $4\frac{4}{15}$

11 **a** 24 **b** $1\frac{7}{8}$ **c** 2 **d** $3\frac{1}{5}$ **e** 10

12 17 days 13 $\frac{1}{6}$ L

Lesson 3 (pages 8–9)

1 **a** $\frac{9}{10}$ **b** $\frac{9}{1000}$ **c** $\frac{9}{100\,000}$
d $\frac{9}{100}$ **e** $\frac{9}{10\,000}$

2 **a** False **b** True **c** True
d True **e** True

3 **a** > **b** > **c** < **d** < **e** >

4 **a** 11.01, 11.34, 12.03 **b** 6.556, 6.87, 6.9
c 1.02, 1.202, 12.02 **d** 5.7, 5.701, 5.72
e 91, 91.002, 91.202

5 **a** $\frac{3}{5}$ **b** $\frac{1}{4}$ **c** $\frac{71}{100}$ **d** $\frac{1}{250}$ **e** $\frac{1}{8}$

6 **a** 0.375 **b** 0.7 **c** 0.12 **d** $0.8\dot{3}$ **e** $0.\dot{7}$

7 **a** < **b** > **c** > **d** < **e** <

8 **a** Paulus **b** 0.05 minutes

9 Teacher to check reasons.
a Accurate **b** Could use an estimate
c Accurate **d** Could use an estimate
e Could use an estimate

10 **a** 40 **b** 45 **c** 120 **d** 4 **e** 20

Lesson 4 (pages 11–12)

1 **a** 9.24 **b** 31.55 **c** 21.62
d 448.69 **e** 98.741

2 **a** 29.1 **b** 88.2 **c** 3.8 **d** 3.765 **e** 1.03

3 **a** 51.57 **b** 34.6 **c** 37.682
d 41.58 **e** 33.653

4 **a** 78 300 **b** 0.356 **c** 1240
d 0.0457 **e** 82.2 **f** 0.0145

5 **a** 348.5 **b** 34.85 **c** 34.85
d 0.3485 **e** 0.034 85

6 **a** 13.6 **b** 273.9 **c** 57.5
d 2.48 **e** 0.34 **f** 0.276

7 **a** 20.5 **b** 0.204 **c** 0.194
d 39 **e** 0.7 **f** 1.2

8

Fraction	Recurring decimals
Thirds	$0.\dot{3}, 0.\dot{6}$
Sixths	$0.1\dot{6}, 0.\dot{3}, 0.\dot{6}, 0.8\dot{3}$
Sevenths	$0.\overline{142857}, 0.\overline{285714}, 0.\overline{428571}, 0.\overline{571428}, 0.\overline{714285}, 0.\overline{857142}$
Ninths	$0.\dot{1}, 0.\dot{2}, 0.\dot{3}, 0.\dot{4}, 0.\dot{5}, 0.\dot{6}, 0.\dot{7}, 0.\dot{8}$

Lesson 5 (pages 12–13)

1 K52.50 2 K30.46

3 50 yam plants 4 $27\frac{3}{4}$ km

5 **a** 20 bottles **b** 15 litres

6 18 times 7 $3\frac{1}{6}$ kg

8 **a** 892 litres **b** 223 days

9 **a** 197.2 mm **b** 57.5 mm
c 49.3 mm per month **d** 53.8 mm

10 **a** Tiffany **b** Christine
c Barbara **d** Sonya and Christine
e 53.86 seconds

Ratio and percentage exercises

Lesson 1 (pages 14–15)

1 D 2 C 3 D 4 D

5 **a** Cyril **b** Luana **c** Kenai

6 C

7

5%	$\frac{1}{20}$	0.05
10%	$\frac{1}{10}$	0.1
20%	$\frac{1}{5}$	0.2
25%	$\frac{1}{4}$	0.25
50%	$\frac{1}{2}$	0.5
60%	$\frac{3}{5}$	0.6
75%	$\frac{3}{4}$	0.75
80%	$\frac{4}{5}$	0.8
100%	1	1

8 a $\frac{3}{100}$ b $\frac{3}{10}$ c 3 d $\frac{2}{5}$ e $\frac{11}{20}$
f $\frac{34}{50}$ g $\frac{33}{100}$ h $\frac{11}{10}$ i $\frac{1}{25}$
9 a 0.43 b 0.08 c 1.04 d 0.48 e 0.21
f 0.35 g 0.66 h 1.25 i 0.01

Lesson 2 (page 16)

1 a 34% b 46.5% c 5%
d 0.2% e 200% f 570%
2 a 7% b 14% c 7.5%
d 4% e 125% f 302%
3 a 14% b 230% c 4%
d 66% e 35% f 2450%
g 22.5% h 100% i 332%
4 a i 0.225 ii $\frac{9}{40}$ b i 0.065 ii $\frac{13}{200}$
c i 0.0025 ii $\frac{1}{400}$ d i 1.205 ii $\frac{241}{200}$
e i 0.875 ii $\frac{7}{8}$
5 a 34% are boys b $\frac{33}{50}$ are girls c $\frac{17}{50}$ are boys
6 a Jemima b Aaron
7 80% 8 Lini

Lesson 3 (page 17)

1 a 15 b 2000 c 32 d 2.7 e 10
f 1.4 g 13 h 3 i 36
2 a K4.50 b K10.50
3 9 school children 4 72 kaukau, 108 taro, 60 yams
5 114 villagers 6 12 left-handed students
7 K13 440
8 a 2% b K7500 c K1500 d K7300
9 1998 people 10 253 entrants

Lesson 4 (page 18)

1 a i 80% ii 8% iii 0.8%
iv 0.08% b 0.0008%
2 a 32.5% b 25% c 72.5%
d 87.5% e 20% f 25%
g 36% h 12.5% i 50%
3 a 35.29% b 66.67% c 2350.00%
4 a 75% b 8% c 1.25%
d 13.33% e 2.08%
5 84.8% 6 125% 7 Miron, by 4.4%

Lesson 5 (page 20)

1 a 1 : 2 b 1 : 3 c 2 : 1 d 4 : 1
e 1 : 3 f 400 : 1 g 16 : 3 h 6 : 7
i 8 : 5 j 11 : 20 k 1 : 115 l 4 : 1
2 a 1 : 2 b 1 : 3 c 2 : 3
d 1 : 3 e 1 : 15 f 5 : 3
3 Teacher to check.
4 a 20 : 4 : 15 b 230 : 120 : 1
c 20 : 14 : 11 d 15 : 42 : 32
5 a 3 : 1 b 2 : 25 c 1 : 25
d 150 : 1 e 1 : 9 f 1 : 30
g 200 : 1 h 8 : 1 i 15 : 73
6 a 2 : 3 b 3 : 1 c 2 : 3
d $1\frac{1}{2}$ hours digging, $2\frac{1}{4}$ hours weeding,
$1\frac{1}{2}$ hours planting, $\frac{3}{4}$ hour cleaning up

Lesson 6 (pages 21–2)

1 a 3 b 6 c 36 d 3
e $1\frac{2}{3}$ f 0.75 g 2 h 4.4
2 a 21 kg b 57 kg
3 55 cm 4 34 kg 5 3 litres
6 33 students
7 4 buckets of cement, 20 buckets of gravel
8 4 kg copper, 1 kg tin
9 a 2000 m b 9 km
10 a 720 cubic metres b 800 cubic metres

Challenge:
a 10 mL b The second glider, by 200 metres
c i Fresh M 2 : 5, Go-mango 3 : 8, Mighty Mango 4 : 9
ii Mighty Mango

Lesson 7 (page 23)

1 a K20 : K30 b K40 and K10
c K5 and K45 d K24 and K26
e K35 and K15 f K7.50 and K42.50
g K32.50 and K17.50 h K18.75 and K31.25
2 a 25 cm and 25 cm b 40 cm and 10 cm
c 17.5 cm and 32.5 cm d 30 cm and 20 cm
e 20 cm and 30 cm f 7.5 cm and 42.5 cm
g 34 cm and 16 cm h 22.5 cm and 27.5 cm
3 2.5 L petrol, 0.5 L oil 4 28 pigs, 98 chickens
5 a 3 : 1 b Tao 36, Kaupa 12
6 6 kg flour, $4\frac{1}{2}$ kg sugar, $1\frac{1}{2}$ kg cocoa
7 a Adults
b i 102 are adults ii 114 are not teenagers

Rates exercises

Lesson 1 (pages 24–5)

1 a cost/weight b volume/time c length/time
d cost/distance e cost/time

2 a 23 toea b K1.85 c 48 d 70 e 7

3 Teacher to check.

4 a $6\frac{1}{4}$ kg per minute b 0.2 cm/m
c 1050 litres/day d K5 per hour
e 48 km/h

5 a 15 km b 7.5 km c $11\frac{1}{4}$ km
d $3\frac{3}{4}$ km e $5\frac{5}{6}$ km

6 a K8 b K4 c K3.60
d K18.40 e K5.80

7 a 40 L b 60 L c 70 L d 15 L e 45 L

8 a 700 hamburgers b 1750 hamburgers
c 2450 hamburgers d 7350 hamburgers
e 127 400 hamburgers

9 a 120 g b 300 g c 750 g
d 369 g e 558 g

10 a 2.8 L b 7 L c 11.2 L
d 70 L e 9.968 L

Lesson 2 (pages 25–7)

1 a 40t b K8.10 c K1.50 d K3.60 e K45

2

Number of hours worked	0	1	2	3	4
Kina earned	0	10	20	30	40

Teacher to check graph.
a K10 b i K15 ii K70
c For every additional **hour** he works, the builder earns **K10**.

3 a Teacher to check graph.
b i 33.75 cm ii 42.5 cm c Teacher to check.

4 a 52 cm per day b Teacher to check.
c 88 cm d No

5 a The temperature is decreasing as the time increases.
b 80°C c 80 degrees
d 2° per minute e −20°C

6 a 100 L b 20 L per minute c 200 L
d 1200 L e 50 000 min (35 days)

7 a Teacher to check.
b

Kilometres travelled	Zippy Vans (kina)	Top Trucks (kina)
5	9	12
10	13	15
15	17	18
20	21	21
25	25	24
30	29	27
35	33	30
40	37	33

c 20 km
d Work out the distance to be covered (in kilometres). If the distance is less than 20 km, use Zippy Vans. If the distance is greater than 20 km, use Top Trucks. If the distance is exactly 20 km, use either.

Lesson 3 (pages 28–9)

1 a Rabbit 112.6 km, Pig 35.4 km, Chicken 29 km, Tortoise 0.54 km, Snail 0.1 km, Cheetah 28 km, Lion 20 km, Zebra 16 km, Giraffe 12.875 km, Elephant 10.1 km
b The animals most likely could not continue to travel at the same rate.

2 a 50 km/h b 90 km/h c 900 km/h
d 5 km/h e 32 km/h

3 a a jet aeroplane 900 km/h
b a person walking 5 km/h
c a PMV 50 km/h
d a car on a freeway 90 km/h
e a heavily loaded truck 32 km/h

4 a A b E c D d C e B

Challenge:
The object could also be moving around you in a circle.

5 D, she will walk home more slowly with the shopping.

6 a 4 min b 3 km c 45 km/h
d From C to D, the graph is steeper, he travels more kilometres in the same time.
e 7 km f 6 min

7 Teacher to check. 8 Teacher to check.

Lesson 4 (pages 30–1)

1 a 3 degrees per hour
b The temperature is not changing.
c 36°C d 11:00am
e 7 hours; The graph slopes to the right
f Between 8:00am and 9:00am; The graph is the steepest
g At 11:00am
h Between 3:00pm and 4:00pm
i 11:00am to midday; 4:00pm to 5:00pm

2 a 23 min b The water level dropped
c Teacher to check.

3 a Jo b Josie's container
c Josie, the graph is steeper; it takes less time
d Peter's graph starts where Josie's starts and is parallel to Jo's graph

4 a Abel
b Abel, his graph does not level off at any stage
c Maybelle, her graph has the longest horizontal section

5 a The graph shows the rate of digging in **m^2** per **minute**.
b Jake c Sung d Katie

Lesson 5 (pages 32–3)

1 a C, because it is narrower and the water will rise more quickly.

b

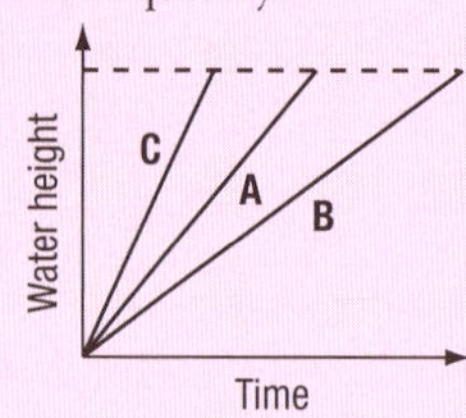

2 a i The water level will start rising more slowly than in Container A, until it reaches the narrow section, then it will rise more quickly.

ii The water level will rise more quickly than in Container A until it reaches the wide section, then it will rise slowly.

b

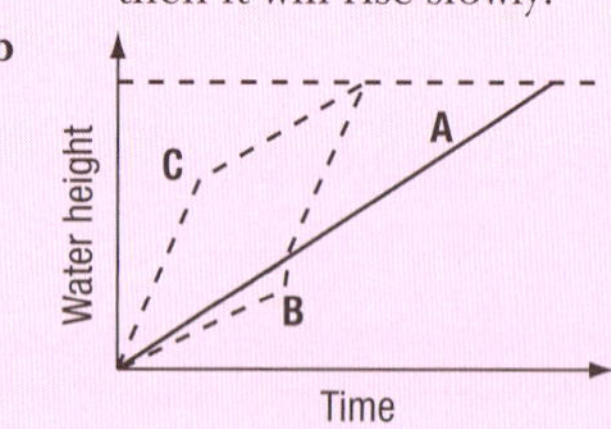

3 a D b C c B d A

4 a

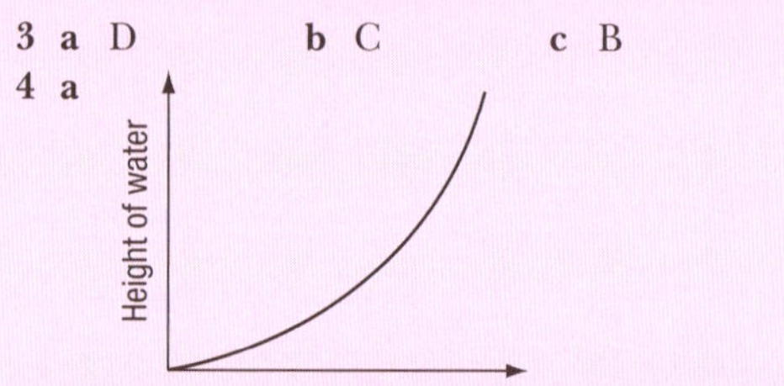

b

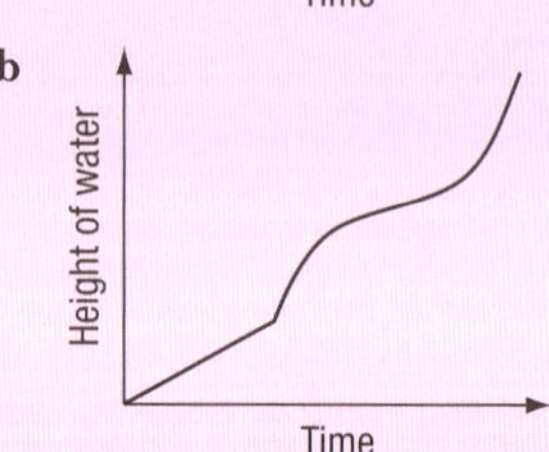

5 a Because the car is already travelling at speed.

b 3 c On the second dip (the lowest point).

d C

Challenge:
Teacher to check.

Money exercises

Lesson 1 (pages 34–5)

1 a K2 b K56 c K5
d K67 e K100 f K120

2 a 85t b K3.70 c K72.40
d K9.10 e K13.45 f K66.20

3 a K5.97 b K14.73 c K11.71
d K14.17 e K79.42 f K408.06

4 Between K6.48 and K6.52

5 K21.90

6 a K6.25 b K4.25 c K4.80
d K5.40 e K6.15 f K4.35

7 a **A** = K8756.00, **B** = K9292.05, **C** = 9381.55, **D** = K9730.20, **E** = K269.80

b K8756.00

c Thursday, the greatest amount was deposited

8 K1.25

Lesson 2 (pages 35–7)

1 a $\frac{1}{10}$ b $\frac{1}{20}$ c $\frac{1}{4}$ d $\frac{1}{2}$ e $\frac{4}{5}$ f $\frac{1}{5}$

2 a K19 b K300 c K28
d K6.65 e K2.33 f K8650

3 a K3.45 b 15t c 40t
d K60.10 e K29 f K52.70

4 a K3.90 b K18 c K495
d K3.18 e K5000 f K4970

5 a K45 b K12 c K200
d K375 e K85.20 f K230.25

6 a K6000 b K234 000

7 a K506 b K552 c K483
d K575 e K496.80 f K920

8 a K85.50 b K90.25 c K23.75
d K38 e K66.50 f K80.75

9 a K12 600 b K26 250 c K618 d K7310

10 a K60 b 5% c i K1323 ii K1389.15

Lesson 3 (pages 38–9)

1 Teacher to check.

2 a i K700 ii K700

b Lollies, Magazines, Savings and Entertainment

c $\frac{1}{2}$ d Clothes e K416

f 'Power, phone and other services' and 'Transport'

3 a K31.50 b Rent c $\frac{1}{30}$ d 5% e 5%

4 a 25 weeks b 5 weeks

5 a K1554 b Food c K1631.70 d Yes

Lesson 4 (pages 39–40)

1 a K21.60 b K11.25 c K26.96
d K157.50 e K7.16

2 a 6 bananas b 5 kg c 2 L
d 5 kg e Both the same

3 The 10% off deal is 90t cheaper.

4 9.09%

5 a K5 b 15%

6 a $45 b 9%

7 a i K1.45 ii 13%
b i K0.65 ii 9%
c i K0.50 ii 6%
d i K0.90 ii 23%
e i K0.30 ii 13%
f i K1.25 ii 21%

8 a K12 b K9.38 c K13.95
d K11.18 e K14.99 f K17.59

9 a K8.93 b K3.99 c K2.85
d K19.81 e K23.70 f K9.45

10 a 22.50 Australian dollars
b 2.50 Australian dollars = K5.56

Lesson 5 (pages 41–2)

1 a K1750 b K350 c 20%
d K1750 e K2012.50

2 a

Profit (K)	% profit
40	8.3
30	6.5
40	8.5

b May c July
d Teacher to check.
e K550

3 a 13 chickens b K35 c 53.8%

4 a K50 b K37.50 c She makes a loss of K2.50.

5 a

Month	Income ('000 K)	Costs ('000 K)	Profit ('000 K)
January	4.8	3.9	**0.9**
February	4.5	**3.2**	1.3
March	3.6	3.4	**0.2**
April	2.9	**2.8**	0.1
May	0.75	2.75	-2.0
June	2.8	2.75	**0.05**

b February, profit of K1300
c Teacher to check. d K19 350

6 a K50 b K26.25 c K120
d K45 e K52.80

7 a 41% b 96% c 54%
d 100% e 35%

Lesson 6 (pages 43–4)

1

Principal	Interest
K200	K60
K150	K120
K40	K12
K3000	K1200
K600	K144

2 a K1500 b K600 c K180
d K90 e K1500 f K892.50

3 a K52 b K152

4 7% 5 12% 6 25%

7 Teacher to check. 8 Teacher to check.

9 K10 112.50

10 a K22.50 b K48 c K80
d K107.25 e K618.75

Lesson 7 (page 45)

1 a K250 b K360 c K391 d K110

2 a K1263 b K600 c 20%

3 K520

4 a 36 b K2109 c 21%

5 a K60 b K360 c K15

6 a K3640 b K10 920 c K4500
d No, he will still have a lot of interest to pay.

7 a K4400 b K1584 c K139

Measurement exercises

Lesson 1 (page 47)

Teacher to check.

Lesson 2 (pages 48–9)

1 a ——— b Teacher to check.
c 100

2 a — b Teacher to check.
c i 10 ii 1000

3 a Teacher to check. b 1000

4 Teacher to check.

5

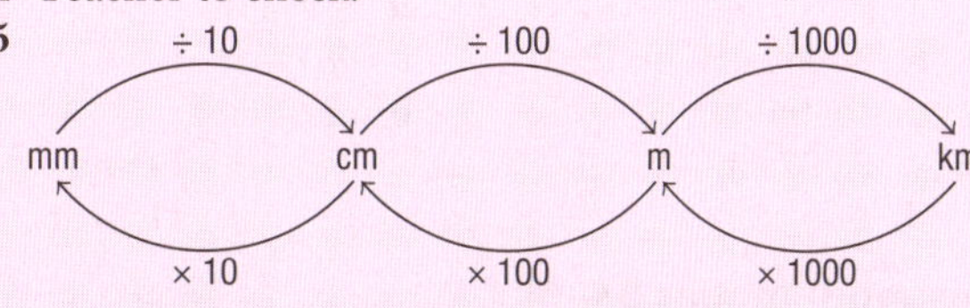

6 a 150 mm b 1.86 km c 2 300 000 cm
d 4.05 m e 2350 cm f 1025 mm
g 65.8 cm h 4750 m i 0.88 m

7 a 158 cm b 4.75 m or 475 cm
c 165 cm d 104 mm or 10.4 cm
e 64 m f 68 cm g 112 cm

Lesson 3 (pages 50–1)

1 a 48 cm b 128 m
c 8.5 cm or 85 mm d 40.8 m

2 a i 13.8 cm ii 3.3 m b 38 268 km

3 a 1980 cm b 15.4 m

4 a 72 mm b 64 mm c 235 cm d 71.2 m

5 a 15.7 cm b 26 cm c 24.75 cm

Lesson 4 (pages 52–3)

1 a

A 1 cm × 1 cm square

b i

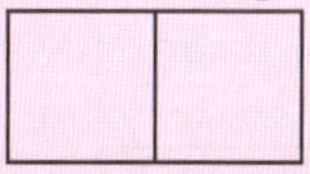

ii One possibility is:

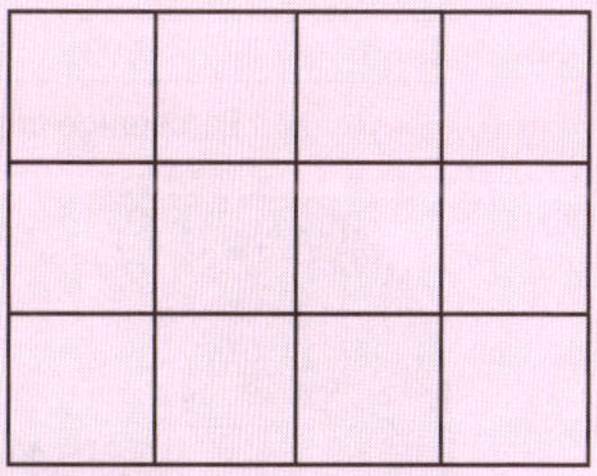

iii One possibility is:

c 100

2 a Teacher to check. b Teacher to check.

c $10\,000\ \text{cm}^2$

3

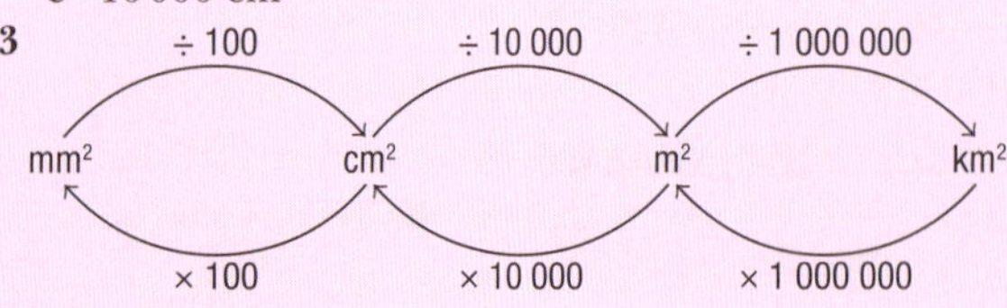

4 a $1.25\ \text{m}^2$ b $563\,000\ \text{m}^2$

c $54.3\ \text{cm}^2$ d $13.4565\ \text{km}^2$

5 a $9856\ \text{cm}^2$ b $6\ \text{m}^2$

c $0.7875\ \text{m}^2$ or $7875\ \text{cm}^2$ d $2.4\ \text{m}^2$

e $20\,000\ \text{cm}^2$ or $2\ \text{m}^2$ f $98\ \text{cm}^2$

g $1549\ \text{mm}^2$ h $5544\ \text{cm}^2$

6 a $28.27\ \text{cm}^2$ b $16\ \text{cm}^2$ c $24\ \text{cm}^2$

d $27.5\ \text{cm}^2$ e $32\ \text{cm}^2$ f $11.17\ \text{cm}^2$

Lesson 5 (pages 54–5)

1 a $80\ \text{m}^2$ b $825\ \text{cm}^2$ c $350\ \text{m}^2$

2 a 45.71 cm, $139.27\ \text{cm}^2$ b 33 m; $71.25\ \text{m}^2$

c88 mm, $308\ \text{mm}^2$ d 85.14 cm, $276.57\ \text{cm}^2$

3 a $824\ \text{cm}^2$ b $340\ \text{m}^2$

c $468\ \text{mm}^2$ d $136\ \text{cm}^2$

4 $1.1522\ \text{m}^2$

Data exercises

Lesson 1 (pages 57–8)

1 a Numerical b Numerical c Categorical

d Numerical e Categorical f Numerical

2 a i Your answer would be one of either Southern, Highlands, Momase or Islands.

ii Categorical

b

Region	Percentage of population	Sector of circle
Southern	20	72°
Highlands	38	**137°**
Momase	28	**101°**
Islands	14	**50°**
Total	100	360°

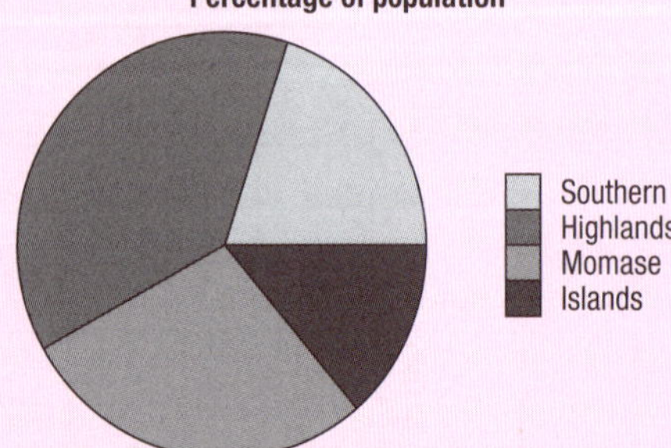

3 a Science b 19 c 20% d 30%

4 a i Banana ii Guava

b 10

c i Your answer would be the name of a fruit

ii categorical

d i

Fruit	Tally	Frequency
Banana	~~||||~~ ||	7
Mango	~~||||~~		6
Pawpaw	||	2
Apple	~~||||~~	5
Guava	||||	4
Pineapple	~~||||~~		6

ii

Favourite fruit

Banana, Mango, Pawpaw, Apple, Guava, Pineapple

1 2 3 4 5 6 7

Frequency

iii Banana iv Guava v 20%

5 a 150 b 33.3%

c

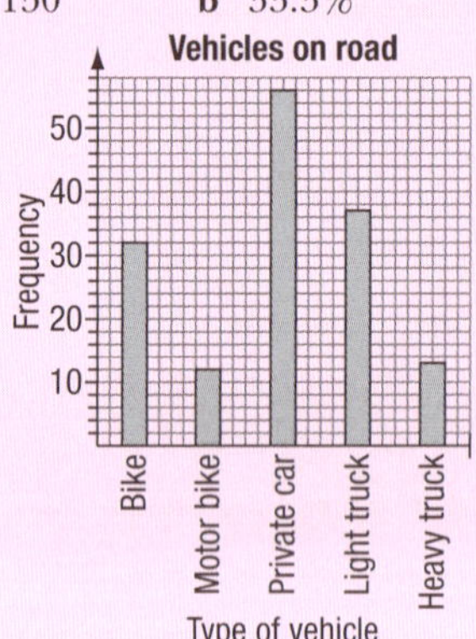

Lesson 2 (pages 59–60)

1 a Continuous b Continuous

c Discrete d Continuous

e Continuous f Discrete

g Discrete h Continuous

2 a 50 b 8 c 14% d 22%

3 a Discrete; the data can only be the numbers 0, 1, 2, 3 . . .

b

Number of pets	Tally	Frequency
0	~~				~~		6
1	~~||||~~ ||	7
2						4
3				2
4			1

c 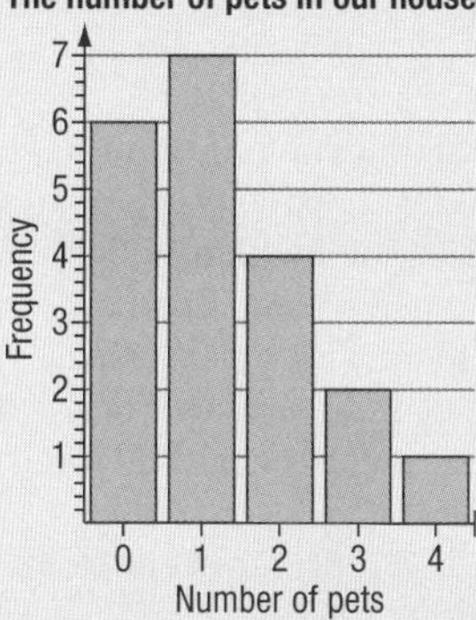

d 1 e 30% f 15%

4 a Discrete data, because the data can only be whole numbers

b

Number of matches	Tally	Frequency
47	|	1
48	𝍸	5
49	𝍸 𝍸	10
50	𝍸 𝍸 𝍸 𝍸 |||	23
51	𝍸 𝍸	10
52	𝍸 ||||	9
53	||	2

c 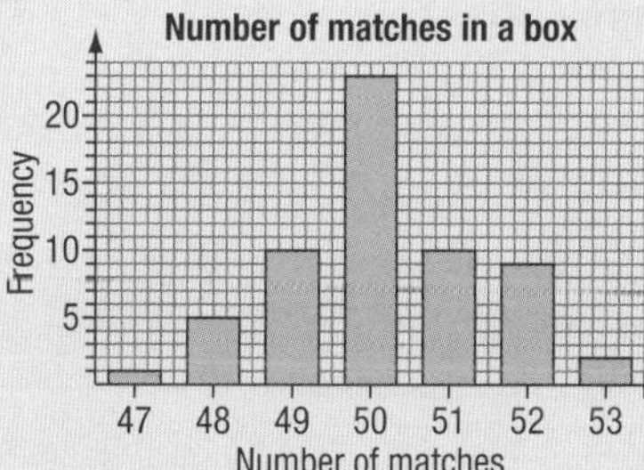

d 38.3% e 26.7%

5 a 80

b The data is discrete, because the number of letters in a word can only be a whole number.

c 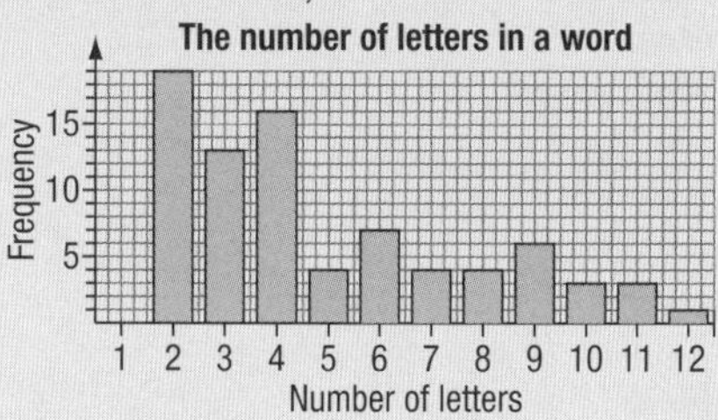

d i 2 ii 28.3% iii 53.3%

Lesson 3 (pages 62–3)

1 Weights of students

Frequency

20 30 40 50 60 70 80 90

Weight (kg)

2 a Rainfall is continuous data because it can take any value, and it is measured.

b

Rainfall (mm)	Tally	Frequency
0–	|||	3
5–	||||	4
10–	𝍸	5
15–	𝍸	5
20–	|||	3
25–	|||	3
30–	||||	4
35–	|||	3

c 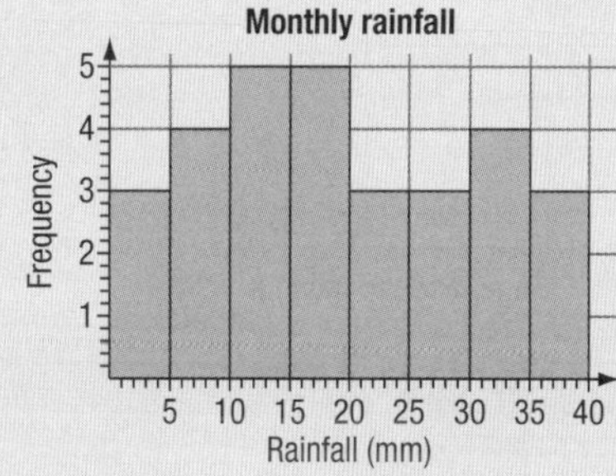

3 a 91 b i 44 ii 50.55%

4 a

Age	Tally	Frequency
0–	𝍸 𝍸 |	11
10–	𝍸 ||	7
20–	𝍸 |	6
30–	|||	3
40–	|||	3
50–	||	2
60–		0
70–	|	1

b

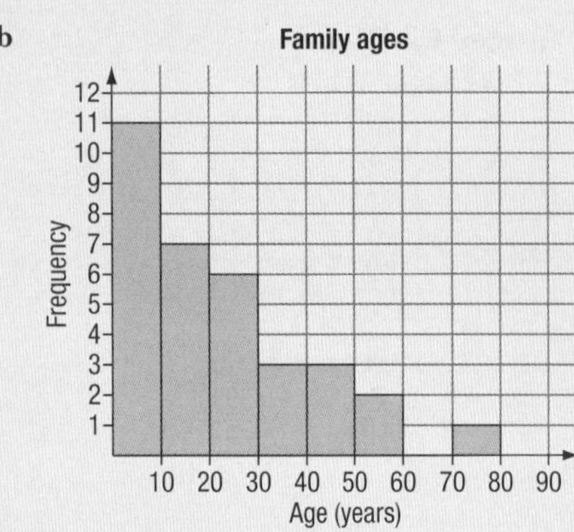

c The '0–' age group is the modal class interval.

d i 54.5% ii 9.1%

5 a Time data is continuous because time can be any value on the number line. In this case, the data has been rounded to the nearest minute.

b

Travel time to school (minutes)	Frequency
0–	2
5–	4
10–	12
15–	14
20–	6
25–	7
30–	6
35–	3
40–	2
45–	4

c

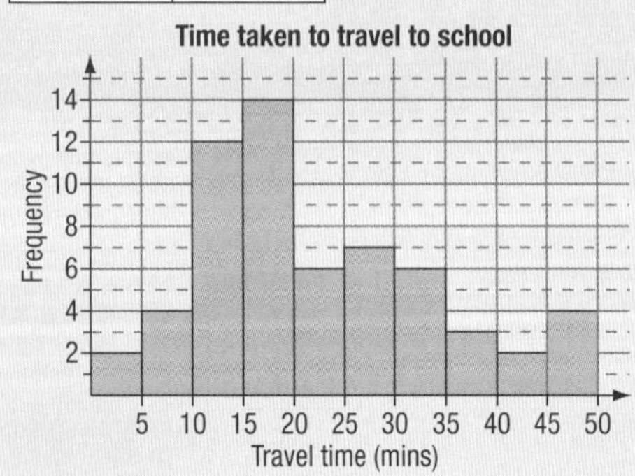

d i 30% ii 25%

Lesson 4 (pages 64–5)

1 a i Mean = 5.61 ii Median = 6
iii Mode = 6

b i Mean = 16.64 ii Median = 17.5
iii Mode = 18

c i Mean = 24.82 ii Median = 24.9
iii Mode = 23.5

d i Mean = 128.6 ii Median = 128
iii Mode =115 and 127 (bi-modal)

2 a An estimate is around 20.

b i Mean = 21.86 ii Median = 23
iii Mode = 18 and 25 (bi-modal)

c The mean is affected by the low value '11'.

d There are two modes.

e Noah's score of 23 is the median, so half the class scored more than Noah's mark and half the class scored less than Noah's mark. Noah's score was slightly more than the average for the class because there were some low marks scored by students in the class.

3 Half the population of Papua New Guinea were less than 21.7 years of age in 2007.

4 a Data set A: 6.46, Data set B: 6.86

b Data set A: 7, Data set B: 7

c Data set A: 7, Data set B: 7

d To find the mean, the data values are added together, so the larger value (15 instead of 10) in Data set B will mean that the total is larger and, therefore, the mean will be larger.
The median is the same for both sets because it depends on position and not the size of extreme values. The mode is the most frequently occurring value and this is not usually an extreme value.

5 a 154 points b 194 points c 16.17 points

6 a 44 goals b 44 goals c 40.2 goals

d i Increase ii 40.33 goals

Lesson 5 (pages 66–7)

1 a Actual values:
i Mean = 5.96
ii Median = 6
iii Mode = 5

b Actual values:
i Mean = 9.1
ii Median = 9.5
iii Mode = 10

c Actual values:
i Mean = 0.64
ii Median = 1
iii Mode = 1

d Actual values:
i Mean = 3.5
ii Median = 3
iii Mode = 2

2 a i Symmetric
Actual values:
ii Mean = 5
iii Median = 5
iv Mode = 5

b i Negatively skewed
Actual values:
ii Mean = 6.33
iii Median = 6.5
iv Mode = 7

c i Negatively skewed
Actual values:
ii Mean = 59.4
iii Median = 65
iv Modal class is 60–

d i Positively skewed
Actual values:
ii Mean = 3.88
iii Median = 4
iv Mode = 4

e i Negatively skewed
Actual values:
ii Mean = 6.18
iii Median = 7
iv Mode = 7

f i Symmetric
Actual values:
ii Mean = 165
iii Median = 165
iv Modal class is 160–

g i Almost symmetric
Actual values:
ii Mean = 50
iii Median = 50.5
iv Bi-modal, the modal classes are 45– and 55–

h i Slightly negatively skewed
Actual values:
ii Mean = 30
iii Median = 30.4
iv Modal class is 30.5–

Revision and assessment (pages 68–74)

Numbers and operations

1 a True b False c False d True
e False f True g True h False

2 a $12\frac{37}{60}$ b $1\frac{11}{23}$ c $3\frac{3}{4}$ d $\frac{4}{5}$

3 a 1447.8 b 1.4478 c 14.478

4 a 137.06 b 95.15 c 211.31 d 950

5 $2\frac{1}{2}$ kg 6 68 traps 7 14.5 containers

8 K147 9 44 km

10 a 13.12 kg b 3.28 kg c 0.46 kg

Ratio and percentage

1 a B b B c C d D
e D f D g A h D

2 a 20% b $33.\dot{3}$% c 0.25% d 7.2%
e 104.29%

3 a 108 trees b 42 trees

4 a 2 : 3 b 10 : 9 c 1 : 3 d 14 : 1 : 50

5 a 1 : 3 b 2 : 3 c 9 : 7

6 21.35 cm

7 a 16 cm and 32 cm
b 8 cm, 16 cm and 24 cm
c 4 cm, 22 cm, 10 cm and 12 cm

8 20 kg

9 a 250 L b 600 L c 570 L
d 142 cans (0.5 L left over)

10 a 3 : 5 b 9 g

Rates

1 a 908 cans/h b 1.5 L/h c K5.50/h
d 15 pizzas/h e 14 km/h

2 a 120 g b K8.76 c 340 seedlings
d 150 kg e 47 km/h

3 a Kala's is the solid line, Kashmir's is the dashed line
b Teacher to check.

4 a After 2 minutes b 2 min
c 4 min d 2 km
e In the first two minutes at the start and, again, after stopping for people (in the last six minutes); 30 km/h

5 a Between midnight and 9:00am; 11:00am to noon; 1:00pm to 4:00pm; 7:00pm to midnight
b 11:00am c 9:00am to 11:00am
d Teacher to check.

6 a Teacher to check graph.
b 6.5 cm/min c 29.25 cm

7 a 500 kL b 120 kL
c Yes, losing 60 kL every day
d 60 kL/day e On the 9th day

8 Teacher to check.

Money

1 K1.35

2 a K5.8 b K2250 c K3.59
d K4250 e K845

3 K8190

4 a K1500 b K11 000

5 K1265

6 a Services b K40
c Food and Rent d 46%

7 a K2.90 b 15.5%

8 a The 185 g tin b The 4 L tin

9 a K12.30 b K22.85 c K21.50
d K152.15 e K315

10 19% 11 K113.10 12 K99.80

13 a K960 b K11 960 c K21.75

14 K457.88 15 12%

16 a K2235 b K660 c 22%

Measurement

1 a 3.45 m b 1654 m c 1.2 m d 256.4 cm
2 a 3.45 cm^2 b 1 200 000 m^2 c 286.5 m^2
3 a 566 cm b 110 cm c 180 mm
4 a 216 cm^2 b 450 cm^2 c 111.22 m^2
5 a 29.75 cm^2 b 170 m^2 c 22 cm^2
6 a 7308 cm^2 b 336 m^2 c 748 cm^2

Data

1 D 2 A
3 a Categorical b Demons
c

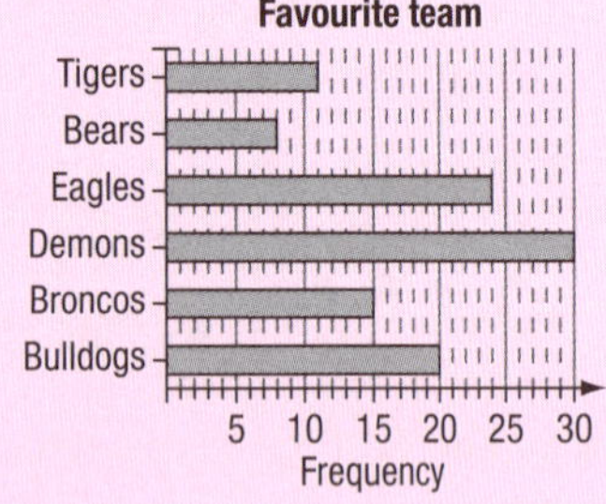

d 22.2%
4 a 5 b 5.5 c 6
5 a 33.4 runs b 31.86%
6 a The data collected is whole numbers.
b 22 c 33.3%
d i 4 ii 5
e Positively skewed
f Larger because the distribution of data is positively skewed and the extra large values would affect the mean.
7 a The '90–' class b 8.3% c 18.3%
d Continuous e About 90 cm

Projects (Pages 75–80)

Teacher to check.

UNIT 2: Patterns of change

Directed numbers exercises

Lesson 1 (pages 88–9)

1 A $^{+}4$ B $^{-}1$ C $^{+}7$ D $^{-}6$ E $^{-}3$
2 a $^{+}7$ b $^{-}5$ c $^{-}9$ d $^{+}25$ e $^{-}13$
3 a $^{-}40$ b $^{+}300$ c $^{-}24$ d $^{-}13$ e $^{+}89$
4 a $^{+}6$ b $^{-}10$ c $^{-}385$ d $^{+}250$ e $^{-}1.8$
5 a $^{+}4$ b $^{-}2$ c $^{+}21$ d $^{-}6$ e $^{-}1$
6 a $^{-}7$ b $^{+}5$ c $^{-}12$ d $^{+}9$ e $^{-}23$
7 a Day 1: $^{-}25$, Day 2: $^{-}10$, Day 3: $^{+}15$, Day 4: $^{+}20$, Day 5: $^{+}25$
b 390 g
8 Week 1: K60, Week 2: K47, Week 3: K37, Week 4: K30, Week 5: K50

9
$^{+}9$ — $^{-}18$
$^{-}45$ — $^{+}\frac{1}{2}$
$^{+}3$ — $^{+}47$
$^{-}\frac{1}{2}$ — $^{-}9$
$^{+}18$ — $^{+}45$
$^{-}47$ — $^{-}3$

10 a $^{-}7$ b $^{+}3$ c 0 d $^{-}5$ e $^{+}1$
f $^{-}5.4$ g $^{+}2\frac{1}{2}$ h $^{-}21$ i $^{+}16$ j $^{+}15$

Lesson 2 (pages 90–1)

1–4 Teacher to check.
5 $^{-}7$ is greater than $^{-}4$. False.
6 a $^{+}9$ b $^{+}5$ c $^{+}6$ d $^{+}6$ e $^{-}7$ f $^{-}4$
7 a 5 b $^{-}4$ c $^{-}5$ d $^{-}9$ e $^{-}12$ f $^{-}8$
8 a False b True c True d False
e True f False g False h True
i True j True
9 a $>$ b $>$ c $>$ d $>$ e $>$
f $<$ g $>$ h $<$ i $<$ j $<$
10 a $^{-}5, ^{-}2, 8$ b $^{-}4, ^{-}3, 0, 4$
c $^{-}11, ^{-}10, ^{-}6, ^{-}5$ d $^{-}9, ^{-}6, 6, 7$
e $^{-}5, ^{-}2, 3, 12$ f $^{-}12, ^{-}1, 0, 7, 9$
g $^{-}3.1, ^{-}1.2, 2.5, 4$ h $^{-}10, ^{-}9.7, ^{-}9.5, ^{-}8.9$
i $^{-}2\frac{1}{4}, ^{-}1\frac{1}{15}, 1, 3\frac{1}{2}$ j $^{-}\frac{7}{8}, ^{-}\frac{5}{8}, ^{-}\frac{3}{8}, ^{-}\frac{1}{8}, \frac{5}{8}$
11 a $0, 3, ^{-}8$ b $4, 0, ^{-}1, ^{-}6$
c $9, 3, ^{-}3, ^{-}8$ d $^{-}2, ^{-}5, ^{-}6, ^{-}9$
e $10, 8, ^{-}2, ^{-}11, ^{-}12$ f $11, 7, 3, ^{-}1, ^{-}4$
g $7.1, 5.2, ^{-}4.9, ^{-}6.3$ h $^{-}1.1, ^{-}1.8, ^{-}3.9, ^{-}5.7$
i $7\frac{1}{3}, 1\frac{2}{3}, ^{-}1\frac{1}{3}, ^{-}5\frac{2}{3}$ j $\frac{5}{12}, \frac{1}{12}, ^{-}\frac{7}{12}, ^{-}\frac{11}{12}$
12 a True b False c True d False
e True f True g False h True
13 a $2, 0, ^{-}2$ b $0, ^{-}5, ^{-}10$ c $^{-}10, ^{-}20, ^{-}30$
d $^{-}8, ^{-}12, ^{-}16$ e $^{-}5, ^{-}9, ^{-}13$ f $^{-}12, ^{-}14, ^{-}16$
g $^{-}5, ^{-}7, ^{-}9$ h $^{-}23, ^{-}28, ^{-}33$
14 a i 15 ii $^{-}1$ iii $^{-}20$
b i 5 ii 6 iii 32

Lesson 3 (page 92)

1 a $^{+}12$ b $^{-}9$ c $^{+}5$ d $^{-}10$ e $^{+}8$
f $^{-}9$ g $^{-}20$ h $^{-}9$ i $^{-}10$
2 a $^{+}16$ b $^{-}4$ c $^{+}2$ d $^{+}4$ e $^{+}3$
f $^{-}4$ g 0 h $^{-}11$ i $^{-}14$
3 a 10 b $^{-}6$ c $^{-}3$ d $^{-}3$ e 14
f 2 g 8 h 0 i $^{-}1$
4 a $^{+}4$ b $^{-}3$ c $^{-}5$ d $^{-}10$ e $^{-}1$
f 13 g $^{-}5$ h 18 i $^{-}5$ j $^{-}11$
5 a 6 b $^{-}11$ c $^{-}16$ d 7 e $^{-}3$
f $^{-}14$ g 14 h $^{-}8$ i $^{-}7$ j 22
k $^{-}11$ l 28 m $^{-}11$ n 30
6 a True b False c True
d True e False f True

7 a −1 **b** −17 **c** 2 **d** −11 **e** 3
f 4 **g** 8 **h** −4 **i** −29 **j** −24
k −6 **l** −7 **m** 30 **n** −36 **o** −54

Lesson 4 (pages 94–5)

1 Teacher to check copied rules.
2 a −35 **b** −18 **c** +6 **d** +28 **e** −40
3 a −8 **b** −7 **c** −48 **d** +36 **e** +16
f −40 **g** +12 **h** +63 **i** −84 **j** −66

4 a

×	+8	−7	−4	+5
−3	−24	21	12	−15
+6	48	−42	−24	30
+9	72	−63	−36	45
−5	−40	35	20	−25

b

×	−9	+5	−6	+7
−1	9	−5	6	−7
−8	72	−40	48	−56
+3	−27	15	−18	21
−11	99	−55	66	−77

5 a −220 **b** −240 **c** 306 **d** −228
e 414 **f** −208 **g** 400 **h** −385
i 315 **j** −1260
6 a 6 **b** −42 **c** 12 **d** −30 **e** 30
f 42 **g** 80 **h** −40 **i** 18 **j** 60
7 a −6 **b** −2 **c** −5 **d** −1 **e** −1
f 10 **g** −3 **h** 2 **i** −2 **j** −2
8 a 15 **b** −2 **c** −9 **d** 4 **e** 6
f −20 **g** −5 **h** −24 **i** 4 **j** 28
9 a −6 **b** 4 **c** −3 **d** −4 **e** −4
f 5 **g** −2 **h** −2 **i** −3 **j** −4
10 a 5 **b** −12 **c** 50 **d** −7 **e** −90
f 10 **g** −9 **h** 33 **i** −3 **j** 25

Lesson 5 (page 96)

1 a −5 **b** 3 **c** 5 **d** −17 **e** −23
f 10 **g** −14 **h** −5 **i** 3 **j** 8
k 4 **l** −15 **m** 0 **n** 7 **o** 3
2 a 53 **b** 2 **c** 0 **d** 52 **e** 10
f −2 **g** −7 **h** 19 **i** 64 **j** 54
k 7 **l** 34 **m** −8 **n** −26 **o** −35

Challenge:
Teacher to check.

Lesson 6 (pages 97–8)

1 a +40, −23, +28, −15, −69 **b** K44
2 a −360 **b** −795
3 a K250 000 **b** K250 **c** K12.50 **d** K2.50
4 a Weeks 3 and 4
b i Week 1 **ii** Week 4
c K2000 **d** K21 300 **e** K61 300
5 a Bahrain, Cairo, Melbourne, Amsterdam and London (equal)
b Stockholm, Helsinki, New York, Oslo
c i 12° **ii** 8° **iii** 9° **iv** 6° **v** 12°
d −6 **e** 95° **f** 106°
6 a 3 bullseyes **b** 10 and two 4s **c** Three 4s
d 10, 4, −5 **e** 4, 4, −5 **f** 10, −5, −5
g 4, −5, −5 **h** −5, −5, −5
7 a 3 **b** 1 **c** 3.5 **d** 3.1

Lesson 7 (page 99)

1 a −1986 **b** 3225 **c** −5852 **d** −707
e 110 **f** 26 197 **g** −66 **h** 58 432
i −996 **j** −821
2 a −337.7 **b** 98.1 **c** −28.1
d 2724.6 **e** 3166.7 **f** 3977
g 5.7 **h** 2 031 203 **i** 5688.8
j −205.1

Indices exercises

Lesson 1 (pages 101–2)

1

n	1	2	3	4	5	6	7	8	9	10	11	12	13	14	15	16	17	18	19	20
n^2	1	4	9	16	25	36	49	64	81	100	121	144	169	196	225	256	289	324	361	400

2 a 121 **b** 3600 **c** 225
d 1600 **e** 3.61 **f** 6400
g 0.81 **h** $\frac{36}{49}$ **i** 0.16
j 90 000 **k** $\frac{81}{16} = 5\frac{1}{16}$ **l** $\frac{1}{625}$
m 81 000 000 **n** 24.01 **o** $\frac{121}{4} = 30\frac{1}{4}$
3 a 8 **b** 10 **c** 4 **d** $\frac{2}{7}$ **e** 1.1 **f** $\frac{9}{11}$
4 a 8 **b** 0.9 **c** 12 **d** 300 **e** 40
f 1 **g** 15 **h** 7 **i** 1.2 **j** 80
5 a i 3 and 4 **ii** 3
b i 9 and 10 **ii** 9
c i 4 and 5 **ii** 4
d i 6 and 7 **ii** 6
e i 6 and 7 **ii** 7
f i 11 and 12 **ii** 11
g i 2 and 3 **ii** 3
h i 9 and 10 **ii** 10
i i 7 and 8 **ii** 8
j i 1 and 2 **ii** 2
6 a 71 **b** 105 **c** 5 **d** $\frac{8}{9}$ **e** 0.69

Lesson 2 (page 104)

1 a i 8 and 9 **ii** 8.8318
b i 7 and 8 **ii** 7.6811
c i 2 and 3 **ii** 2.6458
d i 4 and 5 **ii** 4.6904
e i 12 and 13 **ii** 12.6491
f i 30 and 31 **ii** 30.0666
2 5.57
3 a 1.41 **b** 9.75
4 a 2.236 **b** 8.832 **c** 15.524
d 55.669 **e** 22.869
5 50.21

Lesson 3 (pages 105–6)

1 a D **b** B **c** E **d** A **e** C

2 a 6^5 **b** 3^2 **c** 8^1 **d** 7^8 **e** 12^3

3 a nine cubed **b** six to the power five
c two cubed **d** eleven to the power seven
e eight to the power six

4 a $1 \times 1 \times 1 \times 1 \times 1 \times 1 = 1$ **b** $2 \times 2 \times 2 = 8$
c $3 \times 3 \times 3 = 27$ **d** $4 \times 4 = 16$
e $5 \times 5 \times 5 \times 5 = 625$ **f** $10 = 10$
g $1 \times 1 \times 1 \times 1 \times 1 \times 1 \times 1 \times 1 \times 1 = 1$
h $3 \times 3 \times 3 \times 3 \times 3 = 243$
i $2 \times 2 \times 2 \times 2 \times 2 \times 2 = 64$
j $8 \times 8 \times 8 = 512$

5 a 2^5 **b** $2^3 \times 5$ **c** 3^3
d $2^4 \times 3$ **e** 3×5^2 **f** $3^2 \times 5$
g $2^5 \times 5$ **h** $2^2 \times 5^3$ **i** $2^6 \times 3$
j $2^4 \times 5^4$ **k** $2^2 \times 3 \times 7^2 \times 11$
l $5^3 \times 11 \times 13$ **m** $2^5 \times 5^2 \times 13$

Challenge:
Teacher to check.

Lesson 4 (pages 107–8)

1

a	k^6	$k \times k \times k \times k \times k \times k$
b	y^3	$y \times y \times y$
c	$d^4 \times g^2$	$d \times d \times d \times d \times g \times g$
d	r^2t^4	$t \times r \times t \times t \times r \times t$
e	p^3q^4	$p \times p \times p \times q \times q \times q \times q$
f	qw^5	$w \times w \times w \times w \times w \times q$
g	yz^6	$y \times z \times z \times z \times z \times z \times z$
h	b^3wz	$b \times b \times z \times w \times b$
i	$a^2b^3c^2$	$a \times a \times b \times b \times b \times c \times c$
j	f^7g^2	$g \times g \times f \times f \times f \times f \times f \times f \times f$

2 a 32 **b** 4 **c** 64 **d** 8 **e** 2 **f** 16

3 a 48 **b** 32 **c** 256 **d** 8 **e** 1536

4 a $\frac{1}{4}$ **b** $\frac{1}{4}$ **c** $\frac{3}{16}$ **d** $-2\frac{1}{2}$ **e** $1\frac{1}{2}$

5 a 20 **b** 50 **c** 1000
d 250 **e** 1250 **f** 160

6 a 9 **b** 16 **c** 81 **d** 36 **e** −32 **f** 1
g 20 **h** 86 **i** −54 **j** 12 **k** −6 **l** $5\frac{1}{3}$
m $-4\frac{1}{2}$ **n** $-10\frac{2}{3}$ **o** $\frac{1}{9}$ **p** $-\frac{8}{27}$

7 a i 1 **ii** 1 **iii** 1 **iv** 1 **v** 1
b One raised to any power equals one.

Lesson 5 (pages 108–9)

Index Challenge 1:
a–c Teacher to check.

1 a d^{15} **b** e^3 **c** 6^3
d k^b **e** j^6 **f** 3^3d^5
g $2^7 7^4$ **h** 5^6; 3^m **i** $5^6 r^{13} t^{11}$

2 a t^8 **b** 10^8 **c** y^{11} **d** x^{13}
e a^{10} **f** $20w^3$ **g** p^8q^5 **h** $6a^5$
i $2y^7z^7$ **j** 2^{4n} **k** f^{3+x} **l** 7^{11}
m $32a^6b^9$

3 a $33u^{13}$ **b** $-18z^9$ **c** $-20c^9$
d $60y^{10}$ **e** $\frac{1}{2}a^5$ **f** $\frac{1}{6}a^9$

4 a 12 **b** 27 **c** −216 **d** −216
e 1024 **f** 80 **g** −4 **h** −12

Lesson 6 (pages 110–11)

Index Challenge 2:
a–c Teacher to check.

1 a 8 **b** 10 **c** 3 **d** $b - 1$ **e** 3 **f** 1

2 a d^7 **b** h^2 **c** $4e^3$ **d** $\frac{d^2}{2}$
e $\frac{8f^2g}{9}$ **f** g^4h^2 **g** $\frac{3j}{5k}$ **h** ab^2c^2
i $6r^3$ **j** $\frac{p^3}{2}$ **k** $2t$ **l** $7r^2s$
m $\frac{x^2y^2}{2}$ **n** $6q$ **o** $\frac{16}{5}r^7t^5$ **p** 505^2

3 a 160 **b** 45 **c** 50 421 **d** 40

4 a 6 **b** 216 **c** 42 **d** 30
e 1024 **f** 80 **g** 4 **h** 48
i −2 **j** −32

Lesson 7 (page 112)

Index Challenge 3:
a and **b** Teacher to check.

1 a 6 **b** 9 **c** 3 **d** t **e** $4n$ **f** 2

2 a a^8 **b** c^{24} **c** b^9 **d** 11^{2n} **e** 7^{3t}

3 a $3^3 5^2$, $3(5^2)^3$, $(3 \times 5^2)^3$ **b** $(3 \times 5)^2$, 3×5^2, $3^2 \times 5$

4 a g^{12} **b** h^3 **c** 5^{24} **d** 3^{29}
e j^5 **f** m^4 **g** n^6 **h** $\frac{p}{2}$
i $x^2y^{13}z^{10}$ **j** 3^2s^2t

Lesson 8 (page 114)

1 C

2 a True **b** True **c** False **d** True
e False **f** True **g** False

3 a −4 **b** 4 **c** n^7 **d** p^4q^3

4 a $\frac{1}{m^5}$ **b** $\frac{3}{a^2}$ **c** $\frac{1}{a^2b^2}$ **d** m^5

5 a $\frac{11b^2}{a^6}$ **b** a^9 **c** $\frac{1}{y^5}$ **d** $\frac{2}{5^4}$
e $25w^6$ **f** a^4b^5 **g** k **h** c^7

6 a 1 **b** 3 **c** $-\frac{1}{2}$ **d** a **e** w **f** 4

7 a 7 **b** −2 **c** 10 **d** $\frac{4}{5}$ **e** $1\frac{1}{3}$

8 Polygon

Lesson 9 (page 116)

1 **a** $\frac{10b^2}{7}$ **b** a^6b^4 **c** $c^{15}d^{18}$ **d** $12a^7$
e $4\frac{d^{15}}{g^9}$ **f** $\frac{3m^3}{4n^4}$ **g** $\frac{3}{ab^2}$ **h** $-\frac{g^7h^2}{2}$
i j^8 **j** $\frac{5^{10}h^{33}}{j^6}$ **k** $\frac{1}{k^2m^5}$ **l** 5^{2x+6}
m $\frac{1}{a^2}$ **n** $\frac{b}{a^2}$ **o** $\frac{8a^6}{b}$

2 **a** 2^{14} **b** 3^8 **c** 6^{15} **d** $7^5 \times 5$
e 11 **f** 3

Lesson 10 (pages 117–18)

1 b, e and f

2 **a** 10^{-3} **b** 10^4 **c** 10^0 **d** 10^{-4}
e 10^6 **f** 10^{-1}

3

Number	Calculations	Scientific notation
276	2.76×100	2.76×10^2
8800	8.800×1000	8.8×10^3
5166.66	5.16666×1000	5.16666×10^3
37	3.7×10	3.7×10^1
20.02	2.002×10	2.002×10^1
617.55	6.1755×100	6.1755×10^2
76 000	$7.6 \times 10\,000$	7.6×10^4
400	4×100	4×10^2
1.66	1.66×1	1.66×10^0
8200	8.2×1000	8.2×10^3
3 million	$3 \times 1\,000\,000$	3×10^6
92 330	$9.233 \times 10\,000$	9.233×10^4
0.07	$7 \times \frac{1}{100}$	7×10^{-2}
0.000 54	$5.4 \times \frac{1}{10\,000}$	5.4×10^{-4}
0.008 00	$8.0 \times \frac{1}{1000}$	8.0×10^{-3}
0.001 004	$1.0004 \times \frac{1}{1000}$	1.0004×10^{-3}
0.000 043 333	$4.3333 \times \frac{1}{100\,000}$	4.3333×10^{-5}
0.000 000 150 01	$15001 \times \frac{1}{10\,000\,000}$	1.5001×10^{-7}

4 **a** 8×10^2 **b** 1.23×10^2 **c** 9.8×10^4
d 2.9×10^6 **e** 8.42×10^3 **f** 7.6×10^5
g 7.732×10^5 **h** 9.8765×10^1 **i** 1.76×10^5
j 2.0×10^6

5 **a** 2 000 000 **b** 7900 **c** 90 **d** 5.1

6 **a** 2.3×10^{-3} **b** 4.0×10^{-3}
c 7.8×10^{-8} **d** 4.9×10^{-7}

7 **a** 0.000 82 **b** 0.001 56 **c** 0.09
d 0.403 **e** 0.000 000 7

8 $\frac{7}{10\,000} = 7 \times 10^{-4}$
$\frac{65}{100\,000} = 6.5 \times 10^{-4}$
$\frac{712}{100\,000} = 7.12 \times 10^{-3}$
$\frac{63}{10\,000} = 63 \times 10^{-3}$
In order, from smallest to largest:
$\frac{65}{100\,000}, \frac{7}{10\,000}, \frac{63}{10\,000}, \frac{712}{100\,000}$

Lesson 11 (page 119)

1 **a** 2.8×10^2 **b** 2.1×10^8
c 2.25×10^2 **d** 5.0×10^{-11}

2 **a** 6.0×10^8 **b** 1.2×10^{14} bacteria

3 **a** 6.048×10^5 seconds **b** 1.8144×10^9 seconds

4 **a** 5.67×10^8 kg **b** 5.67×10^5

5 **a** 3×10^5
b **i** 1.8×10^7 km **ii** 1.08×10^9 km
iii 2.592×10^{10} km
c 7.38×10^8 km

6 3.0×10^{-3} mm

7 **a** 2.5×10^7 **b** 2.592×10^{11} blood cells

8 **a** 1.0×10^{-4} g **b** 10 000 000 pins

Lesson 12 (page 120)

1 **a** 1728 **b** 625 **c** 343 **d** 32 768
e 512 **f** 2187 **g** 59 049 **h** 4096
i 14 641 **j** 48 828 125

2 **a** 625 **b** −1331 **c** 0.015 625
d 6561 **e** 2.8561 **f** −40.841 01
g 100 000 **h** −4096 **i** 4492.125
j 28 561 **k** 10

3 **a** 2 800 000 **b** 0.013 **c** 563 900
d 7 203 000 **e** 0.0177

4 **a** 980 000 000 **b** 2 585 714.3 **c** 70 000 000
d 614 000 **e** 27 040 000

Equations exercises

Lesson 1 (page 122)

1 **a** 19, 22, 25 **b** 6, −4, −14 **c** −48, 144, −432
d 8, 4, 2 **e** 18, 22, 28 **f** 15, 18, 13

2 **a** $f = 8$; Add 2 each time
b $y = 9$; Pattern of square numbers
c $q = 32$; Double each time
d $w = 5$; Add 4 each time
e $p = -5$; Subtract 3 each time
f $m = \frac{1}{9}$; Divide by 3 each time

3 **a** 35, 40, 45; Add 5 each time
b −13, −20, −27; Subtract 7 each time
c $\frac{1}{3}, \frac{1}{9}, \frac{1}{27}$; Divide by 3 each time
d −24, 48, −96; Multiply by −2 each time
e $1\frac{1}{2}, 1\frac{3}{4}, 2$; Add $\frac{1}{4}$ each time
f −4.4, −7.2, −10; Subtract 2.8 each time

g 23, 47, 95; Double and add 1 each time

h $\frac{3}{4}, -\frac{5}{8}, -1\frac{5}{16}$; Halve and subtract 1 each time

i −41, −122, −365; Triple and add 1 each time

j $-\frac{1}{25}, \frac{1}{125}, -\frac{1}{625}$; Divide by −5 each time

k 57, 117, 237; Double and add 3 each time

l 1.75, 2.5, 3.25; Add 0.75 each time

4 a 16, 32, 64; Double each time

b 25, 36, 49; Pattern of square numbers

c $\frac{1}{5}, \frac{1}{6}, \frac{1}{7}$; Pattern of reciprocals of consecutive numbers

d 15, 21, 28; Add one more than the last difference each time

e 4, −2, 5; Alternately add and subtract one more than the last difference each time

5–6 Teacher to check.

Challenge:

Teacher to check.

Lesson 2 (pages 124–5)

1 b i Is arithmetic; There is a common difference of −6

ii −6, −12, −8

d i Is arithmetic; There is a common difference of 5

ii 16.2, 21.2, 26.2

e i Is arithmetic; There is a common difference of 2

ii 0, 2, 4

2 a i $a = 4, d = 2$ ii 42

b i $a = 0.5, d = 3$ ii 57.5

c i $a = -8, d = 5$ ii 87

d i $a = 5, d = -8$ ii −147

e i $a = \frac{3}{4}, d = -\frac{1}{4}$ ii −4

f i $a = 34, d = 9$ ii 205

g i $a = 4, d = 0.6$ ii 15.4

h i $a = 30, d = -40$ ii −730

3 a i $t_n = 8 - n$ ii $t_{50} = -42$

b i $t_n = 2n - 11$ ii $t_{50} = 89$

c i $t_n = 7 + 0.5n$ ii $t_{50} = 32$

4 a i 13, 18, 23, 28 ii $a = 13, d = 5$ iii 258

b i 8, 5, 2, −1 ii $a = 8, d = -3$ iii −139

c i 15.5, 16, 16.5, 17 ii $a = 15.5, d = 0.5$ iii 40

d i 13, 2, −9, −20 ii $a = 13, d = -11$ iii −526

e i 180, 160, 140, 120 ii $a = 180, d = -20$ iii −800

Lesson 3 (page 126)

1 a Finite sequence b Infinite series

c Finite series d Infinite sequence

e Finite series

2 a i K33 ii K36 iii K39

b There is a common difference of 3.

c i $a = 33$ ii $d = 3$

d $t_n = 30 + 3n$ e K186

3 a i 2050 ants ii 2100 ants iii 2150 ants

b 4500 ants

4 a i 49.1 cm ii 48.2 cm

b i $a = 50$ ii $d = -0.9$

c 43.7 cm

5 a i $a = 10$ ii $d = -0.2$

b $t_n = 10.2 - 0.2n$ c 4.2 toea

Lesson 4 (page 127)

1 a i 4, 10, 16 ii 2730

b i −10, −9, −8 ii 135

c i 100, 110, 120 ii 7350

2 a i 1 ii 2

b 400 stones

3 540.6 cm

Lesson 5 (pages 128–30)

1 a $m + 6$ b $m - 8$ c $4m$

d $\frac{m}{6}$ e $19 - 2m$ f $3m + 1$

g $2 - \frac{m}{2}$ h $(3 + m) \times 4$ i $11m$

j $\frac{6}{m}$ k $2(m - 4)$ l $\frac{m+9}{2}$

2 C

3 a $5m + 2n$ b $4 + 5y$ c $y^2 + 5y + 6$

d $2mn - 3n$ e $5xy + 6x - 7y$ f $2a^2b - 3ab$

4 a i 13 ii −3

b i 16 ii 0

c i −3 ii −11

d i −6 ii −22

e i 2 ii −14

f i 28 ii 4

g i 15 ii 3

5 b i $2n = 10$ ii 5

c i $5n = 20$ ii 4

d i $19 - n = 6$ ii 13

e i $n - 10 = 6$ ii 16

f i $2n + 1 = 9$ ii 4

g i $3n - 1 = 14$ ii 5

h i $2n + 7 = 13$ ii 3

i i $\frac{n}{5} + 1 = 3$ ii 10

j i $\frac{n+1}{2} = 9$ ii 17

k i $n + (n + 1) = 35$ ii 17

Challenge:

a Yellow b Teacher to check.

Lesson 6 (page 131)

1 a

	x	5
4	$4x$	20

b

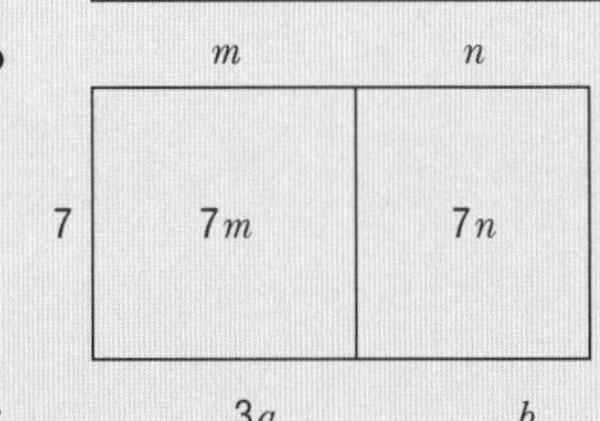

c

	$3a$	b
2	$6a$	$2b$

d

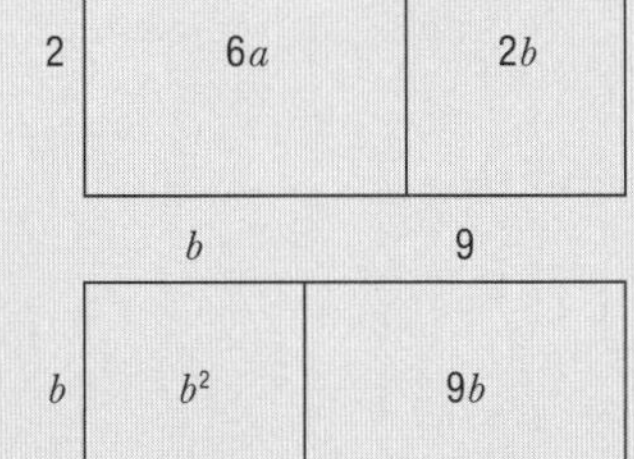

2 Teacher to check product boxes.
a $5w + 15$ b $2t + 2z$ c $12h + 3k$
d $2ab + ac$ e $2b^2 + 2bc$

3 a $4 \times m + 4 \times 11 = 4m + 44$
b $15 + 6y$ c $27 + 9y$ d $40 + 24t$
e $12g + 18$ f $840 + 12b$

4 a $7 \times m - 7 \times w = 7m - 7w$
b $54a - 24$ c $7 - 35x$ d $32p - 8q$
e $18x - 45y$ f $60 - 40w$

5 a $ap + 4a$ b $cg + 11g$ c $5am + 5m^2$
d $4r^2 - 5r$ e $3p^2 - 33p$ f $7a^2 - 49a$
g $2z^2 + 10yz$ h $5b^2 - 15bc$

6 a $16m - 28$ b $8a + 10$ c $13w + 10$
d $4 + 8t$ e $7a + 1$ f $28 + 11p$
g $4t + 4v + 7$ h $10 + 12m + 2n$
i $9a - 6b + 51$ j $27j + 10k$
k $ab - 11a^2$ l $3m^2 - 2mn - 10n^2$

Lesson 7 (pages 132–3)

1 Only final answers given.
a $5x - 15$ b $^-2m + 14$ c $^-3c + 3d$
d $^-40 + 24b$ e $^-6m - 8n$ f $^-6a^2 + 30a$

2 a $^-7a + 28$ b $^-6m - 27$ c $^-15x - 5y$
d $^-18 + 12b$ e $^-2m^2 - 12m$
f $^-18x^2 - 27x + 36$ g $^-24p - 4q + 8$
h $^-18 - 42h + 12h^2$ i $^-28mn - 21m^2 + 14n$

3 a $^-3m - 4n$ b $6 - 7y$ c $^-b + 3c$
d $^-5t - 11$ e $^-9 - 3m + 4m^2$ f $^-xy + x^2$

4 a $^-m^2 - 2mn$ b $^-2h + 3h^2$ c $^-3q^2 + 27q$
d $^-4ab - b^2$ e $^-7h^2 + 5h$ f $^-7w - 5w^2$
g $^-3g^2 - gh + 6g$ h $^-5t^2 - tw + tz$

5 a $^-7a - 28$ b $3p + 56$ c $^-9z - 2$
d $^-3w + 1$ e $^-27t - 40$ f $^-15m + 9$

6 a $7t - v$ b $4p - 6$ c $^-7a - 6$
d $^-r + 35$ e $2h + 9$ f $8x - 3$
g $^-3b^2 - 13b + 1$ h $^-11p^2 + 4pg$
i $14m + 4mn + 6n^2$ j $2y - 9z$
k $^-2m^2 - 10m$ l $4y + 4yz$

Lesson 8 (pages 133–4)

1 a $7m + 19$ b $a^2 + 11a$
c $9 + 16x + 7y$ d $4m^2 - 10mn + 14n^2$
e $^-13a^2 + 46a$ f $7d^2 + 2de + e^2$
g $31w + 25x + 13y$ h $39w - 3z + 21$

2 a $2m + 26n$ b $5x + 7y + 6z$ c $^-3w - 87$
d $2a - 4b + 6c$ e $11t - 34$ f $7r - 9s - 2t$
g 78 h $9m + 5n - 20 + mn$
i $39 - 36t$ j $^-2y^2 - 12y + 8$

3 a $15a - 3a^2$ b $6mn + 20m$ c $21m + 21n$
d $10w - 10y - 22$ e $^-7t + u$ f $6a^2 - 2ab - 6b^2$

4 a i $2[2 + (a + 3)]$ ii $2a + 10$
b i $2[7 + (x - 1)]$ ii $12 + 2x$
c i $2[(m + 4) + (m + 3)]$ ii $4m + 14$
d i $2[x + (3x + 5)]$ ii $8x + 10$
e i $2[y + 2z - 1)]$ ii $2y + 4z - 2$
f i $2[b + (3a + 2b)]$ ii $6a + 6b$

5 a i $\frac{3(x+7)}{2}$ ii $\frac{3x+21}{2}$, or $\frac{3x}{2} + 10\frac{1}{2}$
b i $\frac{4(2m-5)}{2}$ ii $4m - 10$
c i $\frac{m(7m+11)}{2}$ ii $\frac{7m^2}{2} + \frac{11m}{2}$
d i $\frac{x(y+7)}{2}$ ii $\frac{xy}{2} + \frac{7x}{2}$
e i $\frac{2b(a-3)}{2}$ ii $ab - 3b$
f i $\frac{y(3x+y)}{2}$ ii $\frac{3x}{2} + \frac{y^2}{2}$

Lesson 9 (pages 135–6)

1 a False b True c True
d False e False f True

2 a C b B c D d A

3 a $8(m + 2)$ b $2(2a + 5b)$ c $7(k - 21)$
d $5(p - 2q)$ e $a(m + n)$ f $5x(4 + 3y)$
g $6(p - 5q)$ h $5m(a + 5c)$ i $d(c - 8)$
j $8(m + 1)$ k $w(w - 3)$ l $b(1 - 3b)$
m $p(p + q)$ n $3y(2y + 3z)$ o $7t(1 - 3t)$
p $2b(a^2 + 2c)$

4 a i 3 ii $3(x + 3)$
b i 3 ii $3(s - 2)$
c i x ii $x(3 + c)$
d i $3w$ ii $3w(4 + 3s)$
e i 16 ii $16(c - d)$
f i $5g$ ii $5g(2 + p)$

g i 6 ii $6(2 + x)$
h i b ii $b(a + c)$
i i 6 ii $6(2x - 1)$
j i v ii $v(1 + 2x)$
k i 3 ii $3(x - 3)$
l i c ii $c(3a - b)$
m i 4 ii $4(f - 1)$
n i $2d$ ii $2d(2v - 1)$

5 a $a(2 + c)$ b $2p(2 - 3q)$ c $x(x + y)$
d $4h(2 + h)$ e $3w(w - 2)$ f $v(v + 2z)$
g $2d(1 - 4d)$ h $b^2(a + c)$ i $x(7x - 6)$
j $f(f + 2g)$ k $x(x - 9y)$ l $ac(1 - c)$
m $j(j - 4k)$ n $2k(k - 3)$

6 a $^-6(m - 2n)$ b $^-2(x + 6)$ c $^-4h(h - 2)$
d $^-3(p - 1)$ e $^-3a(a + 2b)$ f $^-3y(2y - 3)$

Challenge:
a Teacher to check. b Teacher to check.

Lesson 10 (page 138)

1 a $n = 2$ b $n = 10$ c $n = 23$
d $n = 25$ e $n = {}^-7.2$ f $n = \frac{7}{12}$
g $n = 5.5$ h $d = {}^-3.9$ i $t = 1$
j $r = 4\frac{3}{4}$ k $x = {}^-1\frac{1}{2}$ l $c = {}^-11.1$

2 a $g + 5 = 46; g = 41$ b $g - 23 = 35; g = 58$
c $g + 24 = 99; g = 75$ d $g + 8 = 5; g = {}^-3$
e $12 + g = 9; g = {}^-3$ f $g + 11 = 3; g = {}^-8$
g $g - 7 = {}^-1; g = 6$ h $g - 8 = 4\frac{1}{2}; g = 12\frac{1}{2}$

3 a $n = 8$ b $c = {}^-0.3$ c $r = 3.5$
d $w = {}^-11\frac{2}{3}$ e $x = 3\frac{1}{3}$ f $t = \frac{2}{7}$
g $y = 3\frac{2}{5}$ h $x = 45$ i $w = 8$
j $m = 0$ k $a = {}^-35$ l $k = {}^-8.4$
m $p = {}^-18$

4 a 35 b $3\frac{1}{2}$ c 46 d $^-20$
e $3\frac{1}{2}$ f $\frac{1}{2}$ g $^-0.6$ h $\frac{8}{11}$

5 Teacher to check.

Lesson 11 (pages 139–40)

1 a C b B c C d D e D

2 a $n = 5$ b $c = 6$ c $t = 5$ d $x = {}^-22$
e $d = {}^-8$ f $x = {}^-5$ g $w = 0$ h $x = 3$
i $x = {}^-8$ j $c = {}^-44$ k $d = 25$ l $x = {}^-5$
m $x = {}^-7$ n $x = 1$ o $c = 0$ p $x = 2$
q $x = 60$ r $s = 4$ s $d = 10$ t $x = 6$

3 a 4 b $-1\frac{1}{4}$ c $^-10$ d 3 e 24
f $^-2$ g $^-7$ h 3.8 i $^-40$ j $-1\frac{4}{5}$

4 Teacher to check.

Lesson 12 (pages 141–2)

1 Imelda

2 a $c = 1$ b $n = {}^-5$ c $x = 3$ d $t = -\frac{1}{2}$
e $d = {}^-1\frac{1}{2}$ f $w = 8$ g $x = \frac{3}{4}$ h $m = {}^-1\frac{1}{2}$
i $t = {}^-1$ j $x = 1$ k $m = {}^-1$ l $y = 1$

3 a $v = {}^-9$ b $x = 4\frac{1}{5}$ c $c = 1\frac{1}{2}$ d $m = {}^-1$
e $w = {}^-11$ f $x = {}^-11$ g $v = 19$ h $t = 3$
i $y = {}^-4$ j $m = 6$ k $k = 1\frac{1}{2}$ l $d = 2\frac{2}{3}$

4 a 5 b 11 c $^-3$ d 37
e $^-32$ f $\frac{2}{7}$ g 22 or 2 h $^-8\frac{1}{2}$
i $^-6\frac{1}{2}$ j 18

5 Teacher to check.

Challenge:
a Teacher to check. b Teacher to check.
c Teacher to check.

Lesson 13 (pages 143–4)

1 a $x = 4$ b $n = 2$ c $e = 8$ d $r = 3$
e $x = {}^-1\frac{2}{11}$ f $m = 2\frac{1}{8}$ g $y = \frac{^-6}{19}$ h $p = {}^-1$
i $y = 5$ j $p = \frac{1}{2}$ k $x = {}^-1\frac{3}{10}$

2 a $d = {}^-15$ b $c = -\frac{1}{2}$ c $h = 6$ d $x = \frac{1}{2}$
e $x = -\frac{1}{2}$ f $b = 2$ g $d = {}^-1$ h $t = 2$
i $s = 2$ j $a = 3$ k $x = 1$ l $x = {}^-18$
m $x = 6$ n $x = \frac{3}{4}$ o $x = 3$ p $y = 5$
q $x = \frac{2}{3}$

3 a $x = 5$ b $x = 4$ c $x = 12$ d $x = 1\frac{1}{3}$
e $x = 13$ f $x = 4\frac{1}{7}$

Lesson 14 (pages 145–6)

1 a $m + 5$ b $4(m + 5) = 48$ c 7 years
2 a $3n + 4 = 13$ b $n = 9$ c K9
3 30 years 4 38 fish 5 9 parrots
6 a i $x + 5$ ii $3(x + 5)$
b Daughter is 7, son is 12; father is 36
7 $m = 15°$ 8 24 kg 9 14
10 3, 6 and 11 11 14 fish and 21 fish
12 60 toea
13 Didya 500 m, Luania 125 m, Yawie 375 m
14 8 yellow, 10 blue, 12 red, 20 green
15 25 and 26 16 26, 27, 28, 29, 30
17 Paulius 7, Caleb 14, Moses 21
18 Rose 24 years, Vera 16 years, Keely 8 years
19 A = 6, B = 12, C = 8, D = 24 20 K10 000

Lesson 15 (pages 147–8)

1 a 104°F b 71.6°F c 24.8°F d 32°F
e 89.6°F f 19.4°F g 113°F h 64.4°F

2 a 177°C
b i 121°C ii 149°C iii 204°C

3 **a** 3450 mL **b** 3200 mL
c The 1.6 m tall man would have a greater lung capacity by 180 mL.

4 **a** **i** 20 m **ii** 37.5 m **iii** 60 m **iv** 87.5 m
b 10.625 m **c** 42 m

5 **a** 89 km
b **i** 61.66 km **ii** 69.49 km
iii 42.42 km **iv** 334.87 km

Lesson 16 (page 149)

1 **a** **i** $w = \frac{P - 2l}{2}$ **ii** $w = 6$
b **i** $h = \frac{3V}{A}$ **ii** $h = 20$
c **i** $s = \sqrt{A}$ **ii** $s = 10$
d **i** $b = \frac{2A}{h}$ **ii** $b = 8$
e **i** $r = \frac{A}{2\pi h}$ **ii** $r = 4$
f **i** $T = \frac{100I}{PR}$ **ii** $T = 1.5$
g **i** $d = \frac{c}{\pi}$ **ii** $d = 28$
h **i** $a = \frac{v - u}{t}$ **ii** $a = 30$
i **i** $h = \frac{A}{2\pi h} - r$ **ii** $h = 20$
j **i** $w = Bh^2$ **ii** $w = 45$
k **i** $h = \sqrt{\frac{w}{B}}$ **ii** $h = 1.6$
l **i** $m = \frac{2E}{v^2}$ **ii** $m = 0.26$
m **i** $v = \frac{\sqrt{2E}}{m}$ **ii** $v = 30$

Revision and assessment (page 150)

Directed numbers

1 D **2** C **3** A **4** A **5** D
6 B **7** C **8** C **9** B **10** A

11 a and b

°C
A
C
E
0
B
D

c 22°, 15°, 12°, -4°, -9° **d** 7° **e** 4°

12 **a** > **b** < **c** >

13 **a** $^{+}14$ **b** $^{+}1$ **c** $^{-}5$ **d** $^{-}15$ **e** $^{-}12$

14 **a** $^{+}9$ **b** $^{+}4$ **c** $^{-}14$ **d** $^{-}12$ **e** $^{-}18$
f $^{+}22$ **g** $^{+}1$ **h** $^{+}3$

15 **a** $^{-}30$ **b** $^{+}28$ **c** $^{+}80$ **d** $^{+}3$ **e** $^{-}4$ **f** $^{+}4$

16 **a** $^{-}18$ **b** $^{+}20$ **c** $^{+}19$ **d** $^{+}6$
e $^{-}28$ **f** $^{-}29$ **g** $^{-}12$ **h** 12
i $^{-}6$ **j** $^{-}5\frac{1}{2}$ **k** $^{-}25$

17 A is Benny, B is Cara, C is Dillon, D is Ernie and E is Alina.

Indices

1 **a** D **b** A **c** E **d** B **e** C

2 **a** False **b** True **c** False
d False **e** True

3 D **4** B **5** D **6** D **7** C

8 **a** 2×7^2 **b** $2^3 \times 3 \times 47$

9 54 **10** 7 **11** $x \times x \times x \times x$

12 **a** 4 **b** 12 **c** $^{-}4$ **d** $^{-}8$ **e** 1
f 4 **g** $-\frac{1}{2}$ **h** $^{-}8$ **i** 16 **j** $^{-}10$

13 **a** $5a^4b^2$ **b** $5a^4b^4$ **c** $5^4a^4b^4$

14 **a** x^{11} **b** $^{-}20a^7$ **c** $6x^4y^7$ **d** b^4
e $7a^4$ **f** z^5 **g** $4ab^2$

15 **a** b^{12} **b** 1 **c** $36d^3e^{12}$ **d** $15g^6$
e e^2f^3 **f** 1 **g** $^{-}6h^8j^3$ **h** $10j^5$
i $6n^8p^8$ **j** $\frac{3m^2n}{2}$ **k** $\frac{a^4}{b^4}$ **l** $\frac{m^7}{2k^2}$
m m^5 **n** $\frac{-8}{9}q^6$ **o** $\frac{m^4}{n^3}$ **p** $\frac{g^4}{j^2h^6}$
q $\frac{2^7}{t}$ **r** $\frac{v^7}{2w^4}$

16 **a** 3 **b** $1\frac{1}{2}$ **c** $48\frac{1}{7}$ **d** $3\frac{1}{2}$

17 **a** 30 010 **b** 0.001 932 **c** 0.000 000 812 4
d 500 600 **e** 0.000 054 32

18 **a** 1.2104×10^2 **b** 4.5×10^{-3} **c** 1.0×10^3
d 7.0×10^4 **e** 3.0×10^{-6} **f** 3.75×10^{-5}

19 **a** 1.1034×10^4 **b** 8.85×10^3

20 **a** $\frac{21}{1\,000\,000}$ m **b** 2.1×10^{-2} mm
c 1.05×10^1 mm

Equations

1 B **2** D **3** B **4** A **5** C
6 D **7** A **8** D **9** A **10** D

11 **a** 25, 31, 37 **b** 216, 648, 1944
c $^{-}18$, $^{-}23$, $^{-}28$ **d** $2, 1, \frac{1}{2}$

12 a and c **13** 20, 11, 2 **14** $a = {}^{-}11$, $d = 6$

15 15, 23, 31 **16** 38.5

17 **a** $9t + 5$ **b** $6p$ **c** $x^2 + 6x + 5$

18 **a** $27m - 99$ **b** $^{-}4p - 12$ **c** $w^2 + 7w$
d $2x - 11$ **e** $m + 5$ **f** $2b^2 - 3b$
g $7h - 57$ **h** $^{-}p + 23q$ **i** $8w - 8v$
j $17g + h$

19 **a** $2(3x + 7)$ **b** $b(ab + 6)$ **c** $7n(m - 1)$
d $^{-}5t(t - 2)$ **e** $5a(a + 2 - 4b)$

20 a LHS $= 2(4 \times 3 - 3)$
$= 2(12 - 3)$
$= 2 \times 9$
$= 18$
= RHS

b LHS $= \frac{3w+2}{4}$
$= \frac{3 \times 1 + 2}{4}$
$= \frac{5}{4}$

RHS $= 7w - 6$
$= 7 \times 1 - 6$
$= 1$

LHS ≠ RHS

21 a $x = {}^-8$ **b** $m = 21$ **c** $t = {}^-1$ **d** $v = 21$
e $t = 4$ **f** $y = {}^-3\frac{4}{7}$ **g** $x = {}^-20$ **h** $m = 39$

22 a $m = 1\frac{2}{5}$ **b** $m = 1$ **c** $t = \frac{2}{3}$ **d** $x = 1$
e $n = 2$ **f** $x = 3$ **g** $x = {}^-6$ **h** $d = 4$
i $y = {}^-1$ **j** $p = 17\frac{1}{2}$ **k** $m = {}^-1$ **l** $z = {}^-2$
m $m = 7$ **n** $y = {}^-14$ **o** $c = 1$

23 a $m + 12$ **b** $\frac{m}{4}$ **c** $2m - 5$ **d** $\frac{m+3}{2}$
e $\frac{m}{2} + 3m$ **f** $m(m+1)$

24 a $x-2, x-1, x, x+1, x+3, x+4$
b $n+2, n+1, n, n-1, n-2, n-3$

25 a $4y = 20, y = 5$ **b** $\frac{y}{6} = 8, y = 48$
c $19 - 2y = 9, y = 5$ **d** $3y + 1 = 22, y = 7$
e $3(y+1) = 15, y = 4$ **f** $\frac{y+1}{3} = y, y = \frac{1}{2}$

26 a $x = 7$ **b** $x = 30, 3x = 90, x + 30 = 60$
c $x = 7$, so width = 20; $y = 7$, so length = 6

27 Flora K28, Jason K16

28 a $P = 10.4$ **b** $A = 251.2$ **c** $C = 26.67$

29 a i $n = \frac{S+4}{2}$ **ii** $n = 8$
b i $s = \sqrt{\frac{A}{6}}$ **ii** $s = 9$

Projects (pages 156–61)

Teacher to check.

Enrichment (page 163)

1 a $\frac{2a}{4} = \frac{a}{2} = \frac{8a}{16} = \frac{2ab}{4b}$ **b** $\frac{x}{2} = \frac{3x}{6} = \frac{5x}{10} = \frac{ax}{2a}$

2 a $\frac{m}{4}$ **b** $\frac{t}{2}$ **c** $2m$ **d** ^-2y
e $\frac{5}{a}$ **f** $\frac{(2t+5)}{t}$ **g** $\frac{b-3}{5}$ **h** 7

3 a $\frac{2p}{3}$ **b** $\frac{7m}{20}$ **c** $\frac{7a+4b}{28}$ **d** $\frac{^-y}{21}$
e $\frac{^-10h+21j}{15}$ **f** $\frac{10}{a}$ **g** $\frac{y+12}{4}$ **h** $\frac{36-x}{9}$
i $\frac{7c+2a}{abc}$ **j** $\frac{5+3d}{d^2}$

4 a $\frac{xy}{15}$ **b** $\frac{ab}{12}$ **c** $\frac{2a}{b}$ **d** $\frac{2}{w}$
e 1 **f** ^-2j **g** $\frac{a^2}{16b}$ **h** $\frac{5}{25^2}$

5 a 6 **b** $\frac{2m}{n}$ **c** $\frac{6p}{5q}$ **d** $\frac{2w}{z}$ **e** $\frac{3}{n}$ **f** $\frac{1}{18}$

UNIT 3: Working with data

Recording data and Sorting and organising data exercises

Lesson 1 (pages 166–7)

1 a Numerical **b** Categorical **c** Categorical
d Numerical **e** Numerical **f** Numerical
g Categorical **h** Categorical

2 a Discrete **b** Continuous **c** Discrete
d Continuous **e** Continuous **f** Continuous

3 a Discrete numerical **b** Continuous numerical
c Categorical **d** Categorical
e Continuous numerical **f** Continuous numerical
g Discrete numerical

4 a Continuous numerical **b** Categorical
c Categorical **d** Continuous numerical
e Continuous numerical **f** Discrete numerical
g Categorical

5 Teacher to check.

Lesson 2 (pages 168–9)

1 a

Sport	Tally	Frequency
Rugby	𝍸 𝍸 ||||	14
Basketball	𝍸 ||||	9
Swimming	𝍸 |	6
Volleyball	𝍸 ||	7
Soccer	𝍸 𝍸	10
Tennis	||||	4

b

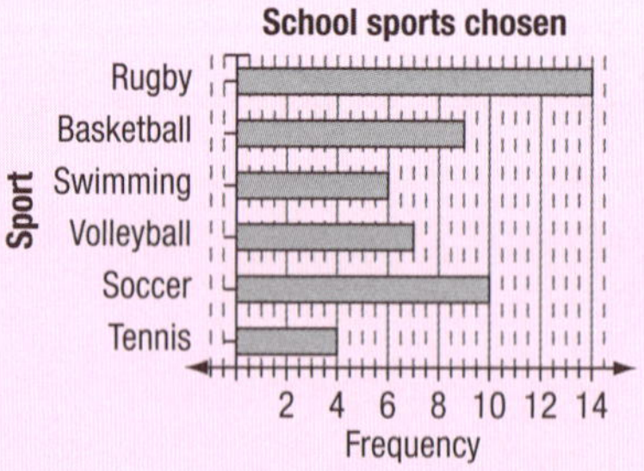

c Rugby **d** 14%

e The most popular sport chosen by this group of male students was **rugby**, with **28%** of the students playing this sport. The least popular sport chosen was **tennis** with only **8%** of students playing this sport. The second most popular sport chosen was **soccer** (**20%**), closely followed by **basketball** (**18%**). **14%** of the students chose volleyball and **12%** chose swimming.

2 a **Weekly expenditure (K)**

7.5%
17.5%
55%
8%
12%
Food
Clothing
Rent
Power
Travel

b A total of **80.5%** of the weekly expenditure was allocated to the essential items of food, rent and power. Rent was the most expensive weekly item accounting for **55%** of the weekly expenditure. Travel accounted for **12%** of the weekly expenditure and **7.5%** was allocated for clothing.

3 a

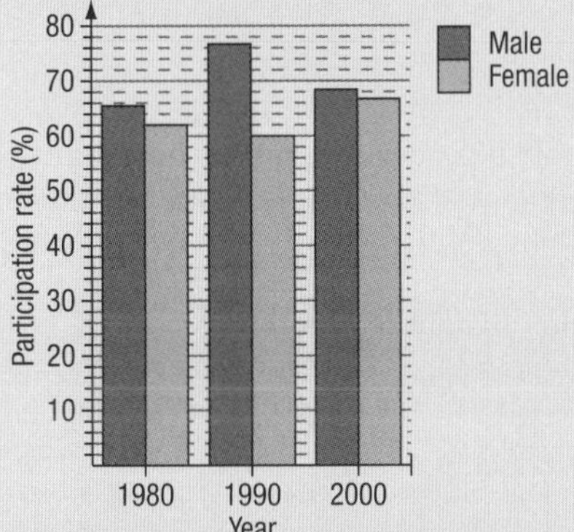

b In each of the three years when data was collected, **males** had a greater participation rate than **females**. The greatest difference between male and female rates was recorded in **1990**, when the difference in participation rate was **16%**. Participation rates decreased for males between **1990** and **2000**. The decrease was **8%**.
In the same period the female participation rate increased by **6%**.

4 a **Childhood malnutrition**

Gulf
Central
Enga
Chimbu
Morobe
Madang
Manus
North Solomons
Region
10 20 30 40
Percentage

b Childhood malnutrition was higher than 15% in all the regions mentioned, with Madang having the highest rate of 40%, and four of the regions having childhood malnutrition rates of more than 30%. Of the regions mentioned Chimbu had the lowest rate of 16%.

c

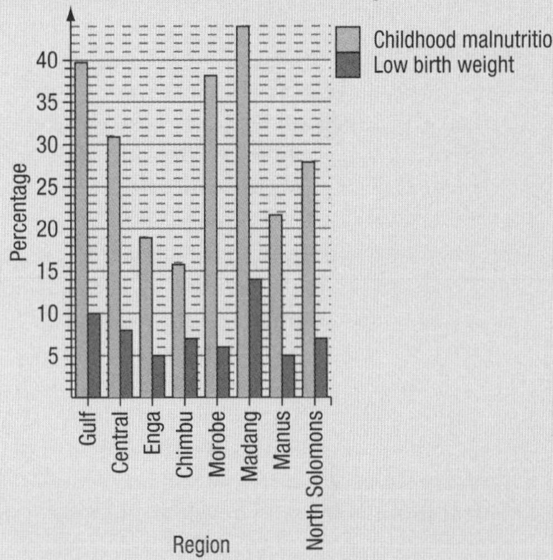

d In all the regions mentioned, the number of children with low birth weight was 5% or more, with the highest incidence of low birth weight occurring in Madang. The rates of low birth weight are highest in those regions with high percentages of child malnutrition.

Lesson 3 (pages 171–2)

1 a

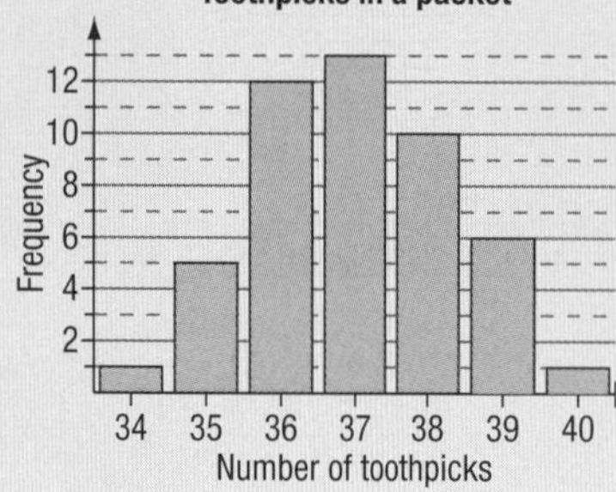

b 37 toothpicks c 12.5% d Symmetric

2 a

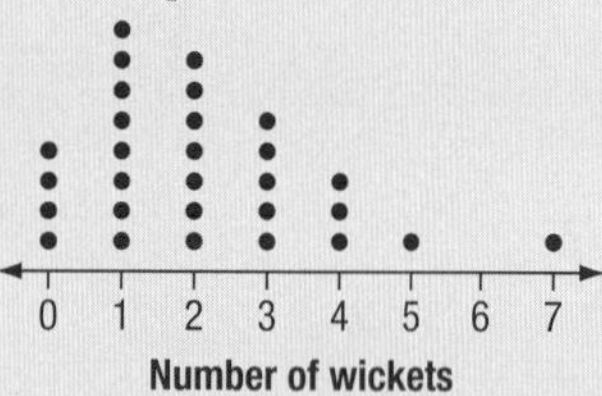

b Positively skewed

c Most often, Leo took **1** wicket in an innings but, in **23%** of the innings he took two wickets. There were **5** innings when Leo did not take a wicket but, on his best performance, he took **7** wickets.

3 a Symmetric

b The maximum number of chickens kept by a household in the survey was **14** and **2.5%** of the households did not keep any chickens. The most frequently occurring number of chickens kept by a household was **7**, and **74%** of the households kept between 4 and 9 chickens inclusive. **25** of the households kept 10 or more chickens and **14%** kept 3 or fewer chickens.

4 a A dotplot or column graph would be suitable.

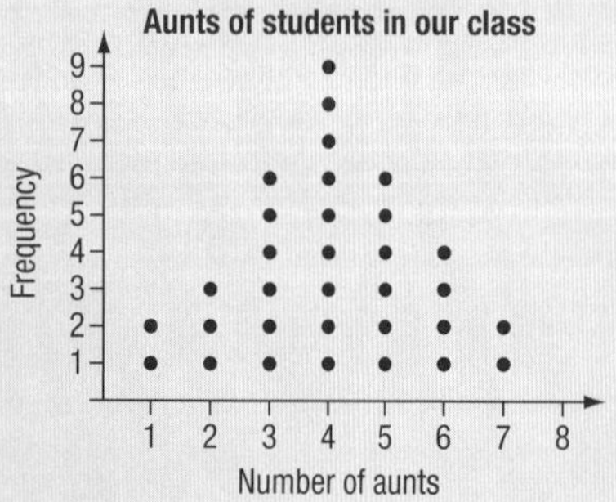

b Symmetric

c Of the 32 students surveyed, the most frequently occurring number of aunts is 4 with 28% of the students having 4 aunts and 66% of the students having 3, 4 or 5 aunts. The maximum number of aunts for the group was 7 (6.25% of the students) and the minimum number was 1 with 6.25% of the students having 1 aunt.

Lesson 4 (pages 174–5)

1 a **Winning basketball scores**

Stem	Leaf
3	5 9
4	2 5 6 6 7 7 8 9 9
5	0 1 3 3 4 4 4 4 4 5 6 8 8
6	1 2 2 4 6 6 6
7	1

4 | 6 means 46

b Symmetric

c Most frequently, the winning teams scored between **50** and **60** goals. The minimum winning score was **35** and the maximum was **71**. **25%** of the winning teams scored more than 60 goals and **12.5%** of the winning teams scored 45 or fewer goals.

2 a Ordered

b i 2 phonecalls ii 55 phonecalls

c 48% d 24% e Negatively skewed

3 a **Test scores**

Stem	Leaf
3	4
4	3
5	0 1 3 6 7 9
6	0 1 4 5 5 6 9
7	0 0 1 1 2 2 5 5 5 6 6 8
8	1 2 3

b 10% c 93% d Negatively skewed

e 34 could be considered an extreme value because it is a lot less than the lowest values for the rest of the data.

4 a and b

Number of vehicles per day

Stem	Leaf
8	8
9	4 4 6
10	3 6 7
11	0 1 9
12	1 2 7 8 9
13	0 1 2 3 5 6 6 8
14	0 3 5 5 7 8
15	4

11 | 9 means 119 vehicles

c Maximum 154; minimum 88

d 130 to 140 vehicles. Most often there were between 130 and 140 vehicles per day along this road.

e 13% f Negatively skewed

g On half of the days there were 130 or more vehicles on the road.

5 a **Number of mangoes per tree**

Stem	Leaf
6	8 9
7	4 5 7 7 8 9
8	0 1 2 3 3 4 4 4 5 6 7 7 7 7 8 9
9	0 1 2 3 4 4 4 4 5 5 6 6 6 6 8 9 9
10	0 1 1 2 3 4 5 6
11	4

9 | 4 means 94 mangoes

b The data is slightly negatively skewed, with the trees most frequently producing between 90 and 100 mangos. The least number of mangos was 68 and the most number was 114. There were 66% of the trees producing between 80 and 100 mangos, with 18% producing 100 or more mangos.

Lesson 5 (pages 176–8)

1

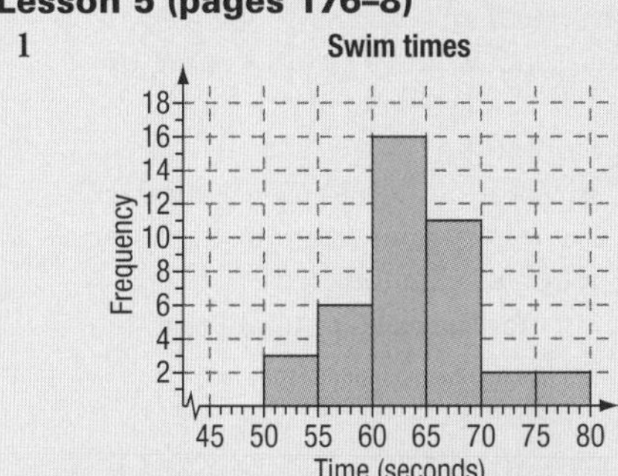

2 a 710 vehicles b 39.4% c 25.4%
d 15.5% e K18 000

3 a

Temperature (°C)	Tally	Frequency
20–	𝍸 ǀǀ	7
25–	𝍸 𝍸	10
30–	𝍸 ǀǀ	7
35–	𝍸 ǀ	6
40–	ǀ	1

b

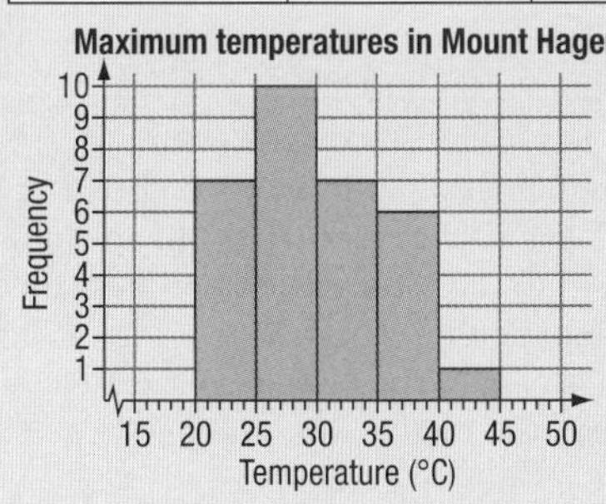

c The temperatures are slightly positively skewed.

d **Daily maximum temperatures (°C)**

Stem	Leaf
2	3 4 3 2 1 4 3
2*	5 6 9 6 5 7 7 5 6 5
3	4 1 2 2 0 4 0
3*	8 8 5 6 5 6
4	1

e The actual data values are all visible on a stem-and-leaf plot.

4 a **Weights of loaves of bread**

Stem	Leaf
87	
87*	8
88	2
88*	5 8 9 5 6 9
89	3 4 2 1 4 4 3 4
89*	5 6 8 5 8 6 7 9
90	1 4 1 3 4 1 1 4 3 3 3 0 1 4
90*	7 7 5 7 9 8 6 8
91	3 0 3 3 0
91*	5 7 6
92	4 1 4
92*	7 8
93	1

b 60% c 13.3% d Symmetric

e Eight of the loaves weigh less than 890 g and six of the loaves weigh more than 920 g, so 14 of the 60 loaves (23.3%) have a weight outside this range, and 46 of the 60 loaves (76.7%) have a weight within this range.

5 a

Weight (kg)	Frequency
0–	3
5–	4
10–	5
15–	5
20–	3
25–	3
30–	4
35–	3

b

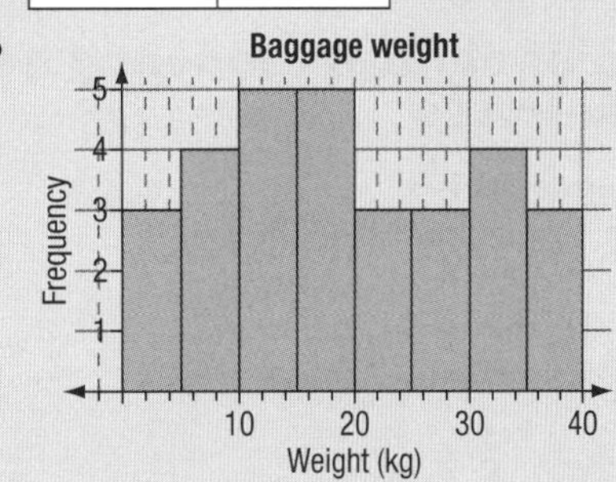

c The 'leaf' is always the last digit and, in this case, this is the number in the first decimal place. This would mean that we would need 'stems' of 0 to 39 (too many stems).

Lesson 6 (pages 180–2)

1 a 420 b 80 c 96
2 a 20 b 97 c 105 d 98
3 a 50
b

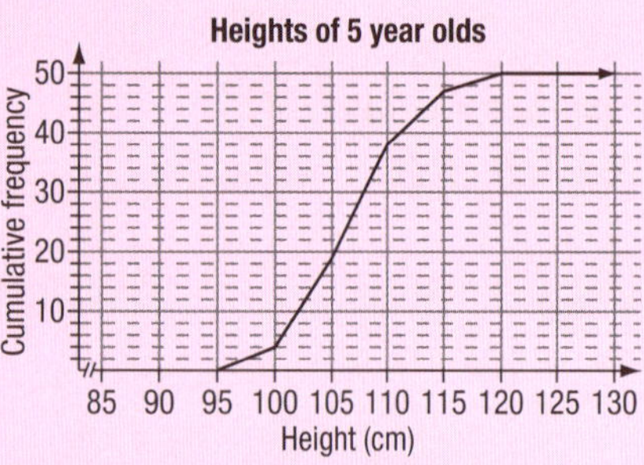

c i 120 cm ii 107 cm
iii 102 cm iv 111 cm, 120 cm

4 a

Number of phonecalls	Cumulative frequency	Percentage cumulative frequency
< 10	6	12
< 20	14	28
< 30	24	48
< 40	37	74
< 50	45	90
< 60	50	100

b

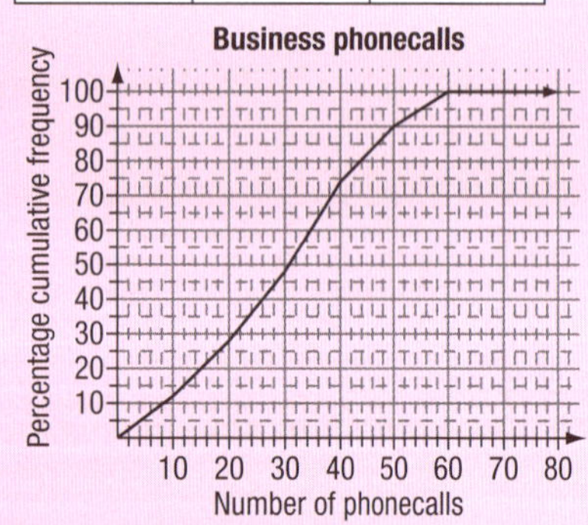

c i 31 ii 44 iii 60 iv 20

Measures of central tendency and spread exercises

Lesson 1 (pages 184–5)

1 a Mean = 5.5, median = 6, mode = 6
b Mean = 13.3, median = 13, mode = 10
c Mean = 6.9, median = 7, mode = 7
d Mean = 4.9, median = 5, mode = 5

2 a

Rooms in houses

Frequency: 2, 4, 6, 8, 10, 12

Number of rooms: 2, 3, 4, 5, 6, 7, 8

b Positively skewed.
c Mean = 3.97, median = 4, mode = 4
d **33.3%** of the houses in the survey had 4 rooms and the median number of rooms was **4**. **11.1%** of the houses had more than 5 rooms and **11.1%** had 2 rooms. The maximum number of rooms for houses in this survey was **8**.

3 a Mean = 21.9, median = 23, 18, mode = 25
b The extreme data value (11) would drag the mean towards it, so the mean is not a good representation of the centre.
c There are two modes in this case (bi-modal); neither is a good representation of the centre.

4 a Mean = 3, median = 2, mode = 3
b

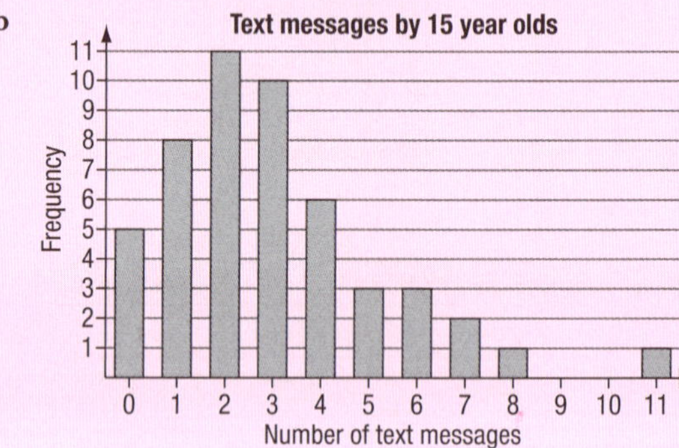

c Positively skewed
d The mean is dragged towards the extreme value (11) whereas the median is not affected by extreme values.
e The median or the mode.

5 a The distribution of the data is symmetric, so the mean, median and mode would be 8.
b The data is positively skewed, so the mean would be dragged towards the extreme values; about 6. The median and mode are 5.
c The data is negatively skewed, so the mean will be less than the median and mode. Mean = 26, median = 28, modal class 30–35.

Lesson 2 (pages 187–8)

1 a i 6 ii 7 iii 4 iv 7 v 3
b i 17 ii 18.5 iii 14.5 iv 14 v 4
c i 5 ii 6 iii 4 iv 4 v 2
d i 5.5 ii 7.5 iii 4 iv 11 v 3.5

2 a 0, 0, 0, 0.8, 1.4, 1.5, 1.6, 1.9, 2.1, 2.2, 2.7, 3, 3.4, 3.6, 3.8, 3.8, 4.5, 4.8, 5.2, 5.2

b i 2.45 **ii** 3.8 **iii** 1.45

c i 5.2 **ii** 2.35

d 50% of the waiting times were greater than **2.45** minutes and 75% of the waiting times were less than **3.8** minutes. The minimum waiting time was **0** minutes and the maximum time was **5.2** minutes The waiting times were was spread over a range of **5.2** minutes.

3 a 1, 1, 2, 2, 2, 3, 3, 3, 3, 3, 3, 4, 4, 4, 4, 4, 4, 4, 4, 5, 5, 5, 5, 5, 5, 6, 6, 6, 6, 7, 7

b 4, 5, 3 **c** 6, 1

d 50% of the students had more than **4** aunts and 75% of the students had fewer than **5** aunts. The minimum number of aunts for students in this class was **1** and the maximum was **7** aunts. The number of aunts was spread over a range of **6**.

Lesson 3 (pages 189–90)

1 a 20 **b** 29 **c** 11.5 **d** 39 **e** 17.5

2 a i 42 **ii** 50 **iii** 31

b Half of the households have more than **42** plants. 75% of the households have more than **31** plants.

c i 52 **ii** 19

3 a i 124 **ii** 130 **iii** 116

b Half of the children are more than124 cm tall. 75% of the children are less than 130 cm tall.

c i 29 **ii** 14

d The children in the survey have heights spread over a range of 29 cm and the middle 50% of the children have heights spread over 14 cm.

4 a i 30 **ii** 40 **iii** 17

b i 53 **ii** 23

c Half of the people surveyed made 30 or more PMV trips per month, with 75% making more than 17 PMV trips per month. The spread of the number of PMV trips per month was 53, with the spread of the middle 50% of PMV trips per month being 23.

5 a Minimum daily temperature in Port Moresby in July

Stem	Leaf
18	3
19	4 7
20	1 6 7 9
21	1 1 2 2 4 6
22	0 1 8
23	0 1 1 2 2 4 5 5 9 9
24	2 6
25	0 2 9

21 | 6 means 21.6°C

b i 22.88 **ii** 23.53 **iii** 21.19

c i 7.61 **ii** 2.34

d 75% of the minimum daily temperatures were greater than **21.19**°C. The minimum daily temperature, in Port Moresby, for July, was centred around **22.83**°C and was spread over **7.61**°C.

Lesson 4 (pages 192–3)

1 a i 36 **ii** 78 **iii** 12 **iv** 52 **v** 26

b i 66 **ii** 26

2 a 97 **b** 25 **c** 70 **d** 84 **e** 55, 84

3

0 10 20 30 40 50

Number of cars

4 a and b

140 150 160 170 180 190

Height (cm)

c range = 37, interquartile range = 16

d The heights of 16-year-olds are centred on 162.5 cm and spread over 37 centimetres. The data is positively skewed. 75% of the 16-year-olds have heights of 152 cm or more and the middle 50% of heights are spread over 16 cm.

5 a i 905 **ii** 913 **iii** 900 **iv** 927 **v** 883

b

880 890 900 910 920 930

Weight (g)

c i 14 g **ii** 44 g

d i Half of the loaves of bread weighed **905** grams or more.

ii **25%** of the loaves weighed 899 grams or less.

iii The masses of the middle 50% of the loaves in this sample were spread over **12** grams.

iv The lightest 25% of the loaves weighed **899** grams or less.

e Symmetric

6 a i B **ii** A **iii** A

b i 43 **ii** 14

c i 75% **ii** 50%

d i Symmetric **ii** Positively skewed

e The students in class **A** generally scored higher marks.
The marks in class **B** were more varied.

Lesson 5 (pages 195–6)

1 a 7 mm **b** July **c** 7 months

2 a About 3100 **b** Increasing

c About 1600 more **d** About 3500

3 a
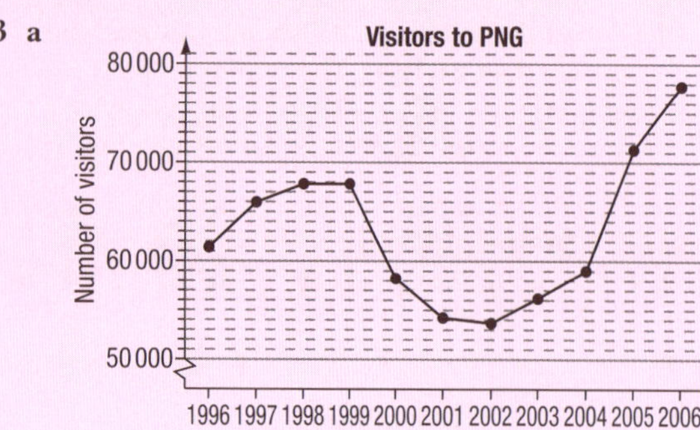

b There is an increasing trend. From 1996 to 1998 the number of visitors increased, then it remained steady for 1999 and then decreased until 2002. From 2002 the number of visitors increased each year, with large increases in 2005 and 2006.

c About 82 000 visitors

4 a April b August and September

c Yes, there is a yearly seasonal pattern.

d Rainfall in Daru is seasonal with the most rainfall falling in the month of **April** and the least amount of rain falling in the months of **August** and **September**. From the months of June to November the average daily rainfall is less than **5** mm.

5 a 2400 million kina b 5200 million kina

c Increasing d Decreasing

e The difference (Exports – Imports) is increasing.

f About 6000 million kina

g About 2000 million kina

Misleading statistical graphs exercises

Lesson 1 (page 199)

1 a The vertical scale does not start at zero, so it gives an exaggerated picture of the increase in 'Sales' over the four years.

b The columns on the graph have been given the appearance of volume and this makes the number of 'Large' drinks sold appear to be about eight times the number of 'Small' drinks sold, instead of twice the number which it actually is.

c The columns are not all the same width so that the frequency of, say, '3 children' appears to be about four times the frequency of '2 children', instead of twice the frequency which it actually is.

d The vertical axis is not evenly scaled, so the comparison of values is distorted.

e The trees used for the columns of the graph invite comparison of areas not just heights, making the frequencies of the larger trees appear greater.

f The columns on the graph appear as if they have volume, making the volume of tank B appear to be eight times the volume of tank C, instead of twice the volume which it actually is.

g The horizontal scale does not start at zero so that the comparison of the frequencies of types of trees is distorted.

h The 'dots' on the dotplot are not evenly spaced so it is difficult to compare the frequencies.

i The horizontal axis is not evenly scaled so that some of the columns, which should all be the same width on this histogram, have a larger appearance.

j The columns on the graph invite comparison of area, so that the area of the larger plots is greatly exaggerated.

Experimental probabilities and Probabilities based on symmetry

Lesson 1 (pages 201–3)

1 Teacher to check.

2 Sample answers given. Teacher to check individual answers.

a Certain b Impossible c Possible

d Very unlikely e Possible f Unlikely

g Possible h Likely

3 a i 0.44 ii 0.16 iii 0.69

b By subtracting from 1 the probability that a child is malnourished.

4 a i 0.15 ii 0.01 iii 0.075

b No

5 a i 0.34 ii 0.16 iii 0.18

b No

6 a i 0.12 ii 0.28 iii 0.04 iv 0.22 v 0.48

b Very accurate in 2000, not so accurate now.

c i 0.14 ii 0.09

7 0.16

Lesson 2 (pages 204–5)

1 Teacher to check.

2 a i The experiment is tossing two coins together.

ii There will be 100 trials.

iii 'Two heads', 'two tails' or 'a head and a tail'

iv 'A head and a tail'

v Divide the resulting number of 'a head and a tail' by 100.

vi Teacher to check.

b Teacher to check.

Challenge:

Teacher to check.

Lesson 3 (page 206)

Teacher to check.

Lesson 4 (pages 207–8)

1 a {head, tail}, equally likely
b {1, 2, 3, 4}, equally likely
c {1, 2, 3, 4}, not equally likely.
d {red, blue, yellow, green}, equally likely.
e {yellow, red, blue, green}, not equally likely.
f {blue, red}, not equally likely.

2 a $\frac{1}{52}$ b $\frac{1}{2}$ c $\frac{1}{\text{the number in the class}}$
d $\frac{1}{7}$ e $\frac{1}{365}$ f $\frac{1}{4}$ g $\frac{1}{2}$
h $\frac{1}{\text{the number of tables in the classroom}}$

Challenge:
Teacher to check.

Lesson 5 (pages 209–10)

1 a i {1, 3, 5} ii $\frac{1}{2}$
b i {5, 6} ii $\frac{1}{3}$
c i {3, 6} ii $\frac{1}{3}$
d i {1, 2, 3, 6} ii $\frac{2}{3}$
e i {2, 3, 5} ii $\frac{1}{2}$

2 a $\frac{1}{13}$ b $\frac{1}{4}$ c $\frac{4}{13}$ d $\frac{1}{2}$
e $\frac{3}{13}$ f $\frac{1}{52}$ g $\frac{3}{13}$

3 a $\frac{1}{4}$ b $\frac{1}{4}$ c $\frac{1}{8}$ d $\frac{3}{8}$
e $\frac{5}{8}$ f $\frac{7}{8}$

4 a $\frac{5}{12}$ b $\frac{1}{3}$ c $\frac{7}{12}$ d $\frac{2}{3}$

5 a i $\frac{1}{2}$ ii $\frac{1}{2}$ iii $\frac{3}{10}$
b i An odd number
ii A number greater than 5
iii A factor of 8

6 a $\frac{18}{37}$ b $\frac{18}{37}$ c $\frac{1}{37}$ d $\frac{18}{37}$
e $\frac{3}{37}$ f $\frac{12}{37}$ g $\frac{18}{37}$

7 a $\frac{1}{11}$ b $\frac{2}{11}$ c 0 d $\frac{2}{11}$ e $\frac{7}{11}$ f $\frac{4}{11}$

Lesson 6 (pages 211–12)

1 a 12
b i $\frac{1}{12}$ ii $\frac{1}{12}$ iii $\frac{1}{4}$ iv $\frac{1}{3}$

2 a

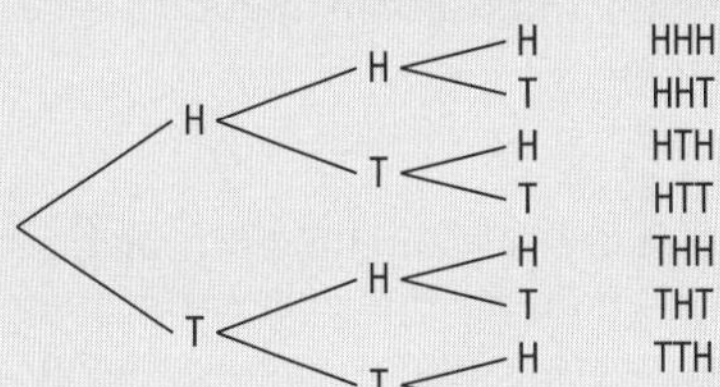

b 'Three heads', 'two heads and a tail', 'one head and two tails', 'three tails'.
c i $\frac{1}{8}$ ii $\frac{3}{8}$ iii $\frac{3}{8}$ iv $\frac{1}{8}$
d 1

3 a

		Spinner 1			
		1	2	3	4
Spinner 2	1	1 and 1	2 and 1	3 and 1	4 and 1
	2	1 and 2	2 and 2	3 and 2	4 and 2
	3	1 and 3	2 and 3	3 and 3	4 and 3
	4	1 and 4	2 and 4	3 and 4	4 and 4

b 16
c i $\frac{1}{16}$ ii $\frac{1}{8}$ iii $\frac{1}{4}$ iv $\frac{3}{4}$
v $\frac{1}{4}$ vi $\frac{1}{4}$ vii $\frac{1}{4}$

d i

		Spinner 1			
		1	2	3	4
Spinner 2	1	2	3	4	5
	2	3	4	5	6
	3	4	5	6	7
	4	5	6	7	8

ii

Event: 'sum of'	Probability
2	$\frac{1}{16}$
3	$\frac{2}{16}=\frac{1}{8}$
4	$\frac{3}{16}$
5	$\frac{1}{4}$
6	$\frac{3}{16}$
7	$\frac{1}{8}$
8	$\frac{1}{16}$

iii 5

Lesson 7 (pages 213–14)

1 Teacher to check.

2 a

		Die 1					
		1	2	3	4	5	6
Die 2	1	1 and 1	2 and 1	3 and 1	4 and 1	5 and 1	6 and 1
	2	1 and 2	2 and 2	3 and 2	4 and 2	5 and 2	6 and 2
	3	1 and 3	2 and 3	3 and 3	4 and 3	5 and 3	6 and 3
	4	1 and 4	2 and 4	3 and 4	4 and 4	5 and 4	6 and 4
	5	1 and 5	2 and 5	3 and 5	4 and 5	5 and 5	6 and 5
	6	1 and 6	2 and 6	3 and 6	4 and 6	5 and 6	6 and 6

b 36 c Yes

Event 'sum of'	Nmber of favourable outcomes	Probability (fraction)
2	1	$\frac{1}{36}$
3	2	$\frac{1}{18}$
4	3	$\frac{1}{12}$
5	4	$\frac{1}{9}$
6	5	$\frac{5}{36}$
7	6	$\frac{1}{6}$
8	5	$\frac{5}{36}$
9	4	$\frac{1}{9}$
10	3	$\frac{1}{12}$
11	2	$\frac{1}{18}$
12	1	$\frac{1}{36}$
Total	**36**	**1**

d Seven e Teacher to check.

3 a i $\frac{1}{36}$ ii $\frac{35}{36}$

b i 1 ii Their probabilities sum to 1.

4 a $\frac{1}{6}$ b $\frac{1}{6}$ c $\frac{1}{2}$ d $\frac{1}{2}$ e $\frac{1}{4}$ f $\frac{3}{4}$

5 Teacher to check.

Option A: Random events and simulation exercises

Lesson 1 (page 216)

1 a Boy or girl

b i Head for boy and tail for girl, or vice versa

ii Any arrangement where three numbers represent a boy and the other three numbers represent a girl. For example: 1, 2 and 3 for boy and 4, 5 and 6 for girl.

2 a i A head can represent a boy

ii A tail can represent a girl (or vice versa)

b Teacher to check. c Teacher to check.

d Teacher to check.

3 a (Other answers possible.) A die could be used, with the numbers 1, 2, 3, and 4 representing a win and the numbers 5 and 6 representing a loss. Or a spinner shaped like an equilateral triangle could be used, with two edges representing a win and the other edge representing a loss.

b If using a die, roll the die a maximum of three times until the result is a loss for Leo. If there are three wins rolled in a row, Leo wins the tournament.

c Teacher to check. d Teacher to check.

Lesson 2 (pages 218–19)

1 a 3, 9, 5, 4, 1

b 97, 52, 90, 92, 96, 84, 65, 51, 79, 89

c 526, 501

2 141, 103, 152, 139, 182, 149

3 a Saturday b 26 March

c Day 288, which is 15 October

4 Teacher to check. 5 26, 25, 42, 40, 20

Challenge:

Teacher to check.

Lesson 3 (pages 219–20)

1 a Teacher to check.

b (Other answers possible.) Odd digits represent heads, and even digits (including zero) represent tails.

c Teacher to check.

2 a The digits 1, 2, 3, 4, 5 and 6 could represent the numbers on a die and the other digits would be ignored.

b Teacher to check.

3 Teacher to check.

Lesson 4 (page 220)

1 a 365

b Use three digits. Look for 30 numbers in the range 001 to 365, ignoring numbers outside that range. Record any repeats.

c Teacher to check. d Teacher to check.

e Teacher to check.

Lesson 5 (page 222)

Teacher to check.

Lesson 6 (page 222)

Teacher to check activity.

Lesson 7 (page 223)

Teacher to check.

Option B: Statistical surveys exercises

Lesson 1 (page 227)

1 a A survey of the whole population

b The government

c Every 10 years

d In the year 2000

e It would be too costly and time consuming.

f It is a legal requirement to answer the questions in a census. You are required to answer them.

g (Other answers possible.) Some people may not have been 'at home' when the census was taken; the location of some households may make it difficult to obtain information.

h Teacher to check.

2 a All the students attending the school
b 120 people c Randomly selected students
d 25 people

3 Teacher to check.

4 a All the people in PNG b Sample
c Teacher to check.

5 a All the people in PNG b Sample
c Teacher to check.

Lesson 2 (page 228)

Teacher to check.

Lesson 3 (pages 230–1)

1 a 3, 9, 5, 4, 1
b 97, 52, 90, 92, 96, 84, 65, 51, 79, 89
c 526, 501

2 141, 103, 152, 139, 182, 149

3 Allocate a number from 1 to 20 to each of the provinces. Choose a starting point and a direction on the random number table. Look at two-digit numbers, ignoring 00 and any numbers greater than 20. Note the first applicable number and locate the province represented by that number.

4 a i The days in June; 30 ii 3
iii 26, 27 and 28 of June
b i Thursdays in June; 4 ii 1
iii The second Thursday (12 June)
c i Groups of four consecutive days; 27 ii 1
iii The group of four days beginning on 26 June (that is 26, 27, 28 and 29 June)

5 a Consider three-digit numbers, ignoring numbers outside the range 001 to 365. Write 30 numbers including repeats.
b Teacher to check. c Teacher to check.

6 Streets have different numbers of houses, so not all houses have an equal chance of being chosen.

Lesson 4 (pages 232–3)

1 a Allocate 'head' to one member of the population and 'tail' to the other.
b Any arrangement in which three numbers represent one member of the population and the other three numbers represent the other member. For example: Allocate 1, 2 and 3 to one member of the population and 4, 5 and 6 to the other.

2 a Allocate two of the six possible outcomes to each of the objects. For example: Allocate 1 and 2 to 'object 1', 3 and 4 to 'object 2', and 5 and 6 to 'object 3'.
b Allocate one of the six possible outcomes to each of the objects. For example: Allocate 1 to 'object 1', 2 to 'object 2', 3 to 'object 3', 4 to 'object 4', 5 to 'object 5' and 6 to 'object 6'. Allocate the outcomes to the objects before you roll the die.

3 a The spinner must be a regular octagon, or be a round spinner with eight equal sectors.
b Teacher to check. c Teacher to check.
d Teacher to check.

4 a Discard the court cards (Jack, Queen and King). This leaves 40 cards remaining. Allocate each of the remaining cards to one member of the population. Randomly choose five cards and identify the sample.
b Teacher to check.

Lesson 5 (pages 234–5)

Teacher to check exercise.

Lesson 6 (pages 236–7)

Teacher to check exercise.

Lesson 7 (pages 238–9)

Teacher to check exercise.

Lesson 8 (page 240)

Teacher to check.

Lesson 9 (page 241)

Teacher to check.

Revision and assessment (pages 242–9)

Recording data and Sorting and organising data

1 B 2 C

3 a Categorical b 110 trees
c Banana d 20%

4 a Continuous b Discrete c Continuous

5 D

6 a

b 60% c 25 d Symmetric

Measures of central tendency and spread

1 B 2 D

3 a Points scored

Frequency

Number of points

b i 4 **ii** 4 **iii** 5

4 a and **b**

Stem	Leaf
8	3 5 7
9	5 5 9
10	3 4 5 6 8
11	0 3 5 7 9
12	0 1 2 5 6 8

10 | 3 means 103

c i 109 **ii** 43

5 a and **b**

Time taken to fill orders

Stem	Leaf
1	4 8 9 9
2	2 2 2 2 3 4 5 6 8 9 9
3	1 1 3 7 8 8
4	1 1 2 4 4 5 6 6 9
5	2 2 3 5 5
6	2 4 5 7
7	3

3 | 1 means 3.1 minutes

c 3.8 minutes **d** 2.2 minutes

e There are many more ‘waiting times’ greater than 2.2 minutes than there are times less than 2.2 minutes.

f 47.5%

6 a 5 **b** 5 **c** 6

7 a i 6.6 kg **ii** 6.7 kg and 6.4 kg

iii 0.8 kg **iv** 0.3 kg

b

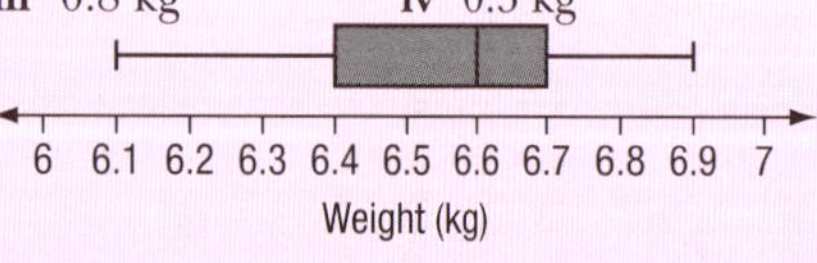

c The weights of bags of fifty apples are negatively skewed about a median of 6.6 kg. The weight is spread over 0.8 kg (range) and the spread of the middle 50% of the weights is 0.3 kg (the interquartile range).

8 60 points

9 a

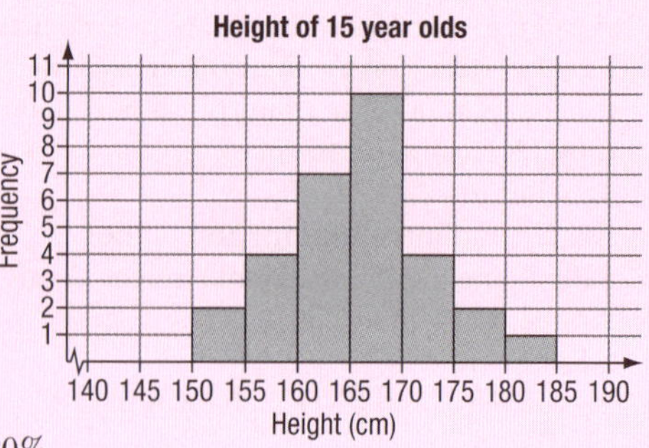

b 20%

c

Height (cm)	Cumulative frequency
< 155	2
< 160	6
< 165	13
< 170	23
< 175	27
< 180	29
< 185	30

d

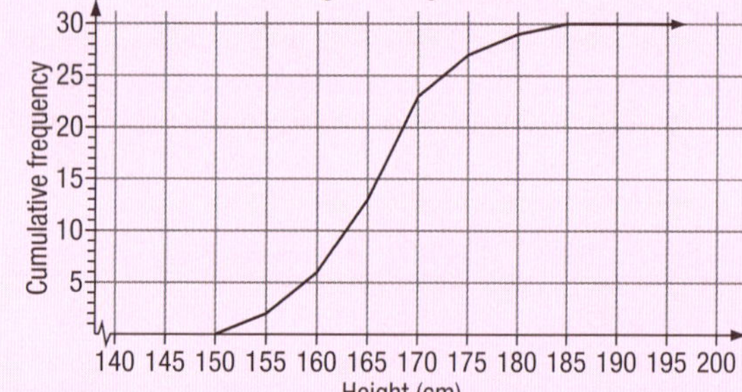

e 166 cm

10 a 40 **b** 24 **c** 11.8

11 a i 157 cm **ii** 172 cm **iii** 30 cm **iv** 19 cm

b Positively skewed

c i 150 **ii** 150 **iii** 164

12 a Day 7 and day14 **b** 4 days

c Saturday **d** 65 customers

e Seasonal with a 7-day repeat

13 a A line graph or time-series graph.

b 2.2 m **c** 1999

d The height of the tree is increasing with time; the increase is greatest in the first five years.

e 2.5 m

Misleading statistical graphs

1 a The horizontal scale is cut off.

b

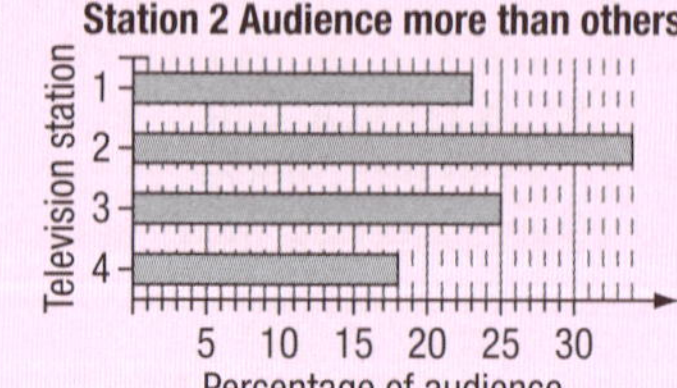

Experimental probabilities and Probabilites based on symmetry

1 a 0.42 b 0.47 c 0.62
2 a 0.55 b 0.2 c 0.25 d 0.5
3 a 0.25 b 0.625 c 0.125 d $\frac{5}{6}$
4 a 0.44 b 0.26 c 0.2 d 0.5
5 a 120 b $\frac{19}{60}$ c More trials
d Probably e No
6 C 7 B 8 A 9 A 10 C
11 a $\frac{1}{2}$ b 1 c $\frac{1}{26}$ d $\frac{1}{52}$ e $\frac{10}{13}$ f $\frac{6}{13}$
12 a {M, A, T, H, E, I, C, S} b No
c i $\frac{2}{11}$ ii $\frac{1}{11}$ iii 0 iv $\frac{4}{11}$ v $\frac{7}{11}$ vi $\frac{4}{11}$
13 a

Disc 1	Disc 2	Outcome
G	G	GG
	R	GR
	B	GB
	Y	GY
R	G	RG
	R	RR
	B	RB
	Y	RY
B	G	BG
	R	BR
	B	BB
	Y	BY
Y	G	YG
	R	YR
	B	YB
	Y	YY

b i $\frac{1}{16}$ ii $\frac{1}{16}$ iii $\frac{3}{4}$ iv $\frac{1}{8}$
14 a {1, 2, 3, 4} b No c $\frac{2}{5}$ d $\frac{3}{5}$
15 a $\frac{1}{12}$ b $\frac{1}{6}$ c $\frac{1}{3}$ d $\frac{1}{2}$ e $\frac{5}{12}$
16 a {2, 3, 4} b Yes
c

		1st spin		
		2	3	4
2nd spin	2	2 and 2	3 and 2	4 and 2
	3	2 and 3	3 and 3	4 and 3
	4	2 and 4	3 and 4	4 and 4

d i $\frac{1}{3}$ ii $\frac{2}{3}$ iii $\frac{1}{9}$ iv $\frac{1}{3}$ v $\frac{1}{3}$
17 a

Fair coin	Double-headed coin	Outcome
H	H	HH
	H	HH
T	H	TH
	H	TH

b i $\frac{1}{2}$ ii $\frac{1}{2}$ iii 0
18 a {1, 2, 3, 5} b No c $\frac{1}{2}$ d $\frac{1}{3}$ e $\frac{5}{6}$
19 a $\frac{1}{10}$ b $\frac{1}{20}$ c 0.45 d 0.2 e $\frac{1}{20}$

20 a A larger area of the board scores '1'.
b i 60 ii $\frac{5}{12}$
c i

First dart score	Second dart score	Total score
1	1	2
	5	6
	10	11
5	1	6
	5	10
	10	15
10	1	11
	5	15
	10	20

ii 2, 6, 10, 11, 15, 20 iii no

Unit 4: Design in 2D and 3D geometry

Properties of plane figures exercises

Lesson 1 (pages 254–7)

1 a An equilateral triangle b A square
2 a Convex b Concave c Convex
d Concave e Convex f Concave
3 Teacher to check. 4 Teacher to check.
5 Teacher to check.
6 a i Hexagon ii
b i Octagon ii
c i Parallelogram ii
d i Octagon ii

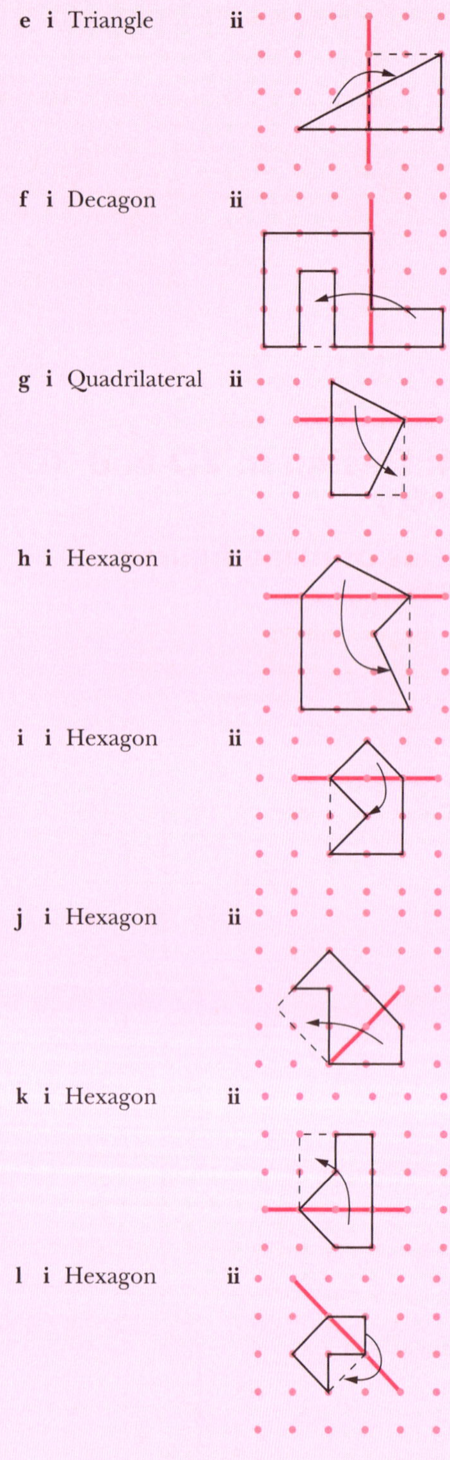

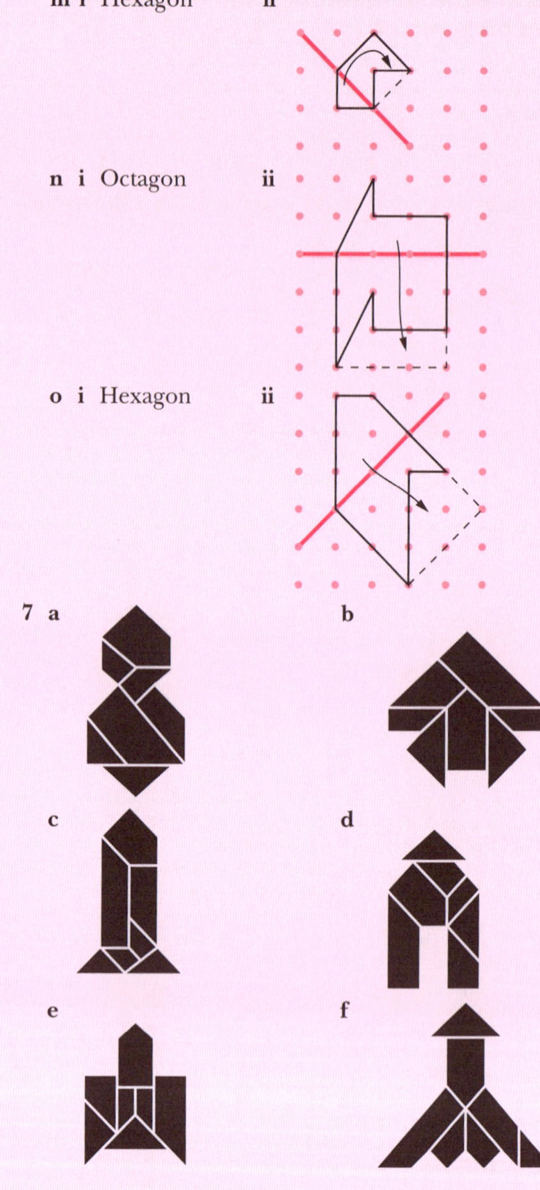

Lesson 2 (page 260)

1 **a** $a = 30$ **b** $b = 58$ **c** $c = 60$
d $d = 155$ **e** $e = 45$ **f** $f = 80$

2 **a** $a = 110$ **b** $b = 120$ **c** $c = 100$
d $d = 105$ **e** $e = 70$ **f** $f = 33$

3 **a** **i** Isosceles **ii** Acute-angled
b **i** Scalene **ii** Right-angled
c **i** Equilateral **ii** Acute-angled
d **i** Scalene **ii** Obtuse-angled
e **i** Isosceles **ii** Right-angled
f **i** Isosceles **ii** Acute-angled

4 Teacher to check.

5 The longest side of a triangle must be less than the sum of the other two sides.

Lesson 3 (pages 263–4)

1 a Square, rhombus b Quadrilateral
c Parallelogram, rhombus, rectangle, square
d Parallelogram, rhombus e Square, rectangle
f Quadrilateral, trapezium g Trapezium

2 Teacher to check.

3 a True b False c True
d True e False

4 Square, kite or trapezium 5 No

6 a 90° b 90° c 106°
d 36° e 212° f 90°
g 222° h 50° i 144°

7 **d**, **e**, **g** and **h** 8 Teacher to check.

Lesson 4 (pages 266–7)

1 a Teacher to check.
b Divide 360° by the number of sides; 360° ÷ n (where n = number of sides)

2 a

Number of sides, n	3	4	5	6	8
Number of right angles, S	2	4	6	8	12

b $S = 2n - 4$
c i 900° ii 1260° iii 1440°
d Teacher to check.

3 $a = 108°$, $b = 72°$, $c = 36°$

Lesson 5 (pages 267–9)

1 a Squares, rectangles
b Rectangles, squares, triangles
c Squares, triangles
d Rectangles, triangles, squares, parallelograms
e Rectangles, squares, triangles

2 to 4 Teacher to check.

Challenge:
Teacher to check.

Lesson 6 (page 270)

1, 2 Teacher to check.

3 a Triangles, squares, and regular hexagons will all tessellate.
b If the size of the interior angle is a factor of 360, the polygon will tessellate.

Lesson 7 (page 271–2)

Teacher to check.

Angles and lines exercises

Lesson 1 (pages 274–6)

1 a 36° b 19° c 60°
d 132° e 148° f 336°

2 a 70° b 30° c 28°
d 133° e 21° f 305°

3 Teacher to check sketches.
a Acute b Obtuse c Right
d Reflex e Acute

4 Teacher to check.

Lesson 2 (pages 276–8)

1 Teacher to check. 2 **b**, **d**

3 a 45° b 74° c 22° d 77° e 65°

4 **b**, **c** and **e**

5 a 80° b 46° c 142° d 137° e 91°

6 a $a = 65$ b $b = 33$ c $c = 14$ d $d = 60$
e $e = 45$ f $f = 26$ g $g = 29$ h $h = 29$
i $n = 43$, $m = 137$ j $y = 92$, $w = z = 88$

7 a $a = 270$ b $b = 92$ c $c = 343$
d $d = 221$ e $e = 177$ f $f = 339$
g $m = 132$ h $l = 127$ i $y = 45$
j $s = 37$ k $m = 73$ l $q = 16$, $t = 254$

Lesson 3 (pages 280–1)

1 a Corresponding b Alternate
c Co-interior d Corresponding
e Corresponding f Vertically opposite
g Vertically opposite h Co-interior
i Alternate

2 a Corresponding b Co-interior
c Corresponding d Vertically opposite
e Corresponding f Co-interior
g Alternate h Vertically opposite
i Alternate

3 a $a°$ and $c°$; $a°$ and $q°$; $b°$ and $d°$; $b°$ and $p°$; $d°$ and $r°$; $c°$ and $s°$; $p°$ and $r°$; $q°$ and $s°$; $d°$ and $p°$; $c°$ and $q°$
b $d°$ and $q°$; $c°$ and $p°$; $a°$ and $b°$; $c°$ and $d°$; $p°$ and $q°$; $r°$ and $s°$

4 a $a = 105$ b $b = 60$ c $c = 110$
d $d = 38$ e $e = 47$ f $f = 145$
g $x = 80$, $y = 100$, $z = 80$
h $x = 135$, $y = 135$, $z = 135$

Challenge (page 282):
Teacher to check.

Lesson 4 (page 283)

1 i c and e are congruent pairs
ii a and d are similar pairs
iii b is neither

2, 3 Teacher to check.

4 a True b Teacher to check.

Surface area and volume exercises

Lesson 1 (pages 285–6)

1 a 4.8 m b 4 m c 1.65 m
d 1.02 m e 3.2 m

2 a 110 cm b 160 mm c 3.1 m
d 4.4 m e 3.1 m f 83 cm

3 a 6.4 m b 168 mm c 13.2 m
d 6.9 m e 4.2 m

4 a 106 cm b 110 cm

5 a i 62.96 m ii 58.96 m

b Using millimetres means that there is greater accuracy and no confusion about where the decimal point is placed.

Lesson 2 (pages 288–9)

1 a 352 cm b 13.2 cm

2 220 mm

3 a 75 cm b 47 m c 34.27 cm

4 a 198 mm b 280 cm c 8.6 m

5 144 cm 6 500 times 7 245 cm

8 a i 123 cm ii 143 cm

b Groups of 4

Lesson 3 (pages 291–2)

1 a 92 m^2 b 1200 cm^2 c 3150 cm^2

d 7 m^2 e 9650 cm^2 f 37.5 m^2

g 292.5 mm^2 h 121.5 cm^2 i 4.8 m^2

2 i 240 cm^2 ii 480 cm^2

Challenge:

a i Teacher to check. ii 6.882 cm

iii 34.410 cm^2 iv 172.05 cm^2

b i Teacher to check. ii 8.66 cm

iii 43.3 cm^2 iv 259.81 cm^2

c Teacher to check.

Lesson 4 (pages 295–7)

1 a 154 km^2 b 962.5 mm^2

c 6.16 cm^2 d 105.6 cm^2

2 1078 cm^2

3 a 2.66 m^2 b 189 cm^2

c 308 cm^2 d 519.75 mm^2

4 a 83.31 m^2 (82.8 m^2) b 1.96 m^2 (1.875 m^2)

c 19.73 cm^2 (22 cm^2) d 122.16 cm^2 (108 cm^2)

5 a 10 345.86 m^2 (10 157 m^2)

b 490.38 m^2 (478 m^2)

6 a 1051.55 cm^2 (1225 cm^2)

b 2796.9 cm^2 (2450 cm^2)

Lesson 5 (page 302)

1 a i hemisphere ii two faces

b i cone ii two faces

c i octagonal prism ii ten faces

d i pentagonal prism ii seven faces

e i right pentagonal pyramid ii six faces

f i tetrahedron ii four faces

2 a i rectangular prism

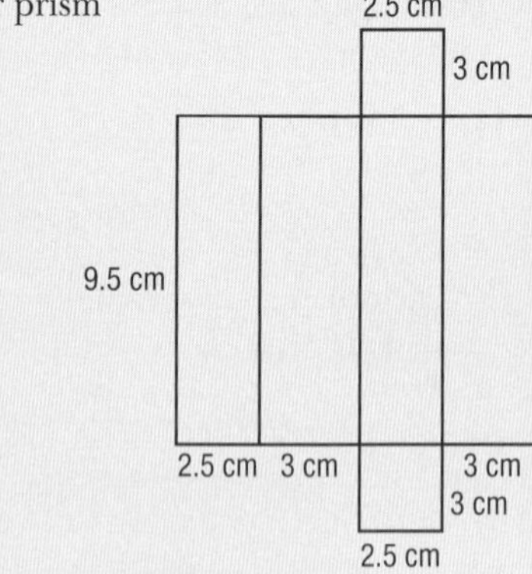

ii 119.5 cm^2

b i cube

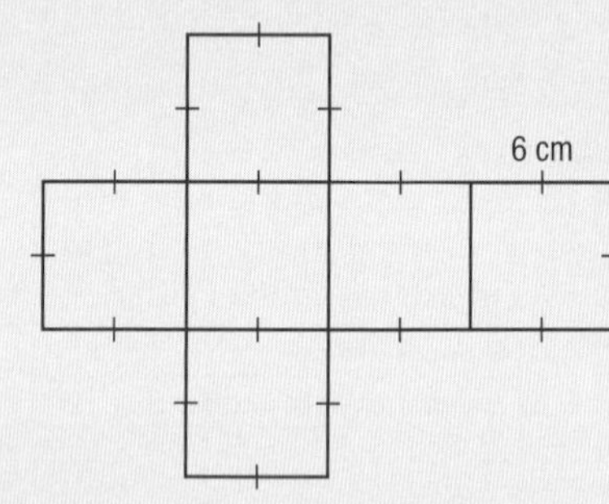

ii 216 cm^2

c i triangular prism

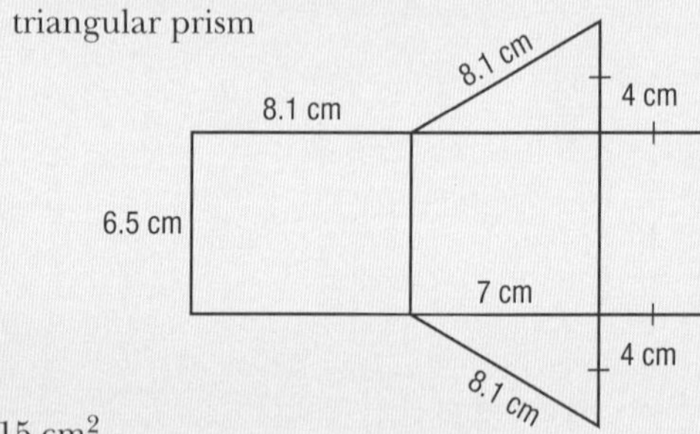

ii152.15 cm^2

3 a 1.375 m^2 b 200 cm^2

Lesson 6 (pages 304–5)

1 a 442.90 cm^2 (422.94 using $\pi = 3$)

b 12.75 cm^2 (12.18 using $\pi = 3$)

2 a TSA = 226.2 cm^2, curved surface area = 169.6 cm^2 (162 cm^2)

b TSA = 226.2 cm^2, curved surface area = 201 cm^2 (192 cm^2)

3 i The cylinder in part **b** is 31.4 cm^2 larger (30 cm^2)

ii Both are 226.2 cm^2 (226 cm^2)

4 a 519.21 cm^2 (503.3 cm^2)

b 1190.31 cm^2 (1161 cm^2)

5 a 5

b

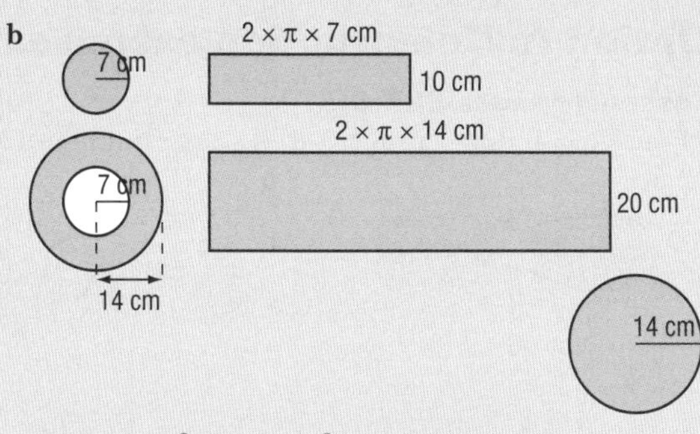

c 3430.6 cm^2 (3276 cm^2)

6 a 19.79 m^2 (18.9 m^2) b 22.1 m^2 (21.6 m^2)

Lesson 7 (pages 308–10)

1 a 819.96 cm^2 (783 cm^2) b 15.19 m^2

2 a 158.34 cm^2 (151.2 cm^2) b 104.64 cm^2

3 a 1256.64 cm^2 (1200 cm^2) b 1.54 m^2 (1.47 m^2)
c 295.56 cm^2 (282.24 cm^2) d 1.54 m^2 (1.47 m^2)
e 103 908.18 cm^2 (99 225 cm^2)
f 115.45 mm^2 (110.25 mm^2)

4 a 1253.50 cm^2 (1197 cm^2) b 740 cm^2
c 0.259 m^2 (0.248 m^2)

5 a 4
b

2 × π × 1.4 mm
3.5 mm
1.4 mm
3.5 mm
15 mm

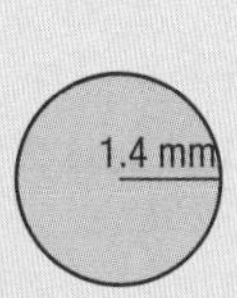

c 247.4 mm^2 (236.25 mm^2)

6 44 m^2 (42 m^2)

7 a Truncated cone b Teacher to check.
c Teacher to check.

Lesson 8 (page 312)

1 603.186 cm^3 2 1112 shovelsful

3 a 7600 cm^3 b 5 kL c 3256 m^3

4 200 5 1250 cm^3 6 250

7 a 33.2 m^3 b 33 200 L

Lesson 9 (pages 314–16)

1 a i

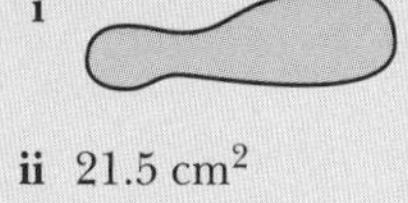

ii 21.5 cm^2 iii 172 cm^3

b i

2 cm

ii 24 cm^2 iii 48 cm^3

c i

0.5 m

ii 0.785 m^2 (0.75 m^2) iii 0.942 m^3 (0.9 m^3)

d i

10 mm

ii 100 mm^2 iii 1000 mm^3

e i

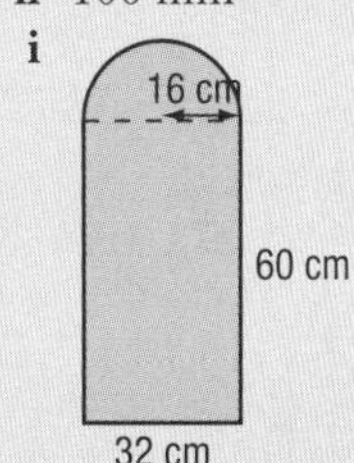

ii 1426.12 mm^2 (1408 mm^2)
iii 85567.43 cm^3 (84480 cm^3)

f i

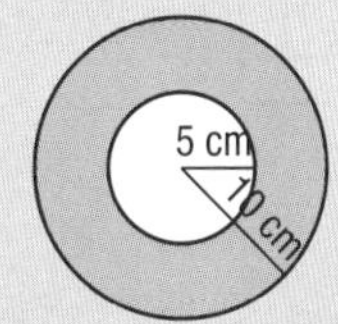

ii 235.62 cm^2 (225 cm^2)
iii 5890.49 cm^3 (5625 cm^3)

g i

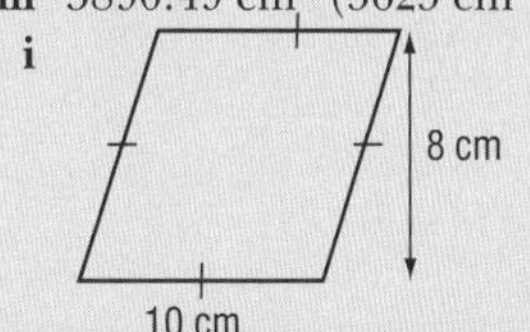

ii 80 cm^2 iii 800 cm^3

2 a i

37 mm
41 mm

ii 758.5 mm^2 iii 15170 mm^3 = 15.17 cm^3

b i

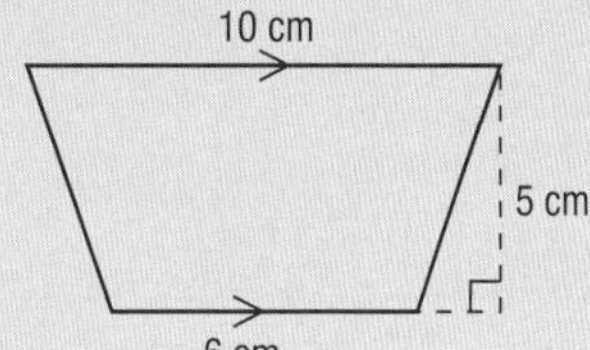

ii 40 cm^2 iii 320 cm^3

c i

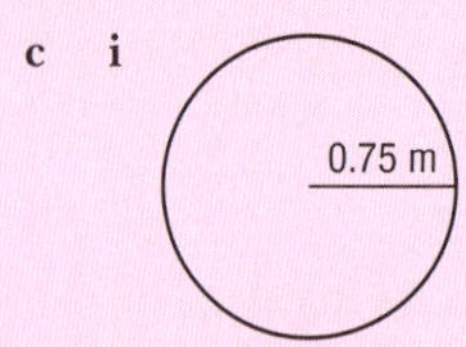

ii 1.767 m^2 (1.6875 m^2)
iii 1.4137 m^3 (1.35 m^3)

3 25 L

4 a

25 m
1 m
2 m

b 37.5 m^2 c 375 kL d 300 kL

5 104.68 L (99.96 L) 6 23.07 L (22.03 L)

7 18.04 cm (18.9 cm)

8 a

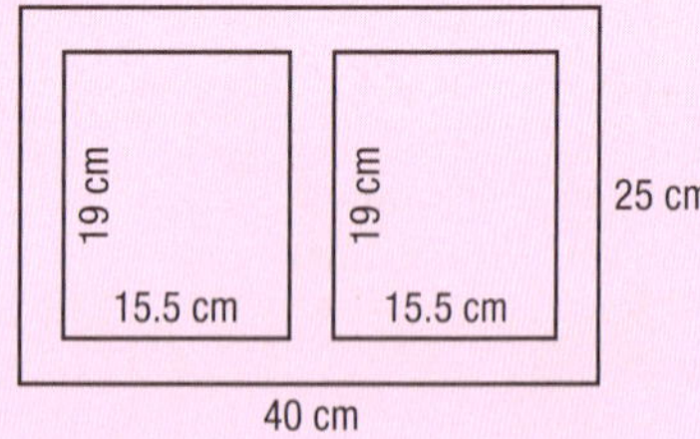

b 705.5 cm^2 c 14110 cm^3
d i 20000 cm^3 ii 70.6 %
e 14.11 m^3

9 5.94 m

Lesson 10 (pages 318–19)

1 a 0.009 m^3 b 5311.1 cm^3 c 75.4 cm^3

2 12800 mm^3

3 a 44602.2 mm^3 (42592 mm^3)
b 883.6 cm^3 (843.8 cm^3)
c 1809.6 cm^3 (1728 cm^3)

4 a 575.2 L (549.3 L) b 0.1732 L
c 5.747 L (5488 L)

5 8.5 cm

6 a 913.2 cm^3 (872 cm^3) b 524 mL (500 mL)

Lesson 11 (pages 320–2)

1 a 320.7 cm^3 (306.3 cm^3)
b 1309 cm^3 (1250 cm^3)
c 3230 cm^3
d 9771.5 cm^3 (9331.1 cm^3)
e 1439.5 mm^3 (1375 mm^3)

2 234.57 cm^3 (224 cm^3)

3 Volume is 1071.7 cm^3 (1023.4 cm^3)

4 a 81.78 cm^3 (78.1 cm^3) b 452 mm^3 (432 mm^3)

5 22.87 m^3 (24 m^3)

6 Teacher to check parts **a** to **d**.
e i 4 ii 43 = 64 iii 44800 cm^3

Lesson 12

Teacher to check Challenge questions.

Option A: Construction exercises

Lesson 1 (pages 327–9)

1 a 1 b 4 c 0 d 6 e Infinitely many

2 a

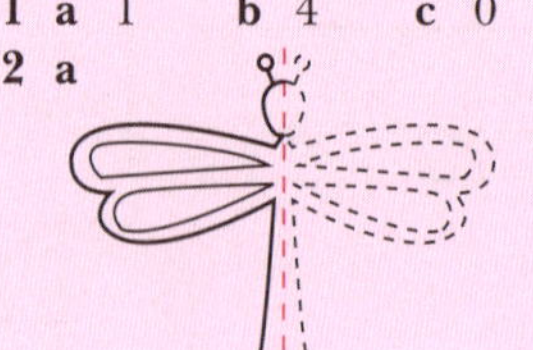

b

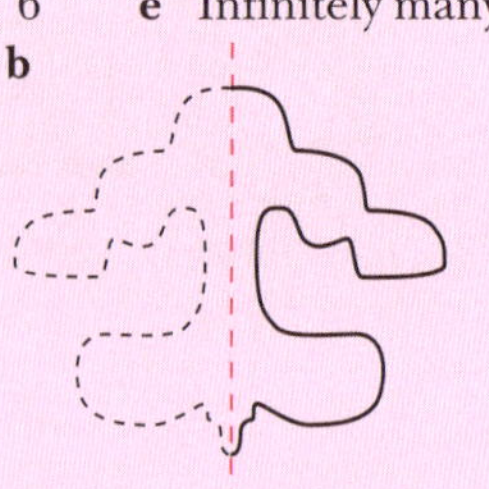

c

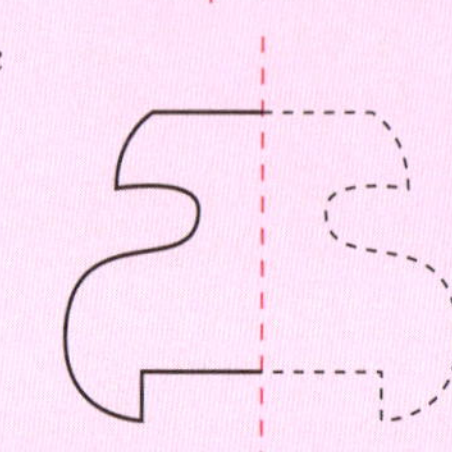

3 Teacher to check.

4 a 4 c 3 e 4 f 2

5 Teacher to check.

6 a **A, B, C, D, E, H, I, K, M, O, T, U, V, W, X, Y**
b **H**, 2; **I**, 2; **O**, infinitely many; **S**, 2; **X**, 2; **Z**, 2
c Teacher to check.

Lesson 2 (pages 331–2)

Teacher to check.

Lesson 3 (pages 334–5)

Teacher to check.

Lesson 4 (page 338)

1 Teacher to check.

2 a

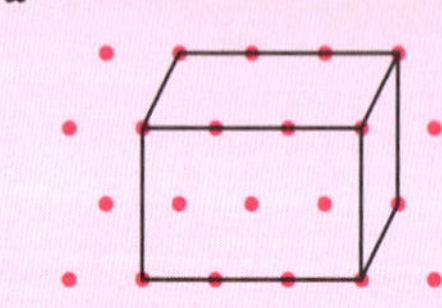

b

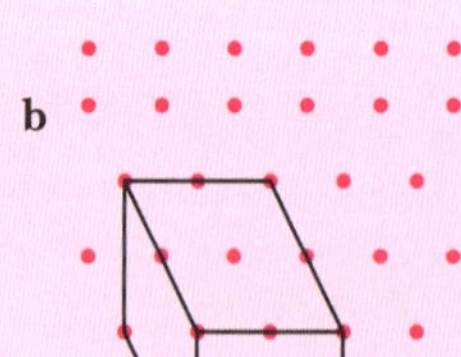

c

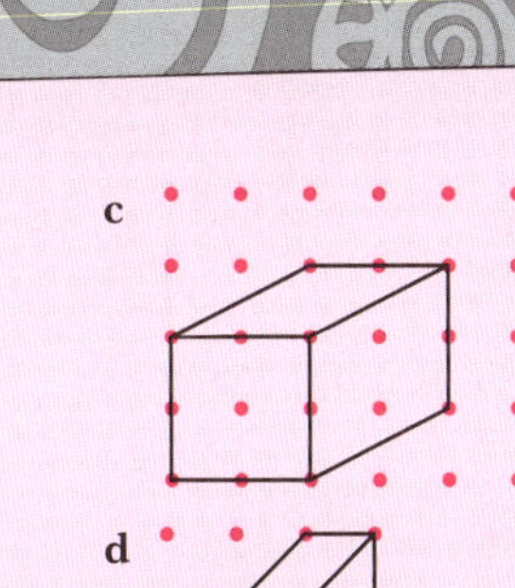

d

Lesson 5 (pages 338–9)

Teacher to check.

Lesson 6 (pages 341–2)

Teacher to check.

Lesson 7 (pages 343–4)

1–3 Teacher to check.

4 a 9 m height; 4 m width b 2.8 cm

5 a 3.2 m b 1 m

6 a 230 km b 420 km c 1370 km

d 450 km e 700 km

7 420 km

Option B: Deductive reasoning

Lesson 1 (pages 350–2)

1 a The line segments **AB** and BC intersect at an angle of **50°**. $\angle ABC = 50°$.

b The **reflex** angle formed by the line segments PQ and **QR** is 220°. The obtuse angle $\angle PQR =$ **140°**.

c $PQRS$ is a **quadrilateral**. Angles QPS and **QRS** are equal in size, and angle **PQR** is a right angle.

$\angle QPS = \angle QRS$ $\quad \angle PQR = 90°$

d Triangle XYZ is a **scalene** triangle. A line segment has been drawn from the point **Y** to the side XZ, bisecting side **XZ** at the point M. The lengths XM, **MY** and **MZ** are equal in length.

$XM = MY = MZ$

e AB and **CD** are a pair of **parallel** lines. $AB \parallel CD$

f A circle with **centre** O has a radius of **5** cm. The **diameter**, AB, is drawn. $AO = OB = 5$ cm

$AB = 10$ cm

g $PQRST$ is a **regular** pentagon.

$PQ = QR = RS = ST = TP$

$\angle PQR = \angle QRS = \angle RST = \angle STP = \angle TPQ$

h AB is an **arc** of a circle of radius **10** cm. AB subtends an **angle** of **40°** at the centre, **O**, of the circle. $\angle AOB = 40°$

$OA = OB = 10$ cm

i **PQ** is a perpendicular bisector of **AB**.

$AQ = QB$ $\quad PQ \perp AB$

j $ABCD$ is a **trapezium** (or **quadrilateral**), with side **BC** parallel to side **AD**. Side AB is **perpendicular** to both BC and AD.

$BC \parallel AD$. $\quad AB \perp BC$, and $AB \perp AD$.

2 a Triangle ABC is an isosceles triangle. $AB = BC$.

b $PQRS$ is a parallelogram.

$PQ \parallel SR$ $\quad QR \parallel PS$

$PQ = SR$ $\quad QR = PS$

c $PQRS$ is a rectangle.

$PQ = SR$ $\quad PS = QR$

d The line AB bisects the line PQ at the point O.

$PO = OQ$

e PQ is a transversal of the lines AB and CD.

f $PQRS$ is a kite.

$PQ = PS$ and $QR = SR$. $\angle PQR = \angle PSR$.

g The lines AB and BC intersect at an angle of 80°.

$AB = BC$ $\quad \angle ABC = 80°$.

h The line LM is perpendicular to the line KN and intersects with KN at the point M.

$LM = MN$ $\quad \angle LMN = 90°$

i A sector formed by the arc AB subtending an angle of 70° at the centre, O, of a circle. The radius of the circle is 25 m.

$OA = OB = 25$ m $\quad \angle AOB = 70°$.

3 Teacher to check.

Lesson 2 (pages 354–5)

1 a 69° b 7° c 79°

2 a 51° b 144° c 80°

3 a $\angle COD$ and $\angle DOE$

b $\angle BOC$ and $\angle COE$, or $\angle AOC$ and $\angle COD$, or $\angle BOD$ and $\angle AOB$, or $\angle AOB$ and $\angle AOE$, or $\angle BOD$ and $\angle DOE$

c $\angle AOB$ and $\angle DOE$

4 a $a = 39$; Angles in a right angle sum to 90°

b $d = 331$; Angles at a point sum to 360°

c $e = 35$; Angles in a right angle sum to 90°

d $g = 32$; Vertically opposite angles are equal

e $h = 117$; Angles on a straight line sum to 180°

f $i = 226$; Angles at a point sum to 360°

5 a $3x + 2x + 4x = 360$; Angles at a point sum to 360°; $x = 40$

b $2x + 10 + 3x + 30 = 180$; Angles on a straight line sum to 180°; $x = 28$

c $x + 5x - 8 + x = 90$; Angles in a right angle sum to 90°; $x = 14$

d $2x + 15 + x + 30 = 180$; Angles on a straight line sum to 180°; $x = 45$

e $3x + 3x + 30 + x - 31 + 67 = 360$; Angles at a point sum to 360°; $x = 42$

f $15x + 2x + 5 = 90$; Angles in a right angle sum to 90°; $x = 5$

6 Teacher to check. 7 Teacher to check.

Lesson 3 (pages 357–9)

1 a

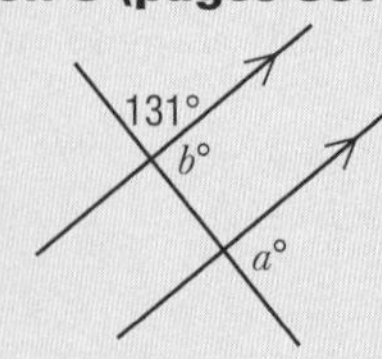

$b = 49$; Straight angle
$a = 49$; Corresponding angles

b $b = 37$; Corresponding angles

c

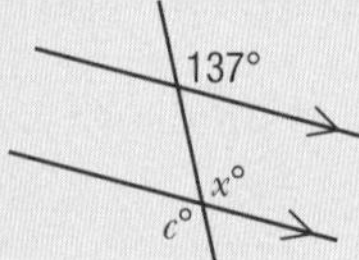

$x = 137$; Corresponding angles
$c = 137$; Vertically opposite angles

d $d = 90$; Corresponding angles

e

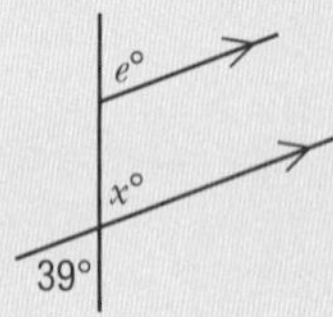

$x = 39$; Vertically opposite angles
$e = 39$; Corresponding angles

f $f = 123$; Co-interior angles sum to 180°

g

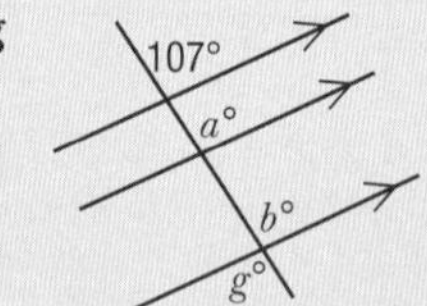

$a = b = 107$; Corresponding angles
$g = 107$; Vertically opposite angles

2 a $x = 50$ **b** $x = 66$ **c** $x = 115$
d $x = 49$ **e** $x = 106$ **f** $x = 30$

3 a $2x + 30 = 130$; $x = 50$
b $3a + 126 = 180$; $a = 18$
c $2c - 30 + 82 = 180$; $c = 64$
d $5p + 38 + 20 + 32 = 180$; $p = 18$
e $3x = 4x - 50$; $x = 50$

4 a Yes **b** No **c** Yes

Lesson 4 (pages 360–1)

1 a $a = 61$ **b** $b = 52$ **c** $c = 130$ **d** $d = 65$

2 a $a = 120$ **b** $b = 128$ **c** $c = 38$ **d** $g = 87$
e $h = 120$ **f** $a = 82$

3 a $x + 6 + 2x - 10 + 43 = 180$; $x = 47$
b $5x + 3x - 8 = 108$; $x = 14.5$
c $2x + 20 + 6x - 10 + 3x + 5 = 180$; $x = 15$

Lesson 5 (pages 363–4)

1 **a** and **d**, **h** and **j**, **b** and **e**, **c** and **g**, **f** and **i**

2 **a** and **n**, **c** and **f**, **d** and **k**

3 a A and C (SAS) **b** B and C (ASA)
c A and B (RHS) **d** A and C (SSS)
e B and C (ASA) **f** A and C (SAS)

4 a $x = 29$, $y = 5$ **b** $x = 30$, $y = 2$
c $x = 104$, $y = 38$, $z = 7$ **d** $x = 117$, $y = 27$
e $x = 90$, $y = 5$ **f** $x = 45$, $y = 3$
g $x = 100$, $y = 42$ **h** $x = 90$, $y = 20$

Lesson 6 (pages 366–7)

1 a $d = 90$ **b** $f = 48$ **c** $i = 44$
d $d = 3$ **e** $e = 30$

2 a $\angle CAB = \angle ACB$
$\angle CDB = \angle CBD$
b $\angle NKL = \angle NLK$
$\angle LNM = \angle LMN$

3 a $x = 113$ **b** $x = 37$

4 Teacher to check. **5** Teacher to check.

6 Teacher to check.

Lesson 7 (pages 369–70)

1 Teacher to check reasons.
a $x = 37$, $y = 63$ **b** $y = 90$, $x = 45$
c $x = 62$, $y = 90$ **d** $x = 39$, $y = 141$
e $x = 78$, $y = 78$ **f** $x = 100$, $y = 100$
g $x = 71$, $y = 109$ **h** $x = 118$
i $x = 6$ **j** $x = 90$, $y = 90$
k $x = 109$ **l** $x = 48$, $y = 132$

2 a $x + 3x + x + 90 = 360$, $x = 54$
b $3x - 30 + 95 + 2x + 95 = 360$, $x = 40$
c $3x + 5x + 3x + 5x = 360$, $x = 22.5$
d $6x + 4x = 180$, $x = 18$
e $180 - 3x + 110 + 5x + 60 = 360$, $x = 5$

3 a True (Teacher to check reasons.)
b True (Teacher to check reasons.)
c True (Teacher to check reasons.)

Lesson 8 (page 371)

1 Teacher to check. **2** Sum = $(n - 2) \times 180°$
3 Teacher to check. **4** Teacher to check.
5 Teacher to check. **6** Teacher to check.
7 Teacher to check.

Lesson 9 (pages 374–5)

1 a $x = 90$ (Half the angle at the centre standing on the same arc)
b $x = 40$ (Sum of angles of triangle is 180° and angle at the circle is 90°)
c $x = 45$ (Angle opposite one of the equal sides of an isosceles triangle)

2 a $x = 64$ (Angle at the centre is twice as big as the angle at the circle standing on the same arc)
b $y = 33$ (Angles standing on the same arc)
c $z = 77$ (Half the angle at the centre standing on the same arc)

d $p = 70$ (Half the angle at the centre standing on the same arc, which measures $360° - 220° = 140°$)
e $q = 99$ (Half the angle at the centre standing on the same arc, which measures $360° - 162° = 198°$)
f $r = 54$ (Both angles at the circle standing on the same arc are 38°. The other angle in the triangle is $180° - 92° = 88°$ because it forms a straight angle. Angle sum of a triangle is 180°, so $r = 180 - 88 - 38$.

3 a $x = 80$ (Opposite angles of a cyclic quadrilateral sum to 180°)
b $y = 70$ (Opposite angles of a cyclic quadrilateral sum to 180°)
c $w = 50$ (Half the angle subtended at the centre from the same arc, which measures $360° - 260° = 100°$)
$z = 130$ (Opposite angles of a cyclic quadrilateral sum to 180°)
d $p = 54$ (Half of the exterior angle 108° because it is an isosceles triangle with radii for sides)
e $r = 130$ (Opposite angles of a cyclic quadrilateral sum to 180°)
$s = 102$ (Opposite angles of a cyclic quadrilateral sum to 180°)
f $b = 70$ (Opposite angles of a cyclic quadrilateral sum to 180°)
$c = 110$ (Co-interior angle with b)
$a = 70$ (Opposite angles of a cyclic quadrilateral sum to 180°)

4 Teacher to check. **5** Teacher to check.
6 Teacher to check.

Revision and assessment (page 376)

2D geometry

1 a False **b** True **c** False **d** True
e False **f** True **g** True **h** False
i True **j** True **k** False **l** True
m False **n** True **o** True **p** True

2 Teacher to check. **3** Teacher to check.

4 a $x = 34$ **b** $x = 124$

5 a $d°$ and $h°$, or $f°$ and $j°$, or $e°$ and $i°$, or $g°$ and $k°$
b $d°$ and $g°$, or $e°$ and $f°$, or $h°$ and $k°$, or $i°$ and $j°$
c $d°$ and $e°$; $f°$ and $g°$; $d°$ and $f°$, $e°$ and $g°$, $h°$ and $i°$, $i°$ and $k°$, $j°$ and $k°$, $j°$ and $h°$, $d°$ and $j°$, $h°$ and $f°$, $e°$ and $k°$, $i°$ and $g°$, etc.
Teacher to check combinations.

6 Pentagon, 540° **7** $\angle LNO = 140°$

8 a 51° **b** 113° **c** 180° **d** 32°
e 117° **f** 226° **g** 108° **h** 143°
i 49° **j** 61° **k** 103° **l** 53°
m 61° **n** 52° **o** 65° **p** 60°
q 36° **r** 212° **s** 90° **t** 106°
u 55° **v** $a° = 99°$, $b° = 94°$

Surface area and volume

1 A **2** C **3** C **4** C **5** B **6** B

7 a 8.5 cm **b** 50 000 mL
c 7100 mm^2 **d** 4 500 000 cm

8 a 47.63 m **b** 78 m^2

9 57.13 cm

10 a 325 cm^2 **b** 577.3 cm^2

11 a 54 cm^2 **b** 287.1 cm^2

12 a

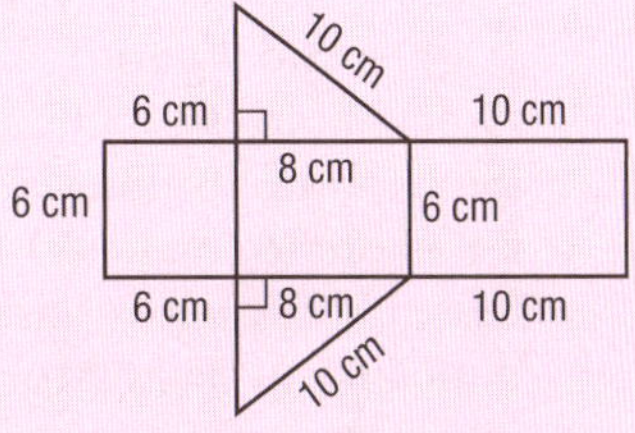

b 192 cm^2

13 a 61 575.2 mm^2 **b** 27 488.9 cm^2 **c** 8.80 m^2

14 a 6000 cm^3 **b** 6157.5 cm^3 **c** $133\frac{1}{3}$ m^3
d 64 654 cm^3 **e** 540 m^3 **f** 2.31 cm^3

15 11.08 kL **16** 10.52 cm^2

17 918.92 kg **18** 1065.69 cm^2

Projects

Teacher to check.

Acknowledgments

The authors and the publisher wish to thank the following copyright holders for reproduction of their material.

AAP/AFP/Adrian Dennis, p. 381 (middle right); AAP/AFP/Kazuhiro Nogi, p. 381 (middle left); Corbis/Hemi/Bruno Morandi, p. 252 (right); Corbis/Wolfgang Kaehler, pp. 34, 82; Corbis/Bob Krist, p. 83 (left); Corbis/David Lees, p. 86; Corbis/Charles & Josette Lenars, p. 380 (left); Corbis/Chris Rainier, p. 37; Corbis/Brian A. Vikander, p. 83 (right); Getty Images/Mark Daswell/Staff, p. 5 (right); iStockphoto, p. 381 (centre); iStockphoto/-Antonio-, p. 157; iStockphoto/GlobalP , p. 160; iStockphoto/guenterguni, p. 148; iStockphoto/humonia, p. 326 (top left); iStockphoto/jgroup, p. 326 (middle left); iStockphoto/mahesh14, p. 381 (top right); iStockphoto/MorePixels, p. 326 (bottom left); iStockphoto/Natural_Warp, p. 327; iStockphoto/nickfree, p. 210; iStockphoto/Pchelarocks, p. 380 (right); iStockphoto/Raycat, p. 159 (bottom); iStockphoto/roptop, p. 274; iStockphoto/Rtimages, p. 21; iStockphoto/stereohype, p. 381 (top left); iStockphoto/Terraxplorer, p. 381 (top middle); iStockphoto/Tjanze, p. 161; iStockphoto/yesphoto, p. 326 (middle right); iStockphoto/typhoonski, p. 267; iStockphoto/Zanthra, p. 159 (top); Lonely Planet/Michael Gebicki, p. 267; Lonely Planet Images/Holger Leue, p. 81; M.C. Escher Company, p. 269 (bottom), for M.C. Escher's 'Simmetry Drawing E18', © 2009. The M.C. Escher Company-Holland. All rights reserved; O'Sullivan Ros, p. 386; Paraide Patricia, p. 75; Pacific Start Ltd PNG, for the graphs on page 164, courtesy of PNG Yearbook 2008 (published by Pacific Star Ltd PNG); Photolibrary/Alamy/VWPICS/Kike Calvo, p. 5 (left); Photolibrary/Tony Bee, p. 326 (top right); Photolibrary/Digital Vision, p. 40; Photolibrary/Fletcher & Baylis, p. 84 (left); Photolibrary/David Gillison, p. 84 (right); Photolibrary/Robert Lawson, p. 304 Photolibrary/Mary Evans Picture Library, p. 345 (right); Photolibrary/Science Photo Library, p. 345 (left); Sawczak Irene, pp. 2, 13, 23, 38, 44, 252 (left), 309, 310, 312, 316, 326 (bottom right), 381 (bottom left & right); Shutterstock/Sybille Yates, p. 270.

The authors and the publisher also wish to express their gratitude to the following:

The Glen Lean Ethnomathematics Research Centre, University of Goroka, for information on traditional mathematics.
Mrs Ellen Naime, Mrs Carol Bisaun and Dr Wilfred Kaleva for their pertinent feedback in reviewing manuscript pages
Mrs Betty Pulpulis, CO Mathematics Secondary, Curriculum & Assessment Division, National Department of Education, for her guidance and encouragement throughout the development of this textbook.

Every effort has been made to trace the original source of copyright material contained in this book. The publisher will be pleased to hear from copyright holders to rectify any errors or omissions.

Notes